TELEVISION
FROM
ANALOG
TO
DIGITAL

Other TAB Books by the Author

No. 1532 *The Complete Book of Oscilloscopes*
No. 1632 *Satellite Communications*
No. 1682 *Introducing Cellular Communications: The New Mobile Telephone System*
No. 1932 *AM Stereo and TV Stereo—New Sound Dimensions*

TELEVISION FROM ANALOG TO DIGITAL

STAN PRENTISS

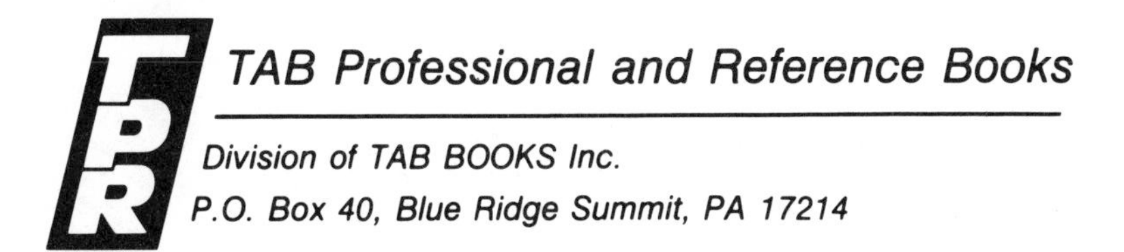
TAB Professional and Reference Books

Division of TAB BOOKS Inc.
P.O. Box 40, Blue Ridge Summit, PA 17214

TRASAR™ of Andrew Corp.
HELIAX™ of Andrew Corp.

FIRST EDITION
FIRST PRINTING

Library of Congress Cataloging in Publication Data

Prentiss, Stan.
Television—from analog to digital.

Includes index.
1. Television. I. Title.
TK6630.P74 1985 621.388 85-22244
ISBN 0-8306-0972-5

Cover photograph courtesy of RCA/Consumer Electronics Division, 600 North Sherman Drive, P.O. Box 1976, Indianapolis, IN. 46206.

Contents

Acknowledgments

I WISH TO EXTEND MY PROFOUND GRATITUDE to all those who have aided in supplying the very valuable material that was used to compile this comprehensive and extensive publication. I am especially indebted to the semiconductor manufacturers both here and abroad for their cooperation, and also to the receiver and transmitter designers and administrators who have contributed both time and excellent engineering information to help make this book much more complete.

Singled out for special tribute are the following: Jack Fuhrer, Dr. Jim Carnes, and Frank McCann, RCA Consumer Products; David H. DeVoe, ITT Seminconductors; Leander H. Hoke, Jr., Pat Wilson, and Ray Guichard, N.A.P. Consumer Electronics Corp.; Jerry Surprise and Hratch Aris, Panasonic; Dan Shockey, National Semiconductor; Marshall Rosen, Motorola Semiconductor; Len Garrett, Terry Burcham, and Ken Umberger, Tektronix; Dave Gray, General Electric; Wendell H. Bailey and Ed Dooley, Natl. Cable TV Assn.; Bruce Romano and Roger Herbstritt, Federal Communications Commission; Edmund A. Williams, Natl. Assn. of Broadcasters; Vern Killion, Andrew Broadcast Antennas; David Kawakami and Neil Cunningham, Sony; Dave Patterson and Jay Yancy, Toshiba; Sy Reich, RCA Solid State Division; D.G. Hymas and D.G. Mager, RCA Transmission Products; John Klecker and Gaylen Evans, Harris Corp.; M.E. Trim, RCA Picture Tube Division; Eric Small, Modulation Sciences; Gus Rose, B & K-Precision Division of Dynascan, Corp.; Mark Sellinger, Avantek; Don Berg and Dave Waddington, Channel Master Corp.; Randy Raybon, Prodelin; Ed Polcen and Jim White, Zenith Electronics.

Introduction

I THINK YOU WILL BE PLEASANTLY SURPRISED at the breadth and depth of this book. In narrative form, but technically applied, it takes you from the beginning of American television right up to the present and beyond, detailing all significant advances along the way. Receivers as well as UHF and VHF transmitters are described, plus antennas and transmission lines, and there's even a chapter on CATV, including the latest head-end and downstream/upstream equipment you may find interesting.

Initially, I had planned a combined chapter on analog (straight signal) and digital (1s and 0s) processing, but ITT Semiconductor supplied so much valuable material that Chapter 5 is devoted exclusively to digital, which, of course, has just been introduced into the U.S. market. Upon reading this chapter, you should develop virtually a complete understanding of the many U.S. and Japanese digital television receivers that will be available to the consumer during the next several years. All these receivers are based on VLSI integrated circuits supplied by ITT, West Germany, and at the end of the chapter there is a discussion of the MAC-B companding digitized system, which eventually should improve our receiver luminance definition by as much as 50 percent once proposed broadcasting begins in the late 1980s.

You will also find new material on cathode-ray tubes, velocity modulation, special tuner controls and applications, chroma and luminance processing, and a special chapter on conventional and multichannel (BTSC-dbx) sound. Of course, sync, high voltage, and power supplies are covered thoroughly, too, as are the host of integrated circuits that made all these excellent television sets possible.

The last book I wrote on color television went into six printings. I hope the carefully selected content of this one finds equal favor both here and abroad.

Chapter 1

Origin and Development of U.S. Television

LET'S BEGIN WITH THE SOMEWHAT TRITE OBservation that the video picture has been a long time coming. But let's augment the statement by emphatically observing that a good and true television picture wasn't available until the middle of 1985. Furthermore, you won't see "out of this world" television reproductions until sometime later in the 1980s or early 1990s when transmissions are fully digitized and transmitted almost exclusively by satellite. Then, by a system of compression and expanding (companding), NTSC (National Television System Cmte.) conventional format should be able to support horizontal dynamic resolutions of 6.3 MHz for luminance and at least 2 MHz for chromance. At the same time, horizontal sync receiver repetition rates should have doubled, so that instead of 525 scanning lines for each frame, of which there are 30/second, you should see 1050 lines of very smooth imagery and an eventual picture rivalling 35 millimeter camera outputs on film.

Even that may not be the climax, because sporadic work still continues on strapping a pair of high-powered satellite channels together and space-beaming very broadband signals down to earth to waiting 2-4 foot antennas, hungry for exceptional, very-high quality pictures on newly-designed receivers that could well be the forerunner of what to expect in the 21st century. This, of course, would not be within the NTSC format that we've been living with since the FCC adopted these NTSC *color* standards December 17, 1953. Prior to that of course, RCA had marketed over 43,000 agc-less 630TS 10-inch black and white receivers in 1946-47, followed by the agc'd 8TS30, and then a number of electrostatically deflected small-tubed receivers by everyone under the sun.

Color sets, however, were to take several different tracks in development before the Federal Communications Commission (FCC) would tender its blessing. And black and white transmissions, based on Paul Nipkow's spiral-holed rotating scanning disc of 1884, actually began both in the United States and England during the years 1925-1926. By 1928-30, there were regularly scheduled TV programs broadcast from New York (Schenectady), Maryland, Massachusetts, New Jersey, and several stations in Chicago, Illinois. The first public service telecasts, however, actually began on April 30, 1939 by NBC. Earlier, in 1935, General David Sarnoff announced general field tests of receivers and transmitters with 343-line, 30 frames/sec., interlaced, at a video bandwidth of 1.5 MHz. By 1939, these parameters had improved to 441 lines at 4.2 MHz with vestigial (modulation) sideband transmission (VSB). Actually, Philco was the first to demonstrate VSB at full bandwidth—a parameter the industry was to adopt shortly thereafter (all electronics now, no more whirling discs for a while).

World War II, of course, interrupted much of this ongoing effort, but even with the delay, New York City, Schenectady, Philadelphia, Chicago, and Los Angeles all had television stations on the air. And by 1949, almost 4 million receivers had been produced (10 million by 1951), and there were 108 TV stations actively broadcasting. Competition, of course, was building rapidly and the country was ready then—as we are today—for new and better developments, and they weren't long in coming.

COLOR TELEVISION

Beginning September 26, 1949 and continuing until May 26, 1950, the Federal Communications Commission received thousands of pages of industry response to its feasibility investigation into the possibility of introducing color television to the United States.

Initially, three systems were proposed: the revolving disc field sequential method by CBS; the line sequential system by Color Television, Inc., and the dot sequential technique by RCA. Because of its simplicity and timely performance, the field sequential system was selected but never widely implemented because of its mechanical and forced-color aspects. A monochrome camera and color scanning disk, synchronized with vertical sweep rates, picked up red, blue, and green light through successive filters, one field at a time. This was transmitted and a black and white-tube receiver with another rotating color filter disk similarly synchronized produced what, at that time, were considered acceptable pictures. However, it was soon discovered that field sequential processes were not compatible with regular black and white TV sets and that rapidly moving scenes and subjects would result in mixed or just blobs of indistinct color.

Meanwhile, RCA began compatible color broadcasts in September, 1949 on its NBC Washington, D.C. station, WNBW, Channel 4. Three-tube image orthicon cameras were used plus rapid time-division multiplexing for the red, blue, and green channels at a sampling rate of 3.8 MHz, and later, 3.6 MHz. Receivers also had three tubes and images were merged with dichroic mirrors in a large cabinet with flip-up, glass-reflective lid. It was not until March 1950 that RCA was able to design and demonstrate a shadow mask color picture tube with tri-dot phosphor structure, three guns, all in a single (in this instance) 16-inch metal-cone CRT. However, even with these improvements, the FCC, in its *Second Report on Color TV*, adopted the CBS system, only to be promptly sued by RCA; but all in vain because of adverse court decisions. So in June of 1951, CBS began color broadcasts on five east coast stations. This didn't last long because of high receiver costs, short programming, and lack of sponsors. The final commercial CBS colorcast was said by Edwin H. Reitan, Jr. in the IEEE Consumer Electronics Transactions to have to have taken place on October 20, 1951 with the coverage of a Maryland-North Carolina football game.

Mr. Reitan also relates that industry work continued rapidly on an all-electronic color-compatible system and that during 1950, Hazeltine demonstrated a "constant luminance concept," RCA offered improvements that reduced dot moire patterns, General Electric showed a frequency-interlaced system, and Philco offered compatible color with wideband luminance and two color difference signals. Color phase alternation was also considered, but later dropped because of edge flicker, according to Mr. Reitan.

By February 1953, the second NTSC color standard was published, the color subcarrier set at its present 3.579545 MHz, and the I-Q chroma bandwidth proposal by RCA was approved. All this went to the FCC and was formally adopted on

December 17, 1953, as related by Mr. Reitan. Thereafter, the race was on to make and sell TV sets.

Briefly, that's a compact history of monochrome and color television as it relates to concepts and developmental struggles. Today, without the hindrance, heat, and power drains of vacuum tubes, tremendously improved in-line, striped phosphor cathode-ray tubes, integrated circuits, and computer-aided engineering, modern transmitters and receivers have realized the full 4.2 MHz luminance bandwidths originally intended, have expanded chroma reproduction to 2 MHz (1.5 MHz for I and 0.5 MHz for Q), and even begun to equip many top-of-the-line receivers with stereo and SAP (second audio programming) multichannel sound. All this should add bountiful dimensions to our viewing pleasures, with promises of much more to come as we approach the year 2,000. Perhaps we'll even have a flat panel picture-on-the-wall by then without shadow mask and other vacuum-tube limitations. If so, we can then proceed to 20 MHz bandpass video equipment having little or no high voltage and fully digitized, including most, if not all, the usual rf and baseband analog functions. In the words of many an entertainer, especially Mammy singer Al Jolson, "you ain't seen nothin' yet!" And when you consider that removal of switching power supplies, high voltage, and picture tubes would downsize receivers to the dimensions of one or two fists, then the late Mr. Jolson would be eternally right since we've already reached the stage of 1-IC throw-away black and white receivers. Considering that mounting repair costs and production-saving curves may eventually cross, you could reasonably project a time when modestly-priced color receivers would become disposables also. In the meantime, a central control for the entire house or business is currently in the works, and may include multi-screen pictures as well as slave cathode-ray tubes (or equivalents) placed wherever you want them. Should this come to pass, it would make sense to market power supplies with each kinescope and pipe audio/video baseband around wherever you wish—a substantial incentive to cable and satellite earth station users who "want to see it all." We might add that some relatively flat display tube-like devices are available now, but the price continues to be momentarily prohibitive. Obviously they'll have to be in large-number production before being offered to the public. The integrated circuits needed for smaller chassis are virtually available now, although there will have to be modifications to existing units before compacting. Optical fibers could also be used to pipe many channels of digitized video/audio around to the various viewing and listening locations using either multiplexing or rf techniques. And phase-locked loop, frequency-synthesized, 100-plus channel tuners will remain in vogue for a long time to come.

Considering that teleconferencing has now become international, direct view picture sizes measure as much as 26 and 35 inches, projection sets offer images in terms of square feet, and banking, mail, and even some shopping is on the way to becoming at least partially electronic. There's no way but up for video communications, be they local or worldwide. This is why we are going to undertake more than usual care in explaining major topics throughout the entire text so that the reader may understand both the electronics as well as foreseeable implications. From our standpoint, they're huge!

So saying, the next logical progression would involve colorimetry and the chromaticity diagram.

Colorimetry

Color combines light or brightness (luminance) with hues (actual colors) and saturation which describes their purity (fully saturated) or pastels that include degrees of white additives. Unlike an artist whose primary colors are blue, red, and *yellow*, and when mixed, actually *absorb* various hues and are called subtractive color mixing, color television uses the *additive* method where luminance (Y) is added directly to primary reds, blues, and greens. These colors, of course, are all in the visible light range between 400 and 700 nanometers and have respective wavelengths of 680, 470, and 540 nanometers. They were deliberately selected to provide the greatest spread of usable coloration. In Fig. 1-1, a half-and-half mixture of these colors

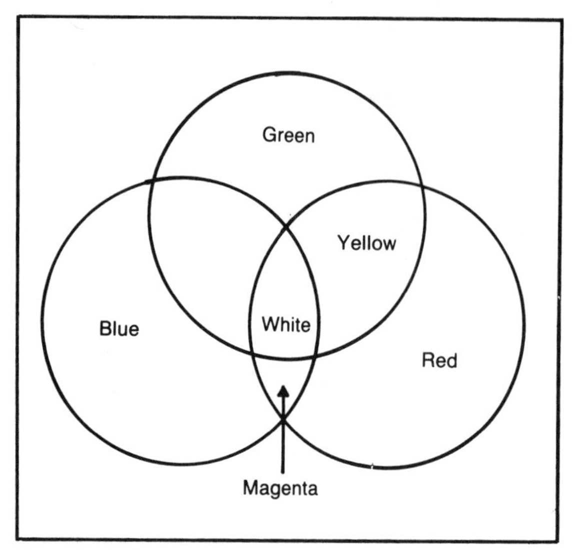

Fig. 1-1. Red, green, and blue primary colors when mixed form magenta, yellow, and white.

without luminance produces yellow, cyan, and magenta, while all three combine and emerge as white. These, of course, constitute your NTSC pattern (color bar) generator output that is standard for the entire United States and probably much of the world. The white raster is useful in color purity setups and establishes the condition for light meter measurements of cathode-ray tube (CRT) color temperatures in degrees Kelvin, varying usually between 6800 and 9000° K, depending on the manufacturer.

While not totally pure colors, these six red, green, blue, yellow, cyan, and magenta are really 75% saturated, with white (75%) and black (7.5%), respectively, on the leading and trailing portions of the display. Including, sync pulse, they consist of a standard 1-volt composite video signal divided into 140 IEEE (Inst. of Electrical and Electronic Engineers) units between −40 (the sync pulse negative-going tip) to +100, which is saturated white. The white portion between 7.5 IRE (or IEEE) and 100, represents the luminance or brightness of the video signal, as we will demonstrate shortly.

In truth, completely saturated colors seldom ever appear in nature and so most colors are less blue, red, and green than full saturation would require; therefore, the 75% saturation compromise just described. Interestingly enough, the eye can recognize over 30,000 colors, and this permits wide latitudes in phosphor lighting on the inside face of the cathode-ray tube. However, for such colors to appear real, a great deal of work has been expended in choosing picture tube phosphors and designing the three guns and backup electronics that excite them. For instance, it was only during the summer of 1984 that full 1.5 MHz I and 0.5 MHz Q/chroma demodulation was announced by (guess who?) RCA's Consumer Products Division in Indianapolis, Indiana. Can you see the difference? There's certainly less chroma noise and little background wiggles, and direct comparison with conventional 0.5 MHz I and 0.5 MHz Q or R-Y, B-Y chroma processors will certainly point to previously constricted color reproduction.

In the meantime, let's look at the NTSC system we've worked with over the years since 1953 and study its composition, along with a brief explanation of French Secam and German PAL derivations that serve Europe and some other continents also.

NTSC Color

NTSC color was established by ongoing efforts of the National Television Systems Committee which was reorganized in mid-1951 and produced a U.S. color standard during the summer of 1953. Upon this, in one way or another, all the other color systems of the world are based, even though the French have their own unique approach and, as is usually customary, promote FM color signal modulation, and cannot use the principle of monochrome and color frequency interleaving—a subject we'll discuss next.

Because of the enormous number of black and white (monochrome) receivers already (or about to be) operating in the United States before the coming of color, a monochrome-chroma compatible system had to be devised that would serve both types of receivers and still remain within the 6 MHz per-channel UHF/VHF frequency allocations mandated by the Federal Communications Commission. Further, color information reaching B/W receivers should not disturb or distort motion or motionless images, even though the eventual result is slightly more contrasty pictures which everyone now seems to accept since there are virtually no remaining black and white transmissions except for old war and other ancient movie flicks. This meant that luminance should not be visibly affected by color information transmitted within the same bandpass yet be pleasing to homo sapiens viewers.

To the everlasting credit of the NTSC group, a system of frequency interleaving was devised so that color information could be "slotted" in groups between monochrome intelligence, permitting smooth and harmonious composite video to be modulated on the various U/V and CATV (cable) carriers between allocated frequencies of 54 and 806 MHz (channels 2 through 69) and CATV from 54 to 500 MHz. Channels 70 through 83 are now assigned to land mobile, although a few TV repeaters still occupy some of the re-allocated spectrum.

Video monochrome signals contain components between about 30 Hz and 4 MHz, horizontal sync and blanking pulses at 15,734 Hz (changed from 15,750 Hz to accommodate color), and vertical sync (with blanking) occurring at 59.94 Hz (formerly 60 Hz before color). Video, then, appears as clusters of energy about the line scanning frequency of 15,734 hertz/second, with each cluster separated from the other by the field frequency rate of 59.94 Hz. In Figure 1-2 it's symbolically evident that spaces exist between each energy group and also that the highest multiple of the line scanning frequency determines the maximum video bandwidth of the system. A *full* 4.2 MHz system would add another several clusters, approaching the 267th multiple.

NTSC members discovered that if color information could be interleaved between the monochrome clusters, then all could be modulated on the same video carrier. This, then, was accomplished at multiples of half the line scanning frequency, permitting a satisfactory luma-chroma-sync transmission and reception fit. Color burst (sync) and audio had to be added to the video components, and we'll get to that shortly as the full line signal develops.

Color Sync and Signals

If you multiply the horizontal scan rate of 15,734.64 by 455/2, a color subcarrier frequency of 3.579545 MHz results, which relates to both vertical and horizontal frequencies and is also the center of color information that continues at intervals up to 300 kHz away from the 4.5 MHz sound carrier. In transmission, however, the 3.58 MHz subcarrier is cancelled by balanced modulators and only chroma is transmitted along with V/H sync, luminance, and sound. In this way, picture elements on one scanning line interleave with those on the next, reducing the appearance of dot-like interference.

The carrier signal from one of these balanced modulators is then transmitted on the back porch of the horizontal sync pulse and used as 8-9 cycles of *burst* to synchronize the receiver's crystal-controlled chroma reference oscillator. In this way, the suppressed color subcarrier is resurrected in the receiver, and chroma detection may proceed as

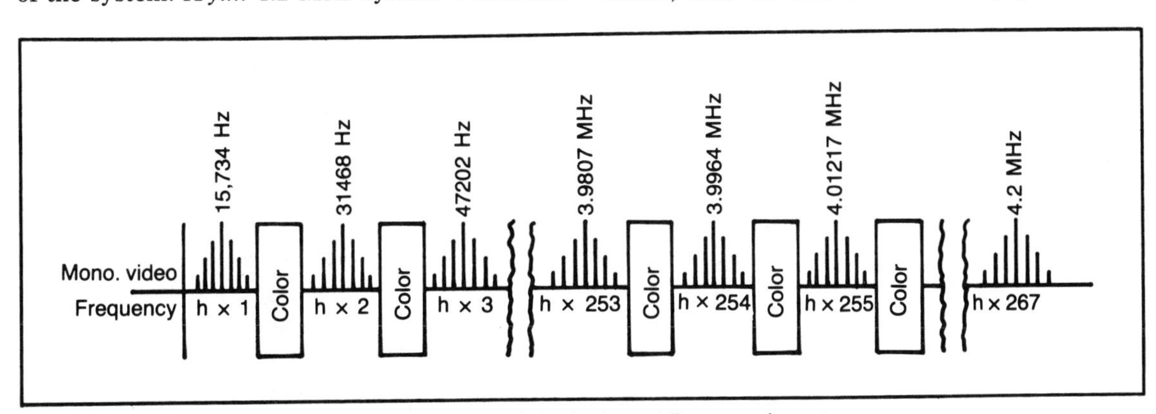

Fig. 1-2. Clusters of video energy about multiples of the horizontal line scanning rate.

though the carrier had been transmitted all along. However, since much of this chroma information appears at the receiver between 3 and 4 MHz, an efficient means had to be developed to separate luminance and chroma to take advantage of extended horizontal resolution as well as avoid the inevitable chroma-luma crosstalk . . . and this was accomplished in the 1970s with introduction by Magnavox of the *comb filter*, a development which made possible RCA's full spectrum receiver, appearing in June 1984.

To transmit red, blue, and green primaries, along with their complements and various stages of saturation, two color mediums were developed called I and Q. The I signal by itself can produce colors in hues from cyan (bluish green) to orange, while the Q signal contains yellowish green to purple. Together, and with the addition of luminance (Y), they handle the entire color television spectrum. These are color difference signals, however, and are derived accordingly:

$$I = 0.74(R-Y) - 0.27(B-Y)$$
$$Q = 0.48(R-Y) + 0.41(B-Y)$$

Where

$$R-Y = 0.96I + 0.63Q$$
$$B-Y = -1.11I + 1.72Q$$
$$G-Y = -0.28I - 0.64Q$$

and an idealized $Y = 0.30R + 0.59G + 0.11B$.

If now you'd like to know how the "burst" or subcarrier is derived,

$$\text{Subcarrier} = 455 \times 15{,}734.262/2 = 3{,}579{,}545 \text{ Hz}$$

Which rounded off equals 3.58 MHz, the chroma burst frequency ordinarily used. As you will learn later, when burst and chroma are precisely in phase, you then have correct color. Further, the color green is never individually transmitted, and usually results from a detected matrix of red and blue outputs, along with luminance added subsequently. This subject is to be covered thoroughly later in the section on chroma demodulation, almost universally executed now by synchronous detection.

Chromaticity

Now that you know how subjective color signals are formed and how they're transmitted, it's time to look at a chromaticity diagram and visualize what the eye can actually see. This particular display (Fig. 1-3) depicts colors in terms of hue (the actual color) and saturation (its purity and intensity), with saturation measured from the white dot just below the CIE illuminant and radiating outward—all in terms of the customary X (horizontal) and Y (vertical) coordinates used in most electronic descriptive material and measurements. Numbers surrounding the X-Y plot are wavelengths in nanometers of the various visible colors. Reds are to the right, blues below, and greens extend vertically, graphically illustrating the 400-700 nanometer numbers given earlier as the wavelengths spanning visible light. The three red, blue, and green primaries are indicated in the diagram, with other intermediate hues called out across the spectrum.

The Federal Communications Commission has specified the values of red, blue and green, according to given coordinates between 0.000 and 0.900 as:

Red = 0.67X and 0.33Y
Blue = 0.14X and 0.08Y
Green = 0.21X and 0.71Y

With all these specific standards, color transmission and reception now must meet such requirements to present the best possible and most acceptable pictures. Camera manufacturers among the broadcasters and chroma demodulation and picture tubes among the receiver group have presented the greatest challenge. Since about 1979, that challenge has been met and substantially overcome by both segments of the industry, until today when we have considerable improvements in the transmission and reception of composite video and high quality sound, with much more to come. Let's also fervently hope that videocassette recorders and consumer product cameras soon begin to show

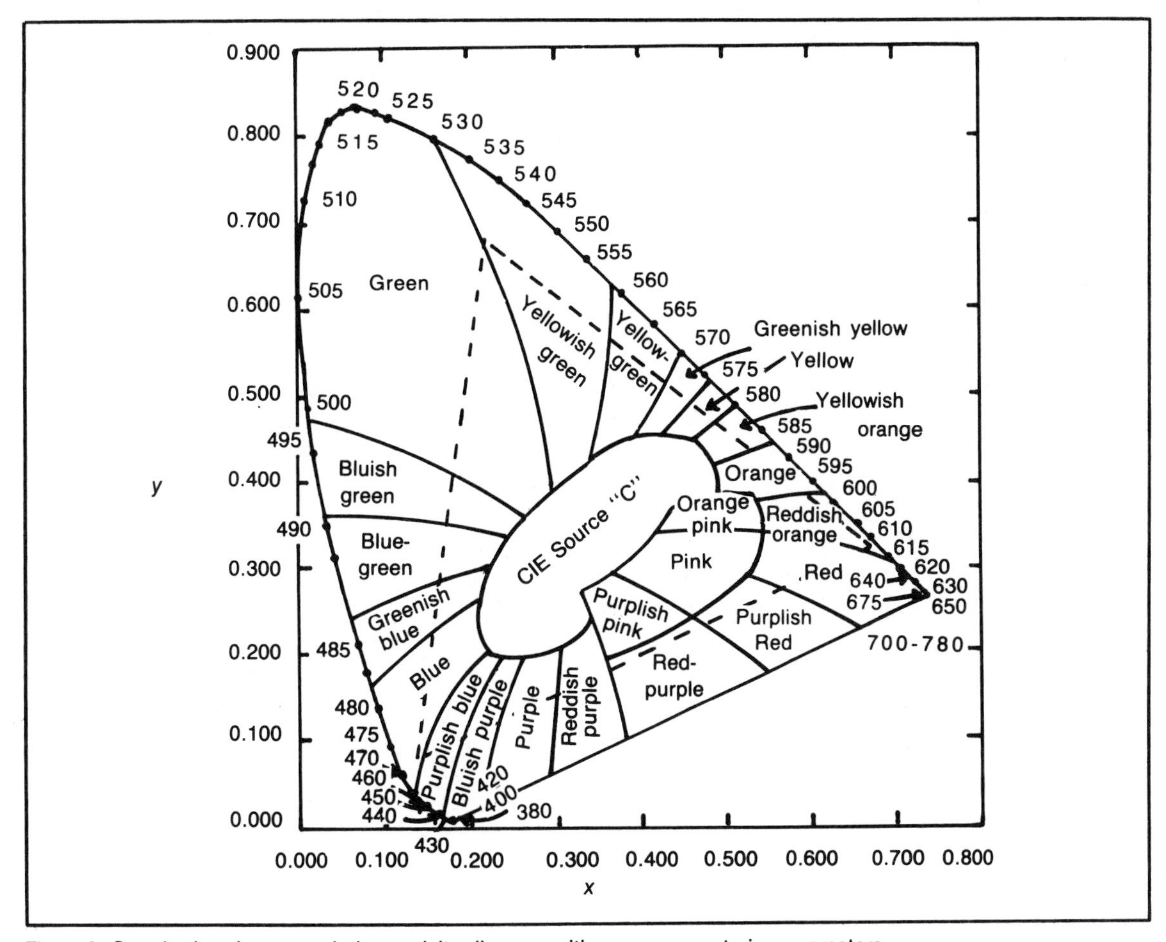

Fig. 1-3. Standard and approved chromaticity diagram with measurements in nanometers.

equal promise. Upgraded audio isn't especially difficult, but widening video bandpasses and processing more saturated colors in dimmer lighting is, and at the moment the home helical-scan recorder and its associated camera aren't really doing the job. Between 1985 and 1990 we hope there'll be some substantial changes for the better as sales of these items continue to boom. Fortunately or unfortunately, we'll have to lean wholly on Japan to do the job since most U.S. types show little or no interest in advanced research and establishment of the necessary facilities to cleanup big bucks with these electro-mechanical monsters.

Meanwhile, picture-tube manufacturers and the important advent of TV/monitor combinations have wrought wonders for receivers in analog reproduction, with immensely greater rewards to come as digital transmissions and reception move ever closer to the state of operational reality.

VIDEO-AUDIO CHARACTERISTICS

Now that you've had a preliminary introduction to the state of television, especially color of yesterday and today, it's time to investigate audio and video signals together and work through the complete and fascinating process. Included, of course, will be the vertical and horizontal synchronizing pulses that ensure coordinated timing with broadcast transmitters wherever they are.

Logically, then, let's begin briefly with the transmitter, say a word or two about antennas, and then devote a fair amount of space to receivers, which must now accommodate pure video/audio (baseband inputs) as well as broadcast signals, cable television (CATV) and, in some instances, red, green, and blue (RGB) information from office equipment displays, games, and even sophisticated computers. For although this chapter is really only an introduction, good reporting does require emphasis on the more relevant subjects as the description progresses. We will also stress a new and most important change in the original composite signal, that of multichannel TV sound, which was approved by the Federal Communications Commission on March 29, 1984 and commercially implemented in 1985.

The Composite Signal

As broadcast, the composite video/chroma/sound signal appears within an allocated 6 MHz bandspread as illustrated in Fig. 1-4, just as it always has. Vestigial sideband filtering eliminated the lower (left) sideband, leaving a 0 to 4.5 video/chroma MHz bandpass with ± 25 kHz of audio that's located exactly at 5.75 MHz of the transmitted envelope. With the coming of multichannel sound, however, the FCC now permits sideband modulation to 50 kHz, with total baseband limiting to 120 kHz, and doublesideband modulation of the AM stereo-suppressed subcarrier as L − R (left minus right) information.The original monophonic L + R intelligence remains in standard FM modulation, as it always has been. This multichannel sound is not a simple subject as you must realize, so we'll have to refrain from a full explanation until reaching the separate chapter on sound. There you'll be given "the works," including stereo sound, SAP, and professional channel detection methods, which are rather intricate, to say the least. SAP stands for *separate audio program*, used primarily for bilingual translations.

Although not this time FCC-directed, the other change affecting the composite signal diagram is the first use of full 1.5 MHz I chroma sideband (by RCA) as well as the 0.5 MHz Q sideband, now totalling 2 MHz of full color reproduction by the receiver. Heretofore, only 0.5 MHz of I and 0.5 MHz of Q had been exploited in the receivers because of luma-chroma crosstalk. With the advent of comb filters, of course, this limitation no longer exists, and RCA's 1985 top-of-the-line receivers now reproduce color just as it is and always has been broadcast from the hundreds of U/V transmitters across the country. Heretofore, the I signal was divided into upper and lower sidebands with only the 0.5 MHz upper sideband receiver-detected. Now with both complete sidebands working, you're going to not only see a greater range of color but less interference visible in the colorations themselves—another recent advance that will have other manufacturers following suit in the near future. We'll also undertake a few rf and i-f sweep alignment explanations to show you precisely where

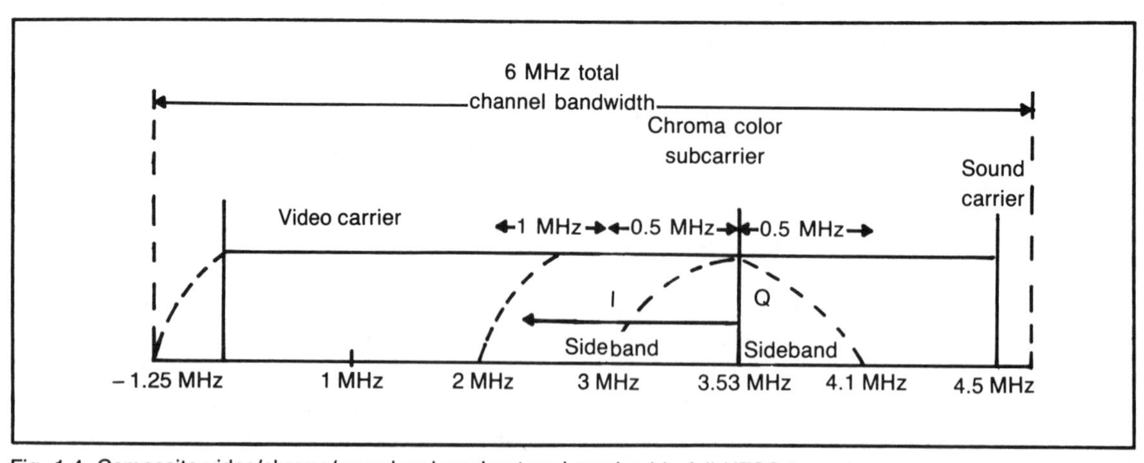

Fig. 1-4. Composite video/chroma/sound as broadcast and received in full NTSC format.

birdie beat markers for all this information must lie within the 4.5 MHz bandpass carrying luminance, chroma, sync, and audio information.

Sync

To keep aural and visual signals in synchronism between television transmitters and receivers, there must be some sort of "trigger" among the electromagnetic energies radiating from the broadcast tower telling the receiver when to stop and start its trace and retrace operations. These are difficult to describe without the partial diagrams, so they'll have to be included even though this is meant only as an introductory chapter. However, we'll keep them as simple as possible for now and give you the full treatment later on when sync circuits are described in detail.

To maintain good resolution and definition, the NTSC system requires two series of interlaced, slightly downward-sloping line traces numbering 262.5 lines, each group called a vertical field (Fig. 1-5). Together they consist of a frame, of which there are 30 frames/second in the 525-line, 60 Hz alternating current (ac) North American system. These lines, by the way, have *nothing* to do with frequency and therefore do *not* vary in application from one brand or manufacturer to another. In addition, they are best resolved at the center of most picture tubes and become less so towards the edges. Cheap receivers and picture tubes, nonetheless, do

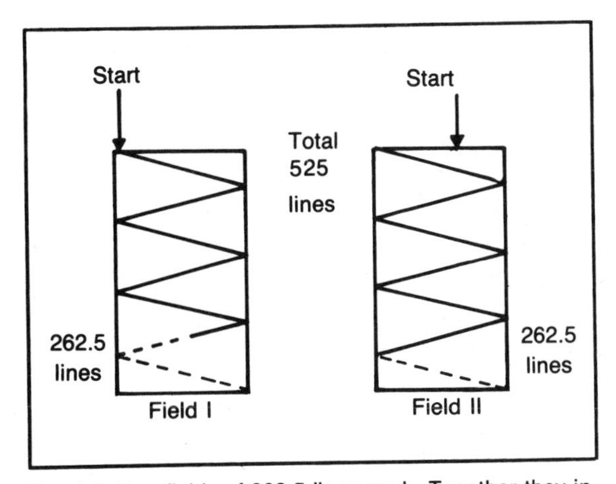

Fig. 1-5. Two fields of 262.5 lines each. Together they interlace in 16.68 milliseconds, constituting one frame.

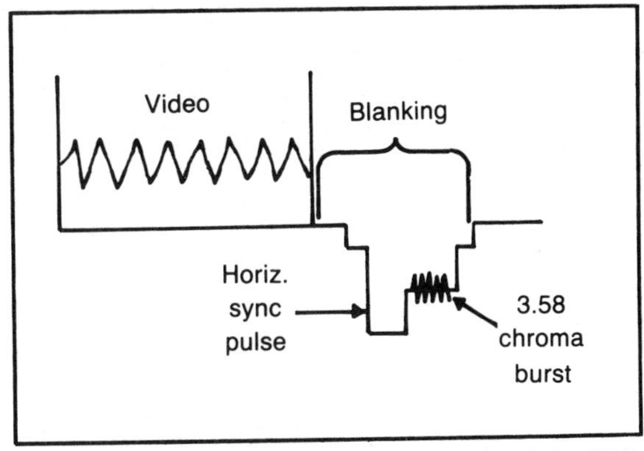

Fig. 1-6. A single horizontal line, completed in 63.5 microseconds, and 525 lines equals 1 frame.

most certainly produce less usable lines of vertical resolution than those from the better manufacturers' top products—so beware.

Viewed individually, each of these "vertical" field and frame lines is also part of the horizontal sweep (Fig. 1-6) and is called a horizontal line, and as such are directly related to frequency. Pulses of test "burst" spaced at intervals between 0.5 MHz and 4.2 MHz accurately define the high-frequency resolution of any transmitted/received image, repeating sequentially throughout the entire 525 lines. In order that the trace beam has time to recover after completion of both vertical and horizontal trace periods, precisely-timed blanking intervals have been injected to separate single and groups of lines from each other. In these blanking periods have been inserted horizontal and chroma sync (following line trace) and vertical sync as well as very useful VITS and VIRS signals, primarily for transmitter testing, but still used by Panasonic and General Electric for automated color control in their better receivers. Special line-selective oscilloscopes can resolve VITS (vertical interval test) and VIRS (vertical color reference) signals, along with some high-quality service scopes, and these—though not always broadcast—can be helpful in determining linearities and bandwidths in the receiver. They appear on lines 17 through 19 in the vertical blanking interval. However, only line 19 remains *reserved* for VIRS, whereas lines 14 through 18 and 20 may also have Teletext (14-16 exclusively), and lines 17, 18 and 20 can show test, que, control, and program que, control, and program indentification as well. Line 21 has been reserved as captioning for the deaf. In 1988, lines 10, 11, 12, and 13 are also scheduled to be added to the approved list carrying Teletext. Other data may be offered on some of these lines when and if the FCC so rules—but that may take a while (Fig. 1-7).

AUDIO

In the past it has been customary to treat television audio like a stepchild, doing no more than absolutely necessary to squeeze acceptable sound mostly in the 3-5 kHz voice range through the least expensive and most available speakers. Today, the lesser receivers and their manufacturers will continue that practice, but not the better ones. With the coming of multichannel TV sound for stereo, SAP, and professional channels, much of this neglect will change due to public demand and Japanese-U.S. competition. Although the larger surge in this new technology will not be apparent until 1986 because of continuing engineering remodeling and equipment installations, in time it will become even more important than stereo FM and virtually all the larger stations will make it part of their major broadcasts.

In the meantime, conventional FM audio in television is ordinarily detected as 4.5 MHz intercarrier sound, demodulated by quadrature or discriminator phase selection and accompanying integrated circuits or transistors, amplified and delivered to one or more speakers with impedance inputs of between 8-and 16-ohms. In multichannel

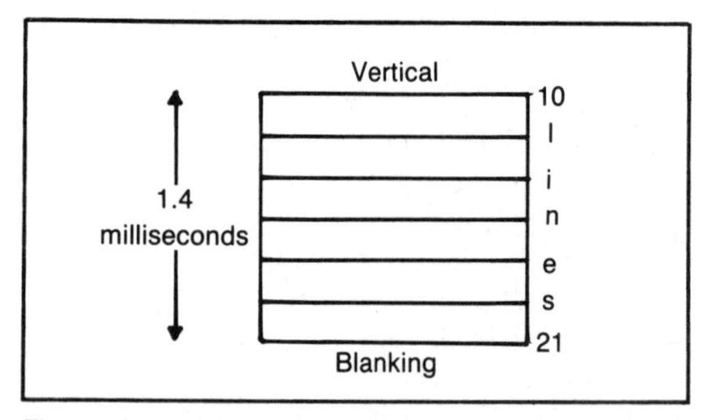

Fig. 1-7. Vertical blanking intervals will soon have 11 useful authorized lines for ancillary information.

sound, this same sort of detection will continue for the L + R monophonic channel, but will differ greatly for the 2 × 15,734 Hz stereo L − R subcarrier and its amplitude modulated sidebands. The bandwidth-diminished second audio program (SAP) will continue as an FM-operated signal whose frequency response cannot exceed 10 kHz and whose carrier is 5 × 15,734 Hz, or 78.670 kHz. The so-called professional channels will *not* ordinarily be involved with television because of their carrier placements and further restricted bandwidths designed for data rather than voice transmissions. However, since the FCC's 1984 decision, a great deal of evolution has still to occur in the use of multichannel sound, with some applications eventually forthcoming that we have yet to consider.

DIGITAL TELEVISION

For starters, digitizing of the analog-processed signal will take place following most synchronous video detectors and the various sound takeoffs normally contained in the intermediate frequency (i-f) and automatic gain control (agc) amplifier integrated circuits. Like any other electronic advance, the new system must start somewhere before it is fully expanded to include most, if not all, of the entire receiver. The first sets to be marketed—and they will include Toshiba, Panasonic, and Sony, probably in that order—will all be basic products of European (West German) Intermetall Semiconductors ITT, who have spent many millions of dollars developing the concept. Named the DIGIT 2000 system, the company claims that a few very large scale (VSLI) integrated circuits can replace "hundreds" of analog- type components, simplifying receiver manufacture and "greatly improving set performance." Claiming greatly reduced noise, flicker-free pictures, extra channel CRT projection, digital comb filters, automatic ghost suppression, a home terminal for video discs, cassettes, Teletext, Viewdata, and computers, the advantages appear considerable. There are, however, some "kinks" yet to be fully overcome, and introductory equipment may not seem all that advanced, especially when the inevitable extra retail selling price has been added.

However, once technical obstacles have been fully overcome, only the very cheapest sets will remain analog, while maximum development should continue with digital receivers, with the expectation that even three-dimensional images may, one day, become a reality. Certainly video and sound will both have greater fidelity, and a great deal of intra-set noise and other objectional features will have been eliminated. At this writing, the development of digital broadcast and receive television is a very exciting prospect, probably for the entire world. For once information is competently digitized, it can be transmitted with considerable fidelity anywhere, and reliably stored on discs, tape, memories, and other media for substantial periods without undue deterioration and in very compact containers. We would all probably be very surprised to know just how many individuals are involved in such concepts all over the globe, for digital data transmission is an electronic marvel whose time has come!

European versions of digital receivers will be designed to process NTSC, PAL, or SECAM for a truly international TV. The block diagram for the ITT set includes many digital terms that may not be immediately familiar, we shall save it for the receiver chapter on digital reception where a full explanation and all the details are given.

VITS AND VIRS

No explanation of video/audio/chroma signals would be complete without some investigation into a pair of test displays that are available from better NTSC color bar generators, especially Tektronix, and are also broadcast, usually by the networks, from time to time as a check on their sync and composite video/chroma signals. Their importance cannot be overstressed because of application to both broadcasters as well as consumer product cameras, discs, videocassette recorders, and all manner of analog television.

If adequately and intelligently applied, these two signals can characterize rather completely any of the video processors whose passbands can and will accept their broadband range of frequencies and measuring symbols. In television receivers, however, only multiburst is really effective beyond the video detector because of blanking, and color bars are only seen as discrete levels at the picture tube, which will vary when adjusted by color, luminance, and contrast controls. But through the video detector and sometimes beyond, all signals are both useful and significant. Considering these circumstances, the reader or instructor may skip this section and pick it up later, if convenient, since we will go into some detail as to appropriate signal analysis. However, in analyzing the various equipment later in the book, most, if not all, the test patterns will bear heavily on the quality and various electrical responses of the singular apparatus investigated. For television receivers, however, most test patterns will be generated by "sidelock" or gated rainbow generators rather than those producing NTSC-type color bars. As for multiburst, cross or dothatch, staircase, and sync, these may be generated by any multi-signal source with good sync countdown and tight crystal control.

VIRS. Since this is the easier of the two patterns to explain and occurs on fields 1 and 2 of line 19, its outline is shown in Fig. 1-8. The display rises from −40 IRE at the horizontal sync tip to +90 IRE at the chroma reference bar, with color burst situated on the back porch of the horizontal sync pulse. A luminance reference then appears at 50 IRE, so that the total amplitude of the chroma reference bar amounts to 40 IRE. At +7.5 IRE you see the black reference, and the waveform then once more sinks into a succeeding horizontal sync pulse, returning to a level of −40 IRE. In all, the chroma reference occupies only 24 microseconds and is somewhat unsteady in the vertical blanking interval when 'scoped in a television receiver. But at the broadcast station, the coincidence of burst and chroma phase is all important since at that time proper color phase is being broadcast. In certain receivers that use the VIR signal for color control, line 19 is detected and, along with well-designed sync logic, this pattern will track color phase and amplitude and keep receivers in the all-important fleshtone range *without* the usual "idiot button" distortion so prevalent in passive RC networks. The luminance and black references are also useful to

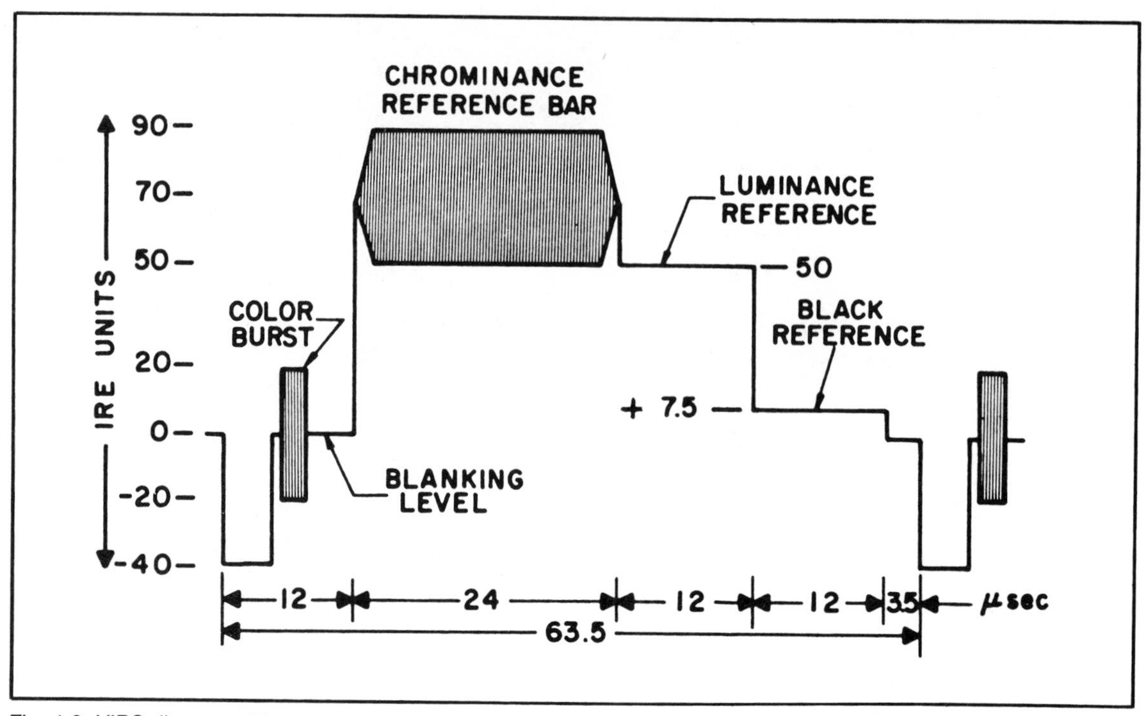

Fig. 1-8. VIRS diagram with color and burst references in terms of IRE amplitude units and microsecond timing.

such receivers and broadcasters in defining the proper picture levels associated with composite video.

VITS. This is both a chroma and luma test signal, but it also checks frequency response in addition to rise and fall times, and gray-scale tracking. The three VITS patterns are illustrated in Figs. 1-9 through 1-11: the first being simple multiburst; the second NTSC color bars; and the third appears as a composite signal on fields 1 and 2 of line 18 which includes modulated staircase, the 2T and 12.5T $\sin^2$ pulses, and an 18-microsecond window. When broadcasting, either color bars *or* multiburst is attached to the composite signal and both appear on lines 17 and 18—nothing more complex than that— simply a pair of electronic test signals in sequence, with VIRS tagging along on line 19.

Multiburst. This is used by broadcasters and TV receivers to chart gain-frequency distortion and burst amplitudes as well as frequency extensions to determine system high-frequency response. Luminance white at 100 IRE is considered saturation and should not be exceeded.

Color bars. These consist of the usual yellow-blue six "saturated" colors as well as white and black on opposite ends. Improper amplitudes are termed chroma gain distortion, poor responses may indicate oscillations, and ragged edges identify nonlinear amplification as well as extra color problems. Each bar has a 6 µsec duration.

The *composite pattern* is by far the most complex and is a complement to both color bars and multiburst. Here you may measure luminance nonlinear distortion in addition to differential gain and phase: differential gain meaning gain changes as signal levels rise and fall, while differential phase goes through the same process as gain but is treated as signal leads or lags rather than amplitudes. The staircase has good uses for modulated gray scale evaluation as well as 3.58 MHz full-amplitude response. The 2T and 12.5T spikes of voltage that follow will indicate relative amplitude of luminance information between 1 and 3 MHz, while the modulated 12.5T pulse is useful in establishing chroma-luminance gain and delays—if relevant. Their amplitudes must match that of the staircase or you

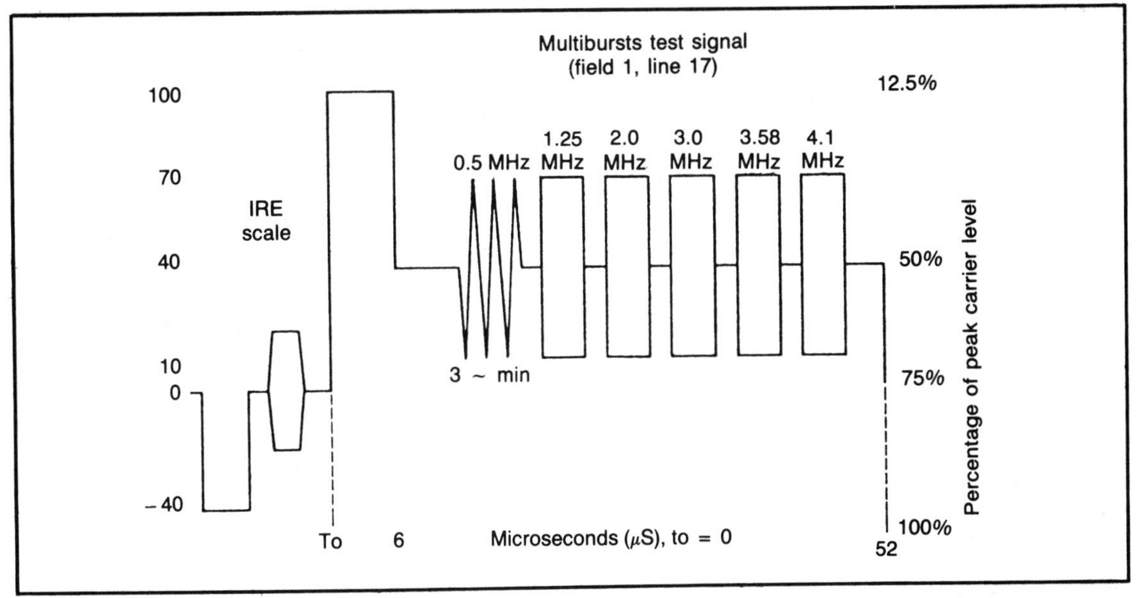

Fig. 1-9. Universally useful multiburst test signal determines high-frequency resolution.

have luma or chroma gain distortion, and should the 12.5T have baseline tilt or seems only partially modulated, chroma lead or lag and color saturation or hue difficulties are often apparent.

Finally, the 18 μsec window following the T-pulses will immediately show low-frequency (top) or high-frequency (sides) rolloff or ringing, suggesting both high- and low-frequency performance for

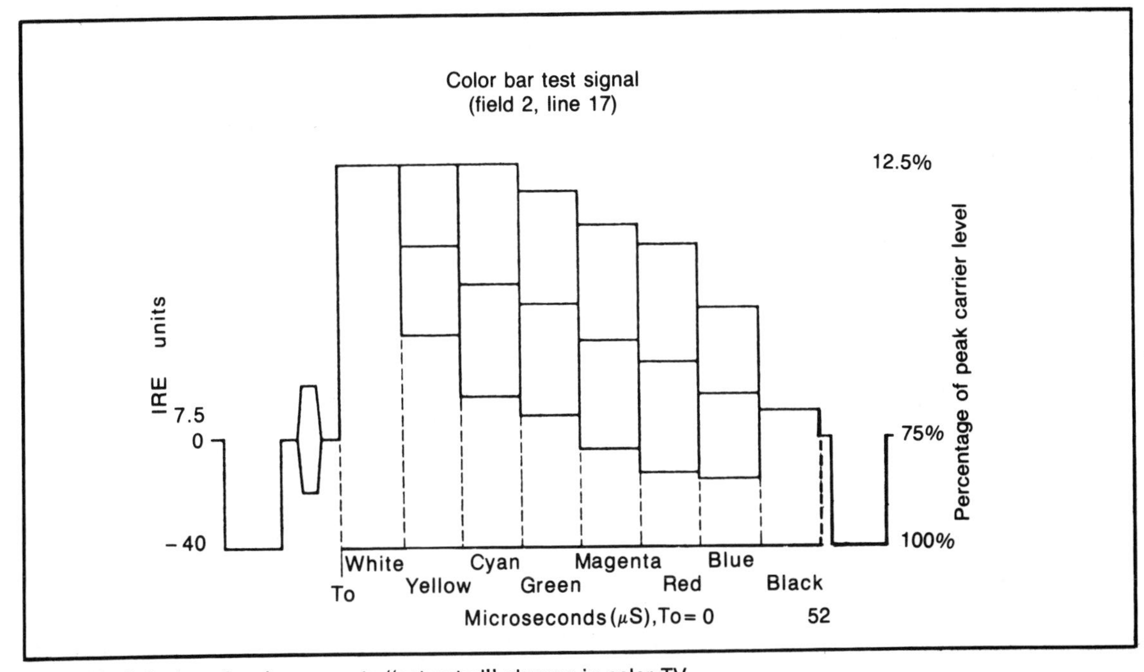

Fig. 1-10. Color-bar signal represents "saturated" chroma in color TV.

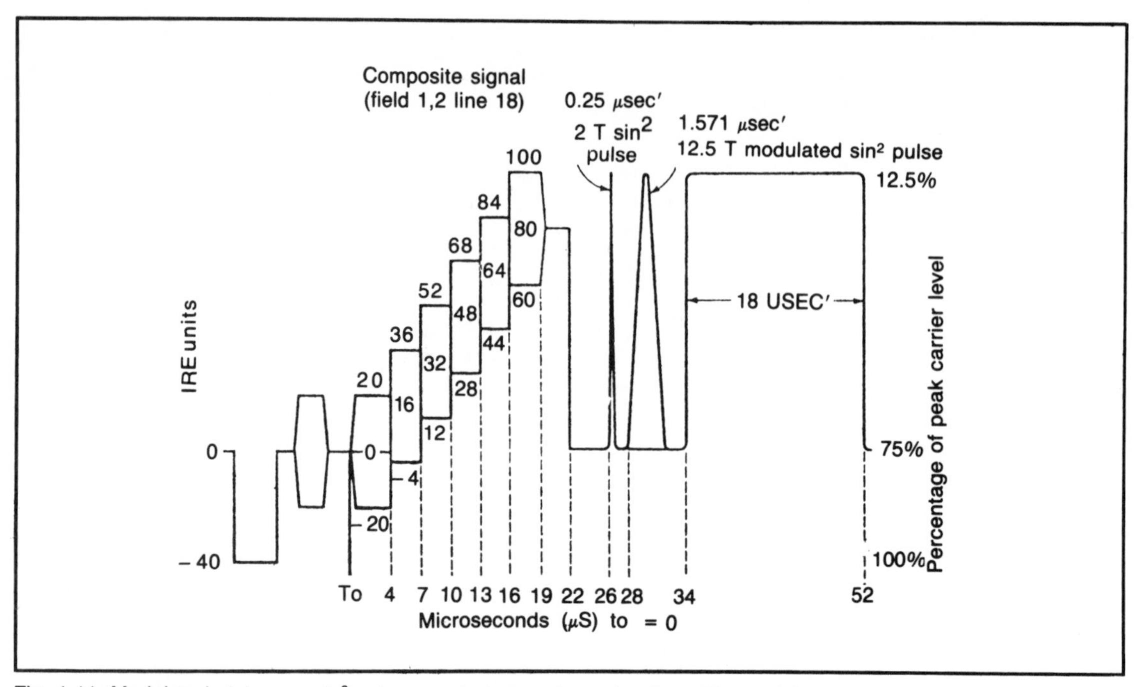

Fig. 1-11. Modulated staircase, sin^2 pulses, and window determine linearities and frequency response characteristics.

the receiver or transmitter. Following the video detector, receivers often exhibit disparities in the amplitudes of these impulses but normally show little chroma or luma lead or lag.

Fortunately for broadcasters, their expensive equipment will generate all these important test patterns with fidelity. But aside from color bars, staircase, and the window, no standard service instrument produces the 2T and 12.5T pulses, even though many more signals are available in the better test gear than we've shown so far. Later, however, especially in the receiver discussions, we will expand the role of test equipment so you can see many possibilities. Additionally, as digital transmit-receivers become available, much of this equipment will change to accommodate that discipline as well.

So, testing accumulates and skills must inevitably grow as we cross the border between signal development and 1s and 0s. Then, paramount concerns won't be just with waveform amplitudes and whether they're horizontal or vertical frequencies, but with logic and signature analysis, rise and fall times, transients, repetition rates, clocks, triggers, and pulse trains in microseconds and possibly even nanoseconds while A/D and D/A converts, NANDS, flip-flops, and one-shots do their stuff. Then add intricate video to sophisticated audio in three or more channels, and you begin to see what's in store.

CHANNELS AND FREQUENCIES

As mechanical television grew into electrical television, and experimental stations sprang up in various parts of the United States, the Federal Communications Commission adopted a ruling assigning 12 VHF (very high frequency) channels to nonconflicting areas throughout the United States, each 6 MHz wide, with no more than seven located in any service area. They occupied frequencies between 54 and 88 MHz for channels 2 through 6, and 174 to 216 MHz for channels 7 through 13. The channel 6-7 break permitted the FM radio band to be located between 88 and 108 MHz, causing little or no interference with TV. Later, the FCC also authorized another 70 UHF (ultra-high frequency) 6 MHz channels beginning with channel 14 (470 MHz) and ending with channel 83 at 884-890 MHz. In all of these frequency designations, note that the video and audio carriers are separated by 4.5 MHz, and that the audio carrier *always* rests at the higher frequency. Recently, however, the FCC assigned former TV channels 70 through 83 to land mobile (mostly cellular phone usage) even though about *500* TV translator stations are still operating in the band. These will gradually phase out with time and cellular radio is scheduled to take over as the service expands and fills the entire 825-845 MHz sector for wireline and 870-890 MHz for non-wireline operations. General Electric is even applying for a Personal Radio Communications Service (PCRS) at 900 MHz as FM CB to fill the need for poor-man's mobile radio with telephone connections. *Maximum* TV transmitter power levels (with certain excessive height restrictions) are:

TV Channels 2 - 6	100 kW
TV Channels 7 - 13	316 kW
TV Channels 14 - 83	5,000 kW
	(except near the Canadian border)

Minimum cochannel separations range between 155 and 220 miles, depending on zones and whether the stations are in the UHF or VHF bands. Usually it's less for UHF because of the higher frequencies and line-of-sight signal transmission/reception characteristics. Low power VHF TV transmitters in the 10-watt power categories have typical contours of 15 to 20 miles with maximum antenna heights that are virtually unrestricted. UHF LP transmitters may have up to 1,000 watts. A listing of broadcast TV channels, including 14-83 follows despite current FCC designations. Sometime in the future these may change as the deregulation stampede pendulum begins to swing back to engineering and not political decisions. Meanwhile, 735 LP stations have been approved and 20,000 applications are pending.

Although not exactly open-air broadcasting, the industry has also assigned a great many cable television (CATV) channels to be included as somewhat offset frequencies between TV channels 2 and 18. Originally termed Community Antenna Systems, the number of channels now in use number in excess of 100 and range all the way from A-1 to the W designations, and as many as 60 may be used by a single CATV head end for subscriber services. Although the FCC does not license such "channels," the Electronic Industries Association, along with the National Cable Television Association, maintain an oversight and coordinating presence in such operations so that cable television does have a unified and recognized position in the video and FM industry. Therefore, television receivers with special wideband and offset frequency tuners can receive these particular rf carrier frequencies as transmitted through the head ends, trunk and distribution amplifiers, and cable-to-home drops, except where scrambled signals require pay-TV decoders. When considering CATV reception, you are advised to consider automatic fine tuner (AFT) ranges of at least ± 1-2 MHz for positive assurance of satisfactory signal pickup, especially in nonstandard (offset) frequency systems. In a number of instances, the cheapies simply won't work and vendor-service people will be asking for trouble—a condition very analogous to computer and VCR display sync problems when this equipment is connected to many of the more inexpensive color television receivers with fixed (non-countdown) vertical sync chains. The image becomes unstable almost immediately and is extremely annoying, especially where there's noninterlaced sync.

NTSC, PAL, and SECAM: THE THREE MAJOR SYSTEMS

Now that you have a good introduction to broadcast and CATV channels, let's wind up with a brief description and technical specifications of the world's three major systems: NTSC, PAL, and SECAM from the United States, Germany, and France, respectively, remembering that we operate at 120 Vac, 60 Hz, while the Europeans are nominally on 220/240 Vac and 50 Hz. This does give them 625 scan lines over our 525, but their power supplies are harder to filter and TV channels considerably fewer. In the past several years, however, conversion electronics compensating for system differences have appeared that permit direct pickup and display of alternate systems, producing pictures and sound approaching that of the original. With the advent of digital TV, this will all be resolved most effectively and you'll not really know the difference, especially if transmissions arrive by satellite which, logically, most are expected to do. PAL translates to *phase alternation line*, while SECAM means *sequential with memory*.

With those words of wisdom extant, time becomes propitious for a 3-system layout so that you may consider the attributes and mediocrities of each national TV arrangement. So without undue influence, the NTSC/PAL/SECAM choice is yours. At least multichannel sound should take care of the audible translations—high notes, too. PAL-

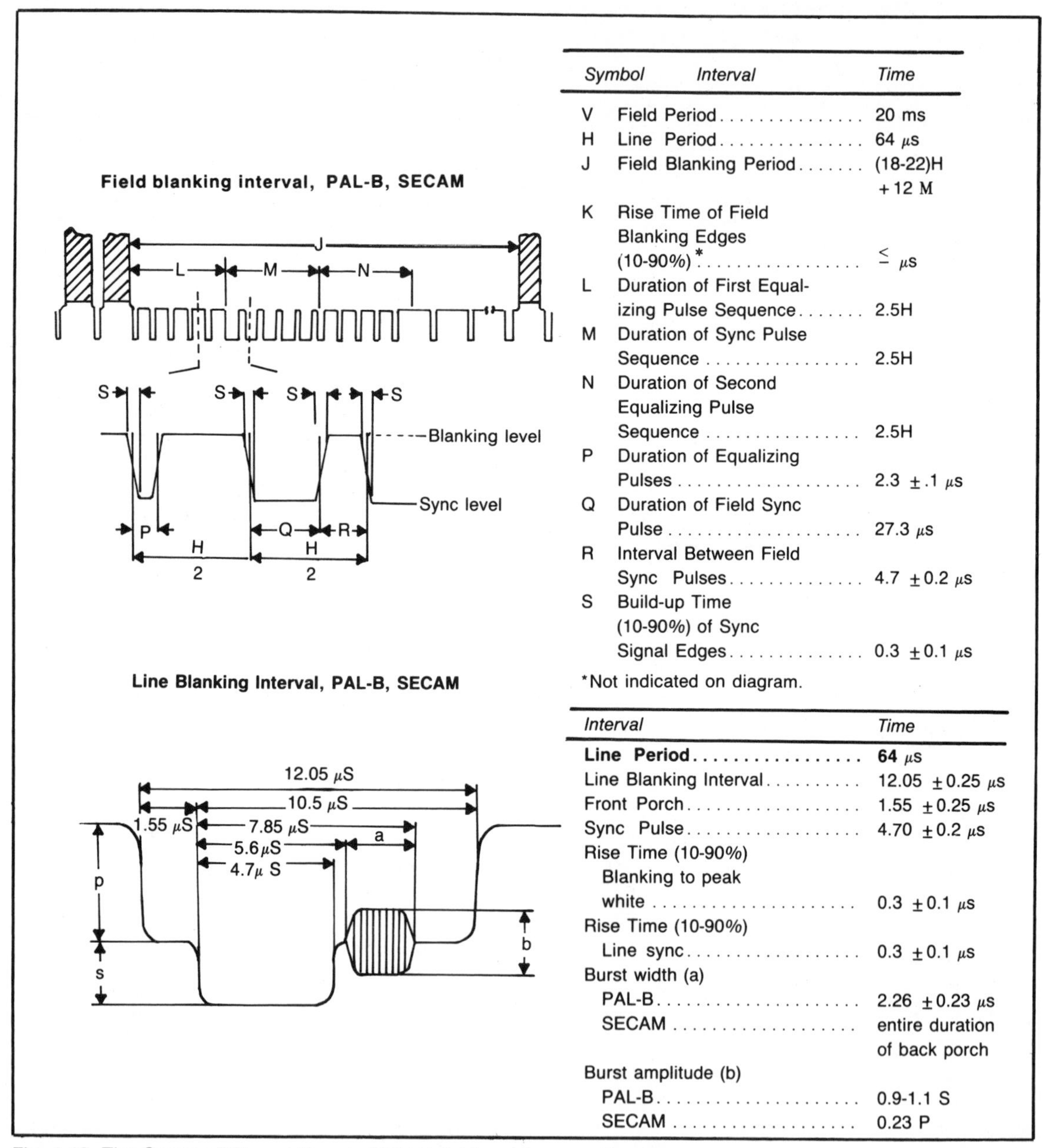

Symbol	Interval	Time
V	Field Period	20 ms
H	Line Period	64 μs
J	Field Blanking Period	(18-22)H + 12 M
K	Rise Time of Field Blanking Edges (10-90%)*	≤ μs
L	Duration of First Equalizing Pulse Sequence	2.5H
M	Duration of Sync Pulse Sequence	2.5H
N	Duration of Second Equalizing Pulse Sequence	2.5H
P	Duration of Equalizing Pulses	2.3 ±.1 μs
Q	Duration of Field Sync Pulse	27.3 μs
R	Interval Between Field Sync Pulses	4.7 ±0.2 μs
S	Build-up Time (10-90%) of Sync Signal Edges	0.3 ±0.1 μs

*Not indicated on diagram.

Interval	Time
Line Period	**64 μs**
Line Blanking Interval	12.05 ±0.25 μs
Front Porch	1.55 ±0.25 μs
Sync Pulse	4.70 ±0.2 μs
Rise Time (10-90%) Blanking to peak white	0.3 ±0.1 μs
Rise Time (10-90%) Line sync	0.3 ±0.1 μs
Burst width (a)	
PAL-B	2.26 ±0.23 μs
SECAM	entire duration of back porch
Burst amplitude (b)	
PAL-B	0.9-1.1 S
SECAM	0.23 P

Fig. 1-12. The German PAL and French SECAM field and line specifications.

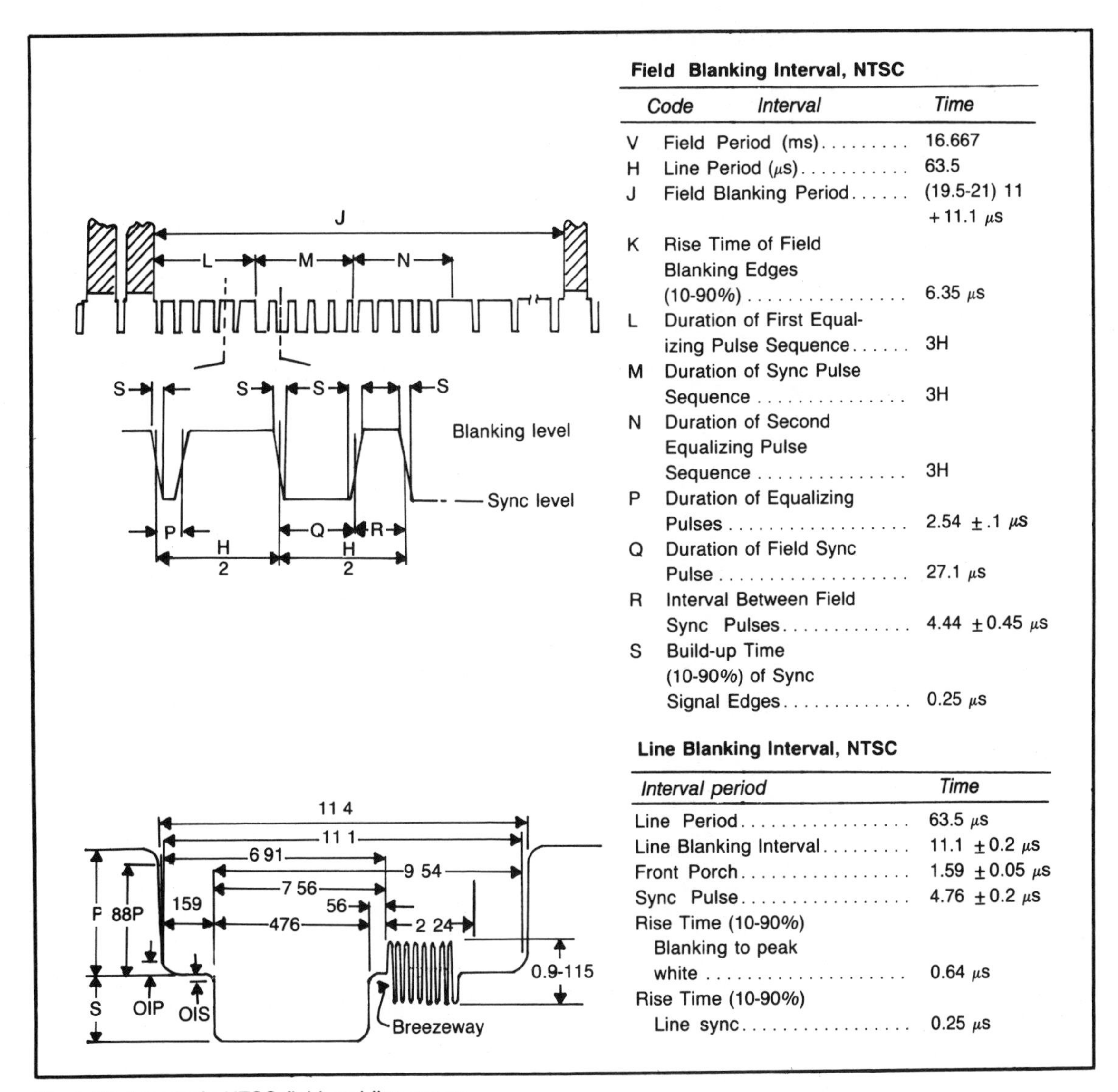

Field Blanking Interval, NTSC

Code	Interval	Time
V	Field Period (ms)	16.667
H	Line Period (μs)	63.5
J	Field Blanking Period	(19.5-21) 11 + 11.1 μs
K	Rise Time of Field Blanking Edges (10-90%)	6.35 μs
L	Duration of First Equalizing Pulse Sequence	3H
M	Duration of Sync Pulse Sequence	3H
N	Duration of Second Equalizing Pulse Sequence	3H
P	Duration of Equalizing Pulses	2.54 $\pm$.1 μs
Q	Duration of Field Sync Pulse	27.1 μs
R	Interval Between Field Sync Pulses	4.44 $\pm$0.45 μs
S	Build-up Time (10-90%) of Sync Signal Edges	0.25 μs

Line Blanking Interval, NTSC

Interval period	Time
Line Period	63.5 μs
Line Blanking Interval	11.1 $\pm$0.2 μs
Front Porch	1.59 $\pm$0.05 μs
Sync Pulse	4.76 $\pm$0.2 μs
Rise Time (10-90%) Blanking to peak white	0.64 μs
Rise Time (10-90%) Line sync	0.25 μs

Fig. 1-13. America's NTSC field and line specs.

B and SECAM are shown as one, while NTSC, which differs considerably from the others, occupies an illustration by itself: see Figs. 1-12 and 1-13.

Observe that line lengths are very similar, but that field periods are 16.67 milliseconds versus 20 milliseconds to accommodate the additional 10 scan lines. Line blanking periods, as you might expect are also very similar, and U.S. sweeps often use a full 12 microseconds to compensate for any horizontal irregularities such as ringing, imprecise design, and component tolerances. European blanking rise times 0.3 μsec versus 0.64 μsec are much sharper, but burst amplitudes and widths vary considerably between PAL and SECAM. There are differences, too, among pulse durations in the field-blanking intervals, but both vertical and horizontal *principles* are exactly the same. So, on the surface, interconversion problems are surmountable. The French, however, do FM their video and

AM their audio, so detection devices compared with NTSC and PAL are reversed. So how do you like your croissants, curled or circular polarized? As favorable/unfavorable comments on both systems, PAL does reverse its subcarrier every other line, cancelling differential phase errors, making a hue control (as we know it in the United States) unnecessary. In transmission, PAL broadcasts only color difference R – Y and B – Y signals as double sidebands with bandwidths of 1.7 MHz below and 1.3 MHz above the subcarrier, which is pegged at 4.443 MHz. There are, however, only 25 frames/second and four fields.

SECAM, on the other hand, transmits its R – Y and B – Y color difference information in sequence, but with a time lag of 64 microseconds (or 1 horizontal line) between them, and color switching at the beginning of each field. Frequency interlace, therefore, isn't part of the system, but subcarrier phase accuracies are not, apparently, important. Fortunately, they're all interesting systems with distinct advantages and disadvantages—too bad they can't be combined with the best features of each included for posterity, but that's probably far too much to expect in this age of furious international competition and overprotective nationalism. Let's hope that impersonal 1s and 0s do the job for us as customs and language translations merge in the never-never land of Karnaugh maps, double negation, and de Morgan's Theorem. A bit of Boolean logic (circa 1854) might do us all good as we near the 21st century.

Chapter 2

Broadcast Antennas and Transmission Lines

THESE SUBJECTS *SHOULD* BECOME VERY dear to both broadcasters and receiver people as high fidelity sound and pictures literally spread across not only this continent but also the entire world. For the better your radiator, the more vivid the picture and greater the sound. This will become especially true in satellite reception and terrestrial propagation as the quality of broadcasting inevitably expands and receivers fully match the trend.

Because of competition and the enormous revenues involved, many television stations will have to install new equipment to satisfy their listener-viewers who will constantly enjoy newer and better receivers, videocassette recorders, cameras, disc players *and* recorders, combination cameras and recorders (camcorders), and multichannel sound. One day, high definition television may require even new methods of electromagnetic propagation and probably additional bandwidth receivers for adequate reception.

Certainly we are now entering the age of CP—known as circular polarization—resulting from broadcast engineering advances, which have found that both vertical and horizontal signal components broadcast together give far better results than just the horizontally polarized signals we have known since TV transmissions first began. A great many television stations are awakening to the many U/V possibilities that better signal dispersal and carrier strength offer (Fig. 2-1).

To some *circular polarization* may come as an unfamiliar term, but those in the VHF television transmitter and satellite downlink business are very much aware of its singular advantages. During the past five years, new and replacement VHF TV transmitters are being installed at an ever-increasing rate using this highly effective dual-polarity method of signal distribution. UHF activity is not as great due to the problem that existing transmitters are already operating at or near maximum output, and doubling their power may neither be economical nor desirable, considering this and also line-of-sight limitations. But, there is distinctive progress here, also.

CP, as circular polarization is familiarly known, is more effective with rabbit ear antennas than

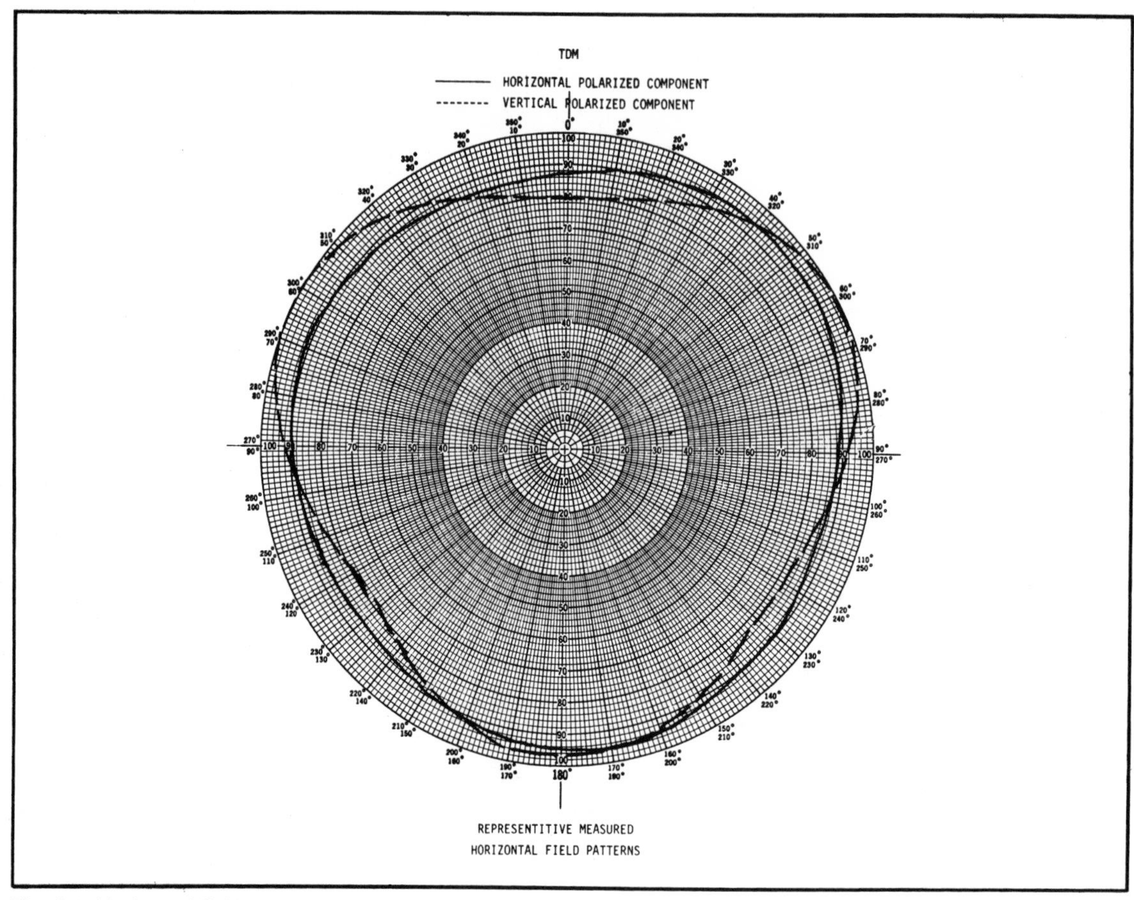

Fig. 2-1. Horizontal field patterns of vertical and horizontally polarized components (courtesy RCA).

standard antennas because these indoor Vees are mainly vertically and not horizontally polarized as are those outdoors. Considering the tiny microvolt and millivolt signals reaching whatever you have in the way of receptors, any useful electromagnetic energy exciting your antenna is important. Therefore, when a broadcasting station reinforces its normally polarized horizontal signal with vertical fare of equal power and intensity, actually doubling the output with an additional transmitter, the "ears" are considerably more receptive. This is especially true for UHF information since it is higher in frequency and usually will readily pass through small window and other entrance areas in residences and apartment buildings.

You might think that only Harris and RCA are the major manufacturers in CP—and that's true for VHF—but Andrew is making considerable headway with UHF by diverting 15 percent of its horizontally transmitted energy to vertical, thereby supplying a great deal of CP coverage we'll be enthusiastically discussing later. Eventually, when broadcasters and the public realize they can expect at least a 3 dB increase in received signal strength, there could be a real awakening demand for *home*-type receptors, especially since considerable ghost cancellation and noise reduction is associated with CP.

GENERAL ANTENNA THEORY

Any transmitting antenna consists of several prime components, one of which is the radiating element. This converts rf current into electromagnetic fields and propagates them into space. Often the radiating element has a number of conductors which may be called an array. In other instances it is a metal horn, a spiral, or parabolic dish—and they all do precisely the same thing—that is, radiating fields of energy directly into the surrounding air. Since they are all radiators, they may receive as well, should such a function be desired.

As current flows through one of these conductors, a magnetic field forms producing lines of force. Conventionally, if you point your left thumb in the direction of current flow, your fingers will indicate magnetic line directions, which are at right angles to the current flow. If an rf generator is connected to parallel elements (as shown in Fig. 2-2) an alternating current can be made to surge and decay in both conductors, building up in one direction, dropping to zero, and doing the same in the other direction.

But when magnetic fields buck one another (as in A) the resulting magnetic flux reduces in intensity. As this magnetic field is "opened" by conductors being moved apart, there is less cancellation and more (B) intensity. In C, current moves away from the generator on top and toward the generator on the bottom so there is minimal cancellation and the field can continue into space as illustrated in D. Of course, as each charge builds up, stops and then decays, magnetic fields build up in opposite directions and electric fields collapse. In E, of course, you see no accumulated charge and current flow once again becomes maximum, completing the cycle.

However, the electric field density is always proportional to the magnetic field density, with fields in space quadrature, expanding at the velocity of light. The frequency of these oscillations, then, is the same as that of the current and voltage in the antenna and when radiated into space is called the radiation field. Initially, this energy radiates as expanding spheres in the form of a wavefront. But at considerable distances from the antenna it becomes a plane at right angles to the direction of transmitted energy. Vertical lines of force lie in a vertical direction and horizontal lines of force are, obviously, horizontal. Magnetic lines of force, naturally, are just the opposite. This brings us to the subject of *directivity*.

An antenna's directivity approximates that of a light source. Maximum intensity occurs at 0° just

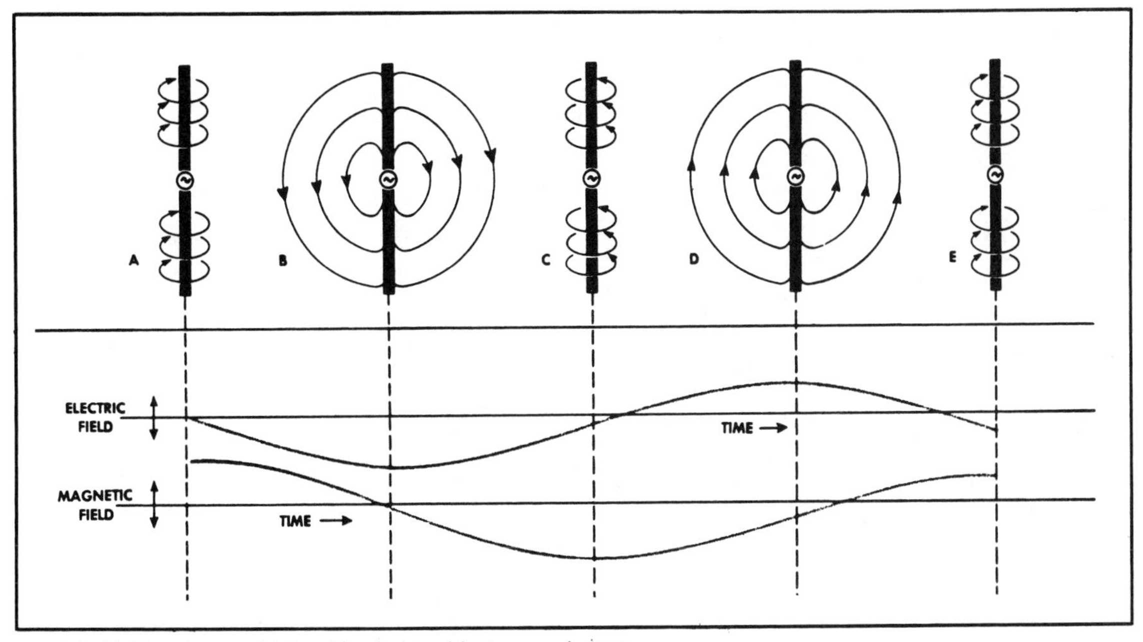

Fig. 2-2. Field and current intensities induced in two conductors.

under the source, then at a fairly uniform intensity for some 30° before quickly dropping off. With a number of concentric circles drawn about this light source, the radius of each circle denotes a certain amount of light or intensity. Between 30° and about 160° relatively little light is measured, but some then reappears relatively faint. This light intensity graphing system has been named one of *polar coordinates*. These, of course, are only two-dimensional portrayals known as the X (horizontal) and Y (vertical) axes. There also needs to be a Z axis, perpendicular to the other two and all three are known as XY, XZ and YZ planes, representing the radiation pattern of some energy source (Fig. 2-3).

That's the general idea of how antennas operate and what makes them function. It does not take into account reflectors, special masts, diplexers (for audio and video) and very-low-loss transmission lines required for maximum power transfer and hopefully small standing wave ratios—those signal-debilitating mismatches and reflections that are prime headaches for any transmitting apparatus. But, these subjects we intend to cover shortly but not in excruciating detail.

TRANSMITTER CABLING

Transmission lines have one prime objective, that of transferring energy from some transmitter to its antenna with minimum loss, maximum efficiency, and utmost reliability over an extended period. Unlike receiver transmission mediums, this line can complete efficient transfer of energy at a single basic rf frequency; but must also handle large amounts of power, maintaining both characteristics under ambient temperature swings that can be enormous. Consequently, very superior transmission lines are required to meet and exceed special requirements, particularly as frequencies increase and dB losses become greater. In dealing with them, you must consider inductance, capacitance, resistance, distributed constants, and the usual voltage/current electromagnetic fields. There's also a characteristic impedance factor usually referred to as Z_0 which identifies the quantity of current flowing in an infinitely long line upon application of a certain voltage. $Z_0 = \sqrt{L/C}$, with L and C representing the inductance and capacitance per unit length of line, depending on conductor size and the spacing between them. Large conductors closely

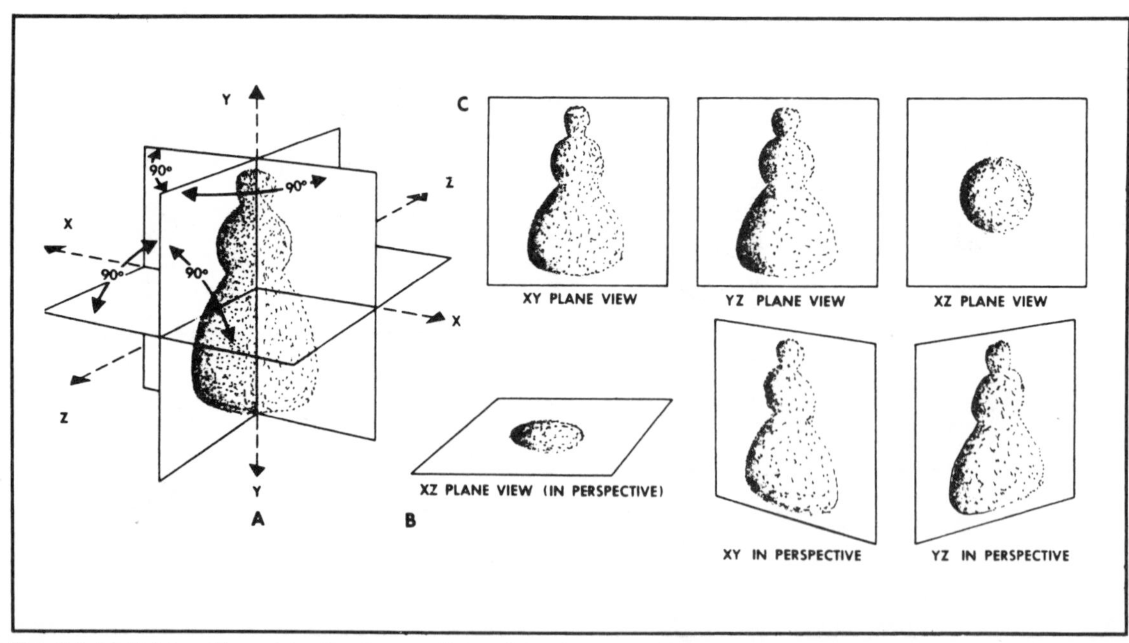

Fig. 2-3. Light intensities relate directly to radiation fields in antenna directivity patterns.

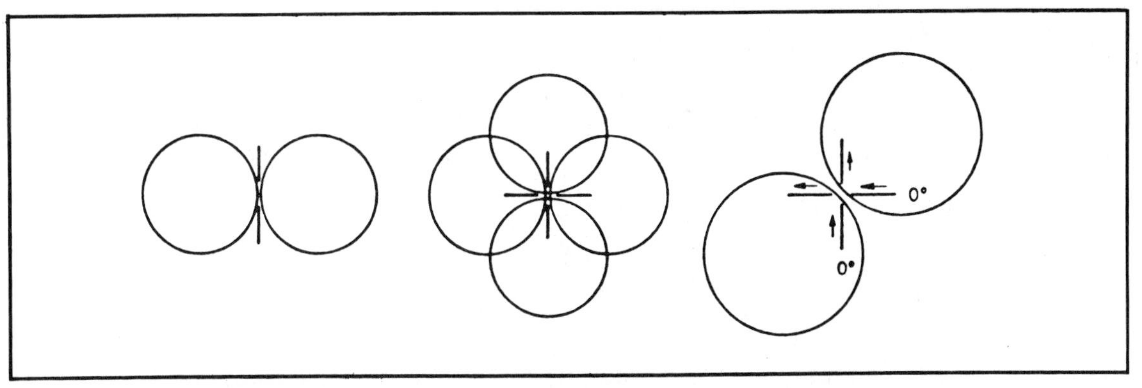

Fig. 2-4. Simple and multiple dipoles produce predictable patterns.

spaced will have a small Z_0, and small conductors widely spaced have large Z_0s. Loss of power in such transmission lines is due to three outstanding conditions: 1) radiation, 2) I^2R loss called heating, and 3) reflection, where power returns to its source. All of this brings us back immediately to two time-proven axioms:

- The impedance of a line at any given point equals voltage divided by current at that point . . . $Z = E/I$.
- Reflected and incident waves of current and voltage exist on an open-circuited line and are called standing waves. Although they appear the same, they are displaced by a quarter wavelength and voltage loops occur at current nodes and vice versa.

With this basic information as backup—and there'll be more in the receiver section—it's well for you to be brought up to date with the most recent developments which you should find rather interesting since relatively few know their way around high-power transmitters.

THE REAL WORLD OF ANTENNAS

Maximum power transfers occur when line and antenna impedances match, and also when the antenna's output impedance approaches that of free space, which is approximately 377 ohms at the point of radiation. Mismatches bring both loss of power and undesirable standing waves.

Various antenna configurations produce different radiation patterns, of which many are not necessarily circular, but can be made so by various mechanical and electrical additions. Where reflections from high mountains or skyscrapers make circular patterns inappropriate, electro-mechanical radiation must be shaped to avoid reflected images.

This is why simple dipoles without reflectors and special harness phasings are usually impractical for either consumer or commercial applications. A good example appears in Fig. 2-4. Here simple dipoles produce a figure-eight pattern; dipoles faced at right angles result in dual figure-eight overlaps (but still only in a horizontal plane); and in-phase feeds of these dipoles reduce the four circles to two in a right-hand diagonal direction but also add as generated patterns, increasing their fields.

In a continuing effort to improve on antenna effectiveness, sophisticated configurations are evolving that do all of the above in various ways and also offer pattern-shaping that's obviously advantageous to existing situations. Design sketches of modern antenna types are illustrated in Fig. 2-5. These include the superturnstile (note dipole positions), the cloverleaf, a triangular loop, square loop, supergain (basket), and wrap-around helical. Four ring antenna elements usually have better circularity than others but early types suffered from reduced bandwidth—a problem factor later models seem to overcome by element-shaping and carefully selected feeds.

Beam shaping also plays a major part in antenna operations, and two of the most important of these are:

Beam Tilts can be developed by instituting specific phase progressions along the antenna

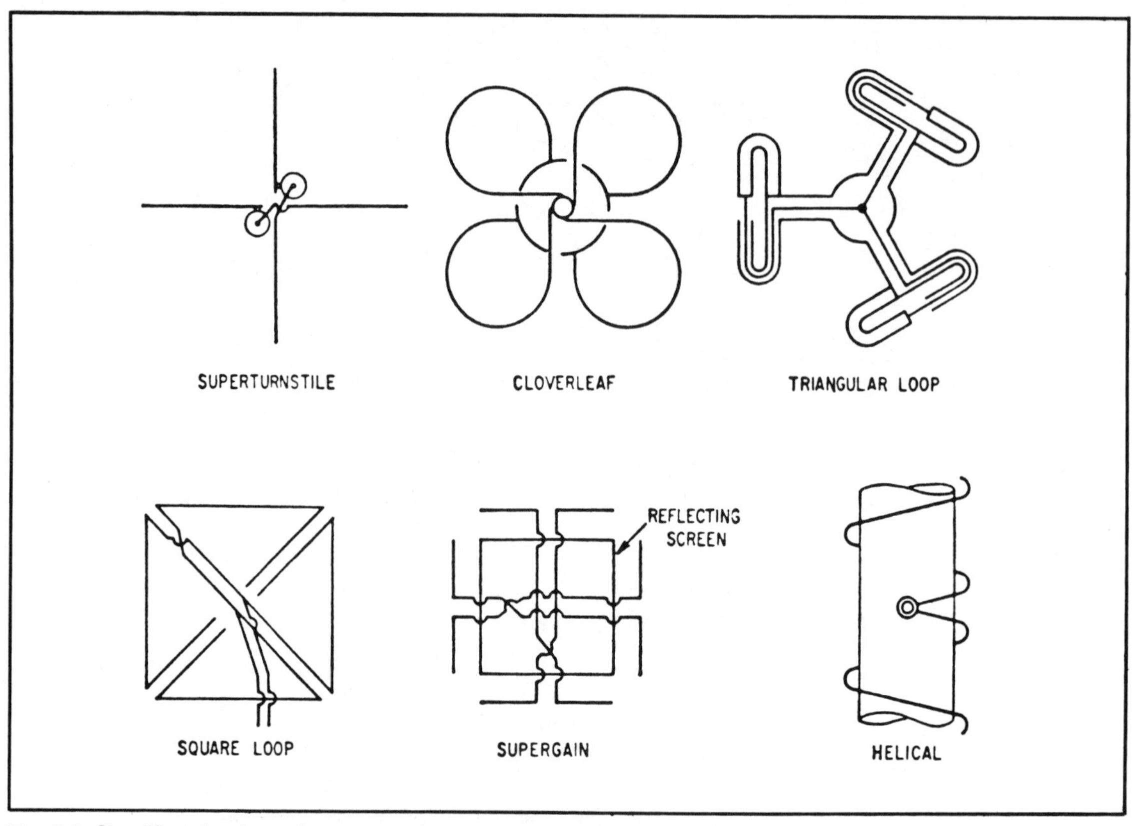

Fig. 2-5. Simplified drawings of representative antenna types (courtesy RCA).

array. The radiated pattern should be orthogonal to a straight line between the transmit radiator and the main beam's midpoint landing.

Null Fills are the result of several techniques required to prevent insufficient signal inputs or outside reflections from interfering with acceptable transmissions. Odd numbered nulls may be reduced or eliminated by supplying two equal antenna segments with unequal power, say 30:70. Amplitudes may also be exponentially tapered across an antenna, much like that of a traveling wave. Second null fills can be accomplished with antenna elements of unequal length, but then some of the signal radiates aimlessly beyond the horizon. Quadratic phase error may be introduced, as well, in addition to phase-shift tilting of the beam. This takes into account relative phase at the array's center, maximum phase errors desired at the extremities, center-edge length, and port distance from aperture center. Unfortunately, null filling occurs usually at the expense of the main transmit beam and, therefore, results in some loss of power.

In evaluating any and all broadcast types of antennas, careful attention should always be paid to vertical and horizontal patterns, rms gains, circularity, VSWR, peak power and de-icing requirements. Axial ratios are especially important in circular polarization and will be discussed separately and at length further into the chapter.

This initial introduction to theoretical and real antennas should serve well as solid foundation for the remainder of this particular discussion. A few of the principles apply to receiving antennas, as well, but since receptors handle little power and are primarily concerned with gain, wind resistance, sidelobes, and beamwidth, their discussion is treated entirely separately, with no references to transmitters.

CIRCULAR POLARIZATION

Used first by FM broadcasters to improve whip-antenna automobile reception during the 1960s, initial tests were only begun for television in 1973 by WLS-TV in Chicago to discover if there could be a comparable effect on television reception. The principle of circularity is rather simple: by propagating two linear fields of equal magnitude in phase quadrature, these fields add vectorially, producing a polarized wave usually designated as right or left hand circular polarization. Horizontal and vertical fields actually traverse the axis of propagation similar to a left-hand screw and appear to rotate clockwise in right-hand circular polarization (Fig. 2-6).

Resulting tests proved the principle and actually doubled the effective radiated power (ERP) in urban areas, delivering considerably greater and more useful signals to whip and rabbit ear antennas; and for suburban/fringe reception, offered considerably improved signal-to-noise ratios along with some increase in received power. Multipath (ghost) reception has also been reduced because delayed (especially bounce) signals tend to reverse polarity and cancel, permitting main signal receive antenna entry with little interference—all this, of course, compared to a comparable broadcast antenna with only the usual horizontal polarity. Naturally, the added vertical component will usually require equal transmitted signals and power, therefore a *vertical* transmitter output stage is also needed to share the load.

Unfortunately, at the moment, very little receive antenna activity has been noted, although a few have been made available by Channel Master, but not in the past several years. If there are any new developments, they will be described at length in the section on standard TV receptors, in addition to the better off-the-shelf antennas available and presently on the market.

The late Dr. Matti Siukola of RCA's broadcasting engineering, Gibbsboro, New Jersey, found that circular polarized signal quality depended largely upon the axial ratio which he described as the ratio of the major and minor axes of the polarized ellipse (Fig. 2-7). In turn, the polarization ratio is also a ratio of the vertical and horizontal voltage as indicated. Should such signals be received by a standard horizontal plane antenna, they will be constant "regardless of the axial ratio." But a circular po-

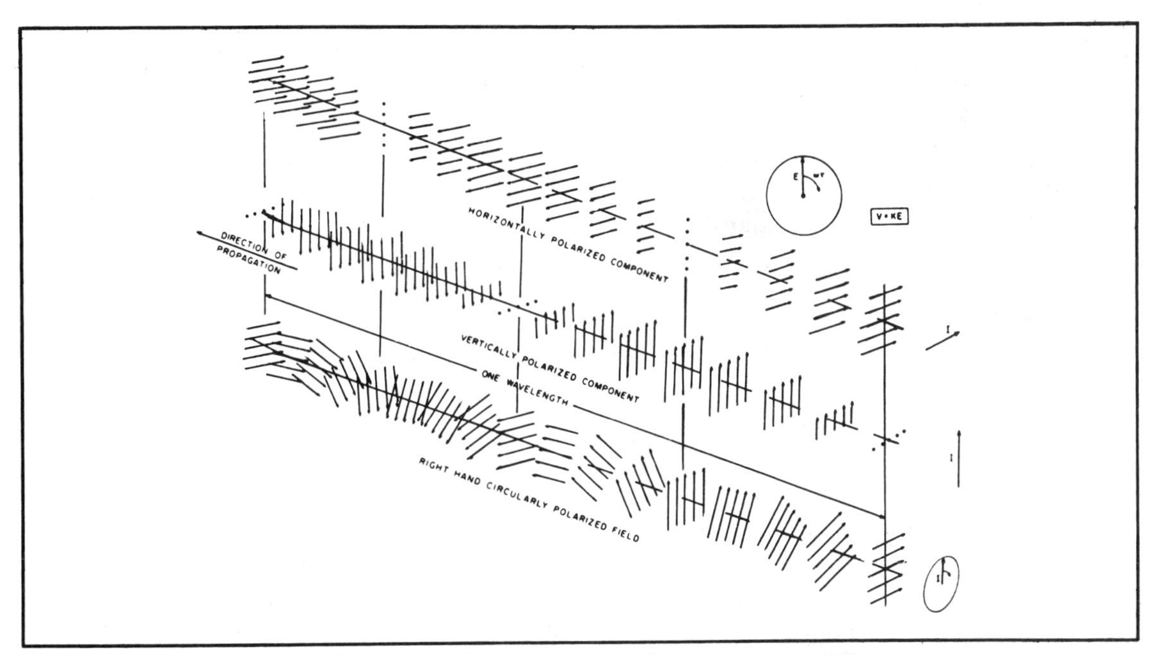

Fig. 2-6. Theoretical diagram of CP with H/V fields in phase quadrature (courtesy RCA).

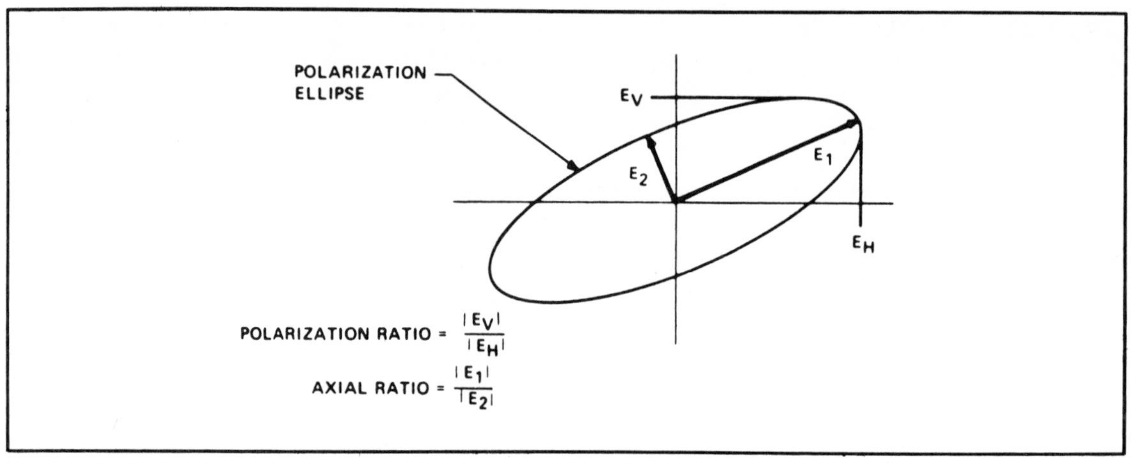

Fig. 2-7. Dr. Siukola's polarization and axial ratio concept (courtesy RCA).

larized antenna would accept the right-hand circular component with the smaller axial ratio and produce a greater receive signal. As Dr. Siukola so aptly demonstrated, a zero axial ratio would completely reject all first order reflected images.

He then selected an average field reflection (ghost) in urban areas of approximately 7%. For a workable picture, you really want no more than 2.5%. So a rejection ratio of 0.36 is required, which can easily be determined from the A_{LH}/A_{RH} rejection ratio curve illustration accompanying the polarization and axial ratio diagram so that all conditions are easily understood. As rejection ratios increase, however, so do the dBs of axial ratios, and ghost problems return to haunt the viewer wherever he may be. In the figure, the required axial ratio amounts to 6.5 dB to reduce the reflection figure of 7% to 2.5%. From these studies, RCA has subsequently developed several types of high-efficiency CP antennas, the latest of which we'll discuss shortly.

The foregoing should have developed a reasonably strong justification for the axial ratio aspect of good circular polarized antenna design. And in consideration of expanded competition, the advent of multichannel sound, full bandwidth television receivers, and continued deregulation policies of the Federal Communications Commission, circular polarization in at least VHF antenna arrays is expected to become a dominant factor in the television broadcasting industry.

Of those named, the Fan Vee seems the more conventional with good electrical and mechanical specs. There are seven layers of dual, interlaced turnstile radiators, with each consisting of one batwing bay for the horizontal signal and four double V-shaped dipoles in a second bay to radiate the vertical signal. All radiating elements are connected by a branch feed system and foam-filled flexible copper feed lines (Fig. 2-8). Four half-inch lines from each Fan or Vee radiators combine in a 4-way junction box, with two being longer than the others by a quarter wavelength. This not only results in an omnidirectional pattern but also improves impedances with phase-error cancellation. The 4-way junction boxes also receive signals from two 3-way cable-connected boxes, and these have transmission lines, which are combined below the antenna.

Used initially in RCA's Superturnstile antennas, the Fan Vee has a rotating phase feed, with the E-W pair of Fan or Vee radiators excited in phase quadrature with their companion N-S pair. The N-S pair are vertically polarized and develop a figure-8 horizontal pattern similar to the E-W horizontally polarized set. As each set becomes excited in phase quadrature on omnidirectional pattern results. With careful vertical and horizontal phase adjustments, a circular pattern emerges as shown in Fig. 2-9.

Fig. 2-8. RCA's effective Fan Vee circular polarized antenna (courtesy RCA).

A notch diplexer at transmitter output combines audio/video rf for passage through the single cable to the antenna. They are then split by a power divider for separate feeds to upper and lower portions of the antenna.

THE RCA CONTRIBUTION

RCA actually built the first CP experimental antenna for initial WLS (channel 7) tests in Chicago which were sponsored by the American Broadcasting Company in the early 1970s. So, it's not surprising that RCA is now a very prominent supplier of CP to the entire western hemisphere, and perhaps offshore, as well.

In the beginning, RCA Broadcast Systems developed four types of CP arrays called the Fan Vee, Tetra Coil, Quatrefoil, and TBJ. Briefly, their descriptions are as follows:

Fan Vee was offered as a 7-layer, top mounting radiator for VHF low band, channels 2 through 6. It consisted of four horizontal and four vertical turnstile and double V radiators. Branch fed, power requirements were 33.5 kW to develop an ERP of 100 kW.

Tetra Coil's design accommodated channels 7 through 13, and appeared as three layers of four interlaced helixes, bottom fed, and terminated by top end loads. An input power of 40 kW is needed for a 316 kW ERP.

TBK's Quatrefoil, first delivered to Miami's WPBT, serves channels 2 through 6. A square basket type one-quarter wavelength deep, it has four round "ring" radiators operating like crossed dipoles. Ring mutual coupling brings broadband response and good impedance matching. Such antennas are especially suitable for large cross-section triangular towers.

TBJ types have also been developed for good circularity with "reasonable" antenna sizes. They're side mounts and permit a "larger tower for a given radiator phase-center to array-center distance in wavelengths." A simple feed system consists of but one feed point for each panel. Both TBJ and TBKs are radome protected from the elements. Examples of the additional three CP antennas are shown in Fig. 2-10.

From the literature at hand, it appears that RCA is now offering three new types of CP broadcast radiators identified as TDM, TCL, and TCP. These abbreviations stand for Dual Mode, Tetra Coil, and Cavity 4-around configurations, the latter especially designed for square or triangular towers. All are, undoubtedly, evolutionary from the original units previously discussed, with refinements

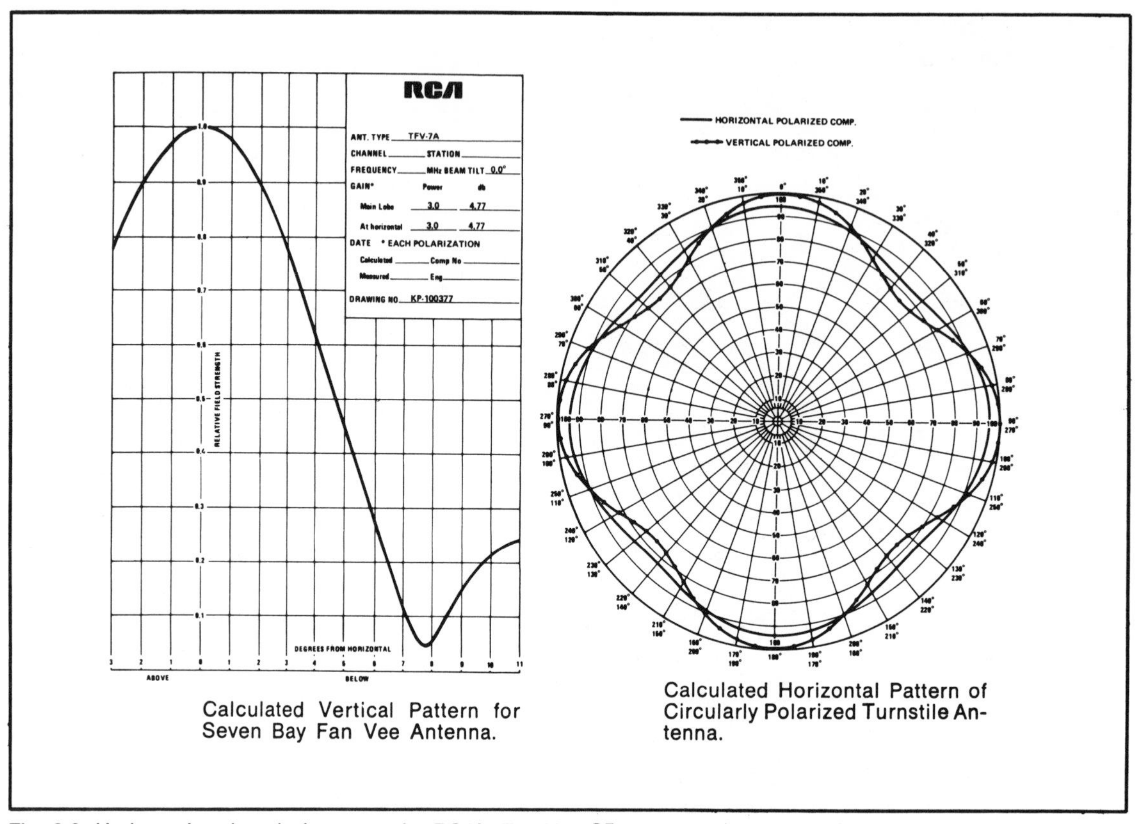

Calculated Vertical Pattern for Seven Bay Fan Vee Antenna.

Calculated Horizontal Pattern of Circularly Polarized Turnstile Antenna.

Fig. 2-9. Horizontal and vertical patterns for RCA's Fan Vee CP antennas (courtesy RCA).

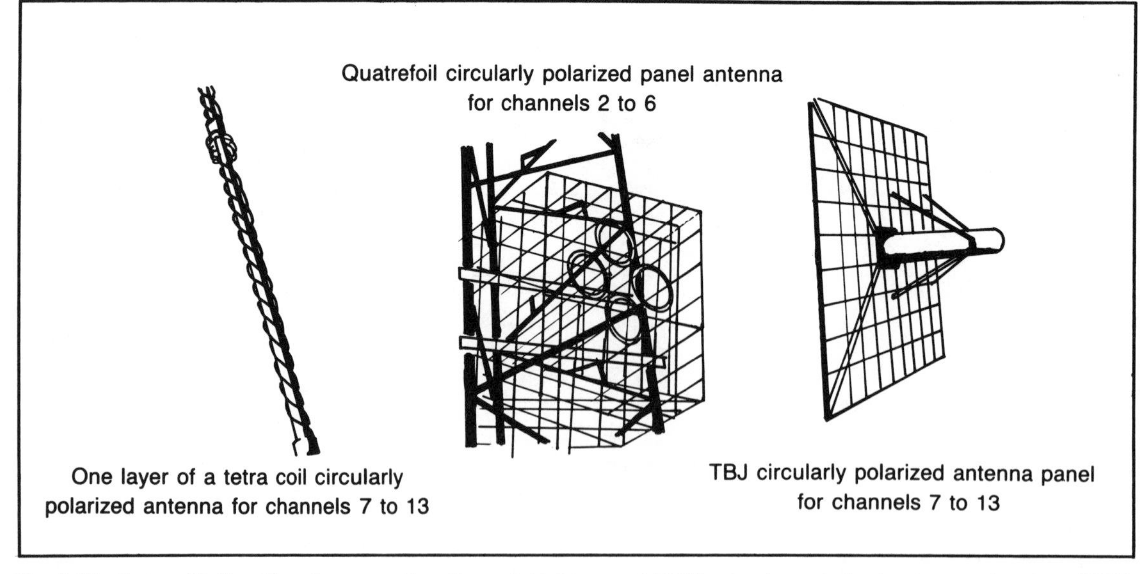

Quatrefoil circularly polarized panel antenna for channels 2 to 6

One layer of a tetra coil circularly polarized antenna for channels 7 to 13

TBJ circularly polarized antenna panel for channels 7 to 13

Fig. 2-10. Along with Fan Vee, these are the other original types of RCA's circular polarized antennas (courtesy RCA).

here and there built on both experience and advanced design practice.

You will also note that all three are VHF types and there is nothing suggesting UHF—a topic we will address in some detail when explaining products offered by one of RCA's competitors located in Illinois and California. By so doing, the entire CP U/V broadcasting band can be covered with worthwhile and contemporary information that should be with us for a good many years.

REC's three circularly polarized antennas have already been identified by name, so it's now a good time to also characterize them functionally.

TDM is a top mount antenna (Fig. 2-11) that will replace Superturnstile radiators having six or more bays on existing towers. It contains seven layers, each consisting of three radiators mounted symmetrically around the mast, with each radiator supplied by a single feedline. One junction box serves the upper four layers and a second connects the lower three layers. According to RCA, this configuration offers low windloading, good horizontal pattern directivity and noninteraction with one another. A branch-type feed system gives good vertical pattern stability, and in the TDM-7A5 unit, an FM channel may also be diplexed to an upper frequency limit of 97 MHz. Unlike the TDM-5a/10A series, there is, apparently, no double stacking for independent 2- and 6-channel operation. Transmitters that combine the video/audio signals require a notch diplexer before the antenna. De-icing power required is 18 kW for channels 2 and 3, and 12 kW for channels 4-6.

Tetra Coil serves high band VHF, consisting of channels 7-13. Good horizontal pattern circularity, excellent axial ratio, and simplified Superturnstile replacement are its primary features. Offered in three types, the TCL-12A, RCL-14A and TCL-16A are constructed of four radiating coils wound around each of the sections of the supporting mast. Their end loads radiate any left over energy and also minimize energy reflections, which could upset the conducting traveling wave during broadcast operations. Phasing between the three vertical sections maintains electrical beam tilt, and power input rating for the TCL-16A amounts to 40

Fig. 2-11. Superturnstile antenna replacement, the TDM (courtesy RCA).

kW. A branching type lightning protector comes with the antenna and all helical radiators are pole-grounded at the bottom of each layer, with the antenna base being grounded to the antenna's top. De-icing power ranges between 12 and 14 kW.

Cavity, or Type TBK Quatrefoil emerges as a side-mounted panel CP antenna has been designed for channels 2 through 6 with both standard and directional patterns and "excellent" horizontal circularity. There are three panel radiators for each layer, supplied with a branch-type feed system. A panel consists of four ring radiators mounted in front of a rectangular pattern-shaping screen—from whence its name derives. The six-bay version has a peak visual power rating of 55 kW and electrically transparent radomes and require no electrical de-icing power. Circularity measures ± 2.5 dB, and axial ratio 1.5 dB average and 3 dB maximum. Figure 2-12 is a photograph of the TBK antenna illustrating three bays in a massive installation.

Fig. 2-12. TBK quatrefoil sidemount CP antenna (courtesy RCA).

CPV AND CBR BY HARRIS

Since the FCC's 1978 approval action, the Broadcast Products Division of the Harris Corp. has introduced two new types of circular polarized antennas called CPV (probably for circular polarized Vee or video) and CBR for cavity backed radiator. Superior axial ratios within 2 dB at any azimuth angle and omnidirectional circularity of ± 2dB is claimed for each. The CBR version has wide bandwidth and allows multiplexing of all VHF high-band TV channels, requires no electrical de-icing when used with radomes, and its wire mesh cavities minimize windloading. Horizontal elements may be driven independently of vertical elements, and all are at dc ground potential for lightning protection.

The equally wideband CPV features top mast loading with windloading comparable to both Batwing and traveling-wave antennas, permitting easy replacement of certain existing antennas with CPV, usually with only minor modifications. Designed for both high and low VHF channel coverage, dual line-feeds separately serve high and low elements. Horizontal/vertical elements are coincident and separate, along with three- or four-side symmetry so that feed systems can deliver a considerable range of horizontal patterns, while low axial ratios tend to reduce ghosting.

Design objectives in these new antennas included reasonable gain equal E and H patterns over the intended frequency band, good axial ratio, minimum susceptibility to ice/snow conditions by good impedance matching, lightning protection, omnidirectional pattern but directional capability, and independent vertical *or* horizontal inputs available during emergencies. One attractive proposal was a basic radiator with arrays and a cavity-backed dipole. Basically, the cavity antenna uses crossed dipoles in quadrature phase set in the aperture of the backing cavity.

Such dipoles excite the cavity with a rotating rf field parallel to the dipoles which can be represented by a constantly rotating vector revolving one rpm/wavelength of propagation distance. Cavity aperture generally determines the radiation patterns, while dipole size and geometry govern antenna impedance and VSWR.

According to Harris, a backing cavity isolates the antenna generally from its tower and other antennas, supplies good beamwidth and additional gain, and excellent pattern control so that horizontal and vertical polarized patterns have nearly equal beamwidth. Cavities may be either rectangular or circular, with rectangular better suited to printed circuit applications and receiving, while circular configurations are more flexible for transmitters.

A special wideband flat dipole had to be designed to excite the cavity and furnish required power levels, with this radiator related to a center-fed cylindrical antenna with 1/a of approximately six for the impedance bandwidth (Fig. 2-13). De-icing provisions included either/both a partial radome cover and/or electrically heated dipole elements. Cavities may be excited vertically, horizontally or circularly, but Harris says they are usually supplied from a power divider and 90° phasing loop for right hand circular polarization, with resulting bandwidths sufficient to allow multiplexing high channel combinations.

For required power gain, a number of vertically stacked arrays are used, with an 8-bay in CP measured at 7.9, a 12-bay at 11.9, and a 16-bay at 15.8. Circularity, VSWR and axial ratio remain constant for all three at 2 dB for the first two and 1.1 for axial ratio throughout, regardless of the number of stacked arrays up to 16. and possibly beyond. Circularity, however, is affected by tower face size, decreasing in both 3-around and 4-around (sides) configurations. Wind loading, according to Harris, is minimized by round frame supports, wide grids, and only bell-shaped radomes cover the enclosures.

LATER CPV VERSIONS

The newest CPV products from Harris (Fig. 2-14) are mainly VHF units, many of which are already installed in more than 20 locations in the United States as well as three countries in Latin America.

In these antennas, cavities have now been supplanted by three crossed Vee poles, separated by an equal number of vertical grids, mounted at 120° intervals about the transmitting mast. The grids isolate their companion dipoles and also aid in horizontal beam shaping, while the dipoles receive signals in phase-quadrature to rotate the transmitted energy in right-hand circular polarization.

Available up to 100 kW power, fiberglass radome covers are supplied to protect all radiating elements, negating the need for electrical deicing equipment, resulting in considerable savings. Single or dual transmission lines may be used, the latter permitting half of the antenna to still receive transmitter signals if the other half develops problems. Azimuth fields patterns for both the channel 2-6 and 7-13 high- and low-band antennas are supplied by Harris as shown in Fig. 2-15. Axial ratios are less than 2 dB, and VSWR of 1.1:1 maximum. Improved ghost reduction and better S/N ratios regardless of the receiving antenna are also claimed by Harris.

HARRIS' TAV-5LE

Designed specifically as a replacement for 6-bay Batwing horizontally polarized antennas, the TAV-5LE fits this space handily and its extended radiation center places it within two meters of existing Batwings. Pattern contouring for beam tilt and null fill may be realized from ordinary phase distribution methods and elevation control does not degrade the 2 dB axial ratio or the 70 kW visual or 20% aural ratings. Power gains with polar-

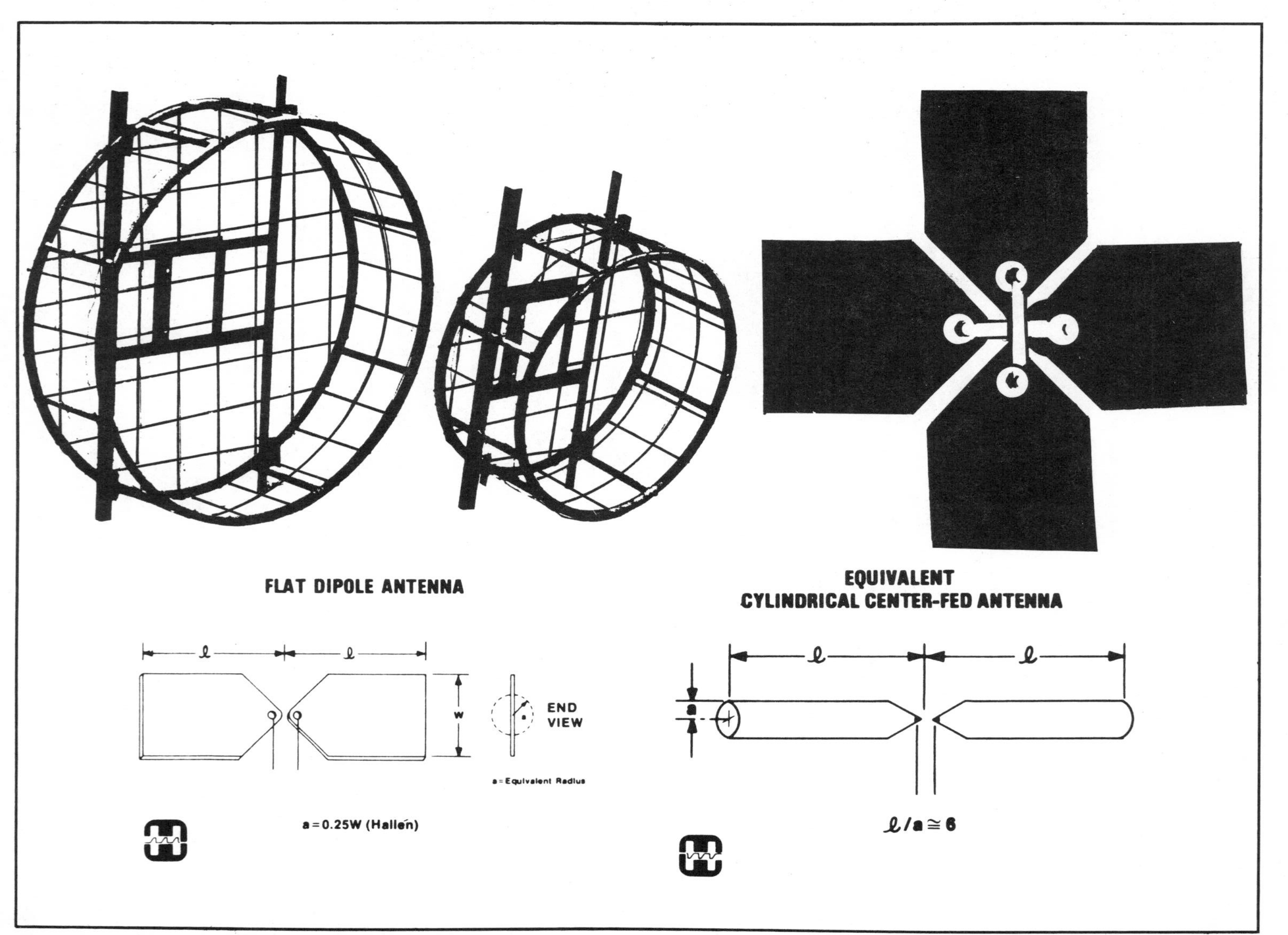

Fig. 2-13. Circular cavities shown with wideband, flat dipole compared to cylindrical dipole (courtesy Harris Corp.).

Fig. 2-14. A new 12-bay CPV circular polarized antenna of latest design (courtesy Harris Corp.).

izations do vary slightly with null fills between 7 and 30%. Antenna appearance is similar to that shown in the previous illustration.

Upper and lower bays are served by independent transmission lines and all masts are hot dip galvanized with stainless steel hardware. For an 85-foot antenna, as an example, the weight for channel 2 of such an antenna amounts to 14,500 lbs, and for a 65-foot antenna serving channel 6, the weight measures 9.200 lbs. So you see as transmission frequencies increase, antenna sizes decrease and weights become considerably less. Patterns, beam tilt, and null fills are required in the manufacturer's ordering information.

Distances and depression angles for various antennas are given in a special chart supplied by Harris for its customers in Table 2-1. You'll find it very easy to use. For instance, an 800-foot antenna would have a horizon distance of 40 miles and a small initial depression angle. But if this angle was specified at 7 °, then distance in miles to the receiving location would become 1.23. So just a small amount of depression angle can mean a great deal in antenna coverage over considerable distances and the obvious loss or gain of large segments of suburban and urban audiences.

UHF CIRCULAR POLARIZED ANTENNAS BY ANDREW

Billed as America's only manufacturer of complete UHF television antenna and transmission lines, the Andrew Corporation also has facilities in Canada, Australia, Brazil, and Scotland. Products include towers, shelters, antennas, waveguide transmission lines, as well as various services.

While not ordinarily manufacturing true all-CP radiators for UHF broadcasters because of several factors including power, cost, and the usual maximum station-authorized user output, Andrew does provide elliptical polarization having 20 percent of the broadcast energy in the vertical plane. This addition is said to reach 4/5 the distance of all horizontally polarized contours offering improved signal strengths and added ghost cancelling.

Named the TRASAR™ for traveling-wave, slotted array design, these antennas are available

Table 2-1. Distances and Depression Angles for Various Antenna Heights (courtesy of Harris Corp.).

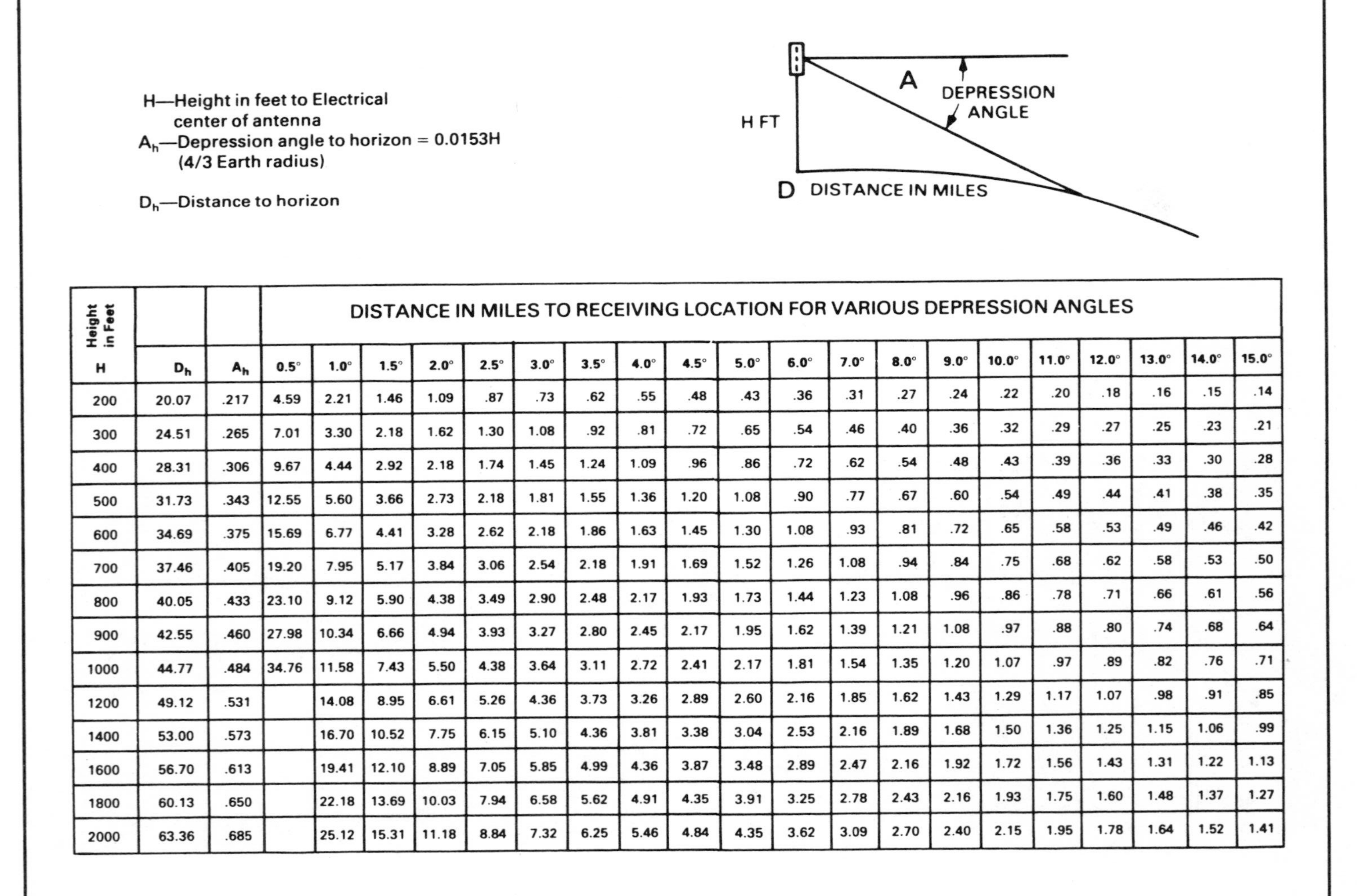

H—Height in feet to Electrical center of antenna
A_h—Depression angle to horizon = 0.0153H (4/3 Earth radius)

D_h—Distance to horizon

Height in Feet			DISTANCE IN MILES TO RECEIVING LOCATION FOR VARIOUS DEPRESSION ANGLES																			
H	D_h	A_h	0.5°	1.0°	1.5°	2.0°	2.5°	3.0°	3.5°	4.0°	4.5°	5.0°	6.0°	7.0°	8.0°	9.0°	10.0°	11.0°	12.0°	13.0°	14.0°	15.0°
200	20.07	.217	4.59	2.21	1.46	1.09	.87	.73	.62	.55	.48	.43	.36	.31	.27	.24	.22	.20	.18	.16	.15	.14
300	24.51	.265	7.01	3.30	2.18	1.62	1.30	1.08	.92	.81	.72	.65	.54	.46	.40	.36	.32	.29	.27	.25	.23	.21
400	28.31	.306	9.67	4.44	2.92	2.18	1.74	1.45	1.24	1.09	.96	.86	.72	.62	.54	.48	.43	.39	.36	.33	.30	.28
500	31.73	.343	12.55	5.60	3.66	2.73	2.18	1.81	1.55	1.36	1.20	1.08	.90	.77	.67	.60	.54	.49	.44	.41	.38	.35
600	34.69	.375	15.69	6.77	4.41	3.28	2.62	2.18	1.86	1.63	1.45	1.30	1.08	.93	.81	.72	.65	.58	.53	.49	.46	.42
700	37.46	.405	19.20	7.95	5.17	3.84	3.06	2.54	2.18	1.91	1.69	1.52	1.26	1.08	.94	.84	.75	.68	.62	.58	.53	.50
800	40.05	.433	23.10	9.12	5.90	4.38	3.49	2.90	2.48	2.17	1.93	1.73	1.44	1.23	1.08	.96	.86	.78	.71	.66	.61	.56
900	42.55	.460	27.98	10.34	6.66	4.94	3.93	3.27	2.80	2.45	2.17	1.95	1.62	1.39	1.21	1.08	.97	.88	.80	.74	.68	.64
1000	44.77	.484	34.76	11.58	7.43	5.50	4.38	3.64	3.11	2.72	2.41	2.17	1.81	1.54	1.35	1.20	1.07	.97	.89	.82	.76	.71
1200	49.12	.531		14.08	8.95	6.61	5.26	4.36	3.73	3.26	2.89	2.60	2.16	1.85	1.62	1.43	1.29	1.17	1.07	.98	.91	.85
1400	53.00	.573		16.70	10.52	7.75	6.15	5.10	4.36	3.81	3.38	3.04	2.53	2.16	1.89	1.68	1.50	1.36	1.25	1.15	1.06	.99
1600	56.70	.613		19.41	12.10	8.89	7.05	5.85	4.99	4.36	3.87	3.48	2.89	2.47	2.16	1.92	1.72	1.56	1.43	1.31	1.22	1.13
1800	60.13	.650		22.18	13.69	10.03	7.94	6.58	5.62	4.91	4.35	3.91	3.25	2.78	2.43	2.16	1.93	1.75	1.60	1.48	1.37	1.27
2000	63.36	.685		25.12	15.31	11.18	8.84	7.32	6.25	5.46	4.84	4.35	3.62	3.09	2.70	2.40	2.15	1.95	1.78	1.64	1.52	1.41

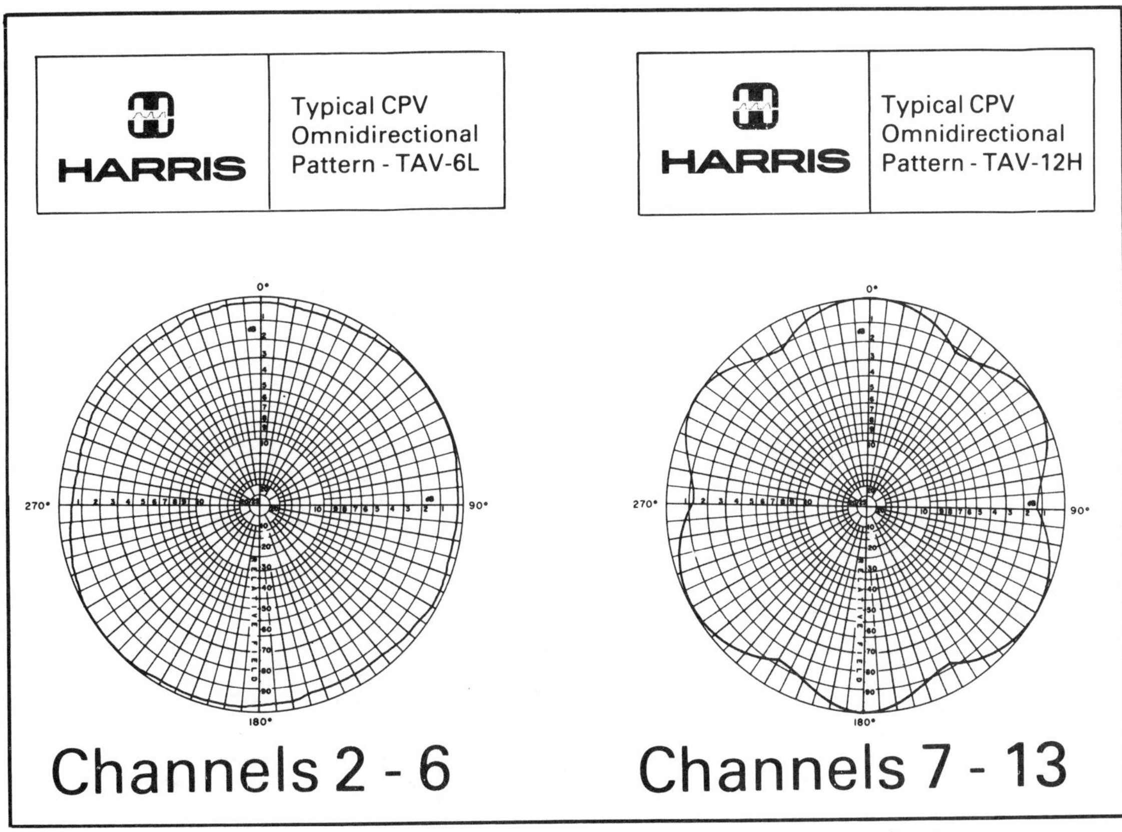

Fig. 2-15. Harris' newest CPV circular polarized antenna patterns for VHF (courtesy Harris Corp.).

in both sidemount and topmount (Fig. 2-16) configurations. They are medium power (500 and 2500 kW) arrays with cylindrical radomes and are suitable for both primary as well as standby installations. Sidemounts produce a skull pattern while topmounts are usually omnidirectional. Sidemounts are available in 2, 8, 16 and 24 elements, and topmounts are stocked for channels 14, 28, 42, 56, and 69. Azimuth examples of each are illustrated in Fig. 2-17.

Topmounts are self supporting and attach directly to the tower's top plate where they have four lightning rods for protection. In addition to omnis, cardioid, peanut, and trilobe patterns are available in both high and medium gain units. Sidemount antennas may also be purchased in cardioid, peanut, and butterfly patterns, all with heavy first null fill. TRANSAR input power ratings are limited by the characteristics of the input transmission

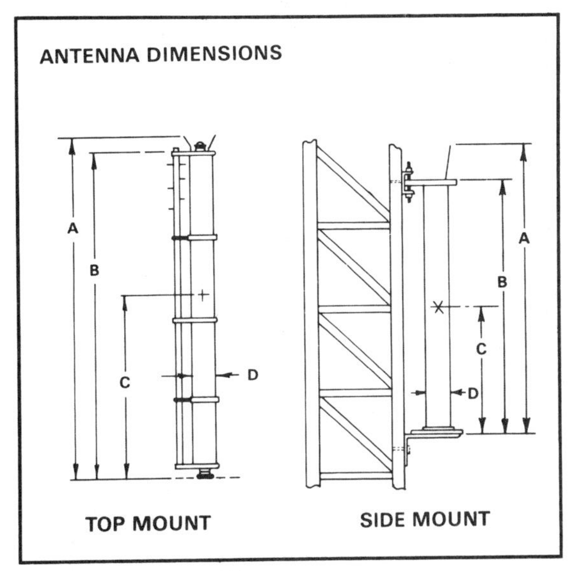

Fig. 2-16. TRANSAR™ top and sidemount slotted array traveling wave UHF antennas (courtesy Andrew Corp.).

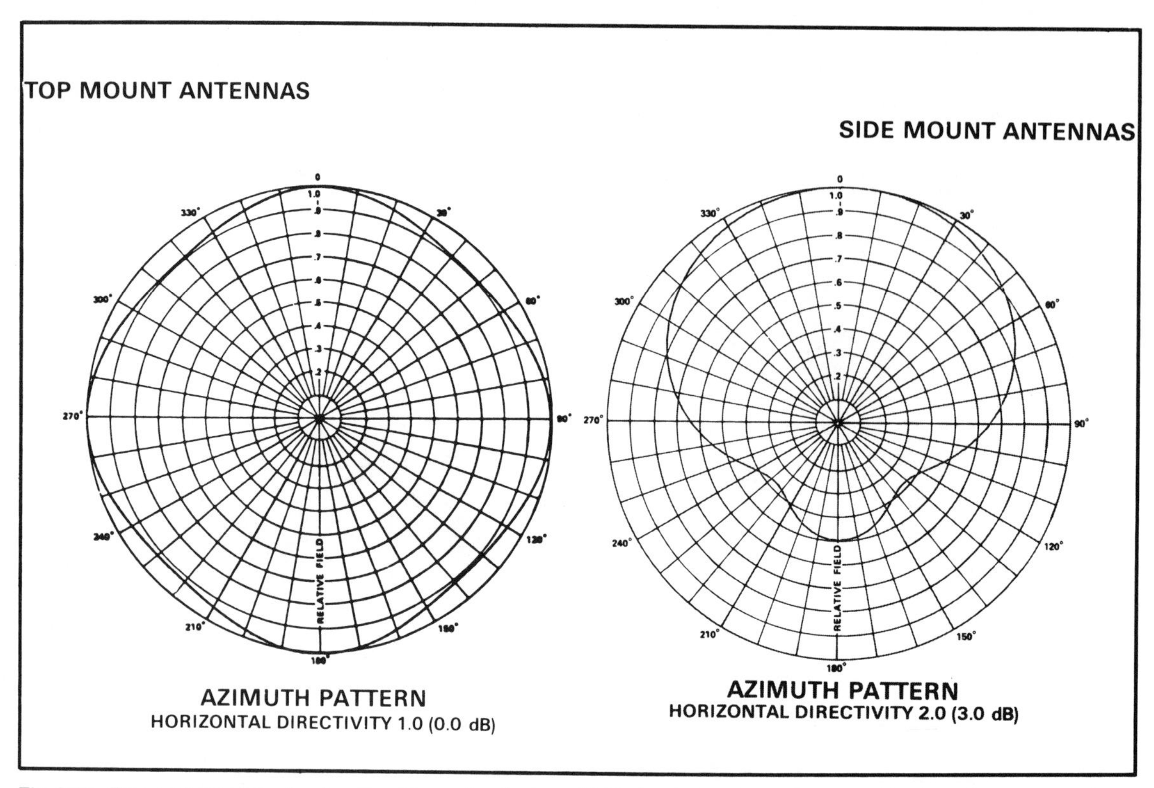

Fig 2-17. Top and sidemount azimuth patterns for Andrew VHF circular patterns (courtesy Andrew Corp.).

lines, which are available with 3 1/8- to 9-inch coax and waveguide connections. These depend, of course, on system design. According to Andrew, the standard topmount array for channels 14 through 50 has an identical power rating as the 8 3/16-inch transmission line. Electrical de-icing equipment is usually unnecessary since a "rugged" pressurized fiberglass radome surrounds the entire radiating antenna surface. Sidemount antennas are less rigid and require a 60-percent lighter antenna structure, but are not recommended for omnidirectional emissions due to tower reflections. Sidemount antennas also require only two lightning rods.

Andrew engineering firmly believes even a partial vertical field addition to UHF signals not only results in better distance coverage but also the ability "to penetrate built-up suburban and city areas at close in distances." With 15 percent of the antenna input power diverted to vertical polarization, it was discovered in one test record that only a 4 percent area coverage reduction in ERP resulted, but considerably better coverage for most of the same or greater area and little or nothing given up in the horizontal field in return.

It might be interesting for some of the receiver antenna makers to undertake similar tests and at least come up with a partial CP *indoor* receptor that would take advantage of these apparent gains in electromagnetic propagation at UHF. Eventually we may all benefit. Meanwhile, Andrew reports 52 TRASAR sidemount and 32 topmount antennas now in operation throughout the United States and Taiwan as of October 1984. Andrew also has 4-bay and appropriate multiples with coaxial power dividers for low-power television (LPTV) operating at UHF.

FEEDERS

This is the connecting feed system between the antenna terminal and its various radiating elements, with some power losses to be expected. Generally, there are three of these feed systems in use: branching, traveling wave, and standing wave.

Branching goes with the Superturnstile antenna, junction boxes, and various signal divisions.

Traveling Wave feeds are common to cylindrical slot antennas and work through gradual attenuation of powered signals as they spread across the aperture of the antenna. Rod radiators in spiral configuration can offer a linear propagation and phase relationship, resulting in desirable lags. Rotating phase, of course, produces a rotating field and, therefore, an omnidirectional pattern.

Standing Wave feeds usually work with slot-cylinder UHF antennas which are the TRASAR antennas we have just described.

NOTCH AND BRIDGE DIPLEXERS

Antennas with single inputs must have a network of some description to combine the modulated video and audio signals. This is called a diplexer and, in this instance, a notching diplexer. There are also bridge diplexers for two-input systems in quadrature, and these are represented as a Wheatstone bridge having two antenna terminations and two bridge elements as balances. These usually work with Superturnstile antennas with visual signals and aural signals balanced and unbalanced with respect to ground, respectively.

Filterplexers are often combined with notch diplexers to absorb lower TV sidebands. Their more common name is vestigial sideband filter. It has two pairs of cavities as rejectors with such wasted energy dissipated by load termination.

TRANSMISSION LINES

Outside, and sometimes even inside the television transmitter, this is one of the more misunderstood segments of the broadcasting industry, and one of the most critical. It is the all-important link between power electronics and the electromagnetic radiating elements that propel audio- and video-modulated signals into space. Its characteristics and constant performance can mean the difference between station coverage, acceptability reliability, and economic success or outright failure. Engineering evaluation and acceptance can never be too cautious or careful since signal losses, changes with temperature, and day-to-day reliability are overriding factors in the broadcast service.

Transmission lines are ordinarily selected for efficiency (meaning least attenuation), power handling, durability, and mechanical suitability. The better the transmission, coupled with a highly efficient antenna, the less power required of the transmitter—an important factor to bear in mind. In addition, standardized EIA-approved products are usually preferable since conductor diameters, line flanges, and connectors from different manufacturers will mate successfully. Ordinarily, VHF lines have 50-ohm impedances, with a few special UHF types manufactured for 75 ohms (see Fig. 2-18).

Four Categories

Transmission lines may be identified generally as occupying four categories:

Flexible Solid Dielectric Coax. RG-213/U and RG-218/U are the solid dielectric cables having solid or stranded center conductors, plastic insulators, and braided copper outer conductors beneath a plastic outer covering. These are *not* usually recommended for permanent use since they have relatively high attenuation losses, age quickly, absorb moisture, and contribute inadequate shielding against outside interference. Temporary, relatively low-power hookups, sampling lines, and jumper connections are their prime uses.

Semiflexible Foam Dielectric Coax. Better temperature and loss figures characterize this cable, but it is mainly used in AM and FM applications rather than television. It has a copper wire or tubing inner conductor, foam polyolefin dielectric, and a flexible outer conductor of either copper or aluminum, and the cost is moderate. Rf leakage, common to braided outer conductors, is eliminated by the solid outer conductor.

Semiflexible Air Dielectric Coax. Available in diameters from 0.25 inch to 8 inches, these air dielectric semiflex cables are quite popular due to easy installation, low VSWR, high efficiency, and power handling ability. Used in AM/FM/TV broad-

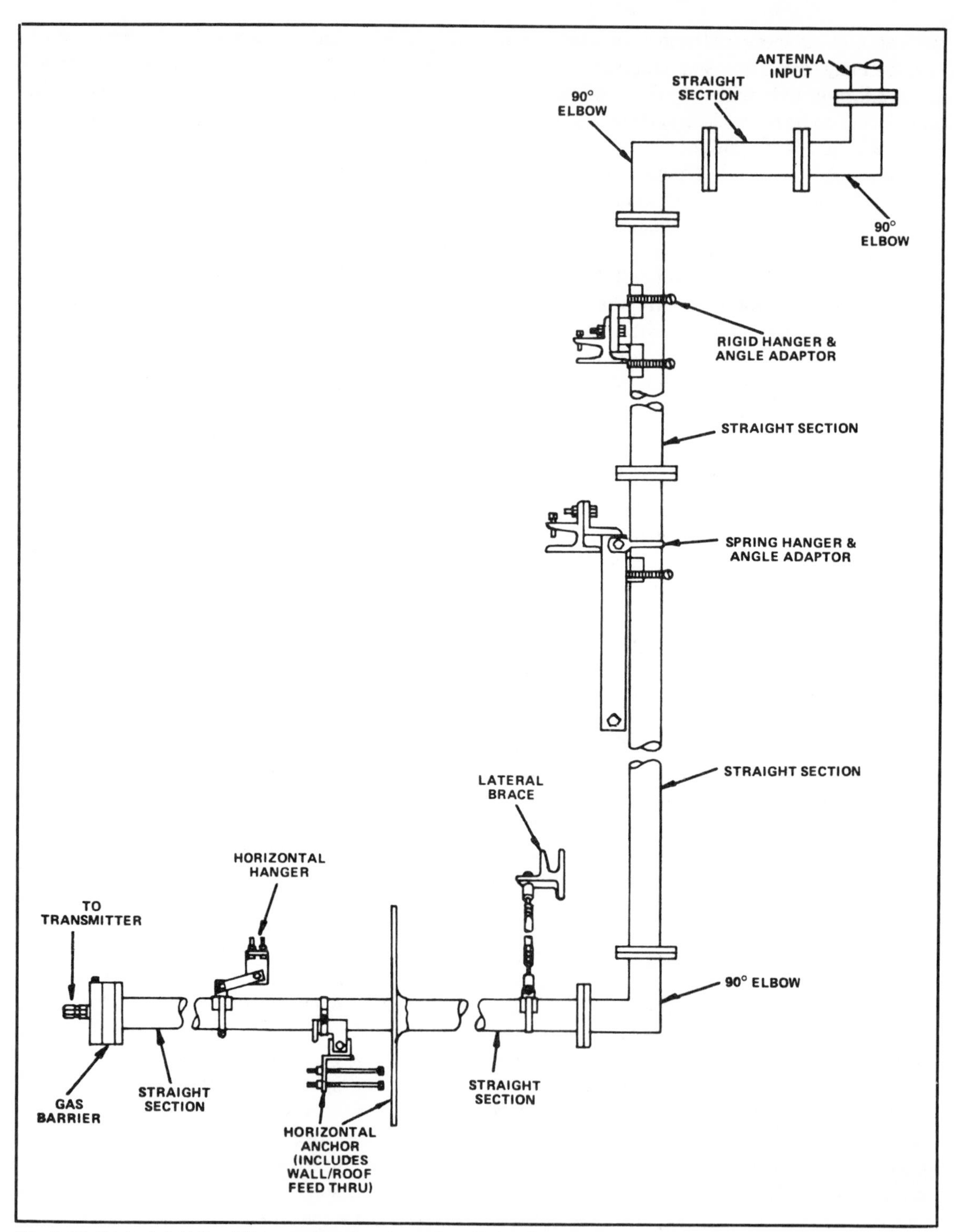

Fig. 2-18. Typical rigid line TV transmitter-to-antenna installation (courtesy Andrew Corp.).

casting, their continuous lengths avoid electrical discontinuities (no joints), and performance can be measured prior to factory shipping. Jacketing and bending characteristics should receive reasonable scrutiny.

Rigid Coax Transmission Line. Manufactured in 20-foot flanged lengths, this rigid, hard-tempered copper tubing features low attenuation and VSWR, with the inner conductor supported by either teflon discs or equivalent standoffs.

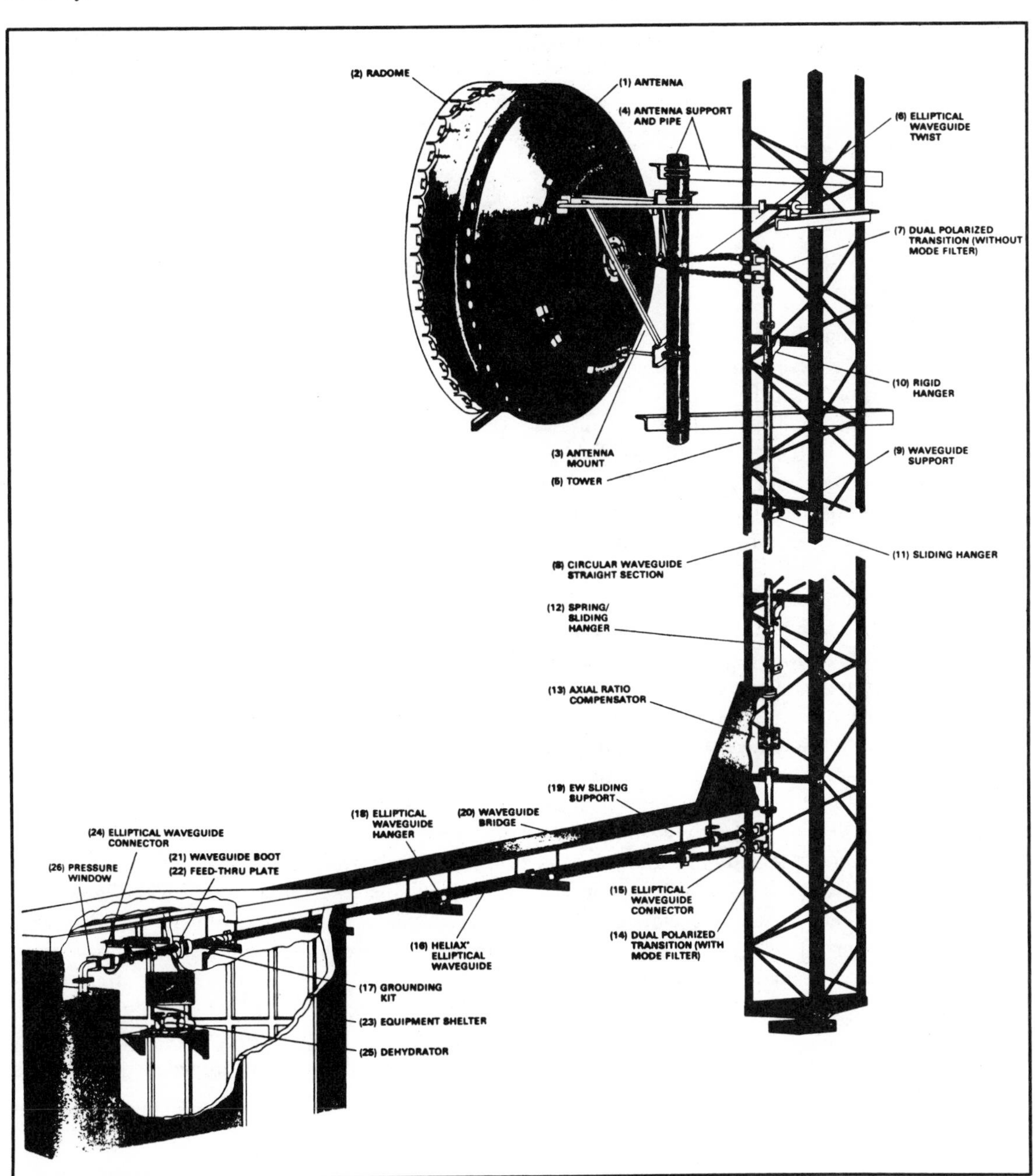

Fig. 2-19. Circular waveguide antenna system with dual polarity (courtesy Andrew Corp.).

Used primarily for high-power applications, you may see it in transmitter interconnections and main feeder lines to the tower, diplexers, and dummy loads. Various diameter sections are available, depending on applications. Air dielectrics, of course, are much less lossy than solid dielectrics.

Transmission line efficiency may be calculated as follows:

$$\text{Efficiency} = \frac{1/(\text{dB attenuation total})^{-1}}{10} \times 100 \text{ (in percent)}$$

where -1 represents the antilog (10^x).

NEWER TYPES

One of the best known and most reliable manufacturers of broadcast transmission lines is the Andrew Corporation which makes HELIAX™ cables for the entire industry. These include air dielectrics; foam dielectrics, elliptical waveguides, rectangular waveguides, and circular waveguides in most or all appropriate and convenient sizes.

Circular waveguides range in diameter from 13.5 to 17.5 inches and are supplied in 12-foot lengths for channels 14 through 69. They are especially suited for lengthy vertical runs to tower-mounted antennas. New high power filters eliminate secondary images, reflections, and any picture smear, and an exclusive 90-degree bend allows usage for both vertical and horizontal runs permitting continuous circular waveguide connection between transmitter and antenna. It may also be air-pressured to 2 lbs/in^2, permitting an air path for whatever reason to the antenna. Andrew says that windloading on the circular shape is 33% less than for an equivalent rectangular waveguide.

With low attenuation and high power capabilities, circular waveguides may also carry two polarizations with 30 dB minimum isolation between the two. A typical circular waveguide system appears in Fig. 2-19. Note dual polarization (with mode filter), waveguide hanger, and its exit compartment from the transmitter. You're looking, of course, at a radome, three-sided mast and antenna pipe support. Attenuation figures per hundred feet range between 0.0300 and 0.0521 at average powers of 294 to 670 kW, depending on channel number and waveguide diameter. Try matching that with solid conductor, sheath cables and see what results.

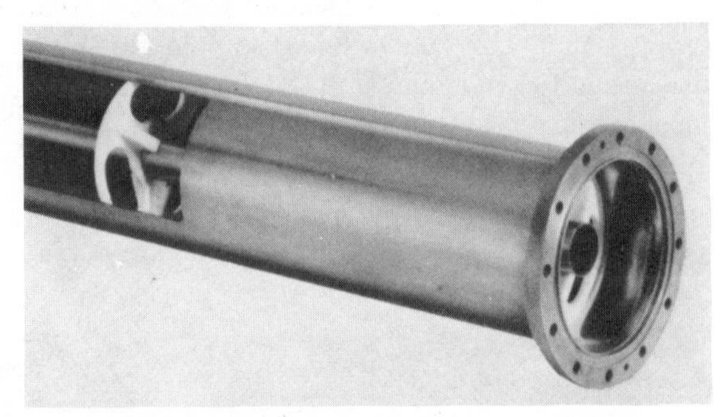

Fig. 2-20. Differential expansion between inner and outer conductors ensures long life and excellent line characteristics (courtesy Andrew Corp.).

Andrew also is offering a new 6 1/8″ 75-ohm rigid coax line (Fig. 2-20) having a special inner conductor support, permitting differential expansion between inner and outer conductors. With 75-ohms impedance and a maximum frequency handling ability of 900 MHz, its rated velocity is 99.8 percent and peak power rating 1000 kW. Attenuation per hundred feet ranges between 0.1599 and 0.0416, once again depending on the channel and power output. It is available in standard 19 1/2, 19 3/4, and 20-foot lengths and some special (custom) spans as well. Fixed male inner connectors are at both ends and may be used as gas inlets.

An "exclusive" bellows expansion device completely eliminates sliding contact wear, which can result in failures of anchor connector galling, slippage and sliding. There are also ultra heavy duty Teflon disk insulators, welded flange attachments to the outer conductor, and a heavily reinforced unequal 90° leg elbow with captive inner conductors guaranteeing constant impedance and maximum mechanical strength.

Chapter 3

Receive Antennas and Transmission Lines

WITH THE VERY RAPID DEVELOPMENT OF broadband television receivers and TV/monitor combinations, plus coming advances in videocassette recorders and video discs, those in the service industry had better tend to their knitting and supply the public with what they need. Understanding the theory *and* applications of both antennas and transmission lines can aid those involved considerably and bring to all of us the kind of reception these new video entertainment equipment deserve and require. Never before have full NTSC frequency chroma, luminance, and BTSC-dbx multichannel sound been offered to the public, and they should receive all of it for maximum enjoyment.

Unfortunately, the underskilled and underpaid television service industry is shrinking every year as old timers retire or die off, new people with little training filter into the depleted ranks, and only the almighty dollar remains as a mark for survival. For although VCR and receiver sales are expanding at a tremendous pace, fewer and fewer independent technicians remain to render service, and the major load for all repairs is rapidly devolving on factory or distributor-operated installations throughout the United States. These people, of course, are specialists and normally work only on their own captive equipment, consequently narrowing equipment maintenance to a relatively few brands.

As for antenna installers, their numbers have depleted also, and only the largest and (we hope) most adept have survived. Normally these people are not technical, have little extensive instruction, and work with only the materials assigned. We hope this will change, but if installation people *continue* to put up antennas and transmission lines with cheap products and/or insufficient knowledge and skills, the entire TV industry is bound to suffer, as it has for over 30 years.

RECEIVING ANTENNAS

Today's ordinary antennas have principally originated from either the free air Hertz 1/2 wavelength dipole or Marconi's 1/4 wavelength design. The latter has one end grounded or attached to counterpoise wires measuring a second 1/4 wavelength. The Hertz (also the new designation for cycles/second) antenna is our main interest since all receive television antennas consist of various dipoles coupled or summed to simulate a *half* wavelength. A wavelength, of course, is defined as the distance between two same-phased wavefronts and is mathematically known as lambda (λ), where:

$$\lambda = \frac{\text{Velocity (wavelength in similar time units)}}{\text{Frequency in Megahertz}}$$

with velocity (and lambda) either in terms of meters (300) or feet (984), relative to frequency in MHz.

As current becomes maximum in any Hertz antenna, impedance goes to minimum, measuring 73 ohms, while end impedances are 2400 ohms. Its half wavelength above 30 MHz is calculated by lambda/2 = 492/f(MHz), and below 30 MHz, lambda/2 = 468/f(MHz) because of a 5 percent reduction in electrical length. Pairs of these half wave units folded in parallel with center connections constitute a folded dipole having center impedances of 4 × 73 ohms, or just under 300 ohms. If folded all the way into a circle such dipoles are still useful and all three configurations have *identical* gains. Remember UHF loop antennas?

Antenna elements, of course, may be tuned to various frequencies by trimming them to certain lengths, connecting them in certain ways, and adding or subtracting any reasonable number of elements. When these elements are added to U/V antennas, the short or round ones in front are known as *directors* and those in the rear as *reflectors*. Such additions are specifically designed to increase the directivity of an antenna since directors are always forward of signal-amplifying *driven* dipoles with reflectors behind (Fig. 3-1). Reflectors also shield out unwanted rear interference, including cochannel signals from a same-frequency distant station. When directors and reflectors are added, their orientation, length, and mounting positions determine many antenna characteristics, especially in Yagi arrays usually designed for high gain and specific directivity.

Electrically, the dipole may be represented as an open circuit length of quarter wave line excited by a generator (Fig. 3-2). Each half of the line then

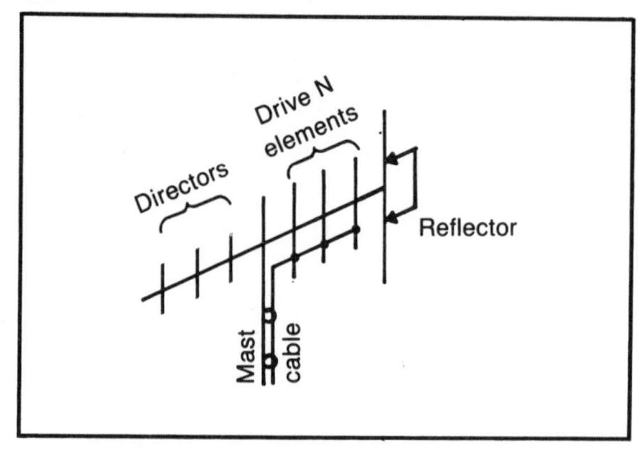

Fig. 3-1. Makeup of a modern antenna. Driver elements connect directly to transmission lines.

constitutes a resonant circuit when the generator is tuned to that specific frequency. When energized, the generator output induces high circulating currents, setting up two fields about the antenna, one electric (E) and the other magnetic (H) (for oersteds of magnetomotive force field intensity). When the two wires are moved apart, overall current flow moves in the same direction and the E/H fields aid one another, producing strong antenna fields. Observe that maximum current flows at the antenna's center while maximum voltage appears at the ends. Note that maximum current and H fields are coincident, while minimum voltage and maximum E fields are coincident.

Such fields when aiding the transfer of energy into space are known as radiation fields and decrease linearly with distances from the antenna, and are unlike induction fields which decrease with the square of the distance from the antenna.

So, without repeating germane information dwelt upon at length in the previous transmitter antenna chapter, we trust the foregoing will be sufficient to serve as adequate introduction to the highly relevant subject of receiving antennas—a topic that has always been of singular importance to some 90 or more million Americans.

RADIATORS WITH GAIN

Without amplification an antenna has no gain. But in comparison with a test dipole or isotropic "radiator," said to have equal gain in all directions,

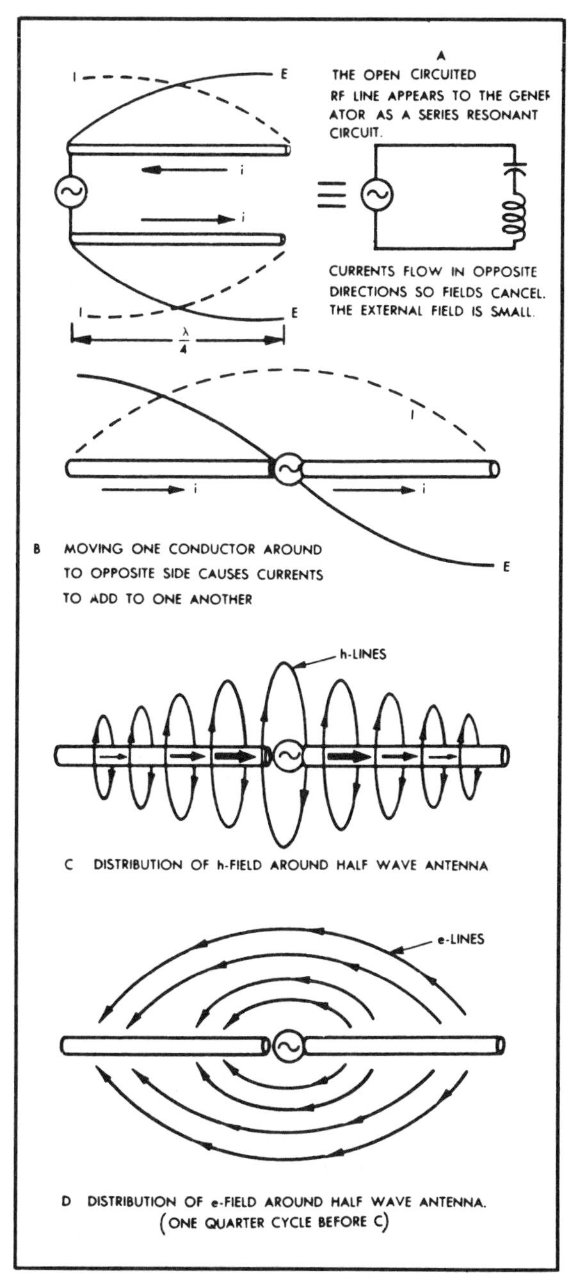

Fig. 3-2. Electric and magnetic fields characterize any operating antenna.

a boom with active and parasitic elements added does have "gain" by analogy. Further, any real antenna never has receive ability from all directions—you wouldn't want that—and is said to be directional, especially with narrow front lobes resulting in diminished beamwidth. So when extra element antennas are compared in signal strength reception with the isotropics and simple dipoles, they are, indeed, found to have gain (or amplification) in terms of decibels (dB) and, depending on the number of signal collecting (active) elements, such gain increases or diminishes accordingly.

However, with increased gains, their beamwidths decrease so that large antennas with many elements are highly directional and often need powered antenna rotors to direct them towards the different television stations, especially if transmit towers are scattered here and there throughout a community. On the other hand, well-designed medium and lower gain antennas traditionally have broad beamwidths (sometimes as much as 60 °) and will pick up both considerable frontal electromagnetic radiation as well as some side signals, too. This is not always desirable, however, since the entry of non-frontal energy can and will result in acceptance of secondary images commonly known as ghosts. Such ghosts are nothing more than secondary signals, arriving a few microseconds after the originals, producing halos or offsetting images annoying to the viewers.

If, therefore, you know the incoming signal strength of some TV station, especially by field strength meter measurement, and you also know the dB gain of your antenna, the two added together will determine if your TV receiver or VCR can handle the information sufficiently to process and reproduce a reasonable picture. Such "reasonableness" must be within TV tuner pickup abilities and should result in a signal-to-noise ratio of much better than 36 dB at the TV's cathode-ray tube, since 35 dB is considered the S/N minimum threshold. Most all the better videorecorder and TV products now are capable of 42-43 dB S/N measurements and, consequently, present satisfactory pictures compared with studio standards of 54 dB. Of course horizontal and vertical resolution also enter into final picture judgments, so we can't arbitrarily state that good S/N ratios mean excellent pictures—they don't. But it's a good beginning and a lot of people are completely fooled by narrow bandpass (2-2.5 MHz) "clean" video with virtually noise-free reproduction. We're saying that consumer VCRs still have a long way to go before they can compete with 4.2 MHz television receivers. Perhaps amorphous (noncrystalline) heads and metal tape will take care of that problem also as the years roll on, but right now, calibrated eyeballs will transfix and suffer as low noise, poor picture video equipment continue their roll. Regardless, BTSC-dbx-equipped VCRs and their U/V/CATV front ends have the same antenna requirements as television receivers, and their tuner pickup sensitivities also have to be taken into careful consideration—possibly more because of the double AM-FM-AM video conversions. Audio will *not* be affected since its bandpass is relatively narrow to begin with, and separate recording/play heads and some faster tape speeds will handle this signal portion easily.

All this, of course, adds up to a sincere admonition to choose all types of test and installation equipment wisely, spending a few more dollars here and there, offering maximum results with fewer time and dollar-wasted callbacks for no other reason than pure negligence. Done right the first time, there's no need for double labor at half the going charge.

Characteristics

Antenna characteristics aren't difficult to understand since they're not really complex, once the fundamentals are mastered and digested. Initial design, however, has been and is almost brutal because of the multiple factors of beamwidth, resonance tuning, U/V matching, the addition or nonaddition of parasitic elements, size, gain, linear electronic amplification, wind resistance, and boom-element strength. Actually, you use the world's most learned equations for most or all these factors, but all fine tuning after initial concept comes from the automated test range with hundred foot towers that may be separated by many thousands of feet, and even miles, depending on antenna types and projected uses.

Gone forever are the days of twin lead and high-low folded dipoles, narrow beamwidth vee-shaped receptors, simple dipoles, and arbitrarily stacked arrays with poorly-matched connecting feeds. Today's antennas are most likely U/V in-lines, reasonably efficient, and coated with a special substance to avoid salt and smog corrosion in attractive gold or blue colors with fairly generous price tags to match. Electrically, the better radiators are mostly log-periodics that have good gains, broad beamwidths, and excellent SWR and front-to-back ratios we'll be discussing next. Good balance takes some strain off the mast, concealed amplifiers in the antenna housing produce even better signal inputs, and undesirable FM signals in the middle of the VHF TV band are often rejected by additionally tuned elements already in place (but hand-removable) as you receive the antenna. Since most are of the "one-piece" concept, just open the antenna box and "click" each aluminum section of tubing into place before final mounting on the mast. But don't be clumsy. Wind, rain, ice, and snow these antennas survive, but not 200 lbs. of man and shoe leather. For once elements are bent or broken, straightening becomes difficult, electrical contacts may be lost, and aluminum welding is very tricky at best. So do handle with respect and TLC, then the antenna should be good for probably 10 years—decidedly *not* 15 or 20. Lead-in life can be approximately the same, but probably less, depending on conductor, insulator, covering materials, and care in installation. Annual visual inspections of both lead-in and antenna should tell the story. Ohmic measurements only indicate an open or short, not impedance or conductivity.

Analysis

Let's start the receive antenna analysis with a simple pattern that's exaggerated, but does make all the significant points. Then we'll continue with various antenna types and their explanations plus gain and polar patterns to substantiate. Remember that cutting a dipole so that it delivers maximum resonance when its measured length is half the wavelength of some incoming signal isn't difficult. But constructing a signal receptor for all channels between 54 and 890 MHz is tough! A great deal of tuning, harnessing, resonance factoring, element diameter size, and just plain luck all accumulate their weighting factors—and they're a load!

The polar pattern shown in Fig. 3-3 is a simple 360° circle marked off in 30° increments for both simple measurement and basic explanation since we need to demonstrate beamwidth, front-to-back ratios, sidelobes, and 3 dB half-power points. As you see, the indicated main lobe of the diagram intersects the 3 dB circle at approximately 17° from 0° which, when doubled, amounts to a 34° signal-acceptance region known as the "beamwidth." In addition to the front lobe, there's also a back lobe, and the extent of this back lobe also determines the front-to-back ratio which, in this instance, amounts to 8:1 and is very poor. A 20:1 FB ratio would be considered very good. Note that pattern sides are relatively clean, and that side signals have about a 90° clear area where interference should not be accepted.

With that abbreviated polar pattern description you should easily be able to read most standard consumer antenna patterns and adapt them to your special circumstances. Usually, however, polar patterns refer to all low or high VHF channels, or UHF and single diagrams for each channel are no longer published. We'll demonstrate . . .

As an example—and a live one—let's consider a Model 973-218 Zenith Chromatenna II antenna designed for U/V/FM suburban reception. This particular antenna has only 3 driven and 9 parasitic elements for low and high VHF and 1 driven and 27 parasitic elements for UHF. This means there are but four driven and 36 parasitic (passive, usually boom-connected) elements on the entire antenna, with a V-shaped UHF corner reflector. In Fig. 3-4 you see the results. For channels 2-6 you have a −3dB beamwidth of approximately 60°, fair sidelobe rejection, but a 65% backlobe which translates to an F/B ratio of 18:1 since 10 log P_2/P_1 gives you the dB figure of 18. In 10% steps, this puts the high VHF at an F/B dB ratio 18.9, and the UHF reading at 19.6:1. Of course, you can take a little of this with a slight grain of salt since all channels do not measure precisely the same, but the idea is there nonetheless. Note, however, the general absence of sidelobes after LO-VHF, and there aren't even any substantial backlobes at UHF.

But how about those diagrams on the left hand side? No calculations here, they're all easily readable and are simply the measured channel gain figures in dB. For instance, the VHF LO-BAND gains for channels 2 through 6 are approximately 4 dB, and that's sufficient for metro stations at VHF. On the other hand, channels 7 through 13 increase from 6 to 8 dB, as they should, while UHF channels go from 10 dB to 12 dB then trail off to 4 dB at channel 70. So above channel 60 you'll have to watch station reception in this instance, making sure it's strong, if not robust. Should all such allocated chan-

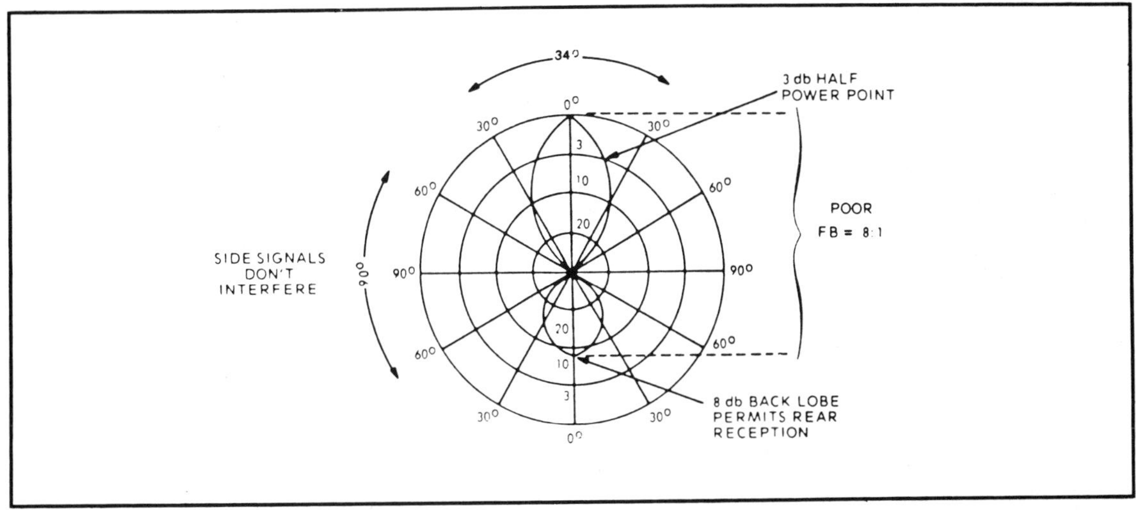

Fig. 3-3. Polar pattern example demonstrates 34° beamwidth, 90° side rejection, but large backlobe and poor 8:1 FB ratio.

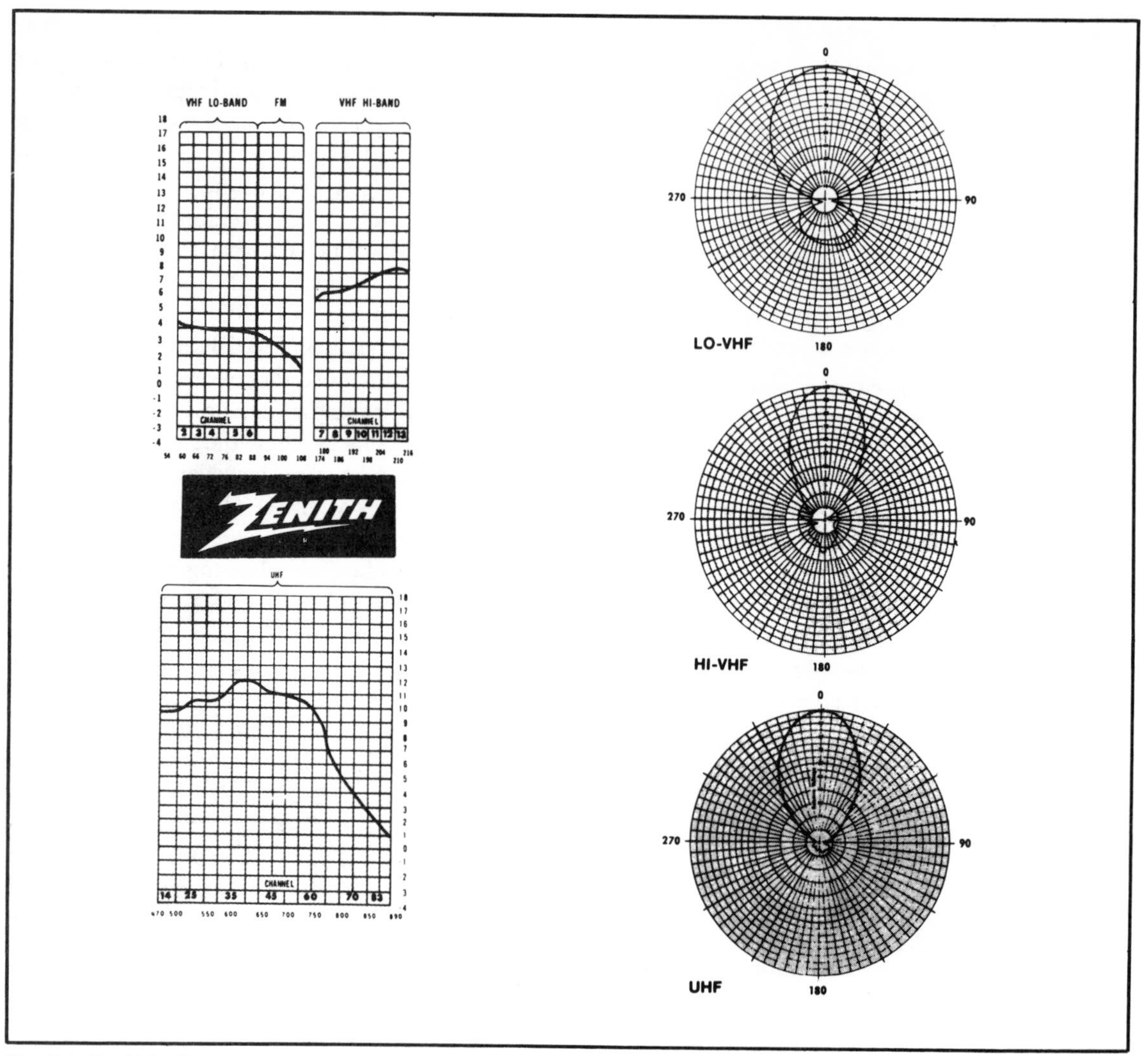

Fig. 3-4. Zenith's Chromatenna II metro antenna with both gain and polar patterns.

nels and their transmitter-antennas occupy one collective location, a simple U/V in-line antenna will receive them all. But if separated, you'll have to use a "hand crank" or ac-powered rotor to do the job. Further, if one of these channels is exceptionally high powered with respect to the remainder, then you may have to sacrifice your broad beamwidth and less expensive antenna for a narrow beam, high gain unit with padders connected to reduce the amount of incoming signal. This can also be a ghost-reduction method using the technique of high directivity. Where no other results are possible, you may have to use a multi- or single-channel Yagi with extremely narrow beamwidth to aid the situation as much as possible, picking up other channels via a second suitable array and either an electronic rf coupler or a simple mechanical throw switch—but in the latter case watch out for added capacitance, and in the former be sure there's enough isolation between the two or more input channels to avoid adjacent channel interference. This can be done with either a sensitive oscilloscope or even more sensitive spectrum analyzer by measuring any signal leakage on the unused channel while delivering full signal to the adjacent one.

Just now we mentioned the antenna designation Yagi. Before RCA sold its receiver antenna business some years ago, that's what it used to design and peddle—end fire Yagis. These are multi-parasitic element antennas with both driven (signal) and parasitic (signal-shaping but dormant) elements, usually with narrow beamwidths but delivering high gains when covering a specific range of frequencies. Veteran TACO/JERROLD, owned by General Instrument Corp. has, in the past, offered cut-to-channel Yagis between channels 2 and 6, or a rugged J-series UHF receptor available between channels 14-27, 28-35, 36-42, 43-50, 51-60, and 61-70. At a 12+ dB gain, TACO/JERROLD says it may be used as a "single-channel yagi" (specify channel) or as a broadband (specify highest channel to be received). As you can tell by the accompanying gain listing, however, it is basically cut for 6 to 10 channels in the UHF band to operate at maximum efficiency. (See Fig. 3-5). Several arrays may also be stacked.

LOG PERIODICS

All the above brings us to the prime topic among receive antennas: that of *log periodics.* Originally discovered by a special research group at the University of Illinois way back in 1954, the prime antennas remaining are log periodics manufactured by General Instrument and Channel Master, although Channel Master does not advertise their Quantum series as such (possibly a patent complication). Those first log periodics were marketed by

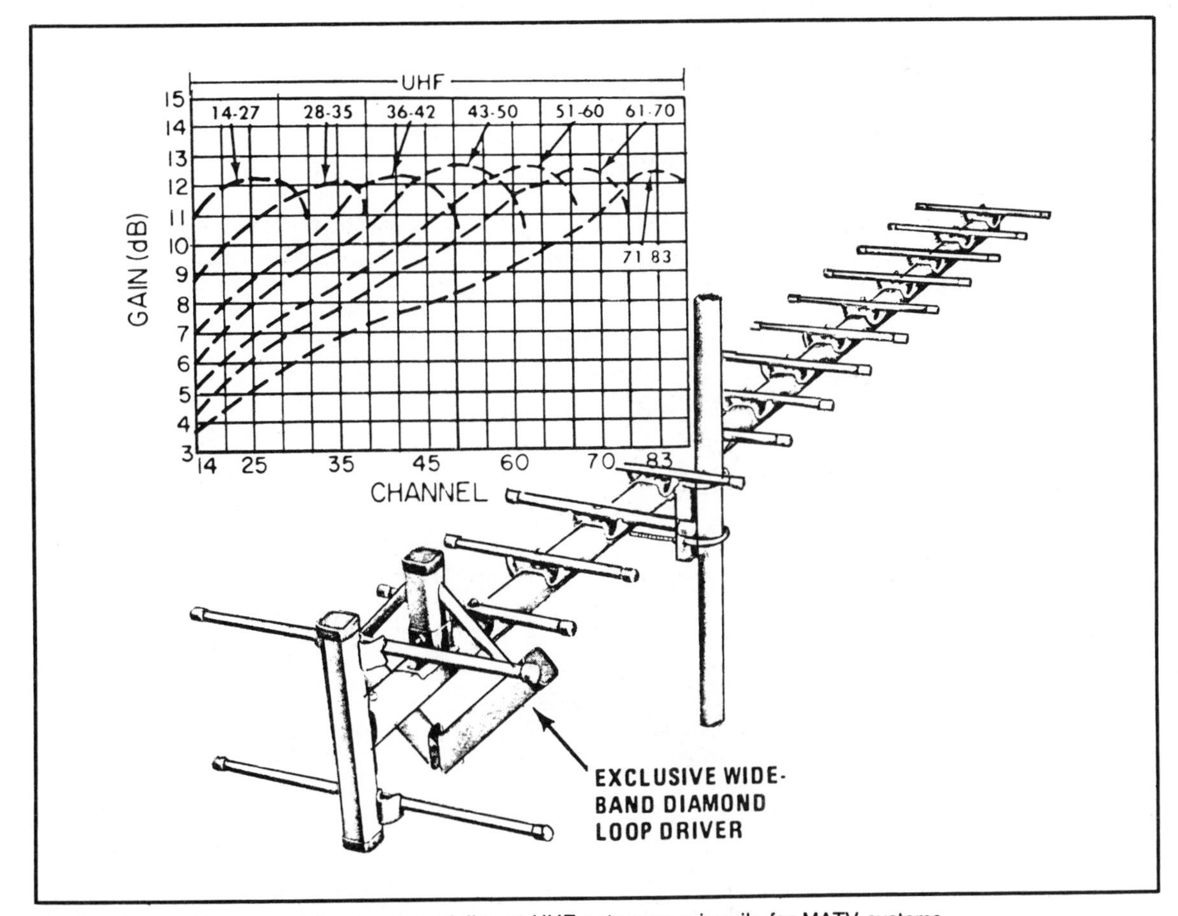

Fig. 3-5. General Instrument J-series especially cut UHF antennas primarily for MATV systems.

JFD (later purchased by Channel Master) and consisted of V-oriented dipoles, cut to a half wavelength of some certain frequency and spaced so that every element depended on the other and all were resonant at some designed part of the spectrum. Later, these forward-looking V-positioned elements had capacitative tuning added, along with circular directors for UHF. With introduction of the Quantum series, however, Channel Master did away with the Veeing and capacitance tuning, added parasitic elements for pattern-shaping, and placed a large number of small horizontal directors in front of the VHF receptor for good UHF response. There they remain, right into 1985, being probably the most popular antenna group made in the United States today. Channel Master remains America's largest home TV antenna supplier, now operating out of Smithfield, North Carolina, the old Sylvania plant.

Channel Master, however, isn't the "pure" log periodic of Illinois days but does have front-fed directors, an efficient rear reflector, and a unique tapered drive system, all producing relatively narrow beamwidths, high gains, and little sidelobe problems. Double booms and "rugged" snap locks supply extra antenna rigidity during ice and snow conditions, and gold-hued alodine coatings that are, in themselves, conductive, self healing and prevent rust. CM claims front-to-back ratios of as much as 35 dB for some of these high gain antennas, of which several models are shown in Fig. 3-6, some-with built-in, weatherproof housings for cartridge amplifiers or special FM traps. Either 300- or 75-ohm transmission downleads may be used but operating voltage still must reach any antenna-mounted amplifier through some coaxial cable center conductor. So in any event, a BALUN—a balanced-to-unbalanced transformer—must be used to match antenna and transmission line impedances while delivering power to any installed amplifier. Otherwise, the antenna's characteristic center impedance is a nominal 300 ohms. We say "nominal" since this impedance is not really a constant because of the extremely wide range of incoming frequencies.

In the drawing you will notice additional elements connected to some of the forward driven elements for tuning these particular dipoles. The short parasitic elements, helpful in directional pattern shaping, are mounted on standoff insulators, while all *driven* elements occupy riveted and locked positions on the twin, electrical harness booms. These are mounted on extremely durable plastic insulators and have no electrical contact with the mounting mast. Each of these antennas has its specialty:

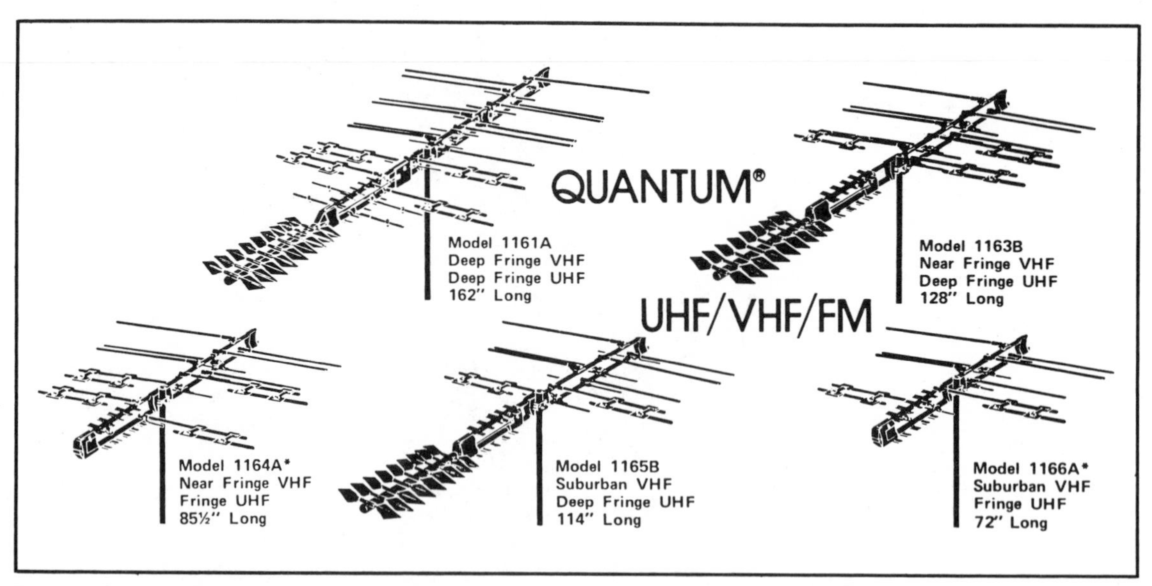

Fig. 3-6. Channel Master's high-gain, superior front-to-back ratio antennas for all terrain.

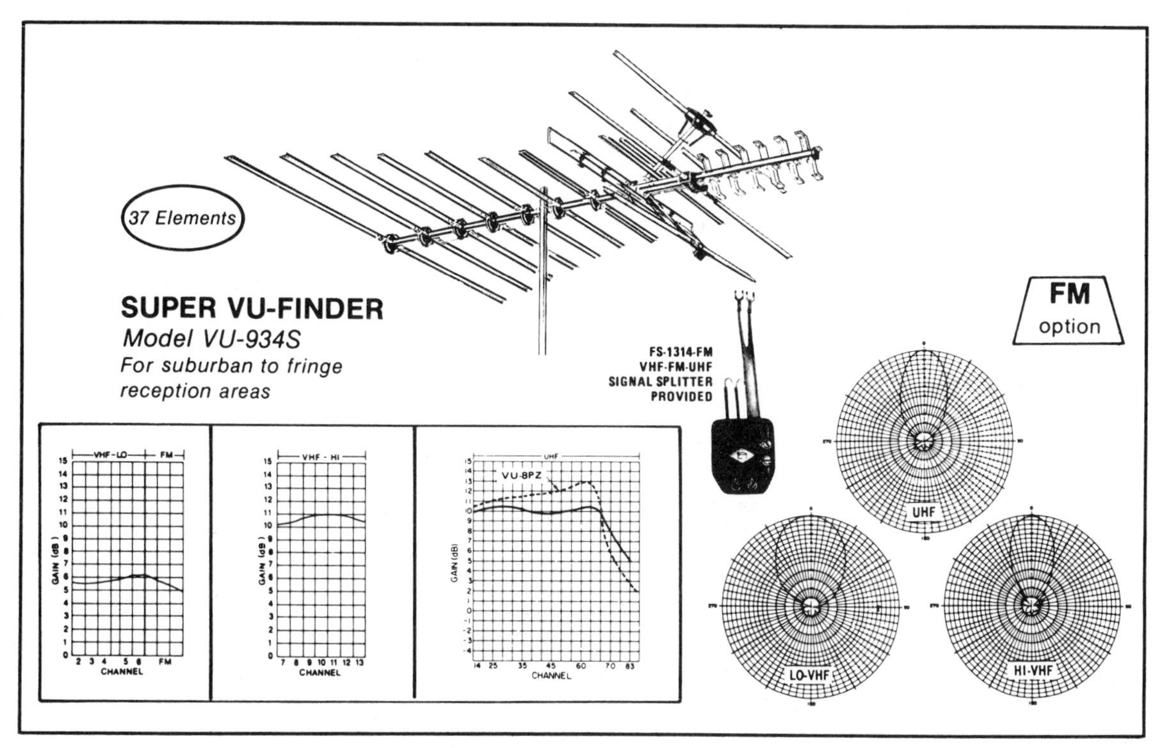

Fig. 3-7. General Instrument's pure log periodic broad beamwidth and efficient U/V antennas with good gain.

fringe, suburban, super UHF/VHF, so you can custom order to suit the circumstances. Are they good? I have two on my laboratory roof right now equipped with heavy duty rotors and they perform exceptionally under all weather conditions.

Dissimilar in appearance but also highly effective in application are the Super Vu-Finder series by General Instrument which also have fine engineering characteristics, slim lines, and well-developed corner reflectors for UHF reception. These feature square booms, alodine finishes, stainless steel down-lead connectors to prevent rust and oxidization, specially stamped termination brackets, aluminum-to-aluminum contacts to prevent corrosion, durable plastic insulators, a special U-wire parasitic about a driven element for higher gain and flatter response for VHF, and positive-locking fold-out elements. Except for two low gain antennas, all others have breakaway blocking signal elements that reduce incoming FM by as much as 12 dB. All are 300-ohm center impedance arrays, have single or double booms and require only 6 inches of mast area for mounting. Beamwidths range from 75° among the smaller antennas to as low as 30° for UHF sections in the deep fringe receptors, while boom lengths measure between 50 inches and 194 inches, depending on the number of elements and intended applications. Front-to-back ratios are listed from 8:1 to 24:1, and there are power zoom attachments for situations needing additional UHF gain. An illustration of Model VU-934S mounted on my roof is shown in Fig. 3-7. Designed for suburban to near fringe, this antenna has a 5.5 to 6 dB gain for lower VHF channels, 10-11 dB gain for 7-13 channels, and a 10 dB gain for UHF channels through channel 70. With the VU-8PZ super UHF power zoom attachment, UHF gains increase as much as 2 dB. Along with polar patterns and gain charts, the antenna and its 37 elements appear as advertised in the figure. Color it blue with fixed installation, no rotor.

SWR

In all this receive antenna discussion we have

yet to mention standing wave ratios (SWRs). The reason is that not much can be done about it among antennas except pay attention to specifications. A low-value SWR is always desirable, a high one means problems, not only with the antenna, itself, but any connected transmission line.

Radio frequency energy in space consists, especially at some frequencies, of both ground and sky waves, travelling their respective paths across *terra firma* and off into free air. As transmission frequencies increase, ground waves tend to diminish, but at low frequencies they may even cancel sky-wave signals. Since television fare moves swiftly over any type of terrain from mountains to skyscrapers, much of the ground wave portion cancels, leaving the 54-890 MHz spectrum largely unaffected by most secondary signals.

You noticed we said "most." This does not include secondary "bounce" images known to the trade as "ghosts." Far from being gentle spooks, these microsecond-delayed signals are fuzzy clones of the original and offer nothing more than contamination of the prime scene. Fortunately, we are well aware of their causes and can deal with them by installing narrow beamwidth antennas, some signal trapping, rotators, lead-in change to better coaxial cable, re-routing lead-ins, and positive separation of the transmission line from all surrounding water and metal, especially if it is unshielded twin lead—for which there's no excuse anyway. Some older shielded twin lead, by the way, seemed to produce its own ghosts by internal capacitative action—at least that's been my experience.

Standing waves may originate as either voltages or currents and result largely from mismatched terminations among and between transmission lines and antennas. They may be represented mathematically as either voltage or current maximums and minimums (E_{max}/E_{min} or I_{max}/I_{min}) divided, as shown, into one another. Since we can do nothing more with antennas other than try to purchase minimum SWRs with each, we'll continue this discussion momentarily upon entering the topic of transmission lines. For here we can make adjustments that will aid many existing SWR problems immediately.

RECEIVE TRANSMISSION LINES

Possibly the least understood and certainly the most abused auxiliaries of received television are transmission lines, commonly called lead-ins. Twin, or bright lead, with 300-ohms impedance, little or absolutely no shielding, skimpy plastic covering, improper separation, and often atrociously AWG-sized stranded conductors is still used by home or novice installers who haven't even the faintest concepts of signal handling between radiator and receiver. And even with fairly recent introductions of better quality coaxial cable, many still believe that average coax attenuations are prohibitive, the cable is expensive, and it's a lot of trouble anyway. Fortunately, none of the coax talk represents fact, which we intend to prove as this discussion continues.

Coaxial cable (cable within a cable) has advanced considerably from the dark times of the '50s when poor dielectrics and largely open basket weave "shielding" did little to aid perpetual waterlogging that plagued almost anyone who attempted to use it. Today, with vast dielectric improvements, reduced capacitance, fast rates of propagation, strong center conductors, and up to 100% shielding, coax has become a highly attractive medium of video-rf processing. For its per-foot losses are at least as good and often better than twin lead, shielding *contains* the circular, interacting fields common to twin lead, costs have been substantially reduced because of intense competition, and the ease of connections and insensitivity to metal areas, extraneous signal pickup, and longevity make it *the* most desirable lead-in available. Furthermore, in- and inter-building signal routing for both audio and video, as well as low noise satellite down-converted signals from dish (antenna) feeds are processed well by this emerging quality medium. At 1 GHz, however, you had best pay close attention to specified losses per hundred feet, and keep lead lengths to a minimum. This is especially true for low noise block downconverters (LNBs) feeding satellite receivers at frequencies between 950 and 1,450 MHz. RG/6U or its equivalent is needed here for minimum capacitance, maximum shielding, and at least an 18 AWG center conductor. In other words, you'll have to use quality, multiple outlet, commercial-type cable. Compared with ordinary RG/59, the results are almost dramatic, especially if said cable is manufactured to tight specifications and has special shielding. A comparison between flat lead and shielded coax is illustrated in Fig. 3-8. With three interactive fields in twin lead and only one in coax, the argument should be relatively self-convincing since fully *contained* signals are always considerably more desirable than those dancing freely on some conductor's external surface and bucking one another. In any two conductor cable, as illustrated, there are always two equal currents flowing in opposite directions, with a third in-between. Even though shielding results in added capacitance and slows, somewhat, the normal flow of electrons, the coax we're suggesting has a nominal propagation velocity of 78% as opposed to 80% for best quality twin lead, although capacitance of 4.5 pf/ft. versus 17.3 pf/ft. doesn't compare. However, if all this does is retard propagation velocity 2%, then the added capacitance is no contest. Attenuation differences are minor, depending on the quality of each as frequency approaches 1 GHz.

So there you have a pretty solid comparison. Will you take signal containment, excellent noise and stray signal shielding, good propagation velocity, and probably 2:1 useful life, or would you rather replace leaky twin lead every five years or less? If you're still adamant, don't forget that twin lead weathering and cracking can reduce incoming signals by 50-80% very easily and in a remarkably short time, especially if conductors are laid down in cheap plastic covering and have any contact with metallic or damp surfaces. Think there's a real choice? Think again—but this time *think!*

STANDING WAVES AND INTERFERENCE

They often are one and the same—or at least wavy lines, ghosting, and herringbone interference seem to be. If you'd like to know the actual frequency of some of the above, just connect an rf signal generator to the antenna terminals through a splitter and match the number of lines, etc. 1:1. When they double up you'll have the culprit redhanded. So what's to be done? Fortunately, plenty. We have such accommodations as notch filters, tunable traps, signal reduction devices, trimmers of all descriptions, and both quarter-wave and half wave stubs—all effective on either coax or (perish the thought) twin lead.

First, however, we'd like to show a well-developed SWR diagram taken from a military manual that succinctly shows what transpires in the realm of standing waves and why. An explanation will be forthcoming as we go along (Fig. 3-9).

At the top is a symbolic rf line with signal source in the middle. Below, lines represent incident (originating) and reflected (returning) currents or voltages moving right and left. Currents and voltages are plotted as nominally being the inverse of one another; that is, when current is low, voltage is high. Remembering that VSWR = E_{max}/E_{min}

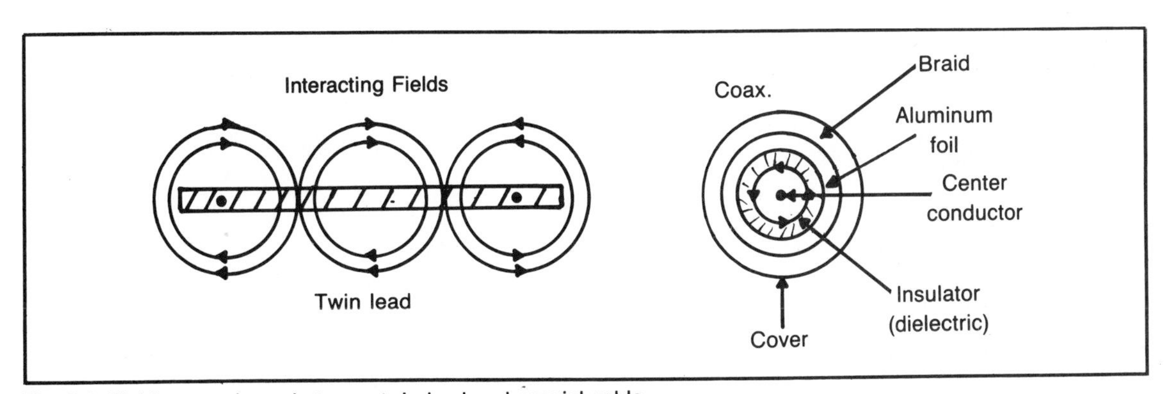

Fig. 3-8. Field comparisons between twin lead and coaxial cable.

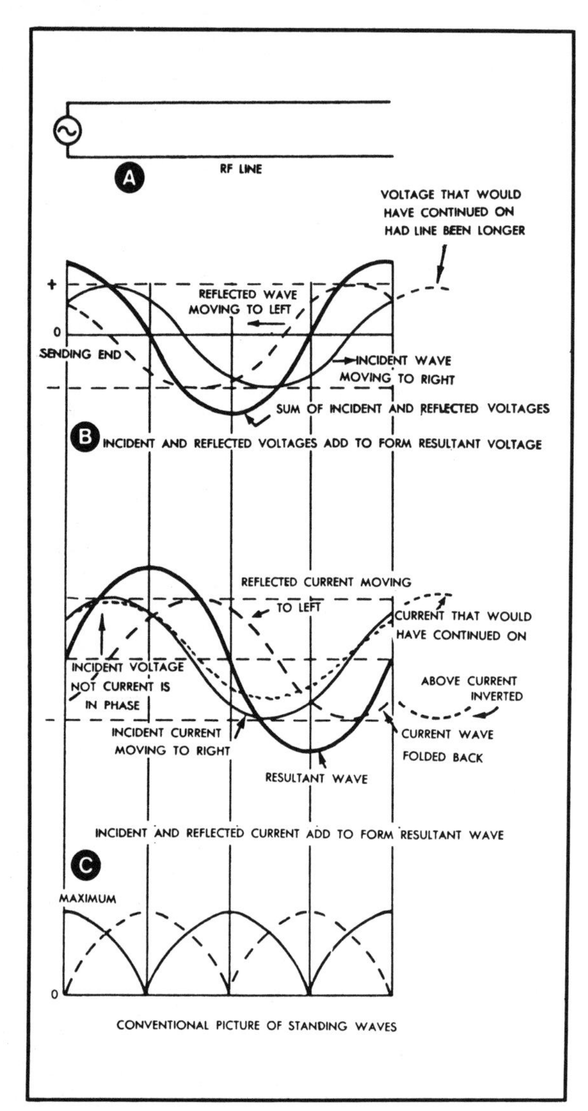

Fig. 3-9. A simple rf line and resulting standing wave formations due to impedance mismatches.

and ISWR = I_{max}/I_{min} we can proceed from there and retrieve some pretty good information.

The open-ended rf line with signal source is identified by the letter A, denoting an alternating mix consisting of both voltage and current. In B, when this voltage reaches the open end, it is reflected back in like polarity and amplitude. If generator and line impedances are equivalent, the reverse energy is absorbed at that point, but incident and reflected voltages otherwise add to form a resultant which, in this instance, is larger than both of them and contains valleys and peaks, as predicted.

In C, currents of this same energy are illustrated moving in incident and reflected directions, with current considerably out of phase, upon reflection, approximately 180°. With their obvious peaks and valleys, the forward and reverse currents add, with a heavy dark line illustrating the resultant. Note the obvious phase difference between the resultant voltage and current. Superimposed below, they show a well-formed pattern of standing waves.

As for shorted lines, currents are reflected in-phase, whereas voltages are returned out-of-phase. Lines terminated in their characteristic impedances minimize or negate completely the effect of standing waves. This is why line and antenna matching are especially important—you always want to reduce any possibility of standing waves and also transfer maximum power between radiator and signal source or receptor. In this regard, coaxial connectors are especially important and should be one piece units to prevent any leakage and resulting circulating currents which are ruinous to low-value signals.

At this point we could jump into a little mathematics to prove the obvious, but any such exercise would be more esoteric than instructive, so we'll just move along into shorted and open stubs as a home-made cure for at least some of these SWR troubles. But you should know at least that actual stub impedance equals the square root of the product of any two impedances to be matched: $Z_{actual} = \sqrt{Z_1 \times Z_2}$ if Z_1 and Z_2 are the to-be-matched impedances. But the main point is how transmission line stubs look and what they do.

CABLE AND TWIN LEAD STUBS

Quarter wave transmission line sections at resonance appear as parallel tuned circuits. Above resonance they are capacitative and below resonance, inductive. Quarter-wave open stubs, on the other hand, become series-tuned at resonance. Below resonance they are capacitative and above resonance inductive. When they become or approach half wave approximations, the shorted stub appears

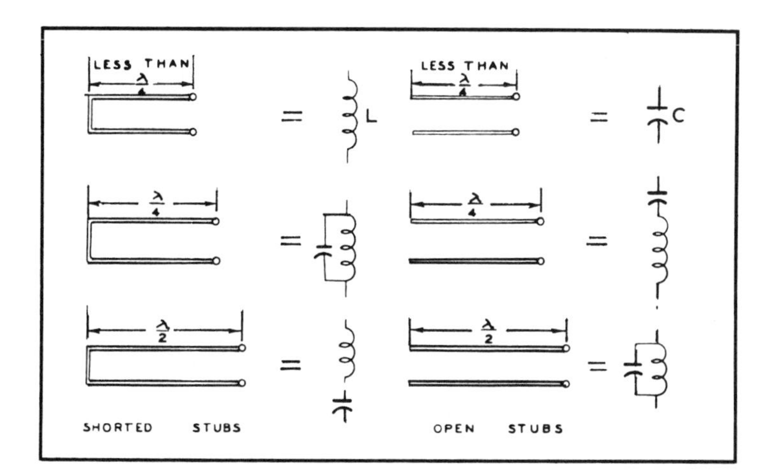

Fig. 3-10. This shows how transmission stubs appear electronically in quarter and half wave reactions.

series tuned and the open stub parallel tuned. All this is shown in Fig. 3-10 where lambda/4 and lambda/2 are the quarter wave and half wave illustrations, respectively.

With that succinct explanation under your belt, look at Fig. 3-11 for practical applications. There you have both quarter and half wave coaxial and twin lead fixed and variable traps to aid your problems. Their various lengths can be determined by either trim and peek methods or by half and quarter wave formulas in the center of the diagram. On the right is a small chart for U/V/FM applications showing channels, frequencies, lengths, and ceramic capacitor trimmer values. Try the signal generator frequency matching system first, then work out your traps accordingly. Remember that although trapping is intended to *remove* the offending frequency, sometimes a quarter wave *open* stub can generate more problems than it solves on adjacent channels that were formerly troublefree. Therefore, use *shorted* stubs or notch filters to absorb offending interference wherever possible and leave the open jobbies alone.

Should your problem be one of excessive signal strength, either balanced (for 300 ohms) or unbalanced (for 75 ohms) will certainly relieve the problem. Being purely resistive, current and voltage will continue in phase so that all you have to do is to use the R-values indicated, pay attention to the A1, A2 input and output voltages, and come out with a winner. A little arithmetic (Fig. 3-12) never hurt anyone when it's for a good cause. The "A" factors represent voltage decreases which attenuate from10-35 dB when using these methods.

If you don't want to experiment, there are plenty of commercial variable and fixed attenuators available at a price. Be sure, however, if you purchase, they are resistive only and have no reactances to add to your already evident problems—a little additive SWR, and things will really begin to pop, and excessive high-frequency attenuation can give UHF a fit!

Now, let's close out the section with a crisp dissertation on that sometimes elusive electrical creature the decibel. Formerly used almost exclusively to express large gains and losses in audio and power, this 1/10 of a Bel symbol has now become universal language when dealing with both large numbers and nonlinear functions. In spectrum analyzers, for instance, amplitudes, signal-to-noise, and power measurements are *all* in terms of decibels.

When dealing with dBs, however, do remember that one set of decibels applies to voltage and current, while the other operates with power. Were you to have a 35% power loss, for instance, a convenient method of expressing this would be in terms of dB. Then $10 \log_{10} P_2/P_1$ would be the equation, and P_2/P_1 would become the simple ratio of 0.35. But wait, this is a *loss* and not a gain, therefore you have to place a minus (−) sign in front of 0.35 and take the anti-log (inverse) for correct calculation. The equation now becomes: $10 \log^{-1} -.53$ (your inversion) and means that you must take the

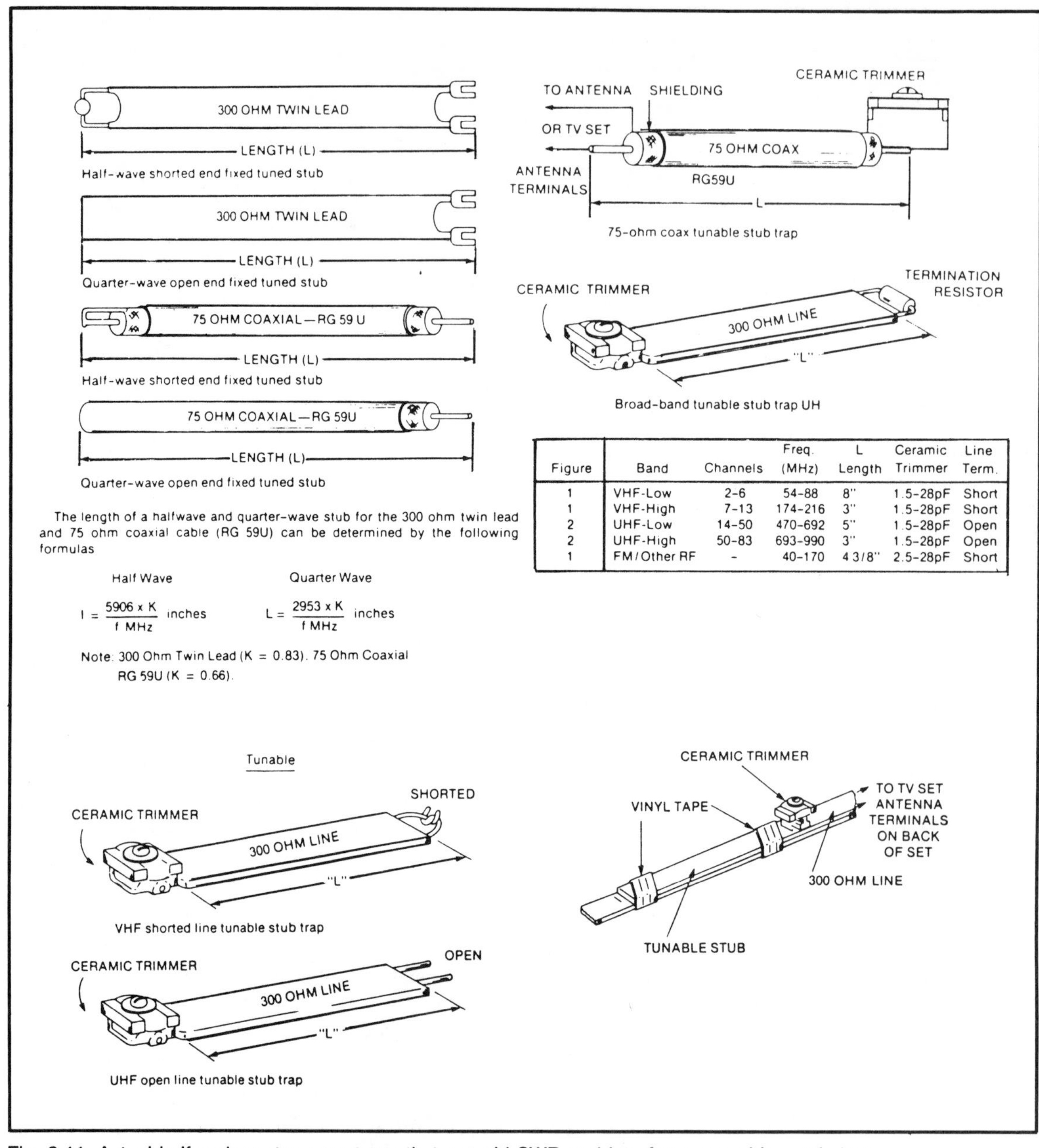

Figure	Band	Channels	Freq. (MHz)	L Length	Ceramic Trimmer	Line Term.
1	VHF-Low	2-6	54-88	8"	1.5-28pF	Short
1	VHF-High	7-13	174-216	3"	1.5-28pF	Short
2	UHF-Low	14-50	470-692	5"	1.5-28pF	Open
2	UHF-High	50-83	693-990	3"	1.5-28pF	Open
1	FM/Other RF	-	40-170	4 3/8"	2.5-28pF	Short

Fig. 3-11. Actual half and quarter wave traps that can aid SWR and interference problem solutions.

anti-logarithm of −.53 and multiply by 10 to solve the power equation. The answer: 35% power loss = 0.44683592 × 10, or −4.47 dB rounded off.

It's all quite simple with any scientific-type calculator and you should use it whenever necessary to understand these essential working terms. Why were negative (loss) numbers chosen first, because most of the "great" mathematics books around won't talk losses (they're tough to explain), only gains that can be done with your eyes closed.

As example, suppose you wanted to double *voltage* or *current* gain in some CATV distribution

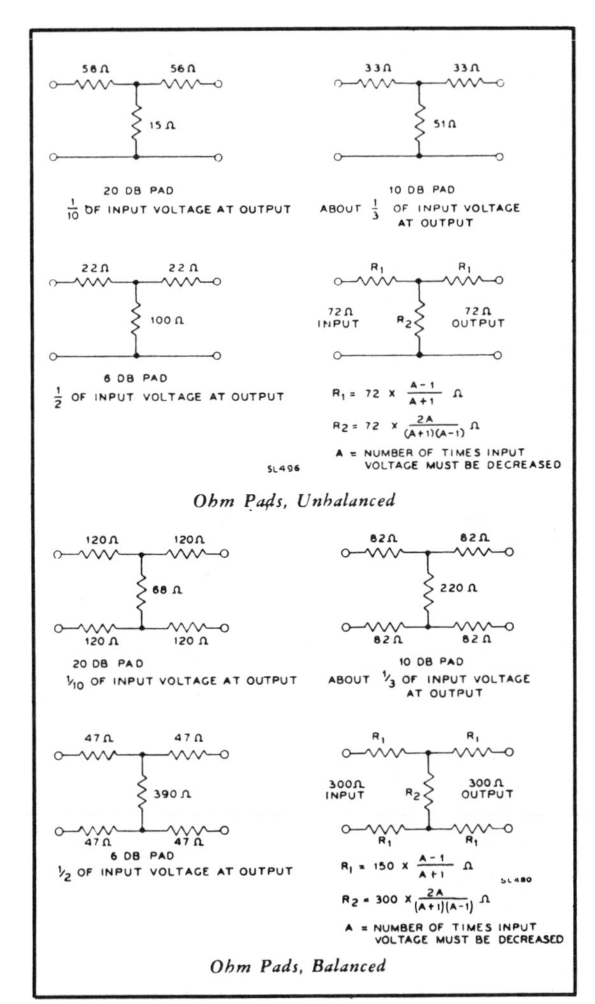

Fig. 3-12. Pads for 75- and 300-ohm lines used to reduce overload signal inputs.

amplifier? Here's how:

$$20 \log 2 = ?$$

But now you don't have the problem of percentages, just standard digits. So the anti-log disappears, logarithms reappear and

20 log 2 (but no minus sign for loss since this is a *gain*

becomes

$$20 (\log) \times 0.3 \text{ or } 6\text{dB}$$

Here, a 100% voltage gain meant *doubling* the figure, not times 100 as some might mistakenly calculate. Just be a little careful and these dBs quickly fall into place. To give you a small hand, here's a voltage/current/power comparison which some may find helpful.

	Voltage/Current		**Power**	
dB	**Gain**	**Loss**	**Gain**	**Loss**
3	1.414	0.707	2	0.5
6	2	0.5	4	0.25
10	3.16	0.316	10	0.1
20	10	0.1	10^2	10^{-2}
30	31.6	0.032	10^3	10^{-3}
40	10^2	10^{-2}	10^4	10^{-4}
60	10^3	10^{-3}	10^6	10^{-6}

If you study this listing, the inversion process is obvious, and only the initial dB figures need be memorized to derive the resulting voltage/current and power gains and losses. One final note, when you see V/I (or E/I) loss listed as 0.707, this means that 70% of the signal remains and 30% is lost. Also note that the voltage/current half (E/I) point is 6 dB versus power's 3 dB. These are highly important differences because transmitter video output stages are stated in dB power, video signal stages are normally in voltage, and audio outputs are in watts. There is, of course, a correlation among the three:

$20 \log P_{out}/P_{in}$ = voltage or current measurements in dB.

$10 \log P_{out}/P_{in}$ = power out in terms of dBm (relative to 1 mW)

and for conversion, dBW = dBm − 30/10 = dB in watts. Translated, dBW = $\log^{-1}$ (dBm − 30)/10, taking the anti-log only after all subtraction and division have been executed.

We could go on, but it would turn into a mathematics course on electronics—and that's not really what this book's about. However, we'll spot you a few equations (not formulas) from time to time just as a reminder that most earth shaking mathematics are often written *after* many "incredible" breakthroughs and discoveries are completed—not before!

THE TV INSTALLER

Usually these are less skilled and lower-paying jobs in the industry that serve as an entranceway into TV-land. Normally, an "apprentice" is given a few days on the job with a more experienced individual, familiarized with whatever antennas and cable the employer offers, then sent out to do the job on his own. He is normally not instructed in normal antenna beamwidth and patterns, area TV station transmission characteristics, cable losses per hundred feet, necessities and advantages of shielding, prohibitive twin-lead problems, parallel radiation signal pickups, and the absolute prohibition of cable contacts with even potentially wet areas such as rooftops, gutters, drains, and rotting wood.

Nor is he usually provided with a workable field-strength meter covering the U/V ranges through channel 69. While this is neither mandatory nor absolutely necessary, familiarity with such equipment makes for a considerably better installer in very short order. In complex, multi-receiver installations, a signal receiver (and often a channel source) are necessities, although a television receiver with known receiving U/V capabilities will also do fairly well except in marginal instances where exact voltage levels are required. The field-strength meter, however, doesn't read secondary image (ghost) factors at all well nor does it indicate when color phase or amplitude is distorted or lacking. Therefore, a good installer would have both a color TV portable and field-strength meter that are adequate and dependable, not a pair of beat-up pieces whose antecedents and calibration are usually suspect. He would also understand that good quality chimney strapping, secure antenna mounting, and even an occasional antenna rotor would aid considerably even in "routine" installations where one or more U/V stations have ghosting or gain problems. If initially an inadequate antenna and transmission line find their way to some unfortunate individual's rooftop, think what will happen when better signals, multichannel sound, and superior receivers try to operate? His squawk should be louder than the Battle Hymn of the Republic, and Mr. installer will have to do the job all over again at considerable cost in time and money. A good combination U/V in-line antenna with good beamwidth, mechanical support, superior electrical characteristics, reasonable sidelobe and gain patterns, and a single U/V downlead that can be amplified, should be adequate for just about any situation. The same goes for well-shielded and braid-covered coaxial cable having a copper-flashed steel center conductor, good conductor insulation and tough external plastic coating. If the loss per hundred feet up to 1 GHz (10^9 Hz) amounts to no more than a few decibels (dB), then many signals from divergent sources are possible, including satellite earth stations. We have more to say about this towards the end of the chapter since not nearly enough attention is being paid to cable operating at 1-2 GHz. Here, agc (automatic gain control) indicators on satellite receivers and especially spectrum analyzers are of considerable aid in tracking signal strengths, although we don't expect TV installers to tote around multi-thousand dollar analyzers for the usual routine installation. These instruments, however, are sure to become considerably more popular as intricate installations and smaller (fainter) signals are recovered. Calibrated vertically in either dB or (to a minor extent) linear volts, and horizontally in fairly accurate frequencies, it's pretty easy to differentiate between transmission/receiver faults, as well as determine shortcomings in either transmitters/receptors or transmission lines. Although sophisticated operationally and decidedly experience-interpretative, spectrum analyzers are going to become considerably more popular as additionally sophisticated equipment is marketed. The same will be true for logic and signature analyzers. And although digital multimeters and oscilloscopes will lose none of their popularity, these old-line instruments have their limitations and additional measuring equipment is needed to characterize and troubleshoot what's currently reaching the consumer. Naturally, more intricate functions mean extra stacks of dollars, but expert and rapid usage may even save considerably more than is initially spent—at least we'd hope so.

Invert "gloom and doom," however, and you can really uncover both city and rural individuals and a few companies that actually revel in working the skylines in search of top reception. These are usually dedicated people with limited electronics experience who, by trial and error, have found some pretty sound solutions to most area problems, or know where to find help if needed. Usually they charge a bit more, but give good *annual* warranties and use fairly decent cable and antennas. If you want the best, they probably can do that too at a somewhat elevated charge and spending a bit longer on the job. Word of mouth, here, probably elicits the best recommendation. Heavy advertisers, unfortunately, are not always among this small, competent group because of frequent personnel changes, fast training, and the necessity of supporting all those radio, TV, and print-press spots. Someone has to pay for notoriety, however bad or good, and the consumer is it! We'd like to see good money drive out bad, in this case, with satisfactory value received.

In all instances, look for name brand equipment such as Jerrold, Channel Master, Belden, Zenith, Winegard, Sadelco, B & K- Precision, etc. Usually they're quality and the manufacturers stand behind their wares. Offshore cable and antennas are often "price merchandise" that won't really cut the mustard in comparison.

GOOD CABLE

For "proof of the pudding" emphasis covering much of what has been expostulated throughout the chapter, let's take a page directly from Belden's Electronic Wire and Cable catalogue for CATV and show some highly desirable statistics (Fig. 3-13). Note that all center conductors are AWG 18, with steel center conductors flashed with copper. Shielding material is both braid and aluminum foil, cellular polyethylene the dielectric, and all of these 6U-type cables are 100 percent sweep tested from 5-450 MHz *before* leaving the factory.

Sure you're going to pay a little more, and certainly the larger diameter wire and its shielding are more difficult to work with, but think of the superb results? When properly installed, there are no callbacks with this quality cable connected to a good log periodic. With that combination you can't miss! Try something like this on your next installation, especially if it's in a real "dog" area and enjoy the results.

CP RECEIVE ANTENNAS

Although there's no substantial development to report anew at this time, at least one or two consumer antenna manufacturers are still aware of the possibilities and we may see some further efforts by them in the near future—especially for indoor UHF units.

The outdoor units to date are somewhat clumsy and their channel number reception limited. But that should be worked out in time, especially considering the very, very good receivers, multichannel sound, the coming of digital broadcasting, and all the other developments pending in this ongoing video revolution that may continue without end.

Because this is another chapter altogether, let's briefly review circular polarization and continue with another few words in the receiver category. Initially introduced to broadcasting during the 1960s to aid frequency modulation (FM) stations in an economic slump, right-hand circular polarization was finally approved by the FCC in 1978 for TV following extensive testing by WLS-TV in Chicago, and KLOC-TV in California. to understand CP's theory, place a pair of orthogonal dipoles in near proximity and excite with equal signals 90 degrees out-of-phase. A constant magnitude field then emerges that rotates either right or left, depending on induced phase of the two signals. Any antenna, therefore, that receives such a field, accepts a constant signal level regardless of horizontal or vertical direction (orientation). Both fields maintain separate vertical and horizontal directions by traveling along the same axis, and since each is equally strong and independent, separate transmitters are needed for their outputs. Besides less

6/U Type

6/U Drop Cables are 100% Sweep Tested from 5-450 MHz, SRL 26 db min.

Duobond® II/Braid

Duobond II plus Braid has proved to be an economical solution to many CATV installation problems — even in high-density metropolitan areas. The aluminum-polyester-aluminum foil is bonded directly to the cellular polyethylene core and provides 100% coverage, seals out moisture and assures easier, more reliable connector installation.

Trade Number	Standard Lengths		AWG (Stranding) & Material		Insulation Nom. Core O.D.		Shield Material	Jacket	Nom. O.D.		Nom. Imp. Ohms	Nom. Vel. of Prop.	Nom. Cap.	
	Feet	m	Inch	mm	Inch	mm			Inch	mm			pF/ft.	pF/m
18 AWG														
9114	1000 U-1000 (UNREEL)	304.8 U-304.8 (UNREEL)	18 (solid) .0375 Copper covered steel	.95	Cellular Polyethylene .180	4.57	DUOBOND II + 40% aluminum braid	Black, Beige PVC	.275	7.00	75	78%	17.3	56.7
9115	1000	304.8	18 (solid) .0375 Copper covered steel	.95	Cellular Polyethylene .180	4.57	DUOBOND II + 40% aluminum braid	Black PVC	.275 × .395	7.00 × 10.03	75	78%	17.3	56.7
	MESSENGERED — *.051″ (1.3mm) Galvanized steel messenger*													
9123	1000	304.8	18 (solid) .0375 Copper covered steel	.95	Cellular Polyethylene 180	4.57	DUOBOND II + 40% aluminum braid	Black poly-ethyl-ene	.275 × .395	7.00 × 10.03	75	78%	17.3	56.7
	MESSENGERED — *.051″ (1.3mm) Galvanized steel messenger*													

9116	1000 U-1000 (UNREEL)	304.8 U-304.8 (UNREEL)	18 (solid) .0375 | .95 Copper covered steel	Cellular Polyethylene .180 | 4.57	DUOBOND II + 61% aluminum braid	Black, Beige PVC	.275	7.00	75	78%	17.3	56.7
9117	1000	304.8	18 (solid) .0375 | .95 Copper covered steel	Cellular Polyethylene .180 | 4.57	DUOBOND II + 61% aluminum braid	Black PVC	.275 × .395	7.00 × 10.03	75	78%	17.3	56.7
	MESSENGERED — ***.051″ (1.3mm) Galvanized steel messenger***											

Duobond® II/Braid/Duofoil®

Duobond II plus Braid plus Duofoil is a step up in shield integrity. This multi-layer shield configuration provides excellent protection against ingress/egress problems.

18 AWG

9118	1000 U-1000 (UNREEL)	304.8 U-304.8 (UNREEL)	18 (solid) .0375 | .95 Copper covered steel	Cellular Polyethylene .180 | 4.57	DUOBOND II + 61% aluminum braid + Duofoil	Black PVC	.275	7.00	75	78%	17.3	56.7
9119	1000	304.8	18 (solid) .0375 | .95 Copper covered steel	Cellular Polyethylene .180 | 4.57	DUOBOND II + 61% aluminum braid + Duofoil	Black PVC	.275 × .395	7.00 × 10.03	75	78%	17.3	56.7
	MESSENGERED — ***.051″ (1.3mm) Galvanized steel messenger***											

Fig. 3-13. Belden's electronic cable data.

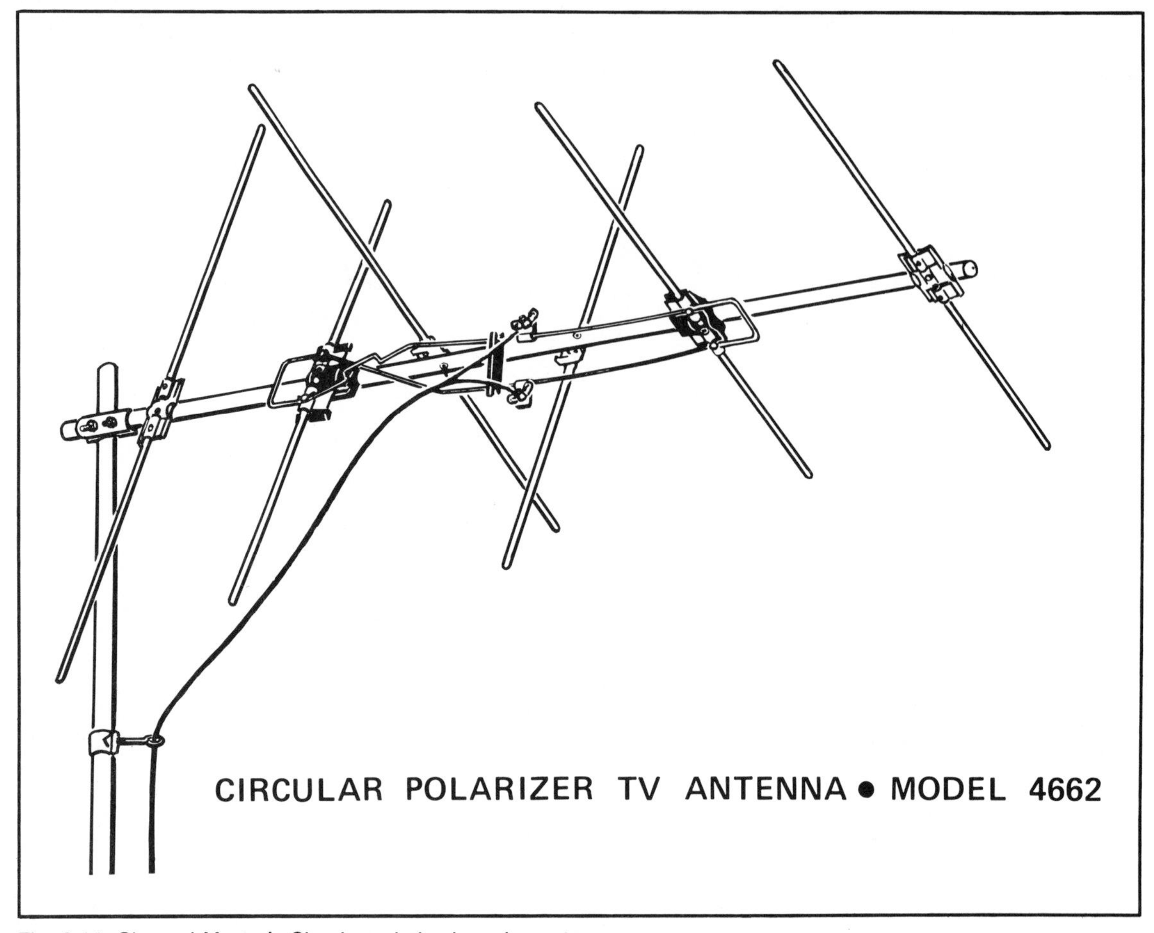

Fig. 3-14. Channel Master's Circular polarized receive antenna.

ghosting, CP should generate better signal-to-noise conditions for fringe sectors, reduce cochannel or adjacent channel interference, decrease critical receiving antenna orientation, and offer better distant and local pictures. Considering that we now have full-dimension luminance, chroma, and sound for the first time in history, it would seem like a very good year to take advantage of CP. To be sure the receive antennas (Fig. 3-14) will cost more initially, but then doesn't everything when first introduced? The only conceivable drawback we can visualize at the moment is their size—you will need somewhat more installation room for the added vertical elements. Standard antennas, of course, will still receive full CP horizontal transmissions, but the vertical portion is lost, negating all the advantages obviously available.

Chapter 4

VHF and UHF Transmitters

USUALLY THIS IS NOT A TOPIC CONTAINED IN contemporary reference or textbooks since most readers are interested in the final picture appearing at the receiver rather than details of origin. In order to view and hear any broadcast information, it must have a transmitter, and that's what Chapter 4 is all about. While we can't and won't print every minute detail, overall operations for the several types of transmitters is included insofar as available information permits. For this we are indebted to Andrew, Harris, and RCA Corporations for the wealth of material graciously provided on their most recent units. Naturally we'll concentrate on circular-polarized equipment since there is every indication the entire industry has adopted both concept and hardware installations. Also, as you are aware, by now, CP simply adds another hefty dimension to the medium delivering equal power to both horizontally and vertically polarized signals for more uniform power, reduced noise, and ghost-cancelling transmissions.With the increased involvement of broadband satellite carriers for maximum video and sound, analog and digital transmissions should soon be producing the best audio and picture information we've ever known. So, if this chapter can give aid and comfort tofurther knowledge and exemplary results, it will!

TRANSMITTERS IN GENERAL

Obviously, the mission of any television transmitter is to accept camera, tape, remote, or studio live performance activity, amplify and modulate luminance, chroma, and sound for some specific broadcast modulator and put the composite audio/video signal on the air. In order to do this, sound and video have to be amplified, delay equalized, filtered, modulated, carefully controlled according to FCC specifications, and then power amplified for the transmitting antenna.

Naturally there must also be vertical and horizontal sync signals accompanying, as well as 3.579545 MHz chroma sync burst in the form of 8-9 cycles once each horizontal line to maintain color phase and red, blue, and green registration. Should any of these synchronizing signals drift or be forced out of strict FCC specifications, thousands to millions of receivers would be unable to reproduce transmitted pictures. Fortunately, sync problems rarely occur unless there is a catastrophic transmitter failure or occasional network problems, therefore they are but minor considerations. Probably the worst fault of all lies with all those television receivers of poor design, outdated electronics, or those that are improperly serviced, and there have to be hundreds of thousands of these, not to mention owner mistuning, especially of color hues. Unfortunately, flesh tones are difficult enough to reproduce without deliberate operator control distortion. Gremlins, not people, have red, green, or lavender faces, and should remain in comic strips.

CHANNELS AND POWER

As you probably have already learned, each television channel (Fig. 4-1) has a luma/sound bandwidth of 6 MHz, with audio being separated from luminance by 4.5 MHz. On channel 4, for example, the video carrier is pegged at 67.25 MHz and audio at 71.75 MHz— and that's precisely 4.5 MHz separation. Note that audio *always* occupies the higher frequency in any of the assigned channels between 2 and 69 . . . formerly 2 through 83 before the FCC assigned spectrum space between 70 and 83 to land mobile. You should also be aware of the break between channels 2 through 6 (54 to 88 MHz) and 7 through 13 (174 to 216 MHz) to accommodate FM radio service in the very familiar frequency assignment of 88 to 108 MHz. AM radio, of course, remains at 0.535 to 1.605 MHz where it's always been.

As for power, you have to consider two kinds in assessing broadcast muscle. They are, the actual power generated by the transmitter and the effective radiated power (ERP). The FCC, of course, limits low VHF to 100 kW ERP, high VHF to 316 kW ERP, and UHF to 5,000 kW ERP. So you may actually operate your transmitter *below* rated power to cover FCC-assigned territory at rated ERP which, by the way, is the transmitter output *multiplied* by antenna gain.

Consequently, we really can't honestly offer an average or maximum of power operating percentage VHF stations usually radiate, but most of them do generate maximum power. Not so with UHF since antenna gains increase with frequency (15 to 25 dB at UHF), and full power costs are extremely high. Under such considerations for UHF, then, you can see why CP via the Andrew method could be more than casually attractive. (See Chapters 2 and 3.)

In the past, gridded tubes have been used in

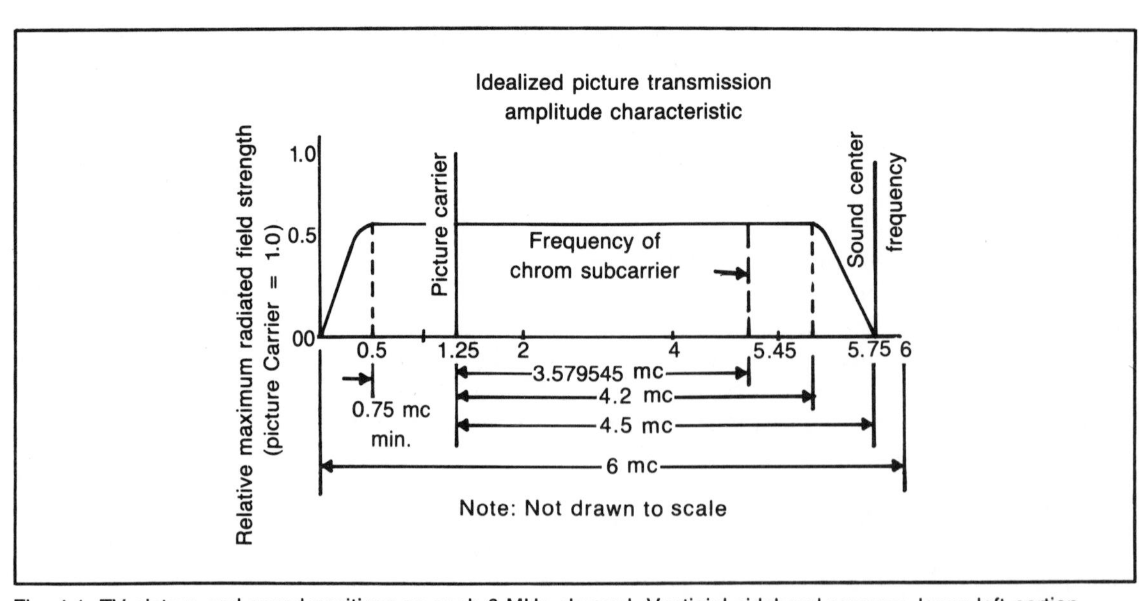

Fig. 4-1. TV picture and sound positions on each 6 MHz channel. Vestigial sideband removes lower left portion.

power stages, while today considerably more use is being made of klystrons, especially at UHF. Generally, all new transmitters are solid- state for better reliability, and more linear signal processing up to the final amplifier. There's less heat, reduced power, greater component longevity, and considerably better use of space, along with excellent remote controls. Actually, many transmitters may be left unattended, subject only to remote monitoring and periodic checkups—a condition not possible until the full application of solid-state signal conditioning and amplification. With further FCC deregulation now in progress, Federal licensing is a thing of the past, and competition alone forces adherence to government regulations set up long ago when FCC engineering controlled the industry. Now, despite some private licensing-type quizzes and certificates, every station is really on its own.

Fortunately, however, enough common sense and built-in nostalgia remains to continue successful operations at least through this generation of employees. If low power (100W) TV ever gets off the ground, this may not be the case—we'll have to wait and see. It may be that Government second thoughts resulting from this *laissez faire* philosophy may reverse the trend before there's chaos in the industry. In the meantime, the Electronic Industries Association and the National Broadcasters Association have a big and important job to do policing their own.

A good example of continuing change has now eliminated detailed TV station logging, bypassed consideration of community needs, and nullified any program percentage relationship between entertainment and nonentertainment video fare. Under multiple ownership rules one may now have a dozen FM or AM stations, and these numbers may also shortly apply to television station ownership, with the possible addition of a cap on VHF audience coverage and/or exceptions for UHF. Newspapers, however, may not own radio/TV broadcast facilities in their immediate area *unless* grandfathered under previously existing statutes. In addition to the above, the FCC in late 1984 released a number of technical changes; the more important ones we'll note where they affect television broadcasting.

FCC RECENT RULE CHANGES

This Federal Communications Commission Report and Order was adopted November 8, 1984 and released on Pearl Harbor anniversary day, December 7, 1984. As with most FCC regulations these days, there's further deregulation accompanied by increased dependence on commercial competition to keep the airways "clean." Overall results at this time are indeterminate, but less FCC engineers and more lawyers do build the paper factory a mite higher and longer. Regardless, here's what has happened:

Paragraph 73.687 Transmission System Requirements

Subparagraphs (a) (1) monochrome transmitter attenuation characterisitics measurements; (a) (2) color transmissions with 3.58 MHz sinewave measurements; (a) (9) white and black level transition spacing linearity; (b) (1) aural transmitter deviation relative to 100% modulation; part of (b) (2) audio bandpasses from 50 to 15 kHz internally to transmitter with preemphasis (now same as FM); (b) 3 through 5—harmonic measurements over 50-15,000 Hz, transmitter output noise level at FM and AM in the audio range "have been deferred to the marketplace." The 75 μsec preemphasis, however, has been *retained* as it was before.

The above, of course, refers to television transmitters. In considering the assorted list of changes, however, *we decided to print the entire group since there may be certain information contained referring to other than TV that may be highly useful to others involved in, say, AM and FM broadcasting also.* Appendix C, therefore, of the FCC Dec. 7, 1984 changes are printed verbatim as issued.

THE 1984 REPORT AND ORDER CHANGES

A. Title 47 of the Code of Federal Regulations, Part 15, is amended as follows:

1. Section 15.333 is amended by revising paragraph (b) to read as follows:

§15.333 Operation in the band 72-76 MHz.

*   *   *   *   *

(b) A receiver may be operated as part of an auditory assistance system provided it meets the technical specifications in §15.367 and is certified pursuant to §15.345.

2. Sections 15.361, 15.363 and 15.365 are deleted in their entirety.

B. Title 47 of the Code of Federal Regulations, Part 73, is amended as follows:

1. Section 73.40 is amended by revising paragraph (a) and subparagraphs (b) (2) and (3), and by deleting subparagraphs (a) (1), (2), (3), (4), and (5) to read as follows:

§73.40 AM transmission system performance requirements.

(a) Stations must annually show compliance with §73.44. These emission limitations must be met under any conditions of modulation expected to be encountered by the station.

(b) *   *   *

*   *   *   *   *

(2) For main channel modulation only, the total audio frequency distortion from terminals to antenna output shall not exceed 5% harmonics (voltage measurements of arithmetical sums or r.s.s) when modulated from 0 to 84% and shall not exceed 7.5% harmonics (voltage measurements of arithmetical sum or r.s.s.) when modulating 85% to 95% (distortion shall be measured with modulating frequencies of 50, 100, 400, 1000, 5000 and 7500 Hz up to tenth harmonic or 16,000 Hz, or any intermediate frequency that readings on these frequencies indicate is desirable). Harmonics should be observed to 20,000 Hz. When stereophonic transmission is used, the distortion must be measured in the left and right channels separately using a suitable stereophonic demodulator.

(3) The audio frequency transmitting characteristics for main (L + R), left (L) only and right (R) only modulation shall not depart more than 2 dB from that at 1000 Hz between 100 and 7500 Hz.

*   *   *   *   *

2. Section 73.317 is amended by deleting subparagraphs (a) (1), (3), (4) and (5) and subdivisions (a) (3) (i) and (ii), and by revising subparagraph (a) (2) as subparagraph (a) (1) and redesignating subparagraphs (a) (6) through (14) as subparagraphs (a) (2) through (10) to read as follows:

§73.317 Transmission system requirements.

(a) *   *   *

(1) Pre-emphasis shall be employed as closely as practicable in accordance with the impedance-frequency characteristic of a series inductance-resistance network having a time constant of 75 microseconds. (See upper curve of Fig. 2 of §73.333.)

(2) Automatic means shall be provided in the transmitter to maintain the assigned center frequency within the allowable tolerance (±2000 Hz).

(3) [Reserved]

(4) Adequate provision shall be made for varying the transmitter output power to compensate for excessive variations in line voltage or for other factors affecting the output power.

(5) Adequate provision shall be provided in all component parts to avoid overheating at the rated maximum output power.

(6) Means should be provided for connection and continuous operation of a modulation monitor.

(7) If a limiting or compression amplifier is employed, precaution should be maintained in its connection in the circuit due to the use of pre-emphasis in the transmitting system.

(8) Any emission appearing on a frequency removed from the carrier by between 120 kHz and 240 kHz inclusive shall be attenuated at least 25 dB below the level of the unmodulated carrier. Compliance with this specification will be deemed to show the occupied bandwidth to be 240 kHz or less.

(9) Any emission appearing on a frequency removed from the carrier by more than 240 kHz and up to and including 600 kHz shall be attenuated at least 35 dB below the level of the unmodulated carrier.

(10) Any emission appearing on a frequency removed from the carrier by more than 600 kHz shall be attenuated at least 43 + 10 Log_{10} (Power in watts) dB below the level of the unmodulated carrier, or 80 dB, whichever is the lesser attenuation.

(b) * * *

* * * * *

3. Section 73.687 is amended by deleting subparagraphs (a) (1), (2), (9) (b) (1), (3), (4) and (5), and subdivisions (b) (3) (i) and (ii); by redesignating subparagraphs (a) (3) through (a) (8) as subparagraphs (a) (1) through (a) (6); by revising and redesignating subparagraph (b) (2) as subparagraph (b) (1); and, by redesignating subparagraphs (b) (6) and (b) (7) as subparagraphs (b) (2) and (b) (3), respectively, to read as follows:

§73.687 Tranmsission system requirements.

(a) * * *

(1) The field strength or voltage of the lower sideband, as radiated or dissipated and measured as described in paragraph (a) (2) of this section, shall not be greater than −20 dB for a modulating frequency of 1.25 MHz or greater and in addition, for color, shall not be greater than −42 dB for a modulating frequency of 3.579545 MHz (the color subcarrier frequency). For both monochrome and color, the field strength or voltage of the upper sideband as radiated or dissipated and measured as described in paragraph (a) (2) of this section shall not be greater than −20 dB for a modulating frequency of 4.75 MHz or greater. For stations operating on Channels 15-69 and employing a transmitter delivering maximum peak visual power output of 1 kW or less, the field strength or voltage of the upper and lower sidebands, as radiated or dissipated and measured as described in paragraph (a) (2) of this section, shall depart from the visual amplitude characteristic (Fig. 5a for §73.699) by no more than the following amounts:

−2 dB at 0.5 MHz below visual carrier frequency;
−2 dB at 0.5 MHz above visual carrier frequency;
−2 dB at 1.25 MHz above visual carrier frequency;
−3 dB at 2.0 MHz above visual carrier frequency;
−6 dB at 3.0 MHz above visual carrier frequency;
−12 dB at 3.5 MHz above visual carrier frequency;
−8 dB at 3.58 MHz above visual carrier frequency
(for color transmission only).

The field strength or voltage of the upper and lower sidebands, as radiated or dissipated and measured as described in paragraph (a) (2) of this section, shall not exceed a level of -20 dB for a modulating frequency of 4.75 MHz or greater. If interference to the reception of other stations is caused by out-of-channel lower sideband emission, the technical requirements applicable to stations operating on Channels 2-13 shall be met.

(2) The attenuation characteristics of a visual transmitter shall be measured by application of a modulating signal to the transmitter input terminals in place of the normal composite television video signal. The signal applied shall be a composite signal composed of a synchronizing signal to establish peak output voltage plus a variable frequency sine wave voltage occupying the interval between synchronizing pulses. (The "synchronizing signal" referred to in this section means either a standard synchronizing wave form or any pulse that will properly set the peak.) The axis of the sine wave in the composite signal observed in the output monitor shall be maintained at an amplitude 0.5 of the voltage of synchronizing peaks. The amplitude of the sine wave input shall be held at a constant value. This constant value should be such that at no modulating frequency does the maximum excursion of the sine wave, observed in the composite output signal monitor, exceed the value 0.75 of peak output voltage. The amplitude of the 200 kHz sideband shall be measured and designated zero dB as a basis for comparison. The modulation signal frequency shall then be varied over the desired range and the field strength or signal voltage of the corresponding sidebands measured. As an alternate method of measuring in those cases in which the automatic d-c insertion can be replaced by manual control, the above characteristic may be taken by the use of a video sweep generator and without the use of pedestal synchronizing pulses. The d-c level shall be set for midcharacteristic operation.

(3) A sine wave, introduced at those terminals of the transmitter which are normally fed the composite color picture signal, shall produce a radiated signal having an envelope delay, relative to the average envelope delay between 0.05 and 0.20 MHz, of zero microseconds up to a frequency of 3.0 MHz; and then linearly decreasing to 4.18 MHz so as to be equal to -0.17 μsecs at 3.58 MHz. The tolerance on the envelope delay shall be ± 0.05 μsecs at 3.58 MHz. The tolerance shall increase linearly to ± 0.1 μ0.1 μsec down to 2.1 MHz, and remain at ± 0.1 μsec down to 0.2 MHz. (Tolerances for the interval of 0.0 to 0.2 MHz are not specified at the present time.) The tolerance shall also increase linearly to ± 0.1 μsec at 4.18 MHz.

(4) The radio frequency signal, as radiated, shall have an envelope as would be produced by a modulating signal in conformity with §73.682 and Figure 6 or 7 of §73.699, as modified by vestigial sideband operation specified in Fig. 5 of §73.699. For stations operating on Channels 15-69 the radio frequency signal, as radiated, shall have an envelope as would be produced by a modulating signal in conformity with §73.682 and Figure 6 or 7 of §73.699.

(5) The time interval between the leading edges of successive horizontal pulses shall vary less than one half of one percent of the average interval. However, for color transmissions, §73.682(a) (5) and (6) shall be controlling.

(6) The rate of change of the frequency of recurrence of the leading edges of the horizontal synchronizing signals shall be not greater than 0.15 percent per second, the frequency to be determined by an averaging process carried out over a period of not less than 20, nor more than 100 lines, such lines not to include any portion of the blanking interval. However, for color transmissions, §73.682(a) (5) and (6) shall be controlling.

(b) * * *

(1) Pre-emphasis shall be employed as closely as practicable in accordance with the impedance-frequency characteristic of a series inductance-resistance network having a time constant of 75 microseconds. (See upper curve of Figure 12 of §73.699.)

(2) If a limiting or compression amplifier is employed, precaution should be maintained in its connection in the circuit due to the use of pre-emphasis in the transmitting system.

(3) Aural modulation levels are specified in §73.1570.

* * * * *

4. Section 73.1570 is amended by revising subparagraph (b) (3) to read as follows:

§73.1570 Modulation levels: AM, FM, and TV aural.

* * * * *

(b) * * *

* * * * *

(3) TV station. In no case shall the total modulation of the aural carrier exceed 100% on peaks of frequency recurrence, unless some other peak modulation level is specified in an instrument of authorization. For monophonic transmissions, 100% ± 25 kHz.

* * * * *

5. Section 73.1590 is ammended by deleting subdivisions (b) (1) (i), (ii), (iii), (iv), and (b) (3) (i), (ii), (iii), (iv), and subparagraphs (c) (1) and (c) (6) and marking them [Reserved]; and, by revising subdivisions (b) (2) (ii) and (iii) to read as follows:

§73.1590 Equipment performance measurements.

* * * * *

(b) * * *

(1) * * *

(i) [Reserved].

(ii) [Reserved].

(iii) [Reserved].

(iv) [Reserved].

(v) * * *

(2) * * *

(i) * * *

(ii) Data and curves showing audio frequency harmonic content for 25%, 50%, 75% and (main channel only) 100% modulation for the audio frequencies 50, 100, 400, 1000, 5000, and when attainable 7,500 10,000, 12,000 and 15,000 Hz (either arithmetical or RSS (root sum square)) values up to the 10th harmonic or 30,00 Hz) for equal left and right (L = R), left (L) only and right (R) only signals. A family of curves must be plotted (one for each percentage above) with percent distortion as ordinate and audio frequency as abscissa.

(iii) Data showing percentage of carrier amplitude regulation (carrier shift) for 25, 50, 85 and, if obtainable, 100% modulation with 400 Hz tone for main channel modulation with equal left and right (L = R) signals.

* * * * *

(3) * * *

(i) [Reserved]

(ii) [Reserved]

(iii) [Reserved]

(iv) [Reserved]

(v) * * *

(c) * * *

(1) [Reserved]

* * * * *

(6) [Reserved]

* * * * *

C. Title 47 of the Code of Federal Regulations, Part 74, is amended as follows:

Section 74.750 is amended by revising subparagraph (d) (1) to read as follows:

§74.750 Transmission system facilities

* * * * *

(d) * * *

(1) The equipment shall meet the requirements of paragraphs (a) (1) and (b) (3) of §73.687.

* * * * *

According to FCC Rules and Regulations, Volume III, October 1982, visual transmitter operating power is usually to be determined by the direct method—with a calibrated transmission line meter connected to the rf output terminals at 80-, 100-, and 110-percent of authorized power. Calibration measurements take place at average power into a zero reactance dummy load. Should electrical "devices" be used, their full-scale accuracy must be at least ± 5 percent. The same transmission line meter is also used for aural power measurements without modulation, just like the visual transmitter which, however, must have a standard sync signal with blanking level set at 75 percent of peak amplitude.

There's also an indirect method available where dc input voltage to the final rf amplifier multipled by dc current at the same point is then multiplied by an efficiency factor:

$$\text{Transmitter } P_o = E_p \times I_p \times F$$

Where F is determined by records, measurement data, or manufacturers specifications submitted to the FCC during type acceptance. Figure 4-2 also sets forth power levels with respect to antenna heights in U.S. Zones I, II, and III.

REMOTE CONTROLS

TV broadcast stations authorized to operate by remote control must do and have the following:

- On/off transmitter switch.
- Operating parameter-reading instruments cali-

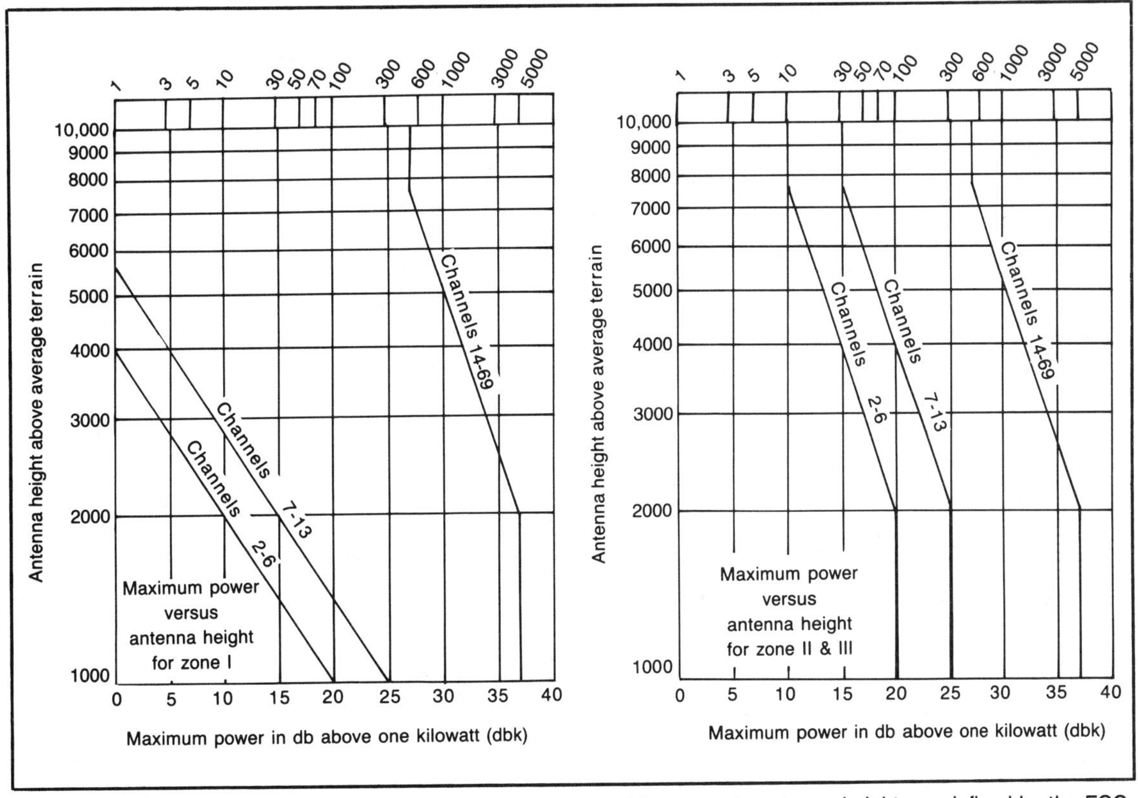

Fig. 4-2. Maximum power diagrams for U.S. Zones I, II, and III, with respect to antenna heights as defined by the FCC.

brated to within two percent of those at the transmitter.

- Remote meters are to meet specs of "regular" transmitter, antenna, and monitor meters.
- Arbitrary scale meters may be used, but their calibration charts and curves must relate precisely to the main meters.
- Sufficient controls to adjust all transmitter requirements for daily operations.
- Equipment to monitor antenna signal radiation such as waveform, percentage of modulation, color phase and amplitude, including full fields and test signal displays on the 21 vertical blanking interval lines.
- A type approved aural modulation monitor and signal frequency amplifier which can continuously (and accurately) show the peak and quasi-peak percentages of modulation.
- Instrumentation for generating particular test signals to adjust and test complete transmission system and to calibrate monitoring equipment.
- Adequate lighting and support tower operation checks.
- Control point to be "under immediate supervision and control of one or more operators at all times" when under remote control.
- Remotes if shorted or open will not "activate" the transmitter and any loss of control will cut off the transmitter.
- Transmitter and control points are to be protected against unauthorized operation.
- Transmitter-control waveform monitors are to read alike or any deviations noted.
- All monitoring equipment shall be calibrated and tested and the broadcast transmitter inspected frequently to insure proper operation.

With a little paraphrasing here and there for plain English translation, that's about the sum and substance of remote operation. You *don't* have to ask the FCC for remote operations permission, but you *must* file a notice with the FCC within three days following the action.

Three are many reasons for tight remote and even master control TV broadcast requirements, especially the quality of signal appearing at the antenna as it is about to be radiated. All transmitters, unfortunately, produce undesirable radiation of one form or another, some coming from the antenna and other types originating in the transmitter. Usually such radiation is harmonically related to the carrier or simply parasitic oscillations—thus, all spurious "glitches" of any description should be identified and, if possible, suppressed immediately whenever said transmitter goes on the air. Prior testing with known waveform characteristics will often recognize these problems and filtering plus shielding can usually eliminate most of them. But in some instances, even entire transmitters or transmission lines in damaged or older categories will have to be replaced to complete the job under critical circumstances. Sorry to say, the commercial broadcast spectrum is just becoming too crowded to permit distorted outputs. Because of the FCC's restricted field operations, bad transmissions may continue for a year, but in the end complaints and mobile monitoring will detect the culprit and he'll have to pay the price—and sometimes it's a big one. Contrary to some wishful thinking, all types of transmissions monitoring do continue with fixed and mobile FCC units, so don't pres press your luck. In addition to harmonics and spurs, transmitters may also generate gamma (X) rays and acoustical noise, and these will have to be measured and certified at the transmitting site rather than under ideal conditions at the originating factory.

Most state-station licensing occurred recently between 1981 and 1984, depending on past issuance, with renewal periods good for five years (seven for radio). Initial or renewals, however may be granted for lesser terms if the FCC finds a "public convenience" factor. The hour of expiration for regular licenses normally occurs at 3 a.m. local time in the morning.

Before we leave the topic of remote controls, let's borrow a worthwhile equation from the Natl. Assn. of Broadcasters handbook and A. Hans Bott of Harris; that is, the reliability factor "R."

$$R = e^{-t/T}$$

where t is the operating period, and T stands for MTBF, or meantime between failures. As Mr. Bott points out, MTBF for 50 kW transmitters is usually about 500 hours. So on the basis of an 18-hour broadcast day:

$$R = e^{-18/500} = e^{-0.036} = 96.46\%$$

which is really a measure of efficiency, and a rather good one at that. The e, of course, stands for the natural logarithm of 2.7183, formally known as Napierian.

LATEST REMOTE RULINGS

The FCC has also issued further rulings on remote controls dated (released) Nov. 21, 1984, FCC 84-549, MM Docket No. 84-110, which should be important to many, if not most station operators and, therefore, will be included as part of this chapter for the record.

The following amendments are made to Title 47, Part 73 of the Code of Federal Regulations:

1. 47 CFR 73.66 is removed in its entirety.
2. 47 CFR 73.67 is removed in its entirety.
3. 47 CFR 73.274 is removed in its entirety.
4. 47 CFR 73.275 is removed in its entirety.
5. 47 CFR 73.574 is removed in its entirety.
6. 47 CFR 73.575 is removed in its entirety.
7. 47 CFR 73.676 is removed in its entirety.
8. 47 CFR 73.677 is removed in its entirety.

9. 47 CFR 73.3548 is removed in its entirety.
10. In 47 CFR, new Section 73.1400 is added to read as follows:

§73.1400 Remote control authorizations.

(a) An AM, FM, or TV station transmission system may be operated by remote control using the procedures described in §73.1410.

(b) No authorization from the FCC is required to operate the transmission system of an AM station operating with a nondirectional antenna, FM station, or TV station by remote control. Authority to operate an AM station using a directional antenna system by remote control is obtained using the following procedures:

(1) An application for a construction permit to erect a new directional antenna or make modifications in an existing directional antenna, subject to the sampling system requirements of §73.68, may request remote control authorization on the permit application FCC Form 301 (FCC Form 340 for noncommercial educational stations).

(2) A licensee or permittee having a sampling system in compliance with the provisions of §73.68(a) must request remote control authorization on FCC Form 301-A, and submit information showing that the directional antenna sampling system has been constructed according to the specifications of §73.68(a).

(3) A licensee or permittee of a station not having an approved directional sampling system in compliance with the provisions of §73.68(a) must request remote control authorization on FCC Form 301-A, and submit information showing that the directional antenna is in proper adjustment and further showing the stability of the antenna system during the 1-year period specified in Section II of Form 301-A.

(c) Whenever a remote control point is established at a location other than at the main studio or transmitter, notification of that remote location must be sent to the FCC in Washington, D.C., within 3 days of initial use of that point. This notification is not required if responsible station personnel may be contacted at the transmitter or studio site during hours of operation when the remote control operator is elsewhere.

11. In 47 CFR, new Section 73.1410 is added to read as follows:

§73.1410 Remote control operation.

(a) Broadcast stations operated by remote control must provide at remote control points sufficient control and operating parameter monitoring capability to allow technical operation in compliance with the Rules applicable to that station and the terms of the station authorization. AM stations that are required to change modes of operation during the broadcast day must provide sufficient redundancy to assure that such mode changes actually occur.

(b) The remote control system must be designed, installed, and protected so that the transmitter can be activated or controlled only by licensed transmitter operators authorized by the licensee.

(c) The remote control and monitoring equipment must be calibrated and tested as often as necessary to ensure proper operation.

(d) The remote control system must be designed so that malfunctions in the circuits between the control point and transmitter will not cause the transmitter to be inadvertently activated or to change operating modes or output power.

(e) Whenever a malfunction causes loss of accurate indications of the transmitter operating parameters, use of remote control must be discontinued within 3 hours after the malfunction is first detected. If the station is found to be operating beyond the terms of the station authorization and such malfunction cannot be corrected by remote control, station operation must be immediately terminated.

(f) AM stations may use amplitude or phase modulation of the carrier wave for remote control telemetry and alarm purposes. FM stations may use aural subcarriers and TV stations may use either aural subcarriers or signals within the vertical blanking interval for telemetry and alarm purposes. Use of such remote control signals must be in accordance with the technical standards for the particular class of station.

12. 47 CFR 73.51 is amended by changing the reference to rule "73.57" in subparagraph (d) (5) to read "73.1410."

13. 47 CFR 73.68 is amended by removing subparagraph (b) (4) and redesignating subparagraph (b) (5) as (b) (4).

14. 47 CFR 73.1225 is amended by removing sub-subparagraph (c) (1) (iv) (A) and redesignating sub-subparagraphs (B) through (D) as (A) through (C).

15. 47 CFR 73.1560 is amended by revising the headnote, redesignating existing paragraph (a) as subparagraph (a) (1), and adding new subparagraph (a) (2) to read as follows:

§73.1560 Operating power and mode tolerances.

(a) *AM stations.*

(1) * * *

(2) Whenever the transmitter of an AM station cannot be placed into the specified operating mode at the time required, transmissions of the station must be immediately terminated. However, if the radiated field at any bearing or elevation does not exceed that permitted for that time of day, operation in the mode with the lesser radiated field may continue under the notification procedures of paragraph (d) of this section.

* * * * *

16. 74 CFR 73.1570 is amended by removing the words "§73.67(b) or" from subparagraph (b) (1) (ii).

17. 74 CFR 1690 is amended by revising the words "See §73.3548" in subparagraph (d) (2) to read "See §73.1400."

18. The alphabetical index to 47 CFR Part 73 is amended removing all references to rule sections numbered 73.66, 73.67, 73.274, 73.275, 73.574, 73.575, 73.676, 73.677, and 73.3548; and the following listings are added in alphabetical sequence:

Authorization, Remote control 73.1400

Mode tolerances, Operating power and 73.1560

Operating power and mode tolerances 73.1560

Power and mode tolerances, Operating 73.1560

Remote control authorization 73.1400

Remote control operation 73.1410

TRANSMITTER OPERATIONS

Since we'll be describing a number of very new and modern transmitters in the next several pages, probably the most illuminating course to take now is a brief discussion of certain transmitter sections and their intended operations so the specific descriptions will have full meaning. Again, we'll borrow some of the forthcoming description from Harris' Hans Bott, who's done an admirable job on TV transmitters for NAB.

VHF TV TRANSMITTERS

Visual exciter/modulators usually have two oscillators, one modulator, certain video processing, an upconverter and a bandpass output filter. The intermediate frequency oscillator begins its operation at 37 MHz, is amplified, and then reaches the usual video modulator. This stage also receives the processed and linearly corrected video signal positioned on its sync pedestal so that linear mixing will take place correctly and can be sustained over a varying amplitude video range without deterioration. Mixing products should also be approximately 100 dB down below carrier level to avoid undesirable beats. When the new Harris transmitters are described, you'll see why this particular i-f frequency was chosen. See Fig. 4-3.

Next, processing power levels are to be established so that a relatively small signal is sufficient that requires few individual circuit adjustments—usually about 1 milliwatt (mW). If both oscillators also generate minimum power, then the additional requirement for built-in spares removes itself. The four (in this instance) printed circuit boards must accept either single- or double-ended inputs, compensate for nonlinearities, and restore dc signal levels. Such circuits are not sensitive to hum, noise, and other buzz-type interference. Differential gain, plus sync stretch is all part of this conditioning, as well as differential phase. In the diagram you also see delay compensation, as well as vestigial sideband, both with switch bypasses, if required. After another gain boost, the 60 MHz rf oscillator is double-multiplied twice, and reaches the upconverter at 240 MHz. Video is now upconverted to its selected channel, bandpassed and delivered as 1 milliwatt (mW) intelligence that, in a 75-ohm system, offers sufficient signal- to-noise (S/N) as well as adequate drive and good linearity. This peak-to-peak signal also matches that of videocassette recorders and other consumer video products. As the diagram shows, however, after bandpassing there is substantial amplification to 1 watt before final power coupling to the antenna.

In the diagram you will notice the output of the second 2X multiplier also connects to the aural

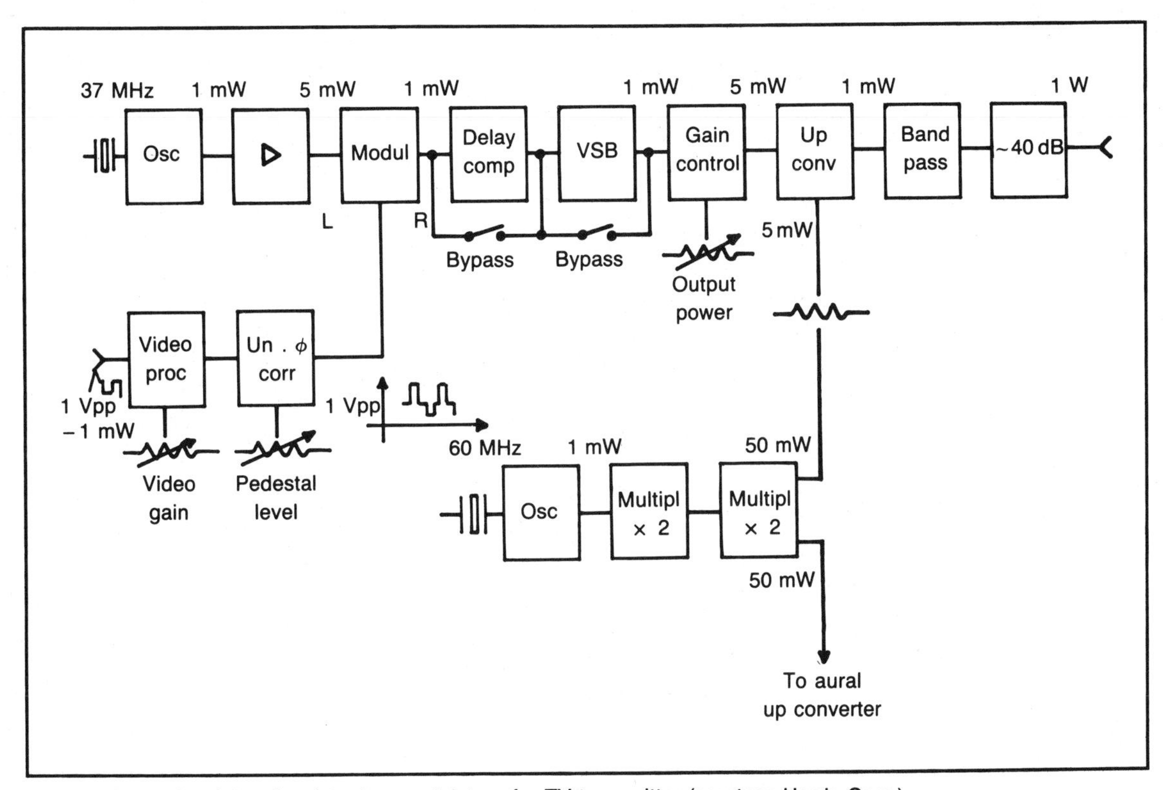

Fig. 4-3. Example of the visual exciter-modulator of a TV transmitter (courtesy Harris Corp.).

(audio) up converter. Figure 4-4 shows a block diagram of this exciter/modulator which also uses an i-f frequency, but this time at 32.5 MHz, with a varactor diode for frequency modulation (FM). An oven with proportional temperature stability and phase-locked loop control maintains this i-f at center frequency by controlling the modulating oscillator via dc corrections from the PLL with its divide-by circuits and crystal-controlled oscillator.

Mixing products are removed by a low-pass filter, and modulated FM then reaches the converter which is energized from multipliers on the visual exciter. A bandpass filter selects the difference frequency and the result is amplified for eventual broadcast. Mr. Bott says that frequency response, FM S/N and distortion are mostly determined by modulating oscillator circuit design. Sync becomes the next subject for discussion since the sync separator has been placed in board No. 2.

This time, however, we'll go directly to FCC specifications and approach the description from this point of view rather than attempting to work down a group of synchronizing chains which are constantly changing with further application of integrated circuits and very tight tolerance controls. One digital method, nonetheless, uses a harmonic of burst and color subcarrier, 3.579545 MHz, to obtain the 15,734 Hz horizontal pulse timing frequency upon which all remaining sync is based. Another system has a 64h (horizontal) 1.0069 MHz crystal oscillator and sampling system to produce the horizontal and 59.94 Hz vertical repetition rate pulses. The end result, of course, is to produce an interlaced raster of 525 lines in two fields of 262.5 lines to develop a full frame of which there are 30 per second (in our NTSC 60 Hz) system. Obviously pulse shaping and time delays are critical and that is why the FCC has issued very specific regulations governing them, the complete layout being shown in Fig. 4-5.

Here, the approved horizontal frequency is 2/455 X burst which calculates to 15734.26374 Hz, with the vertical repetition rate of 59.94 Hz simply a divide-by 262.49583, proving line-field scanning relationship. In transmitters this is normally not especially difficult to do, but in receivers with newly-developed digital countdowns, the procedure is relatively complex because of not only broadcast

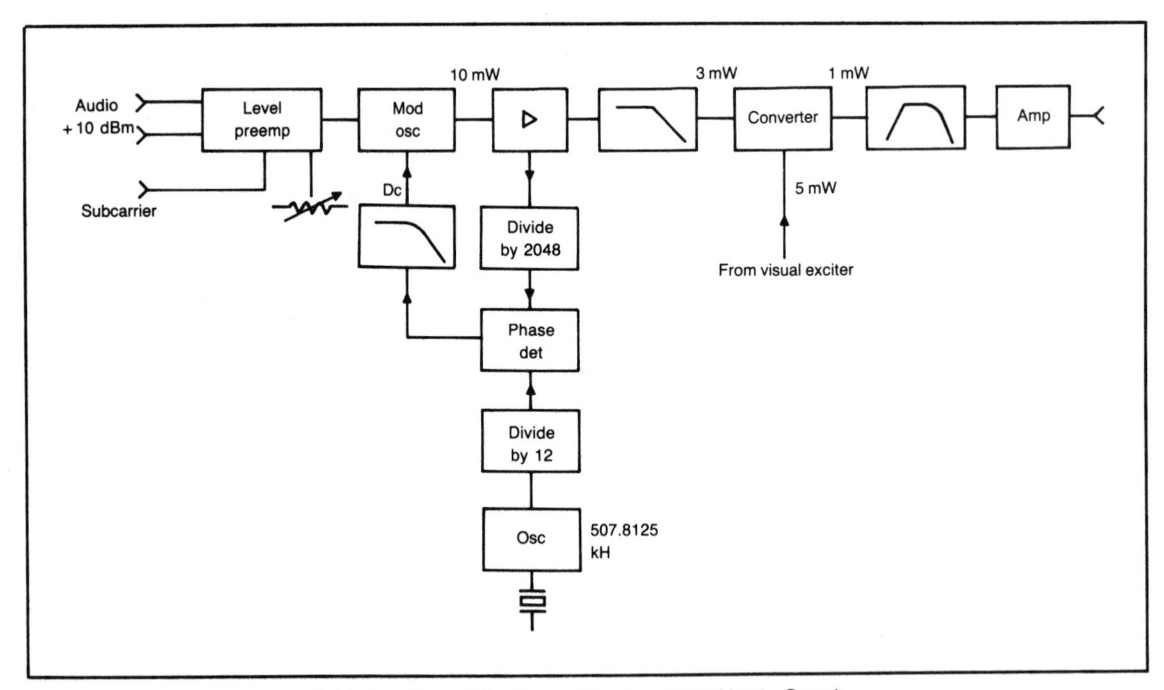

Fig. 4-4. The aural exciter-modulator portion of the transmitter (courtesy Harris Corp.).

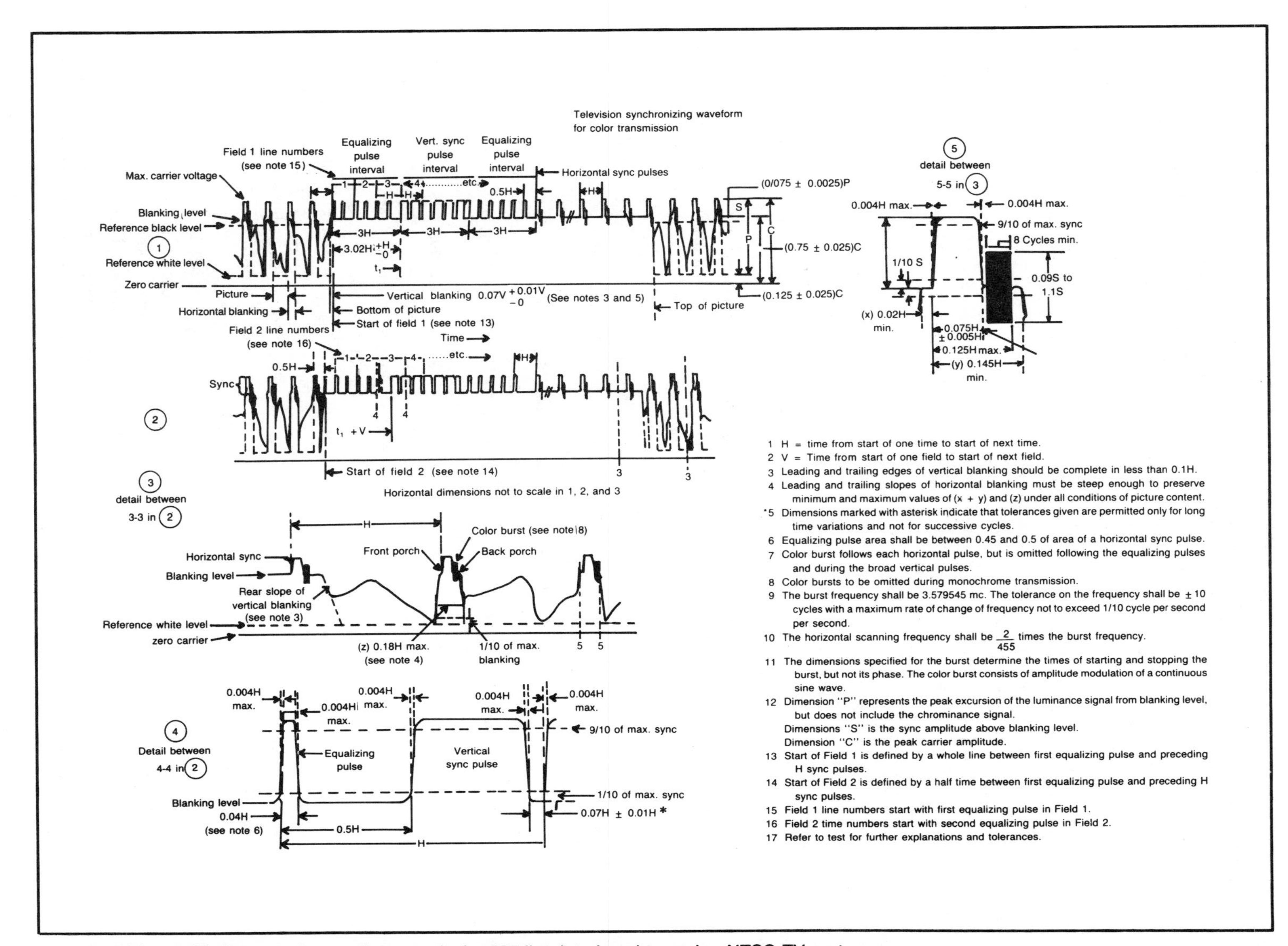

Fig. 4-5. FCC specified sync pulses and sync train for 525-line interlaced scanning NTSC TV system.

signals but inputs from ancillary equipments such as VCRs and video discs which, under certain circumstances can engender problems. In this respect, video readout computers and certain video games are in categories all by themselves—a subject discussed at some length in the receiver portion of the book.

As you can see, all equalizing, vertical pulses and pulse durations are given in terms of a horizontal line. Each horizontal line has a length (with blanking) of 63.5 microseconds, that of the vertical sync pulse has a time duration of 0.5H minus 0.07H times 63.5μsec, or 31.3055 microseconds. Look at detail 4 and you see that 0.07 has to be subtracted from 0.5h before doing the remaining simple multiplication. You can also see in the same detail that one equalizing pulse—of which there are six before and after the six vertical pulses—plus the vertical sync pulse, occupy the time of one horizontal line, and the six equalizing pulses, themselves, take up three horizontal line times.

Below fields 1 and 2 detail is the horizontal sync pulse itself, with color burst on its back porch. The duration of this pulse is 0.075h, or 4.7625 microseconds, as shown in detail 5, along with color burst at 2.24 μsec.

In all television, sync is transmitted during the horizontal and vertical blanking intervals so that it will not appear in the picture. However, slower vertical timing of 16.67 milliseconds per field can and does appear occasionally in audio as a perceptible sync buzz which, if not removed, becomes most annoying, especially with transmitter over modulation. Any 15.734 kHz distortion is usually too high for most ears even though it is mechanically generated in some faulty receivers.

That should about wrap up the simplified transmitter report and introduce you to the very real world of TV radiation that follows. Continued introduction of new transmitters, the extended use of solid state, and remote-controlled units of considerable output contribute to this changing picture, all of it for the better.

RCA's TTG SERIES FOR VHF

Completely solid state to the final amplifier, the TTG series of U/V transmitters by RCA offer improved picture and sound, stable performance, frequency-synthesized exciter, and special acoustical-wave sideband filter. Designed primarily for low/high band VHF, they are available in both single and parallel configurations with picture power from 10 through 100 kW and capable of worldwide transmissions including PAL, SECAM, M, N, B, D, and K1. Diplexing may be either internal or external, and the solid-state broadband driver has eliminated many transmitter tuning adjustments. Automated operation and stabilization requires *no* operator and minimizes most maintenance.

Contained in three compact cabinets with rear-opening full length locked doors are the exciter/driver, audio and video power amplifier, and power-control units. Harmonic and color notch filters are found in external output transmission lines.

The *solid state driver* is the most innovative among the many new features. Power output approaches 1600 watts peak and there is no high voltage, only 45 and 28 volts dc in low- and high-band operation, respectively. Two drawers contain the exciter-modulator, while a third houses the rf signal processor with rf amplifiers separate. All this is done to effectively isolate high and low power divisions of the transmitter to prevent stray pickup or crosstalk, as well as avoid any heat factor, all contributing to better shielding and system thermal reliability.

RCA's *exciter-modulator* is said to have been "meticulously" designed to approach transmitter system "signal transparency" with special high performance circuits, a looped-through video differential amplifier, plug-in equalizer boards for tuned amplifier and notch diplexer. Group delay correction is not required for the phase-linear sideband filter, rf exciter, or rf amplifiers, which are *all* solid state. A block diagram of both the exciter-modulator and rf processor is shown in Fig. 4-6. As you can see, video inputs are immediately subject to amplification, equalization, clamping, phase correction, and white clipping before modulation, and audio undergoes the usual preemphasis before FM excitation. There are also motorized controls in-

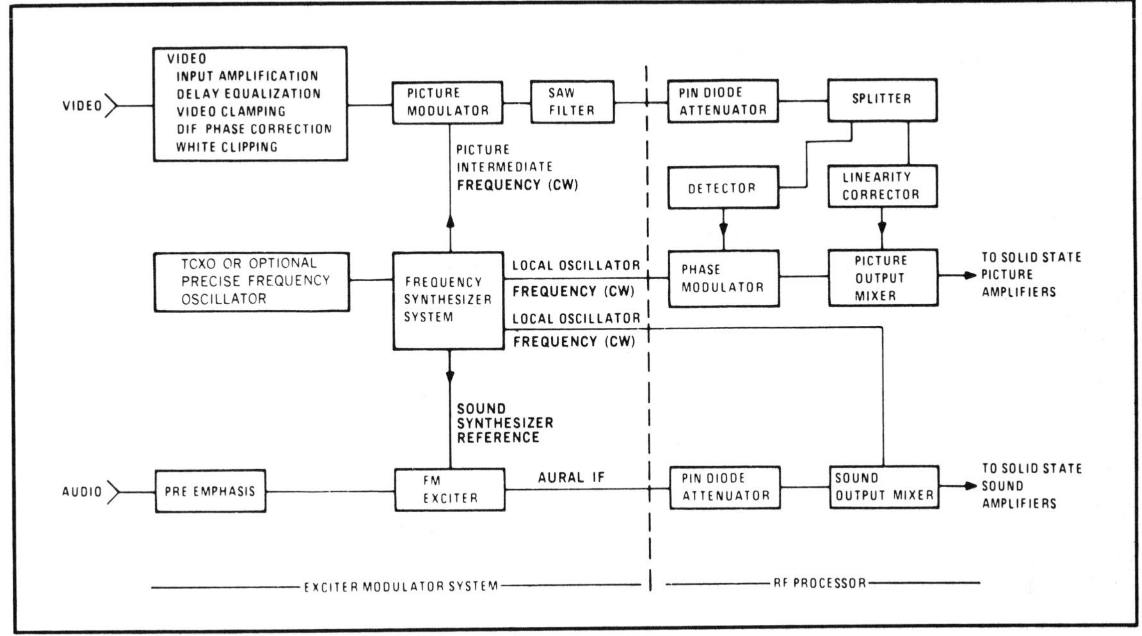

Fig. 4-6. A series TTG exciter/modulator and rf block (courtesy RCA).

cluded for remote adjustment of video levels and sync gain. In the center of the diagram you also see the TCXO or other reference oscillator setting both rate and drive for the following frequency synthesizer which serves as the system frequency standard. Outputs are to the FM exciter, and video modulator, as well as the phase modulator and output mixer in the rf processor.

Video is prepared for transmission by a double balanced modulator delivering signals to system M (FCC) U.S. at 45.75 MHz, or 38.9 MHz for CCIR systems B, D, and K1. By modulating at i-f, all CCIR systems are served by identical low-level, high- precision sideband filters for best picture and accurate response.

Audio, traditionally, frequency modulated, remains phase-locked to the video-audio synthesizer and is shock-mounted to guard against mechanical vibration. It has an extra subcarrier input for optional telemetry transmitter metering.

SAW filtering has also been further developed and temperature stabilized to afford "exceptional" flat amplitude frequency response over the video passband and sharp cutoff skirts. Low ripple and freedom from band edge delay distortions are also characteristic.

Frequency synthesizer, unlike the usual three-crystal oscillator systems, the TTG systems have only one temperature-compensated crystal oscillator (TCXO) of exceedingly high accuracy and standard frequency clocking a phase-locked loop (PLL) frequency synthesizer.

By simply selecting one or more component strapping connectors, any of the common carrier frequencies may be generated without changing the crystal, even including ± 10 kHz offsets. This places audio and video carriers within almost exact tolerance and permits universal interchange of spare TCXOs. In addition, there's also a *precise* frequency control oscillator, oven stabilized, that can maintain the visual carrier within ± 2.5 Hz over 30 days.

The rf processor accepts oscillator/video/audio feeds from the exciter-modulator and executes several operations before delivering outputs to the solid-state picture and sound amplifiers. Signal distortions from power amplification are removed, output power levels controlled, and sound and picture information is upconverted to carrier frequencies. This latter is done by mixing each i-f signal with the PLL local oscillator at carrier plus intermediate frequencies. Incidental phase distortion of the video carrier corrects via phase modulation equal and opposite to the fault, which also corrects color hue errors as well. A sample of the final rf output is detected and compared to sync level and reference voltage. Any resulting error is routed to both the video and audio pin diode attenuators for correction. Should load or VSWR threaten the system, a sensing circuit overrides automatic power stabilization and reduces transmit power to safe levels.

The remainder of this description can probably best be illustrated by one of these transmitters in parallel operation (Fig. 4-7). Here you see the two exciter-modulators delivering their signals to power dividers and the sound A/B and video A/B processors for diplexing and final transmission.

Actually, these are two TTG single units combined and operating as just one transmitter but having double output potential. Along with obvious equipment redundancy, the dual system can absorb any reflected ghosts resulting from poor load impedance matches, and signal quality improves with the averaging out of minor transients and glitches from either transmitter. Combining takes place with 3 dB couplers (combiners) in the audio/video outputs. Should one transmitter fail, total power output reduces to 1/4 of normal level, with half power from the good transmitter going to the load, and the other half to the combiner reject loads. Output switching may be either local or remote.

Naturally, only one exciter-modulator is on the air at a time, with the other in standby. In the event of failure, automatic switchover takes place immediately. The four output modes are: A plus B on the air; A on the air, B test; B on the air, A test; and A + B test. LED indicators deliver reporting transmit status visible through panel openings provided. These include startup confirmation, mode switching, line loss, temperature, power supply fault, overload, and single/multiple fault indication.

Rf amplifiers have separate linearity and incidental phase correction so that each processor may deliver maximum output with minimum distortion and also perform in parallel if necessary. Should the station licensed power not require full parallel output, bilevel switching is available to in-

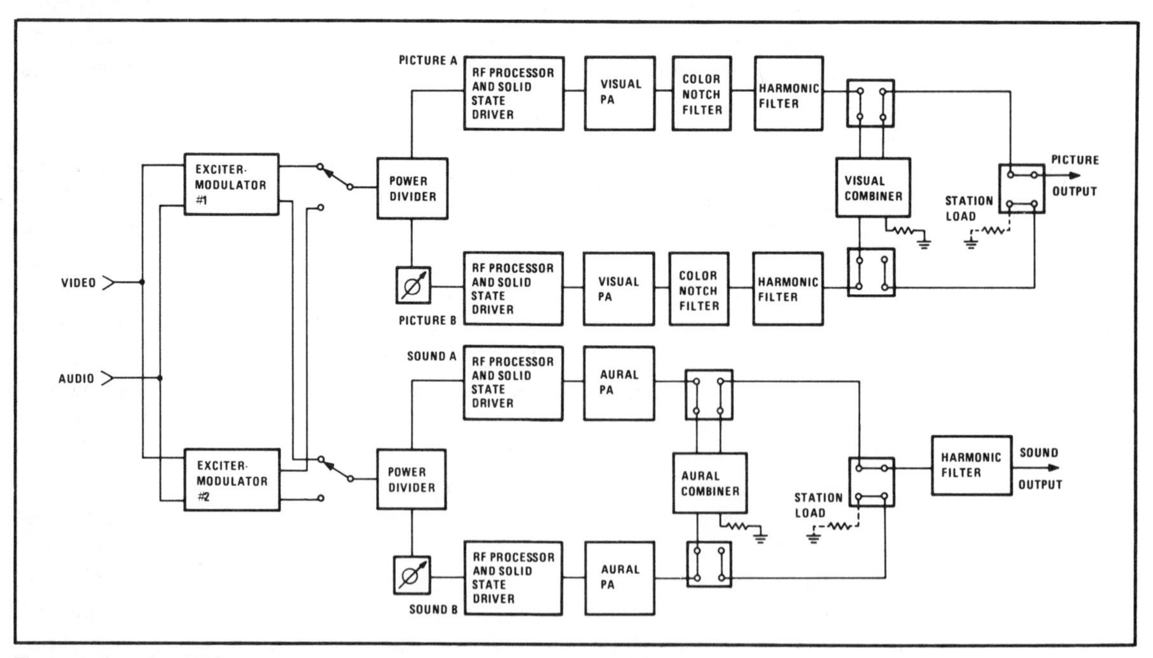

Fig. 4-7. A series TTG audio/video parallel transmitter (courtesy RCA).

crease power output of a single transmitter. RCA also reports that *all* TTG transmitters may be remotely controlled "with a minimum of installation effort" for interconnections with remote equipment.

RCA's TTG-100U FOR UHF

This is a brand new G-Line transmitter RCA has developed for UHF broadcasting using new high power 100-kW klystrons, available in either 100-kW or 200-kW versions. Supplied by Varian, the VKP 7853 100-kW klystron offers regular saturated power output with 55 percent efficiency. RCA, in turn, has now designed a new transmitter including this tube(s) having collector efficiencies of 79 percent with anode modulation sync pulse techniques.

All tubes are liquid cooled and have five integral cavities that operate with variable visual couplers, beam voltages of 33-KV, 6- amperes of beam current, and 50-W of drive power for a 100-kW output. Water cooled, each tube requires collector water flow of 50-gallons per minute with 8.5-inch piping, and 3-gallons per minute at 60-inches for body water flow. The usual klystron protections are needed for body, beam, ion and magnet over/under current overloads, in addition to collector and magnet current water flow. Actually 13 control and protection systems are required to prevent early or catastrophic failures should the occasion arise. Emergency operations consider 80 percent visual power levels and aurals of 5 percent visual and adequate, and multiplexing of visual and aural carriers required. In normal operations visual pulsing and nonpulsing supply a bandwidth of 6-MHz at 105-kW peak, allowing for 5 percent notch diplexer loss. Multichannel sound transmissions occupy approximately 2 MHz. However, intermediate power amplifier (IPA) and Exciter linearity corrections change between the two modes.

When the klystron is sync pulsed, black level anode voltage should be adjusted for minimum beam current for maximum efficiency and Exciter/IPA sync clipping and revised black/white linearity corrections are in order. In nonpulsing, klystron gain becomes greatest with high beam current, along with Exciter/IPA sync stretch to compensate compression effects.

THE TTG-100U TRANSMITTER

A block diagram of the entire transmitter appears in Fig. 4- 8. All subsystems are contained in three main cabinets: 1) Exciter/Control; 2) Aural power amplifier; 3) Visual power amplifier, while the ac cabinet normally occupies a spot between primary power interface and the high-voltage beam supply. Remaining system hardware consists of the heat exchanger and pumps, the notch diplexer, and transmission lines. These units are illustrated at the input and output of the referenced block.

The Exciter/Control cabinet houses its transmitter control, visual and aural exciters, and the exciter power supply, and accepts video, remotes, and multichannel sound. Thereafter, there are interfaces with power amplifier control, intermediate power amplifier (IPA), aural power amplifier, and ac control units. This latter component then supplies ac distribution as well as current for power amplifier logic and the 33 KV beam supply.

The Visual power amplifier contains control logic, IPA, mod anode pulser, IPA power supply, magnet supplies, the klystron and its output harmonic filter. Klystron beam acceleration comes from the 33-KV external supply.

RCA says Aural power amplifier makeup approximates that of the Visual power amplifier except that the anode pulser isn't used and only one magnet power supply and lower IPA power are needed.

IPA driving power for the VKP7853 klystron originates from a broadband, linear UHF transistor amplifier having a 100-W cw output. It has built-in overdrive detection, gain status monitor, and thermal fault sensors designed to limit dc power supplies.

With the loss of either visual or aural power amplification, multiplex electronics restores single-ended outputs and offers emergency backup for partial failures. Visual/aural summing in the Visual Exciter accomplishes this mission, which also requires Notch Diplexer detuning to transmit both carriers on Visual. Aural power amplifier multiplex-

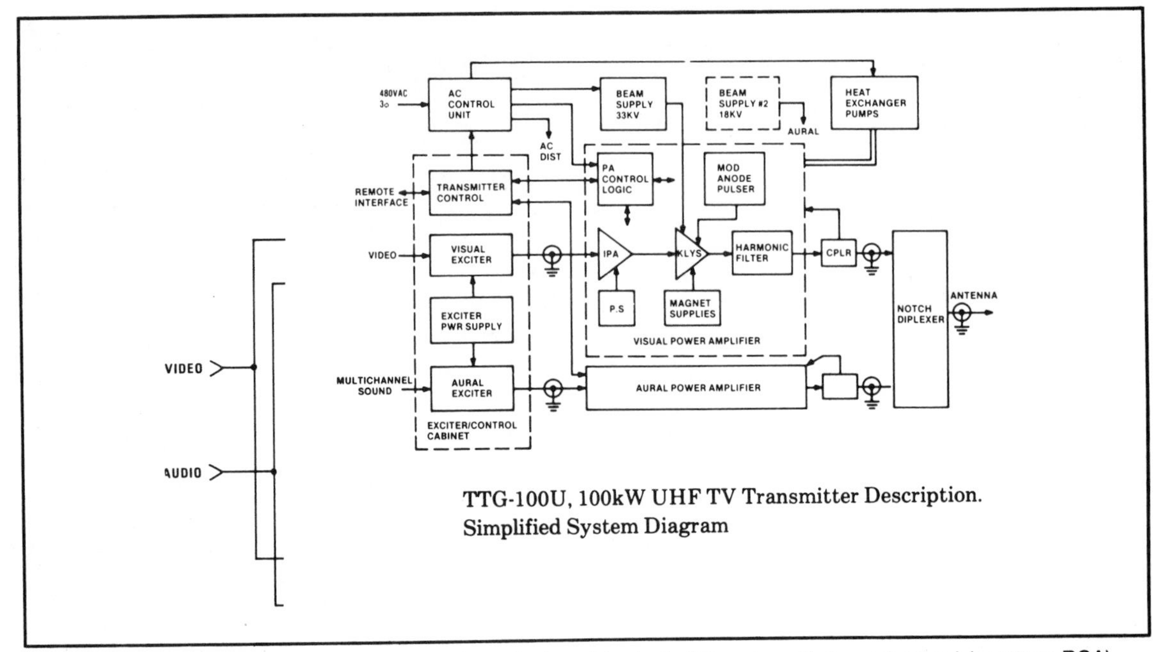

Fig. 4-8. RCA's very new 100 kW VHF transmitter designed for both CP and multi-channel sound (courtesy RCA).

ing does approximately the same thing but with Exciter output/PA input transfer required.

Aural and visual i-f frequencies originate from a single high-stability TCXO providing exact 4.5 MHz intercarrier stability needed for good stereophonic operation at 5 Hz/month tolerance. Precorrection of intercarrier phase modulation continues to be an RCA practice in UHF transmitters, virtually eliminating phase noise, sync transients, random color noise, and improper sync rations.

Vestigial sideband shaping with a SAW (surface wave acoustic filter) offers precision amplitude versus frequency response characteristics with adequate temperature control. Remote control includes 32 status, 19 telemetry, and 22 command functions. Phase sequence sensing inhibits transmitter operations when and if there are excessive klystron currents, arcs or standing waves (VSWRs) exceeding preset thresholds.

HARRIS' CP TRANSMITTERS

This powerful transmitter manufacturing company is also very much involved with circular polarized transmitters models in TVD single- and dual-operating modes. These come in various power delivery systems from 30- to 100-kW, either standing alone or in parallel—so you can take your pick, depending on need and practical limitations. As expected, remote control may be included, as well as Dualtran rf switching systems. We'll select at least two of these VHF units to describe from available literature. Unfortunately we don't have a complete theory of operation, but we'll give you what we can, which should be enough to visualize what the equipment will do.

Dual VHF High Band Transmitters

These are two 70- and 100-kW TVD-70H and TVD-100H color television units with advanced SAW filters, special tube distortion quadrature correctors, hot standby exciters, modulators, and sideband filters, Dualtran output switching for single or parallel operation, and ultra linear drivers with solid-state intermediate power amplifiers (IPAs) for signal transparency and maximum reliability (Fig. 4-9). A notch diplexer is included. They're designed for high VHF channels 7-13 and Channels E5-E12 and consist specifically of two

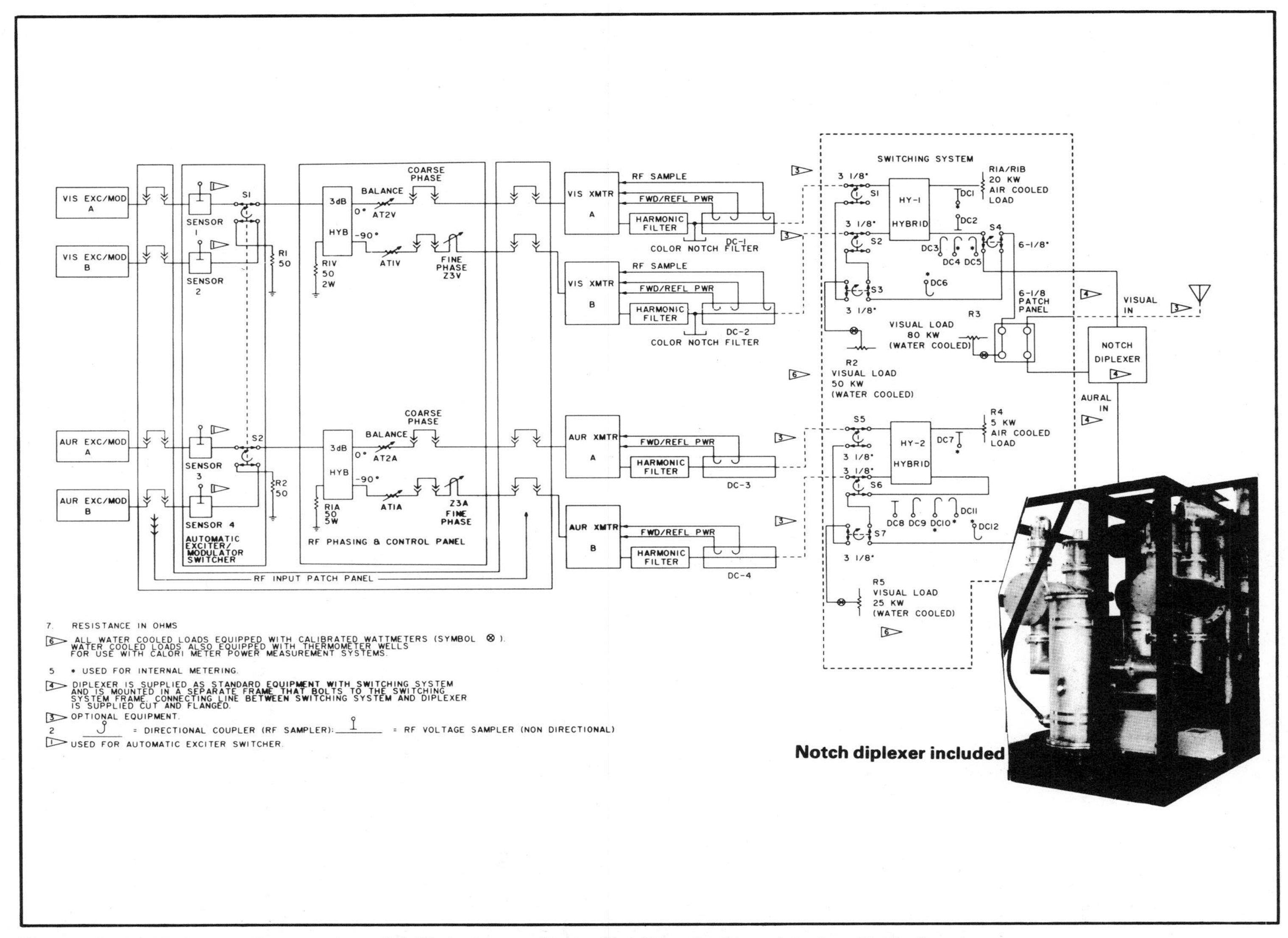

Fig. 4-9. A simplified block diagram of special CP dual transmitters with total redundancy (courtesy Harris Corp.).

separate 50-kW or 35-kW transmitters. Outputs are combined for high power and maximum ERP in radiating circular polarization (CP). You also have built-in redundancy to circumvent any loss of air time, with all low-level processors 100-percent redundant, including video processors, modulators, VSB filters, i-f correctors, oscillators, and upconverters. In the rf portion, the output of one set of exciters splits into two signals paths, passing to both transmitters through splitting and phasing networks, and switching is automatic to reserve exciters if there are problems. An rf input patch panel allows any transmitter to be driven from any exciter during tests or breakdowns.

The MCP-2V visual exciter offers a very linear ring diode modulator with compensation allowing modulation to zero carrier with only negligible phase, gain and differential phase distortions. The MCP aural exciter also delivers wideband low- distortion audio already tested with *all* subcarrier multichannel systems, along with a subcarrier input for telemetry or ENG.

The visual signal is then amplified by an ultralinear visual driver for an 8984 power amplifier tube, minimizing any corrections and permitting Harris transmitters to operate over extended periods without adjustments. There's also automatic gain control of all rf amplifiers, a new vestigial sideband filter for ripple-free amplitude response that's virtually flat between −0.75 MHz to +4.18 MHz, direct drive blower cooling with protected motors, and three water cooled test leads plus directional couplers for internal metering and transmitter monitoring.

In both sets of visual and aural transmitters you see forward reflected power and harmonic filtering, plus color notch filtering and rf sampling for video. In the switching system are the two hybrid switchers and air/water cooled loads. On the combined output you see the 100-kW notch diplexer that's included as part of the Dualtran rf system.

Cyclotran System

According to Harris, this is the most efficient and economical means to enjoy CP—the Cyclo (circular) and tran (transmission) way. According to the company, it's the first complete system specifically designed for circular polarization. Combining the TVD-100H 100-kW high-band transmitter and Harris' CPV antenna, the Cyclotran offers sharper, clearer pictures through ghost reduction, improved S/N, and better fringe area signals even with standard horizontally-polarized receiving antennas.

To save money, the new CP system has been engineered to cover approximately the same floor space as 50 kW transmitters, and the CPV top mount antenna has wind loading specs, equivalent or less than existing Batwings and traveling wave radiators—so there isn't expensive rebuilding and alternations.

The dual 100-kW transmitter consists of two TV-50H 50-W independent transmitters (just described), and has only three tubes, including the 8984 tetrode for the visual power amplifier. With any malfunctions, redundancy keeps you automatically on the air at 1/4 full power and no carrier interruption, or 1/2 power manually in three seconds.

Also included are the Harris transversal sideband filters, i-f modulation, power amplifier linear operation, and solid-state visual and aural exciter/modulators.

TVD-60H Systems

For those who want another configuration and complete transmitter redundancy, Harris also builds dual 30-kW transmitter with three tetrodes: a 9007 in the visual power amplifier, an 8988 in the visual driver and an 8807 in aural. This latter is a "field-proven" tetrode with a Harris built cavity for 3-6 kW output with sufficient bandwidth for hi-fi TV stereo operation with a wide bandwidth of 5.2 MHz. All three tubes have proven reliabilities of between 10- and 20-thousand hours.

In the visual PA, the exciter features i-f linearity correction for amplitude and lower sideband reduction reinsertion. Incidental phase distortions are cancelled, along with intercarrier phase noise for better stereo. TV-30 power consumption is rated at 61-kW for black picture and 55-kW for gray. The overall transmitter consists of a three-bay cabinet housing the exciters and rfs, a separate 8.5/4.25 kV high-voltage power supply, low-pass and color-notch filter, with all monitoring, metering and control circuits designed specifically for remote controls with designated hookup terminals. The TV-30H was first displayed at the 1984 NAB convention.

The TVE-60S for UHF

Not a transmitter designed particularly for CP, but one that will serve well the common cause for greater UHF efficiency at 60-kW with less cost. Starring the new S series, high efficiency 5-cavity klystrons, the TVE-60S offers 5-kW more output but consumes 5-10 kW less power than a Harris UHF transmitter of 1982. Harris also claims 16-30 kW more efficiency that competition and $10 thousand less in annual power bills for the same output. (See Fig. 4-10.)

The high efficiency of Varian's VKP-7550S derives from tuning the 3rd and 4th cavities 12 and 20 MHz above the visual carrier, but reducing gain some 10 dB. So Harris developed a highly-linear solid-state 60-W intermediate power amplifier (IPA) to deliver 50-watts of drive to this pulsed klystron. Also a new linear corrector was engineered having a range of 50-percent differential gain precorrection to compensate for 30-40 percent of differential gain during normal operation. Operating life of this S klystron should match that of the VA-953H and G series tubes, which "live" for at least 22,000 hours. Efficient aural service tuning results from slightly stagger-tuning all cavities except the fourth, which is always tuned some 2 MHz above the aural carrier for maximum saturated power.

When stereo tuned, typical bandwidth is 2.3-3 MHz with drive powers from 0.4 to 2 watts at 10-20 percent aural. There's also a solid-state 6-dB gain amplifier added following the exciter to maximize aural klystron efficiency.

The visual exciter also possesses a SAW filter for a ±0.5 dB transmitter low-ripple passband and better than 15-dB rejection visual at the aural carrier. Corrections include better than 10 degrees of

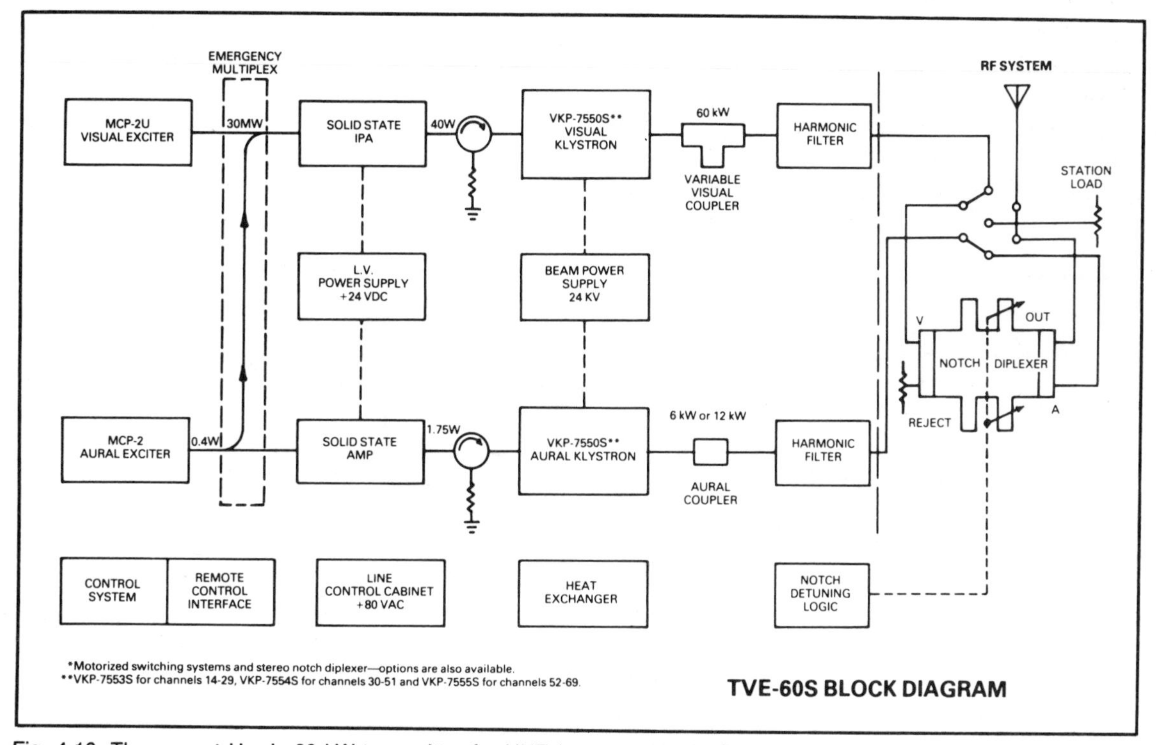

Fig. 4-10. The newest Harris 60 kW transmitter for UHF (courtesy Harris Corp.).

quadrature incidental phase and amplitude nonlinearity of the S klystron.

The 60-W linear IPA consist of four power rf transistors in pairs operating Class A push-pull, with four stages delivering 33-dB gain and AGC action to stabilize gain over temperature and time. The final power amplifier stage is preset to eliminate field adjustments. Rf transistors are cooled with an efficient aluminum heat sink and increases transistor lifetime.

Very low intermodulation distortion arises from the IPA amplifier, permitting emergency multiplexing where the visual klystron can transmit 20-kW multiplexed TV signal with 10-percent aural. At the output, several different notch diplexers are available, some with detunable notches for emergency multiplexing.

SAW Vestigial Sideband Filter

Surface-wave acoustic-wave filters (SAWs) have now become the vestigial sideband filters in the Harris MCP series of visual exciters. As described by Harris' Dr. David Hertling, this is actually a solid-state tapped delay line with very flat passband and steep transition bandwidth. Its phase versus frequency characteristics are almost linear, rejection band attenuation high, and most group delay correction circuits have been removed from the exciter. Given enough taps (Fig. 4-11), Dr. Hertling says it is theoretically possible to synthesize *any* bandpass.

SAW filters, however, have a tendency to undergo bandpass shift with temperature, and this is why Harris has centered its response as well as chosen the 37 MHz i-f frequency since temperature drift in this lithium niobate substrate is proportional to its coefficient of expansion. Harris claims 37 MHz is "the lowest i-f frequency used by a television transmitter manufacturer." We might add that lithium niobate SAW filters have also been used for years by receiver manufacturers to replace heretofore complex LC tuning and filter networks between tuners and the i-fs.

Filter response versus frequency is well illustrated in Fig. 4-12. High rise and fall times, bandwidth, and frequencies covered between 32.3 and 38.25 MHz are evident, with amplitude measurements between −40 and 0 dB. When group delays are integrated in phase-versus-frequency, Dr. Hertling says a "nearly ideal linear phase plot" results. Rapid and periodic fluctuations in magnitudes and phase imply predictable echoes. In this SAW filter, he says that such fluctuations have 100-200 kHz periods corresponding to displaced echoes of 5-10 μsec at 42 dB down. Echoes result in rapid fluctuations but don't cause "undesirable" picture distortions.

Basically, a SAW filter (Fig. 4-13) has two transducers deposited by photolithography on a

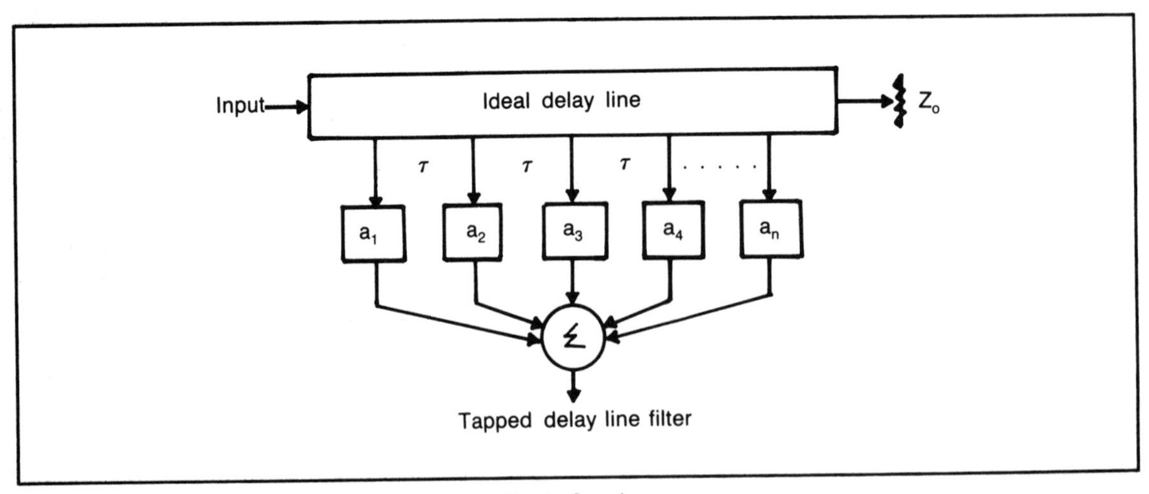

Fig. 4-11. A very linear tapped delay line (courtesy Harris Corp.).

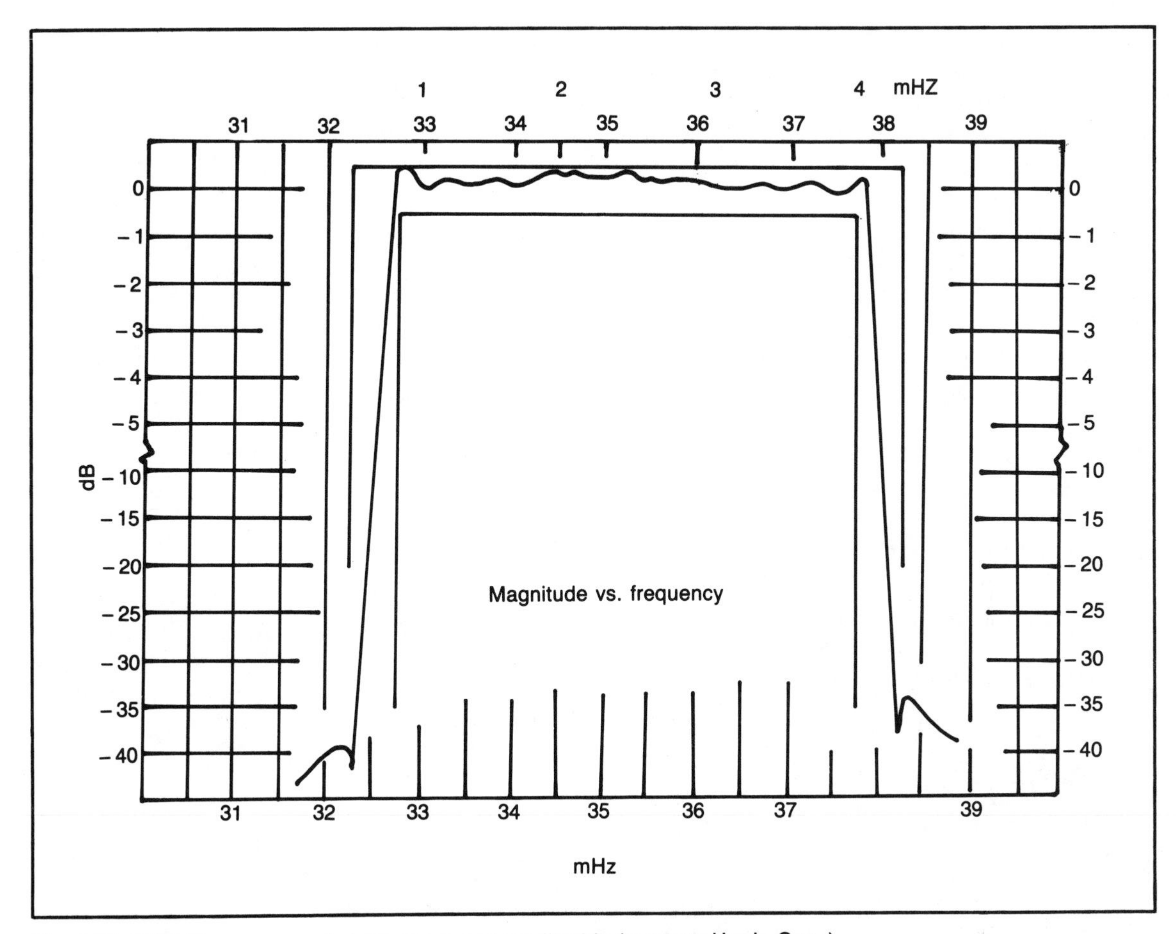

Fig. 4-12. SAW filter response is both sharp and predictable (courtesy Harris Corp.).

piezoelectric substrate, consisting of an interdigital metal electrode pattern having electrode centers spaced at 1/2 wavelength of the operating frequency. An rf potential applied across the input transducer has half power radiating toward the output transducer with the other half radiating to the end of the crystal substrate and is lost. More complicated transducers, however, do eliminate

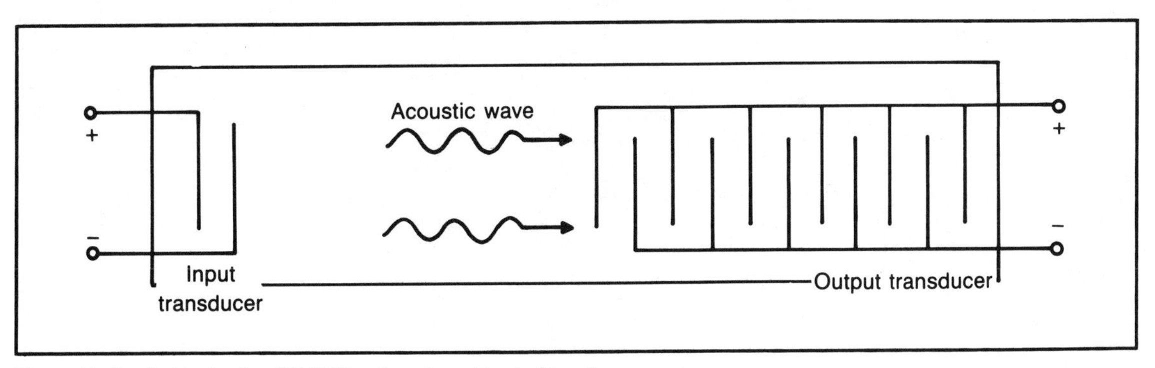

Fig. 4-13. Basic block of a SAW filter (courtesy Harris Corp.).

such bidirectional loss and launch acoustic waves in only one direction where insertion losses may be only about 2 dB or better. Typical production figures of several years ago specified SAW filter frequency ranges between 20 and 700 MHz, with fractional bandwidths from 0.2 to 60 percent, depending on type and usage.

In the Harris filter, maintenance is minimal because of constant alignment, corrective circuits have been eliminated, the output response is considerably better and will offer improved overall system performance.

NOTCH DIPLEXERS

Everyone—or just about everyone—should know what a notch diplexer is, but obtaining specifics turns out to be a mountainous chore. Finally, Mssrs. Mager and Hymas of RCA Broadcast Systems Division supplied the information, along with a photo (Fig. 4-14) and a first class writeup. Here's the story.

As you should know, a properly designed notch diplexer will couple video and audio-modulated signals to a single antenna via a single transmission line, at a constant input impedance for all transmitted frequencies. With the two inputs balanced, aural signals enter a coaxial coupler and divide equally in the transmission lines but separated 90° in phase. As they reach resonant sound notch filters CL1A and CL1B, these appear as short circuits and reflect the signals back to the coaxial coupler and the composite signal now goes to the antenna output.

Video also passes through the other coax

Fig. 4-14. Photo of a notch diplexer (courtesy RCA).

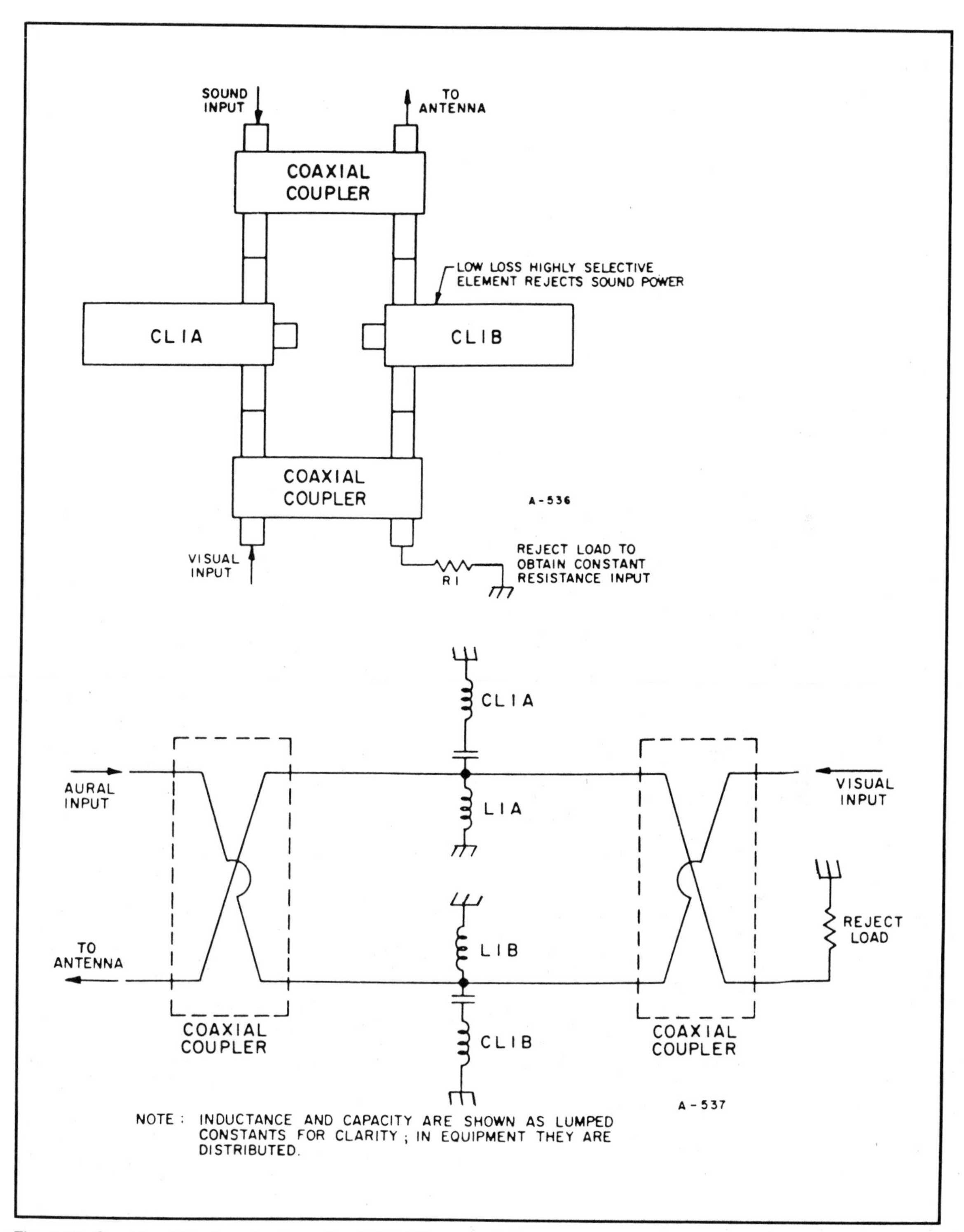

Fig. 4-15. Combined block and schematic of 100 kW notch diplexer (courtesy RCA).

coupler, with CL1A and CL1B offering a good impedance match for the visual carrier. Video then recombines at the other coupler in proper phase so that it, too, reaches the antenna for immediate broadcast.

Pretuned at the factory, these notch diplexers require little in the way of maintenance other than keeping connections secure and free from dust and dirt. A blower used to dissipate heat generated in the sound notching cavities requires little maintenance and its bearings are permanently lubricated. A notch diplexer simplified schematic appears in Fig. 4-15, showing the dual audio-video inputs at opposite ends of the unit and the filters in-between. As you see, when both combine they exit one output to the antenna.

The following definitions are from the FCC. They should help you to understand the technical material presented in this chapter.

TV TECHNICAL STANDARDS

§73.681 Definitions.

Amplitude modulation (AM). A system of modulation in which the envelope of the transmitted wave contains a component similar to the wave form of the signal to be transmitted.

Antenna electrical beam tilt. The shaping of the radiation pattern in the vertical plane of a transmitting antenna by electrical means so that maximum radiation occurs at an angle below the horizontal plane.

Antenna height above average terrain. The average of the antenna heights above the terrain from approximately 3 to 16 kilometers (2 to 10 miles) from the antenna for the eight directions spaced evenly for each 45 degrees of azimuth starting with True North. (In general, a different antenna height will be determined in each direction from the antenna. The average of these various heights is considered the antenna height above the average terrain. In some cases less than 8 directions may be used. See §73.684(d)). Where circular or elliptical polarization is employed, the antenna height above average terrain shall be based upon the height of the radiation center of the antenna which transmits the horizontal component of radiation.

Antenna mechanical beam tilt. The intentional installation of a transmitting antenna so that its axis is not vertical, in order to change the normal angle of maximum radiation in the vertical plane.

Antenna power gain. The square of the ratio of the root- mean-square free space field strength produced at one mile in the horizontal plane, in millivolts per meter for one kW antenna input power to 137.6 mV/m. This ratio should be expressed in decibels (dB). (If specified for a particular direction, antenna power gain is based on the field strength in that direction only.)

Aspect ratio. The ratio of picture width to picture height as transmitted.

Aural transmitter. The radio equipment for the transmission of the aural signal only.

Aural center frequency. (1) The average frequency of the emitted wave when modulated by a sinusoidal signal; (2) the frequency of the emitted wave without modulation.

Blanking level. The level of the signal during the blanking interval, except the interval during the scanning synchronizing pulse and the chrominance subcarrier synchronizing burst.

Chrominance. The colorimetric difference between any color and a reference color of equal luminance, the reference color having a specific chromaticity.

Chrominance subcarrier. The carrier which is modulated by the chrominance information.

Color transmission. The transmission of color television signals which can be reproduced with different values of hue, saturation, and luminance.

Effective radiated power. The product of the antenna input power and the antenna power gain. This product should be expressed in kWs and in dB above 1 kW (dBk). (If specified for a particular direction, effective radiated power is based on the antenna power gain in that direction only. The licensed effective radiated power is based on the average antenna power gain for each direction in the horizontal plane. When a station is authorized to use a directional antenna or an antenna beam tilt, the direction and amount of maximum effective radiated power will also be specified.) Where circular or elliptical polarization is employed, the term effective radiated power is applied separately to the horizontally and vertically polarized components of radiation. For assignment purposes, only the effective radiated power authorized for the horizontally polarized component will be considered.

Equivalent isotropically radiated power (EIRP). The term "equivalent isotropically radiated power" (also known as "effective radiated power above isotropic") means the product of the antenna input power and the antenna gain in a given direction relative to an isotropic antenna.

Field. Scanning through the picture area once in the chosen scanning pattern. In the line interlaced scanning pattern of two to one, the scanning of the alternate lines of the picture area once.

Frame. Scanning all of the picture area once. In the line interlaced scanning pattern of two to one, a frame consists of two fields.

Free space field strength. The field strength that would exist at a point in the absence of waves reflected from the earth or other reflecting objects.

Frequency departure. The amount of variation of a carrier frequency or center frequency from its assigned value.

Frequency deviation. The peak difference between the instantaneous frequency of the modulated wave and the carrier frequency.

Frequency modulation (FM). A system of modulation where the instantaneous radio frequency varies in proportion to the instantaneous amplitude of the modulating signal (amplitude of modulating signal to be measured after pre-emphasis, if used) and the instantaneous radio frequency is independent of the frequency of the modulating signal.

Frequency swing. The peak difference between the maximum and the minimum values of the instantaneous frequency of the carrier wave during modulation.

Interlaced scanning. A scanning process in which successively scanned lines are spaced an integral number of line widths, and in which the adjacent lines are scanned during successive cycles of the field frequency.

IRE standard scale. A linear scale for measuring, in IRE units, the relative amplitudes of the components of a television signal from a zero reference at blanking level, with picture information falling in the positive, and synchronizing information in the negative domain.

NOTE: When a carrier is amplitude modulated by a television signal in accordance with § 73.682, the relationship of the IRE standard scale to the conventional measure of modulation is as follows:

Level	IRE standard scale (units)	Modulation percentage
Zero carrier	120	0
Reference white	100	12.5
Blanking	0	75
Synchronizing peaks (maximum carrier level)	-40	100

Luminance. Luminous flux emitted, reflected, or transmitted per unit solid angle per unit projected area of the source.

Monochrome transmission. The transmission of television signals which can be reproduced in gradiations of a single color only.

Multiplex Transmission (Aural). A subchannel added to the regular aural carrier of a television broadcast station by means of frequency modulated subcarriers.

Negative transmission. Where a decrease in initial light intensity causes an increase in the transmitted power.

Peak power. The power over a radio frequency cyle corresponding in amplitude to synchronizing peaks.

Percentage modulation. As applied to frequency modulation, the ratio of the actual frequency deviation to the frequency deviation defined as 100% modulation expressed in percentage. For the aural transmitter of TV broadcast stations, a frequency deviation of ±25 kHz is defined as 100% modulation.

Polarization. The direction of the electric field as radiated from the transmitting antenna.

Program related data signal. A signal, consisting of a series of pulses representing data, which is transmitted simultaneously with and directly related to the accompanying television program.

Reference black level. The level corresponding to the specified maximum excursion of the luminance signal in the black direction.

Reference white level of the luminance signal. The level corresponding to the specified maximum excursion of the luminance signal in the white direction.

Scanning. The process of analyzing successively, according to a predetermined method, the light values of picture elements constituting the total picture area.

Scanning line. A single continuous narrow strip of the picture area containing highlights, shadows, and half-tones, determined by the process of scanning.

Standard television signal. A signal which conforms to the television transmission standards.

Synchronization. The maintenance of one operation in step with another.

Television broadcast band. The frequencies in the band extending from 54 to 806 megahertz which are assignable to television broadcast stations. These frequencies are 54 to 72 megahertz (channels 2 through 4), 76 to 88 megahertz (channels 5 and 6), 174 to 216 megahertz (channels 7 through 13), and 470 to 806 megahertz (channels 14 through 69).

Television broadcast station. A station in the television broadcast band transmitting simultaneous visual and aural signals intended to be received by the general public.

Television channel. A band of frequencies 6 MHz wide in the television broadcast band and designated either by number or by the extreme lower and upper frequencies.

Television transmission standards. The standards which determine the characteristics of a television signal as radiated by a television broadcast station.

Television transmitter. The radio transmitter or transmitters for the transmission of both visual and aural signals.

Vestigial sideband transmission. A system of transmission wherein one of the generated sidebands is partially attenuated at the transmitter and radiated only in part.

Visual carrier frequency. The frequency of the carrier which is modulated by the picture information.

Visual transmitter. The radio equipment for the transmission of the visual signal only.

Visual transmitter power. The peak power output when transmitting a standard television signal.

(28 PR 13660, Dec. 14, 1963, as amended at 35 FR 5692, Apr. 8, 1970; 36 FR 5505, Mar. 24, 1971; 36 FR 17429, Aug. 31, 1971; 41 FR 56325, Dec. 28, 1976; 42 FR 20823 Apr. 22, 1977; 44 FR 36039, June 20, 1979; 47 FR 35990, Aug. 18, 1982).

Chapter 5

Digital Television Transmission and Reception

As this is written, we are at the early stages of digital television receiver introduction in the United States from both American and Japanese manufacturers, courtesy ITT Intermetall, Hans-Bunte-Strasse 19, D-7800 Freiburg, West Germany. So the secret's out. After spending millions of dollars and many thousands of hours in engineering time, television receiver circuits *following* the video detector may be digitized via fast A/D converters and finally reconverted to analog with the aid of D/A converters just before the picture tube.

Is the picture better? No, not yet, but the flexibility, huge number of options, and the prospects of digital companding in the forseeable future open up completely unexplored avenues of both sound and picture transmissions and reception that will make any hardened cynic capitulate. For openers, how about increasing the luma bandwidth from 4 to 6 MHz, full 2-MHz chroma and audio at any processing stage you may want or wish to pay for. You may also see double or triple noninterlaced scan frequencies, with picture virtually equaling 35-mm film at its best. Remarkably enough, none of this is mere supposition, it's all been proved and needs only commercial and consumer applications to put it on the market.

Will new receivers be required? You'll need a special expander decoder and probably increased bandwidth video amplifiers; but older receivers can pick up this information, too, but at conventional bandwidths and little picture-sound improvement. Consequently, if you want the best, you'll buy. If you're content with yesteryear, you won't. But projected costs seem modest and just a little more will probably buy a lot. Certainly this is the biggest development since color, the comb filter, remote-controlled tuning, multichannel sound or anything else we haven't thought of—it's a *huge* leap forward.

PROGRESS

Does all this occur immediately—not necessarily unless one of the networks and some consumer products manufacturer jumps the gun. Predictions are that the studios, themselves, will have to become totally digitized before going on the air. We do know, nonetheless, there are digital TV systems *in being* that could conceivably start up at any time—and this also may be a means of coding *satellite* commercial traffic so that the general public can't look and listen in without the right equipment. This doesn't necessarily foreshadow program enciphering—such as that used for military or diplomatic traffic—but it certainly could mean regular digitizing with a little sync compression or video inversion thrown in— fortunately not too difficult for entrepreneurs to decode and distribute—for a fee, naturally. As for regular broadcasts, they should remain the same for the foreseeable future without any appreciable change except the welcome addition of multichannel sound, due to give us both stereo and second audio programming (SAP).

In the meantime, while awaiting the networks' levitation from their terrestrial behinds, digitized receivers are scheduled to improve by factors of 10 or more over their 1985-1986 counterparts. Intermetall ITT has enough production now so it can supply at least four or five U.S. and offshore manufacturers with the necessary integrated circuits, and Toshiba, Panasonic, Sony, Zenith and (possibly) others have already demonstrated they're ready to go with the new technology. RCA, NAP (Philips), General Electric, Sharp, Hitachi, and several more haven't yet committed, but probably will as time goes on. Many are awaiting the availability of additional features besides the picture-in-picture addition that most receivers will have initially. This simply means that you may put an extra video signal through baseband and display it as a smaller picture at any one of four positions near the corners of the cathode-ray tube—all in color, of course. In some receivers, the main image may be transposed and become the secondary image, with whatever's entering baseband appearing as the first. You will also find extensive use of digital tuning for both baseband and rf entries, with easy switching between signal sources, characterized by various signature color identifications. Blue for video and green for rf, as an example. The remote control for Toshiba, for instance, has 32 pad keys for every conceivable command, including picture *and* sound. Later, with considerable memory expansion, a half dozen or more pictures from different channels may be sampled and stored for simultaneous display, if you're that curious.

The best innovations of these digitizing efforts will transcend by far those few already indicated. For instance, adaptive noise reduction by correlation or nonlinear processing; flicker-free pictures using storage and increased horizontal and vertical scanning; automatic ghost cancellation; digital comb filters; direct processing of digital audio signals, Teletext and/or Viewdata, Prestel, Antiope, etc., and an excellent terminal for home computers. There are probably more uses besides the perfect vehicle for very large scale integration (VSLI), which will come ever closer to putting most of a television receiver at some future time on one or two ICs. It will also be possible to transmit digital television over cable, especially fiber optics with this development, and that opens up CATV operations to all sorts of possibilities, including advanced two-way at possibly reduced costs and greatly increased information reliability.

The technology was first tested in 1981, TV receiver manufacturers have been evaluating since 1982, and production began in earnest in 1983. The system works with German PAL, French SECAM, and U.S. NTSC—allowing for one or two extra ICs.

After huge financial investments and remarkable engineering diligence, ITT Intermetall has the digital race going, and will probably supply many millions of ICs to the United States and Japan, especially, before another semiconductor house produces a competing product. In the meantime, it's been a real international breakthrough—unfortunately *not* made in America.

HOW IT WORKS

The DIGIT 2000 digital television system now consists of a central control unit (CCU), video codec unit (VCU), video processor unit (VPU), audio A/D converter (ADC), audio processor unit (APU), deflection processor unit (DPU) and a clock generator. This amounts to seven integrated circuits with one, reportedly, having as many as 60,000 transistors.

An RGB digital comb filter will take the place of either analog-applied CCD or glass delay lines, using a RAM for delay and can "be optimized" for best picture. There's also an automatic picture control which analyzes the video signal, then communicates with the central control unit (CCU) to adjust brightness, contrast and color saturation according to comparative values stored in the CCU by the manufacturer. In another year, we're told, these seven ICs will be reduced to three or four, including A/D conversion, as well as additional efforts to be made on adaptive noise reduction, H/V aperture correction, picture tube geometry correction, and noninterlaced pictures. Already a new 8-bit A/D and 10-bit D/A converter designated UVC 3100 has appeared and been advertised as a high speed, flash-type on a 40-pin IC apparently made up of a combination of emitter coupled and transistor-transistor coupled logic (ECL/TTL), with collector implant technology, having both keyed and peak clamping, and its own internal voltage reference. There's also a gated amplifier at the output for analog signals instead of just that from the D/A converter.

A block diagram of a 5-chip system (less the special features) appears in Fig. 5-1. During production, this receiver is automatically and fully aligned by computer control. In it you see a power supply, infrared preamplifier, keyboard, display, tuner interface and the tuner/video/sound i-fs and their respective detectors. Emanating from the tuner are both audio and video baseband, including an input to the MAA2000 central control unit, which is a very important part of the system.

Microprocessor Control

The CCU (Fig. 5-2) is shown in a "typical" ap-

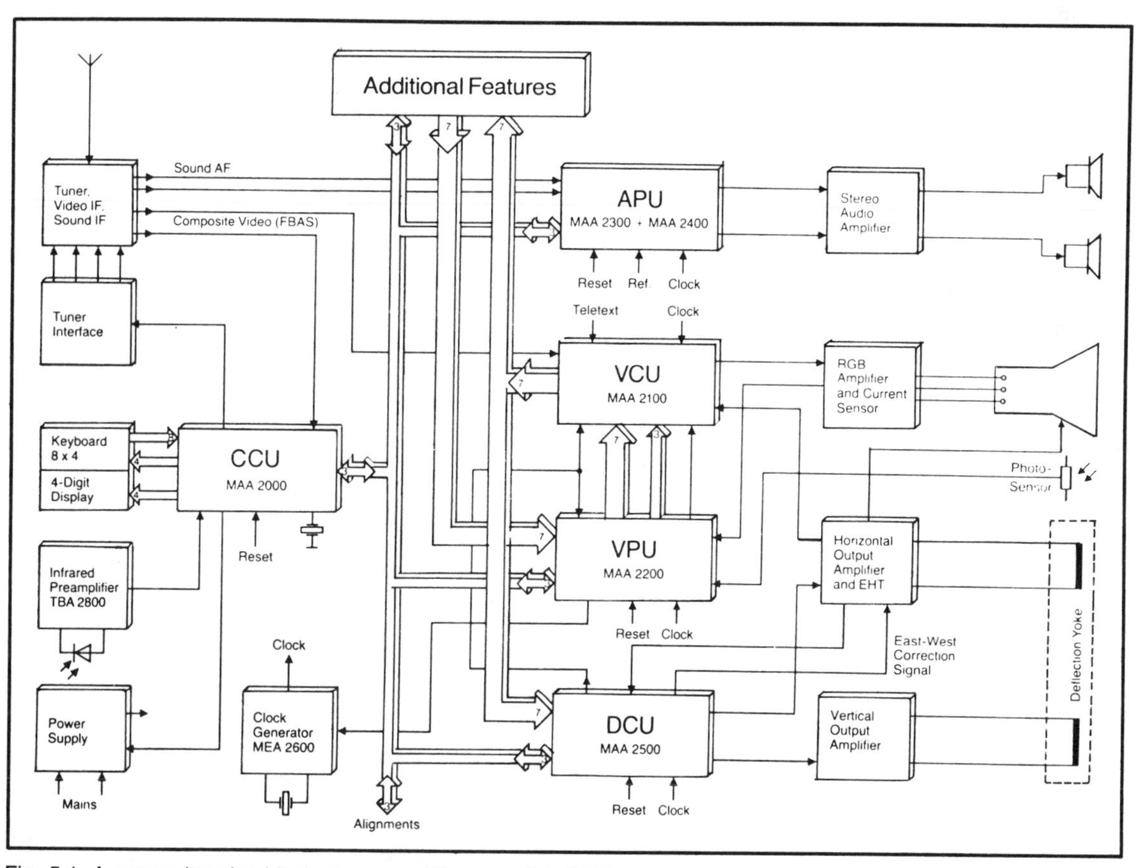

Fig. 5-1. A comprehensive block diagram of the new digital TV system (courtesy ITT Semiconductor).

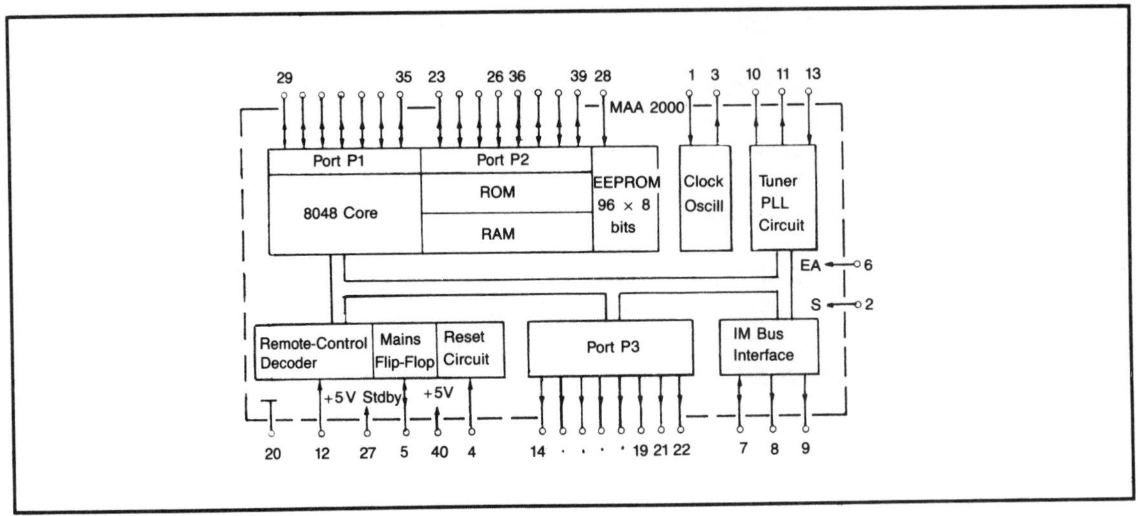

Fig. 5-2. Block diagram of the MAA2000 CCU central control unit (courtesy ITT Semiconductor).

plication circuit and its 8049 microcomputer which has a 8-kilobyte ROM capacity and a RAM of 128 bytes. There's also a 96 × 8-bit EEPROM, three ports, a clock oscillator, tuner phase-locked (PLL) circuit, remote control decoder, power flip-flop, reset circuit, and an IM bus interface. The 8049 and associated logic supplies separate addresses for I/O program information, processes operator commands, controls digital signal direction for video, audio, and deflection, and contains factory alignment information already programmed during receiver production. The clock oscillator operates with crystal control between 3.5 and 4.6 MHz, depending on USA or European systems, and is capable of as much as 11 MHz, if needed.

The remote control decoder receives 10 bits per word, four address and six data bits, and is always powered even in standby. During receiver production, the 128-byte storage EEPROM computer-programs with factory alignment information via the serial IM bus, so that whenever the receiver is switched on, an EEPROM readout flows to all designated parts of the set. The tuning system, also EEPROM-controlled, will store 30 TV channels with 96 eight-bit words. Channel information exits via ports P2 and P3 and port P1 does the bandswitching.

The "mains" flip flop turns off and restores power, as its name suggests, but has a delay of 16 milliseconds to "properly" charge the various stray capacitances before the set can resume operation.

The nonvolatile electrically erasable-programmable read-only memory with its 128 bytes of information, is programmed in two steps: the stored word has to be cleared, then the stored word is written—all assisted by an internal clock-generator of 1 kHz. To keep tuner-programmed channels on frequency, a 64:1 prescaler and phase-locked loop circuit is provided, having an independent crystal-controlled frequency reference of 4 MHz (we presume this is USA, also).

Video Codec (VCU)

Because our prime interest in this digitizing scheme is video processing, the VCU may, appropriately, appear next in Fig. 5-3. This is the high-speed coder/decoder integrated circuit for A/D and D/A video signal conversion in a VSLI 40-pin multifunction package having beam current limiting, noise inversion, color DEMUX, white balance, CRT spot cutoff, black level control, RGB analog matrix, extra analog inputs, and three RGB analog output amplifiers for the cathodes of the picture tube. The large IC is clocked by a 14.3 MHz clock generator from an externally-programmable source.

The A/D converter, ITT says, is the flash type with 2^n comparators in parallel. For slowly varying video signals, 8 bits are needed for the necessary video resolution. This, however, is a 7-bit converter that has its reference voltage changed every other line by a value corresponding to half that of the least significant bit. Thus, a gray value between two 7-bit steps is converted to a next lower value during one line and to a next higher value during the next. The eye averages the two values, resulting in apparent 8-bit resolution. Resolution is 1/2 the least significant bit of 8 bits, and its output is Gray-coded for spike and glitch elimination resulting from various comparator speeds. New video codecs, we're told, have two inputs.

Digitized video proceeds through a noise inverter before entering the video processor and deflection processor. Whenever there are noise spikes from ignition, lawn mowers, trucks, etc., that reach digital step 127, or 7.5 volts, signal levels are reduced to about 40 IRE, which is 10 units under the normal luminance reference and appears gray. Following luma-chroma separation and other video processing, color signals return to the video codec via the color difference DEMUX block, and luminance to the D/A converter. This is a parallel 8-bit signal with 1/2 LSB of 9 bits accuracy sufficient for good limited contrast and also room for peaking filter positive and negative overshoots. Called an R-2R ladder, clock rate of 14.3 MHz is the same as for the A/D previously described.

With R-Y and B-Y signals treated at a bandwidth of 1 MHz, R-Y and B-Y 8-bit converters are also built in as R-2R ladders and clocked at 4 MHz. Here, 4 × 4 bits are processed sequentially producing 16 bits with synchronous coordination, requiring only four lines.

Luminance is now added to R-Y and B-Y analog in the RGB matrix, which also sets the brightness level of the picture via commands from the central control unit. White spot cutoff control develops from a 7-bit D/A converter whose output corresponds to half the luminance range. Dark voltage, then, can be adjusted in one-percent steps and component and picture tube aging may be compensated accordingly. At the same time, data exchanged between the VCU, VPU, and CCU results in control of picture tube alignment, white level and dark current adjustment, and photodetector ambient lighting correction.

Beam and peak beam current-limiting are controlled by reducing reference voltages of the color and luminance D/A converters, and at 0 V, brightness reduces to zero. During vertical and horizontal blanking, all three RGB amplifiers are switched to black or ultra black, depending on the cathode

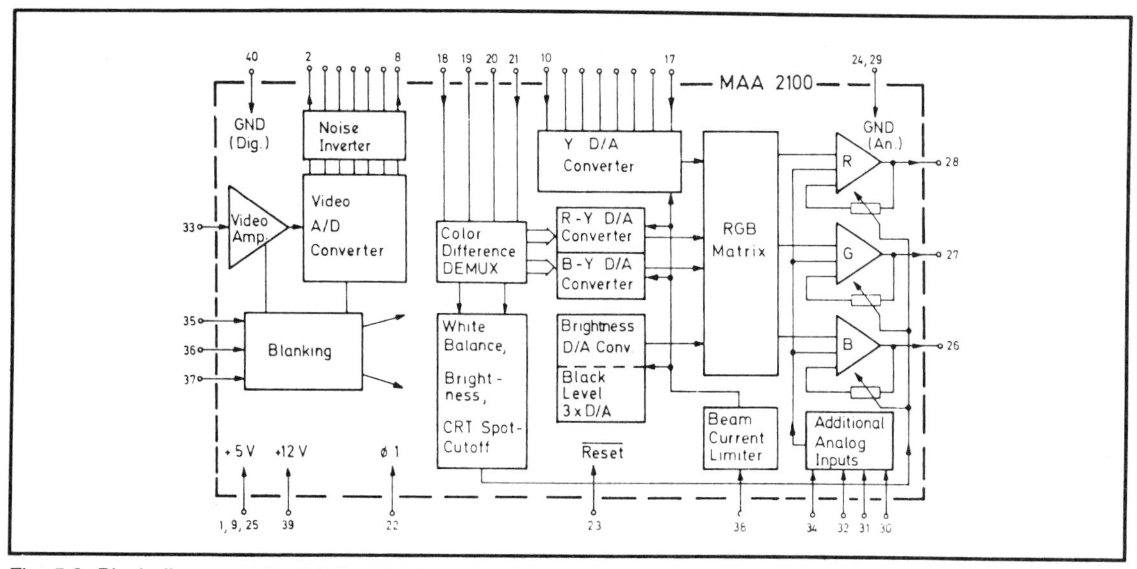

Fig. 5-3. Block diagram of the MAA2100 Video Codec Unit (courtesy ITT Semiconductor).

selected for measurement so that dark current of only one cathode is measured. For leakage current measurements, however, all cathodes switch to ultra black.

Video Processor (VPU)

We come now to the video processor proper which treats real time video in another 40-pin IC N-channel MOS VSLI. It consists of a code converter, chroma trap, peaking and bandpass, the usual auto color control, color killer and decoder, phase comparator, PLL filter, and multiplier and multiplexer for color saturation and hue correction. Analog also has a contrast multiplier and limiter, an A/D converter, IM bus, and chroma-luma data output multiplexer. The IC is illustrated in Fig. 5-4, and converts the parallel Gray code into binary for the luminance channel and offset binary for the chroma channel upon entry into the code converter.

In the luminance portion, a phase-linear filter separates luminance from chroma and peaks luminance around 3 MHz for additional high frequency response. Covering the range between − 3dB and + 5dB, it may be advanced or retarded by the TV operator in eight steps. Luminance is carried in 8-10 bits to the contrast multiplier and luminance limiter. Contrast is both manually and photodiode controlled so that the CCU calculates the necessary contrast, and signals the VPU accordingly. Digitized 10-bit information returns to the VCU where it is absorbed by the luminance D/A converter.

Another phase linear filter appears for color, which may be switched for symmetrical or asymmetrical response, depending on composite video input or i-f amplifier input, respectively. The automatic color control (ACC) then maintains burst amplitude at some value within a control range of 36 dB. Chroma is then decoded for R-Y and B-Y color difference information, while a PAL compensator operates in NTSC as a comb filter (using straight CPU) to relieve cross-color modulation. Digital color difference signals then reach the color saturation multiplier and multiplexer where hue correction takes place as well as operator adjustments for color saturation, which is then coordinated with contrast by the CCU.

Deflection Processor (DPU)

This is the final video-signal processing IC among the five original ICs, with only audio remaining to be described (which will follow). (See Fig. 5-5.)

Another programmable VSLI in N-channel MOS, it contains video clamping, H/V sync separation, east-west correction, sawtooth generation,

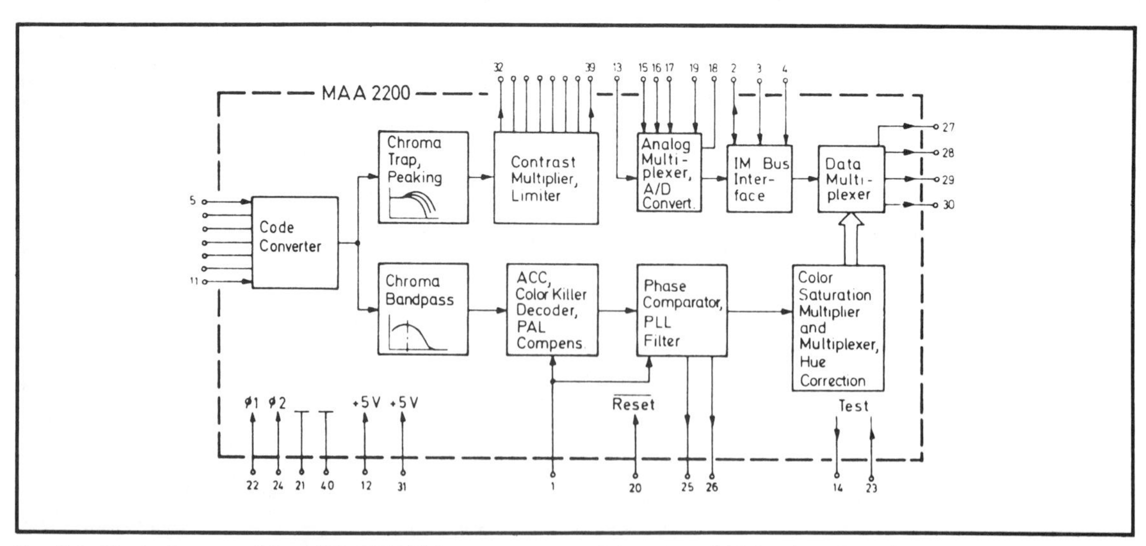

Fig. 5-4. Block diagram of the MAA 2200 Video processor unit (courtesy ITT Semiconductor).

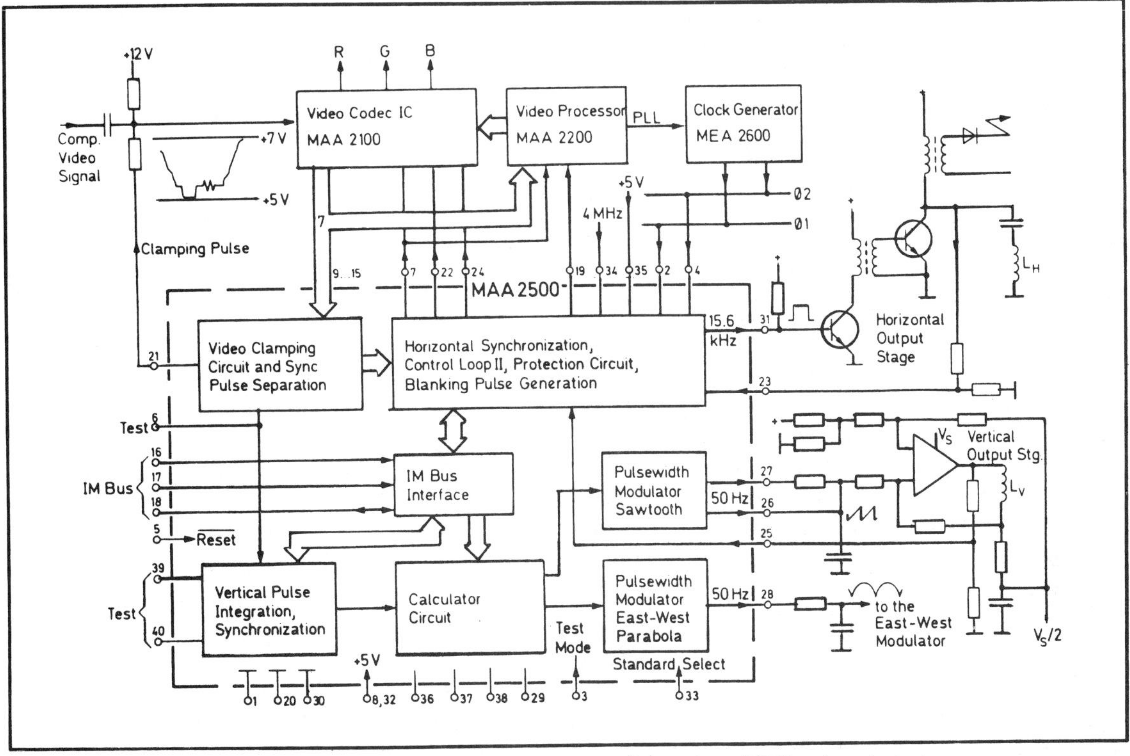

Fig. 5-5. Block diagram of the MAA2500 deflection processor unit (courtesy ITT Semiconductor).

and H/V synchronization. According to ITT, the following functions are programmable (with many more to come):

- Filter time constant for horizontal sync.
- Phase and duration of horizontal output pulse
- Vertical amplitude and position with S-correction
- East-west parabolic and trapezoidal correction, and horizontal width
- Sync switchover characteristics
- PAL, NTSC, or SECAM system selection

As the receiver turns on, the video clamper measures and sets the sync top level at 5.125 volts through received current pulses so the A/D converter can do its job. Horizontal and vertical sync are then separated at a fixed separation level of 5.25 V, permitting horizontal sync to be immediately position-locked and setting its back porch at a constant 5.5 volts for both black level and color burst. With horizontal sync locked, sync pulses are slice-leveled at 50 percent for maximum pulse separation even at small amplitudes.

In nonlocked operation—where there's no fixed ratio between the color subcarrier and horizontal frequencies—the PAL clock is counted down in a programmable frequency divider until the proper horizontal frequency is found. Various key and gating pulses such as color, horizontal blanking and horizontal output control are all derived from the programmable divider's output due to a phase comparator. To equalize horizontal phase changes, a second control circuit takes over where the position of the programmable divider and the output of the flyback (pulse) are measured via a balanced gate delay line. Any deviation phase shifts the horizontal output control and the two are synchronized, along with additional anti-phase jitter control to set pulse phase to within 3.5 nanoseconds for each line.

When phase position has been established in the nonlock mode, the locked-mode programmable divider is set for 910:1 for NTSC and phase comparator No. 1 is disconnected to prevent noise, etc. from interfering with horizontal deflection. But phase corrector No. 2 remains in the circuit to take care of any errors in the horizontal output. There's also an active protection circuit for the horizontal output to ensure any PAL to NTSC clock frequency switches don't vary more than ± 1 kHz and keeps an eye on signals absent from the switch mode. This circuit is only active if the standard input is open-circuited and a 4 MHz clock is automatically applied to the protection circuit. It may also be used as a start oscillator, but certain pins have to be either disconnected or energized to play PAL, SECAM, or NTSC. Higher frequencies are possible, if required.

Vertical sync derives from the separated composite sync signal and continues to the calculator circuit where suitable division takes place to accommodate 625-line PAL and, undoubtedly, the other international systems. Only PAL, however, is described here and the digital countdown approximates the same sort of arrangement we've described several times before in other portions of the text. A nonstandard counting circuit is also included for missing or absent sync pulses. For the PAL system, however, there are three operating modes:

- Nonlocked using a wide trigger window
- Nonlocked using a narrow trigger window
- Locked with standard ratio divider

One might suspect that an equivalent for NTSC has been available for some time but not yet in print. We'll try and cover this later if the information is available in any detail. We're told there's no difference here between PAL and NTSC—so that solves that question. Meanwhile, the calculator circuit and its pulse width modulated signals continue to control both vertical output and the east-west modulator. Receiving adjustment values from the IM bus and CCU stored at the time of manufacture, the signals remain as selectable voltage references. When operations begin, the calculator circuit looks at the vertical trigger pulse and alternately calculates values for the vertical sawtooth and E/W parabola at a clock rate of some 35 kHz. Rate-multiplied, such information is converted for the succeeding clock pulse period into a 560 kHz pulse duty factor signal corresponding to the next computed value of the succeeding 17 kHz period. There are, then, two 560 kHz squarewaves whose duty cycles vary according to programming, which, interestingly enough, are now 448 kHz for our own NTSC.

We now understand that because of strict U.S. radiation requirements the 14.3 MHz system clock has been reduced in amplitude to 5 V p-p and the two-system clock square wave is now single phase and sinusoidal. There have also been some IM clock video changes to slope or lower risetimes so that harmonics, which might fall within the channel 2-5 region would not continue to cause interference. Even Europe, with its two or three available channels per sector, had some problems with one channel, whereas in the United States with a dozen or more active group channels, tuner pickups were inevitable. This is also the reason, presumably, why both Japanese and U.S. manufacturers have been slow in bringing production receivers to market even though they've been showing samples since 1984.

To continue the original discussion, note that in the calculator circuit there's an east-west data register with 8-bit horizontal amplitude processing, another 8 bits for peak value trapezoidal correction, and cushion correction curvature. In the vertical section there are 8 bits each for vertical position and vertical amplitude adjustments to addition to another 8-bit S-curve and symmetry shaper—all considered, a pretty complex piece of sweep parameter adjustment, but one that may have some very useful implications as these digital applications are further imaginatively developed.

The Clock Generator

One of the more important parts of these digital electronics is the MEA 2600 clock generator shown in Fig. 5-6. As you can see it has both data and a data clock input, a control circuit, two shift

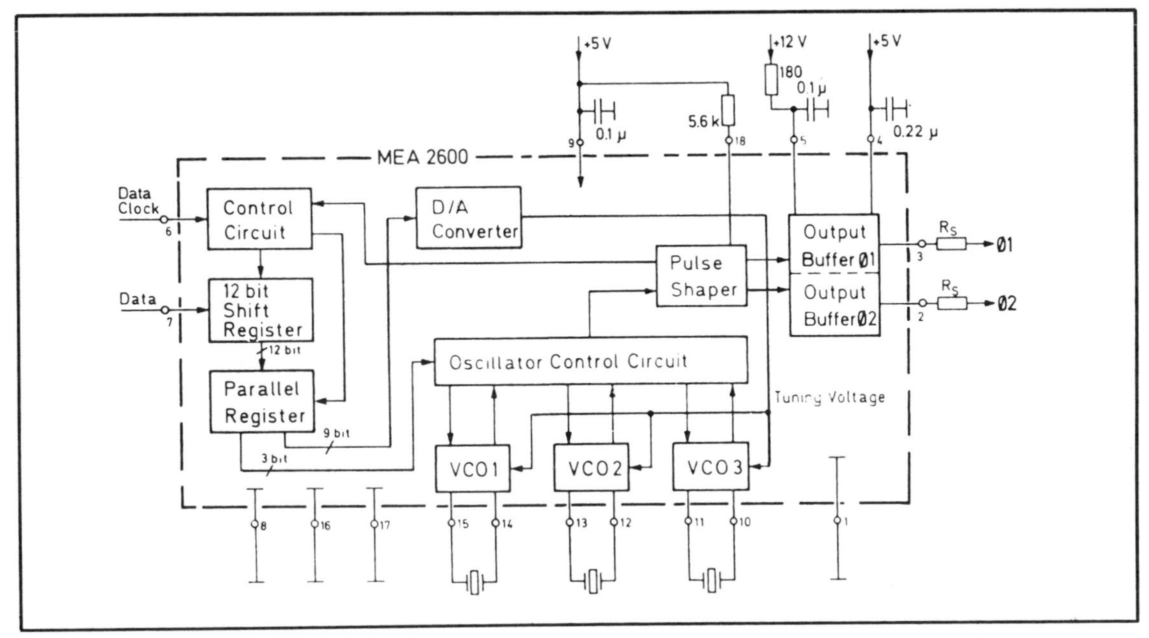

Fig. 5-6. Block diagram of the MEA 2600 clock generator and application circuit (courtesy ITT Semiconductor).

registers, three VCO positions, a D/A converter, pulse shaper, and two output buffers. Each of the VCO crystal slots serves a separate international color television system: one for NTSC, another for PAL, and the third for France's SECAM. They are all a part of a phase-locked loop stabilizing oscillator circuit with phase comparator and filter in the VPU chip.

Upon the data clock signal's negative transition, the shift register receives *write* information and upon positive swing, the loaded register shifts by 1 bit. When 12 bits have accumulated in the shift register and the data clock is high, a delay of one data clock period ensues as the parallel register takes over and passes its information to the oscillator control circuit and a 9-stage D/A converter, producing tuning voltage control for the VCO(s).

Output of the VCO goes to a pulse shaper through the oscillator control which now forms a single phase-buffered clock (another change) which may be converted to dual phase on any IC that requires it. The video processor (VPU) delivers data to the clock generator, with the initial three bits selecting the required VCO frequency, and the ensuing nine bits providing the tuning signal in the form of 2's complement. Where a single standard instead of NTSC, PAL, and SECAM are all combined, the unused VCOs may be externally blocked by a positive 5 volts applied to their crystal-controlling inputs.

Audio Processing

Here, audio must be first digitized and then processed in monophonic as well as stereo, resulting in a fairly complex set of ICs to do the job.

The A/D converter has two pulse-density modulators, five analog switches, a stereo dematrix, and the IM bus interface between this converter and the central control unit (CCU). As you see in Fig. 5-7 there are both auxiliary and TV audio inputs, one of which is the European 2R which we don't use. The five analog switches are two-pole and select all connections between the four analog inputs, the two digital outputs, and the extra analog outputs. The dematrix furnished both 2R and 2L stereo signals at the analog outputs (PAL) and S1 routes the stereo portion to pulse density modulators 1 and II. Their outputs are single-bit data streams at a maximum 4.7 MHz rate, where it is routed to the connected audio processor IC.

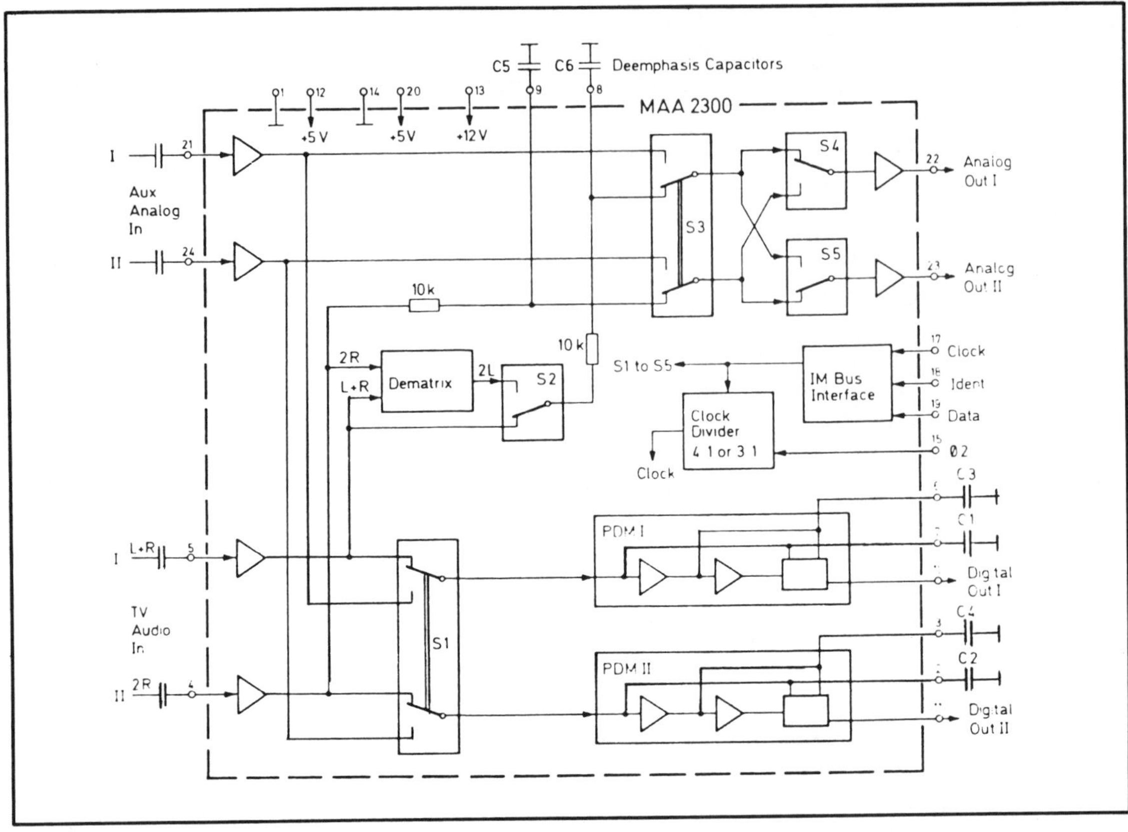

Fig. 5-7. The audio A/D converter (courtesy ITT Semiconductor).

THe IM bus simply links this converter with the CCU unidirectionally from the CCU whose commands are received via 8 data-bits, with NTSC sampling by the main clock frequency assigned.

The audio processor (APU) unit, itself, isn't nearly so simple as the digital converter. Although ITT says the chip's basic functions are mask-programmed, the IM bus signals may be modified so that this processor is not only good for the German 2-carrier stereo but for NTSC and other systems, as well. So in lieu of 17 MHz clocks, think in terms of 14 MHz with some additional programming and the U.S. system will develop practically (Fig. 5-8).

This is another 24-pin N-channel MOS designed to convert pulse-density modulation signals into parallel data at an audio sample rate, dematrixes, deemphasizes, has bass and treble controls, stereo "enlargement" and pseudo stereo. Note that the stereo and pseudo-stereo following the decimation filters and bilingual/dematrix have equal channels through to the PWM (pulse width modulation) outputs.

Upon conversion to pulse-width modulation, the APU can drive output power amplifiers, with programmed operating mode from digitized identification commands. After filtering and mode determination, digital audio filters transform the 1-bit pulse density information into 16-bit words at a sampling rate of some 35 kHz. The digital ID filter separates the AM ID sound signal of channel II, and decides whether transmitted sound is in mono/stereo or SAP. Being cascades of the transversal and recursive lowpass filters, the digital audio filters reduce both word rate and bandwidth and assign word lengths to 16 bits. A parallel-to-serial converter and multiplexer converts audio-filter outputs and identification from parallel to serial and multiplexes so that all information may be transferred serially into the data RAM.

Into the C RAM, the address is transferred first and the data word next in 1-byte increments, with transmission of a single byte requiring 100 microseconds. Readouts are at the same rate of 35 kHz.

When the two digital signals are converted into pulse-width modulation for right and left channels, such signals have broad baseband and there is direct conversion of 4 bits with double-edged PWM. With feedback, any error is corrected so that the signal-to-noise ratio amounts to 75 dB.

LATEST DEVELOPMENTS AND PREDICTIONS

Shortly before press time, Technical Manager David DeVoe of ITT was kind enough to send the 1985 version of the DIGIT 2000 VLSI digital television system down from Massachusetts, along with a listing of several very important developments that will make digital receiver systems highly acceptable to virtually all television manufacturers supplying the United States and Japan, and probably Europe also. Consequently, we are extending the scope of this chapter to include most, if not all, of these developments so you will recognize and understand their extent when appearing in United States TV systems during the next few years. Therefore, you should be covered with the basic features for probably five years to come from publication date of this text. Perhaps then we'll have the urge to supply another worthwhile update that will overlap advances of the mid 1980s. This would be especially true if liquid crystal technology has accelerated enough by that time to actually produce large-screen "picture-on-the-wall" direct view television systems which should, eventually, either supplement to a large extent, or actually supplant the honorable, but venerable cathode-ray tube. By then, also, the first one or two IC analog receivers may be available, and that, in itself, would indeed become excellent meal for the repertorial grist mill. So with these impending and current goals in view, let's return to what's happening at the moment and temporarily leave the remainder to editorial posterity.

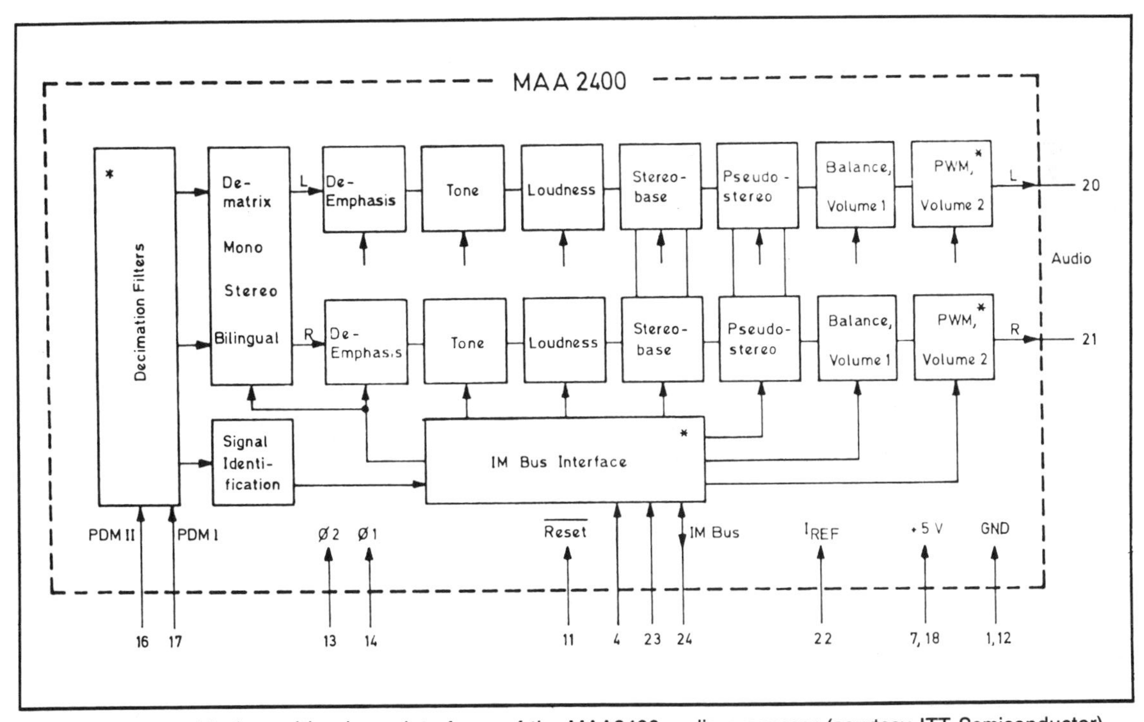

Fig. 5-8. Software blocks and hardware interfaces of the MAA2400 audio processor (courtesy ITT Semiconductor).

The Updated DIGIT 2000 TV System

We'll begin by listing *all* of the various unit ICs involved, then discuss those important ones not already covered in the preceding material. In this way, you'll have *all* available information currently released, much of which is either in the design or application stages of many of the manufacturers' top-of-the-line monitor/receivers right now.

- CCU2000/2030 Central Control
- VPU Video Processor
- APU Audio Processor
- MCU Clock Generator
- CVPU NTSC Comb Filter Video Processor
- SPU SECAM Chroma Processor
- MDA 1024-Bit EEPROM
- MEA Tuner Interface IC
- SAA Infrared Remote Control Transmitter IC
- TBA Infrared Preamplifier IC
- VCU Video Codec
- ADC Audio A/D Converter
- DPU Deflection Processor
- TPU Teletext Processor
- APC Automatic Picture Control
- MEA D/A and Bus Converter IC
- UAA Tuner Alignment IC

We will also print the new block diagram (Fig. 5-9) which contains everything already listed in Fig. 5-1, along with a few additions and some slightly changed abbreviations which the above grouping will identify. Note that the Teletext processor and

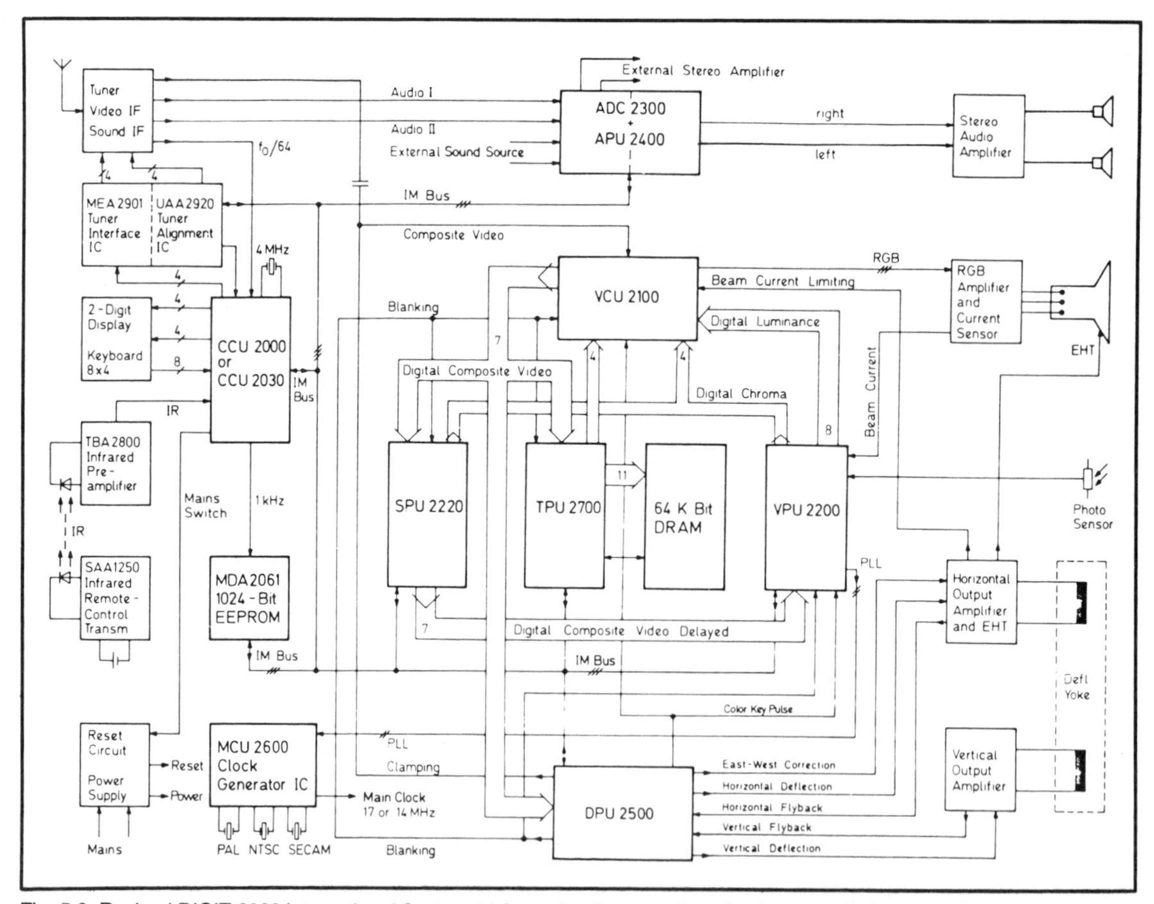

Fig. 5-9. Revised DIGIT 2000 International System. U.S. applications are described separately (courtesy ITT Semiconductor).

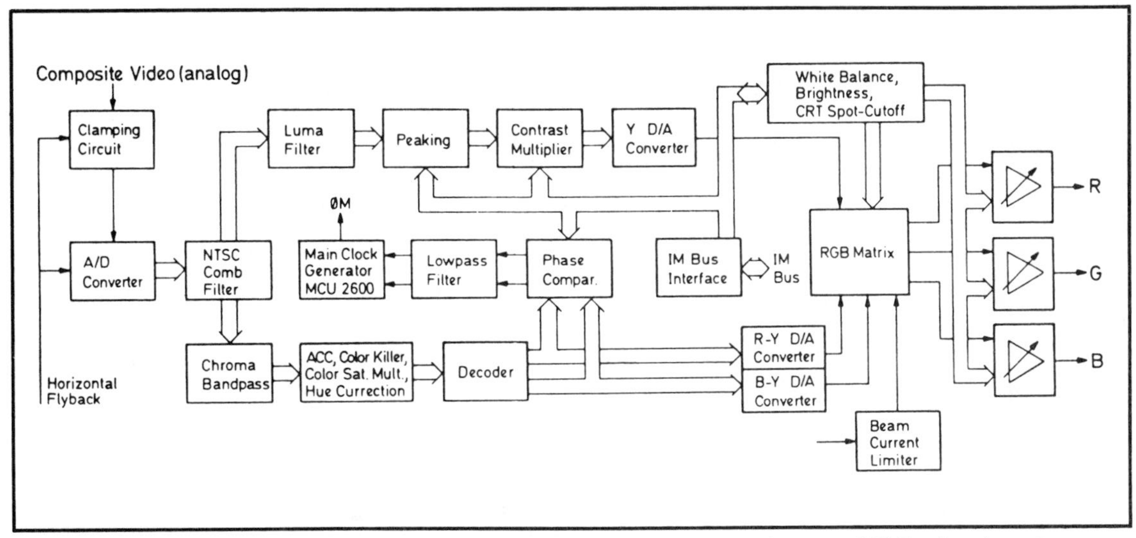

Fig. 5-10. Block diagram of the VCU 2100 and CVPU2210 with comb filtering (courtesy ITT Semiconductor).

its 64 k-bit dynamic random access memory DRAM has now been added along with the MDA 1024-bit EEPROM (previously defined), the SECAM SPU 2220, and the SAA1250 infrared remote control.

Please understand that this is the block diagram of a receiver (primarily European) built "on the DIGIT 2000 concept," and does *not* reflect precisely those now becoming available in the United States. Before the end of the chapter, however, we expect to have one or more blocks of current digital United States receivers imprinted, even if I have to draw the diagrams myself. In other words, ITT Intermetall supplies the basics and different manufacturers do their own applications and innovations thereafter. In such a new and obviously important technology, rapid development could not evolve otherwise. Now, for the recent changes.

CVPU Comb Filter Video Processor

The VPU MAA 2200 has already been discussed previously as well as the VCU video control unit, therefore the best we can do here is to combine the two into a single diagram and add available information on the NTSC comb filter (see Fig. 5-10) as the CVPU2210.

Composite video is first clamped at levels suitable for A/D conversion, then chroma bandpassed, chroma trapped for luminance, then peaked, phase compared and emitted from the D/A converters as separate luminance and R-Y, B-Y information, which is subsequently combined in the RGB matrix.

The luminance filter (when used) has dual bandpasses, with 4 MHz allowed for i-f signals and 7 MHz for baseband inputs from cameras, videocassette recorders, video discs, etc. Eight bits then process luminance for peaking, followed by a contrast multiplier and limiter, which is additionally controlled by the CCU from the IM bus.

Chroma (Fig. 5-11) begins with a phase-linear comb filter, switchable from symmetrical to asymmetrical by the CCU to accommodate baseband or i-f processing. An ACC multiplier adjusts color saturation and hue correction. When composite video inputs have taken place, comb filtering occurs as a RAM takes over the function of delaying a single line of information by 63.5 μsec so that luminance and chroma may be compared and added in time to produce single luminance or chroma outputs into their respective peaking and color control circuits. Naturally there's no chroma trap in luminance when this digital filter separates the two signals. Once chroma is demodulated and becomes color (less luminance), then brightness and color are RGB-matrixed and delivered as RGB composite signals to the picture tube. As usual, phase comparison

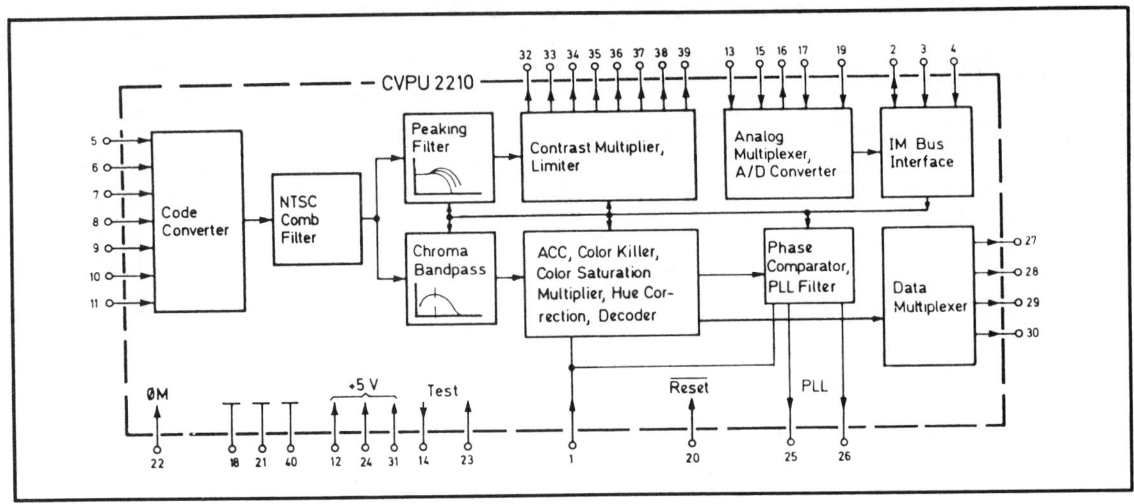

Fig. 5-11. Block diagram of the CVPU2210 Comb Filter Video Processor (courtesy ITT Semiconductor).

sync for the demodulator comes from reference color subcarrier burst at 3.58 MHz. A phase-locked loop circuit does the rest.

SECAM Chroma Processor

An explanation of the French color system processor may or may not be of value to most United States types, but there could be one or two exceptions now and then who want and need the information—so here it is, as illustrated in Fig. 5-12.

This is known as the SPU2220 N-channel VLSI MOS circuit contained in a 40-pin plastic package. It has a code converter, SECAM bell filter, digital FM demodulator with dc offset correction, deemphasis and demultiplexer, digital R/B identification, digital recognition circuit, a color saturation multiplier and multiplexer for color difference signals, and an IM bus interface circuit.

The IC processes SECAM as parallel 7-bit Gray code signals from the VCU. In parallel with the

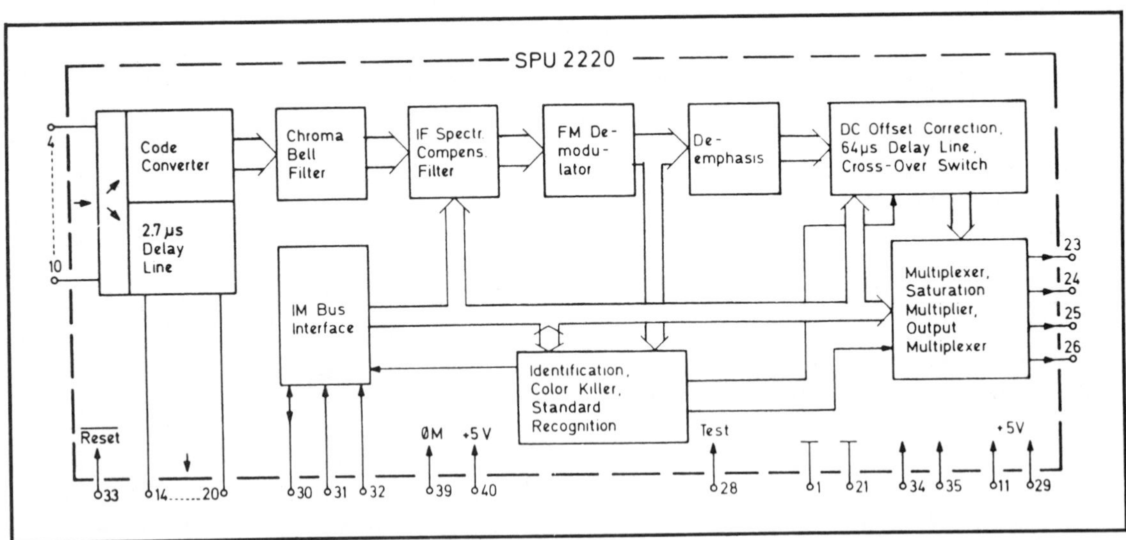

Fig. 5-12. Block diagram of the SPU2220 SECAM Chroma Processor showing how the French do it (courtesy ITT Semiconductor).

VPU, it then separates color from composite video, bell-filters with i-f spectral compensating before reaching the digital algebraic detector. This demodulator then delivers an 8-bit output sampled at 4.286 MHz, whose amplitude is made proportional to the frequency of the received color information, with digital deemphasis before demultiplexing with a 64 μsec delay line. The FM demodulator is a digital FM discriminator, 14 bits wide and covering frequencies between 3.774 and 4.782 MHz. In the following deemphasis filter, SECAM standard sequential chroma difference signals of:

$-1.9(R-Y)$ + Offset and $1.5\ (B-Y)$ + Offset

and any phase position shift is corrected by synchronizing circuits. These signals are now subject to a crossover switch and appear as parallel R-Y and B-Y information, multiplexed with two phases of the 4.286 MHz clock before being introduced to the color saturation multiplier, whose saturation factor may be set in 64 steps. The final step in the French process is transferring the 6-bit color difference information sequentially over four lines to the VCU for final processing. During vertical flyback (blanking) the IC is inactive since the VPU transfers test and alignment information to and from the CCU.

Also included in the SECAM processor is identification logic, a color killer, and a 2.7 μsec coding delay because of the extensive processing for this complex IC. The bell filter removes luminance and compensates for certain transmitter responses. The IM bus function becomes apparent from the diagram.

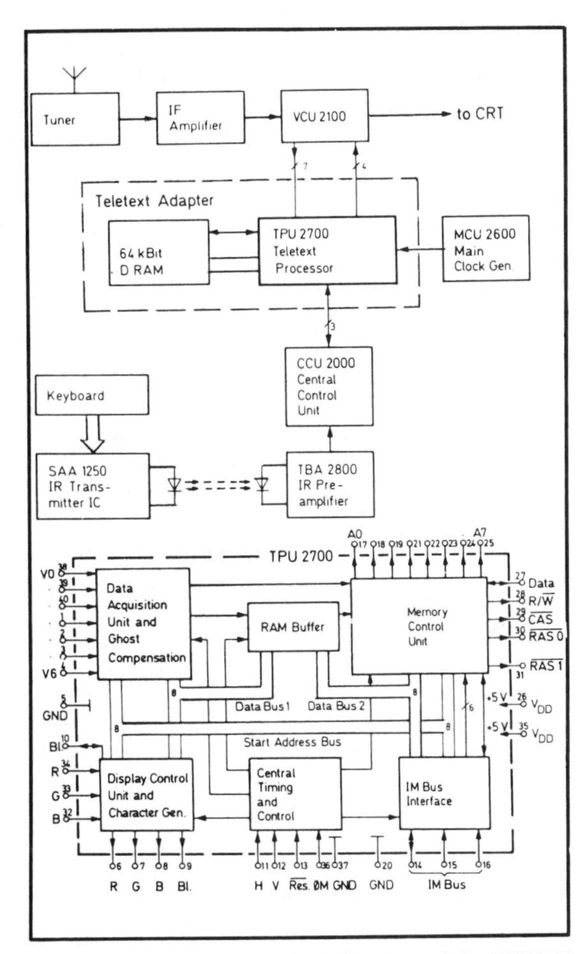

Fig. 5-13. Application and block diagrams of the TPU277 Teletext digital processor (courtesy ITT Semiconductor).

The TPU 2700 Teletext Processor

This particular N-channel MOS VLSI 40-pin IC is designed to process Level 1 Teletext (Videotext in Germany) as transmitted by broadcasters in the U.K., Germany, and other European countries, with modifications for the United States. It has ghost compensation, delivers a 7-bit composite video signal digitized, stores eight pages of text at a time, auto language-dependent character selection, and PAL/NTSC switchover ability (see Fig. 5-13).

Data acquisition begins with line 7 for PAL and line 10 for NTSC, ending at line 22 for PAL and line 14 for NTSC. Inputs pass through a deghosting filter which will compensate for delay time reflections of 0-0.8 μsec for PAL and 0-1 μsec for NTSC.

Incoming teletext is synchronized and identified, and a comparator preselects pages and page numbers required by Central Control. After internal RAM buffering, loads all requested data into the RAM and it is stored. The 8-bit character words are translated into 6 × 10 dot matrix for PAL or 6 × 8 dot matrix for NTSC by a 96-program character ROM and displayed in 24 rows of 40 characters each. Character sets are available for 8 languages and automatically selected, and every 10th PAL or 8th NTSC line is loaded into the RAM buffer. So when the RAM is not accessed, memory control refreshes and fills CCU requests for RAM access. As the diagram shows, the CCU is able to read and write all RAM locations by loading appropriate registers into the RAM. A menu is also available for display. By the use of an external RAM, flexible capacity is available to store 2-, 4-, or 8-pages of menu. External 17/14 MHz clocking and RGB inputs are available, as well as external fast blanking and status output. The IC will also display noninterlaced information and can distinguish between and first and second TV fields. Should single errors appear, they are corrected, while multiple errors immediately stop all Teletext reading. At this point, the error flag is set and TPU attempts to re-read this page when it again appears without clearing the previous data. Should two pages conflict in time, the first page is stored and others must wait for retransmission.

There's also a great deal of additional information in the ITT booklet 6250-11-2E which we won't print, but which we understand will be made available to serious design engineers desiring to use the TPU 2700.

The 1024-Bit EEPROM

This is an N-channel electrically-erasable programmable read-only memory with an 8-bit, 128-word capacity. When reading stored data, memory address must be entered into the memory address register first and then the data is read by entering an IM bus address (Fig. 5-14).

Reprogramming memory requires all bits to be reset to 1, then data is routed into some select memory location through the IM bus with a timed program sequence of some 16 periods via the 1 kHz memory clock. When programming is complete and there are no "busy" signals, memory addresses are restored and normal operations resumed.

Automatic Picture Control

The APC 2230 appears as a real time signal processor that may be inserted between the NTSC comb filter and the VCU video codec (Fig. 5-15). It will control brightness, contrast and color saturation. Various values are measured by the APC, processed by the CCU, and commands generated by the CVPU video processor. During auto fleshtone correction oranges and reds are slightly merged to produce relatively broad areas of fleshtones. There is also a vertical interval (color) reference circuit that activates whenever VIR is broadcast, which is most of the time on the larger stations. This, of course, is an original G.E. development and we presume does digitally what the

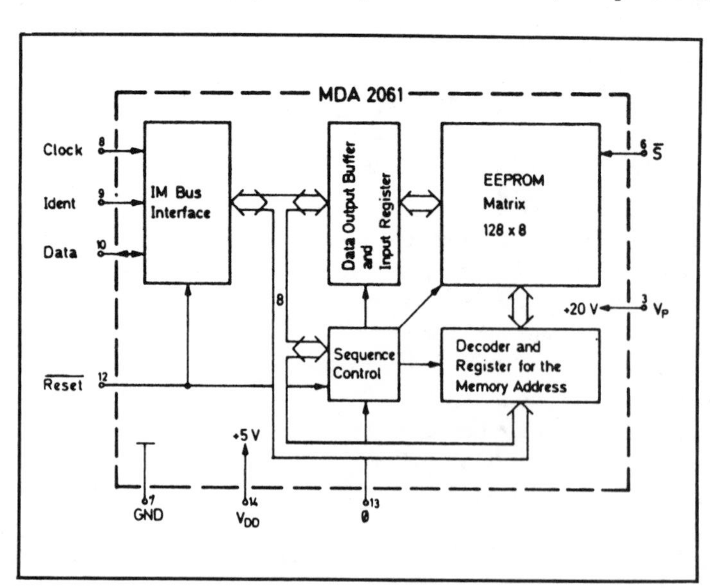

Fig. 5-14. The 1024-bit EEPROM (courtesy ITT Semiconductor).

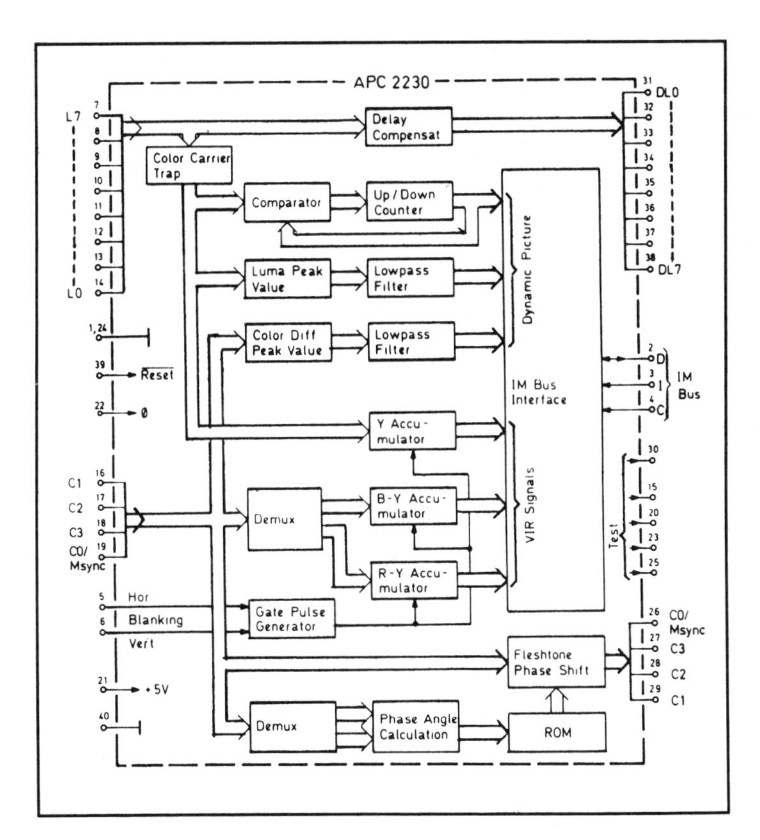

Fig. 5-15. The APC2230 automatic picture processor controlling contrast, brightness, and color saturation (courtesy ITT Semiconductor).

G.E. unit does in analog. A further description of the original VIR is contained in the chapter on Color Processing. Here, oranges and reds are not even slightly merged, and a combination of line 19 signals and sync information with chroma feedback governs fleshtone and other chroma-luma corrections *without* picture distortion.

As ITT states, this is another N-channel MOS, 40-pin IC that supplies automatic brightness, contrast, color saturation, fleshtone correction, and VIR. The dynamic picture portion may either operate continuously or be operator on/off controlled, depending on software programming. During dynamic picture control, digital luminance signals go through the color trap where their black levels and peak values are measured, along with the RB-Y color difference signals. The results are then routed to the CCU for action via the IM bus.

In the black level measurement, an up/down counter becomes an integrator and comparator as it compares momentary luminance points with counts in the up/down counter. Should luminance be higher, the counter counts up, and if lower, the counter counts down at 50 × the speed of up counting so that only the dark portion of the picture is recognized as black level. Brightness is now custom-controlled by both operator and VCU action. Meanwhile, maximum luminance values are stored for every line, with consecutive values going to a low-pass filter having a selectable time constant. Then the CCU reads via the IM bus and cal culates contrast so that with high-brightness portions of the picture contrast is reduced accordingly. Small-brightness picture portions will have no affect.

Chroma, at the same time, has its value derived from nonfiltered color difference information, stored by line and lowpass filtered. The CCU then calculates color saturation and controls the color saturation multiplier in the CVPU so color saturation remains constant.

VIR, briefly, is broadcast normally with color

line 19. An accumulator measures 32 values of luminance and RB-Y during this designated line. These are stored until the next line 19 appears, and then can be read via the IM bus by the CCU which may then adjust color saturation and picture hue. Horizontal and vertical blanking is provided by the DPU which continue during the 32 values of the color carrier, where there is also a B-Y reference. Even in nonlocked sync, the gate pulse generator finds line 19, permitting establishment of the proper operating signal.

A mask-programmable ROM of 256 × 11 bits covers the color range between 30° to 213°, with a step width of some 0.7° per LSB. Such 11-bit data is distributed so that with 8-bit resolution, sine and cosine values of −28° to +28° may be stored.

For the up-to-date information available just before 1986, the foregoing is about it! You now have all the useful information expected in most of the digital receivers during the coming several years. Obviously there's more to come, but it hasn't been announced as yet, and probably won't be until some of what we've just described has been digested and safely tucked into many of the new digital TV models. Obviously we're very much indebted to ITT Intermetall and their able Mr. DeVoe for the information to date and will continuously follow these developments in future magazine articles in some of the popular technical journals such as *Modern Electronics*.

Future Developments

Once again we're indebted to Mr. DeVoe for both information and diagrams of planned improvements in the digital system, which have just been revealed before book publication. Many are planned for soon-to-be-produced receivers, certainly by 1986 or 1987.

Full frame storage, coupled with doubling the line rate should produce flicker-free pictures, full motion and noise elimination are several of the coming innovations, and ITT believes that economic frame stores will be available in 1986. Such a proposal is outlined in Fig. 5-16, which ITT says can be developed using Dynamic Ram Memories (DRAMS), and will also enable the operator to on line 19 of the vertical blanking interval. A color carrier trap is required since the CVPU has no trap and the comb filter has to be switched off during switch between interlaced and noninterlaced frames. In the figure you see a 1-frame memory composed of three banks supplying three multipliers with sequence control from a system clock, along with an image processor and noise detector. When frame store arrives, Mr. DeVoe says, features such as freeze frame, zoom picture-in-picture

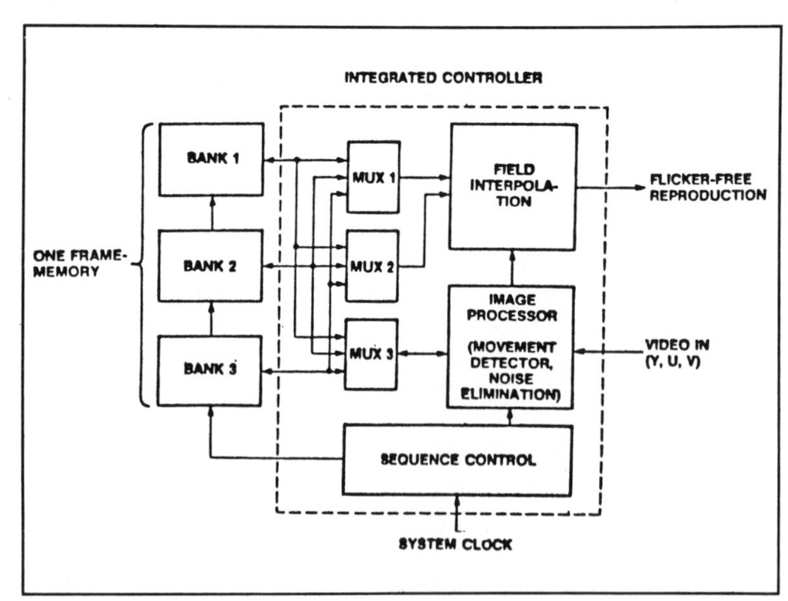

Fig. 5-16. Dynamic RAMs provide full frame storage (courtesy ITT Semiconductor).

and ghost reduction will also be immediately available at *no extra cost.*

There's also a new super telex chip under active design that can be dynamically redefined for character set and ROM-programmable color look-up table. It will be able to handle all levels of teletext in both character set and resolution with 40 characters per line and up to a 12 × 10 matrix. There's also a ghost compensator and bit slicer plus an on board microcontroller decoder with instructions mask-programmed in a ROM. It reads and writes into a RAM and uses memory as a modified bit plane. Graphics are part of the program also.

But the ultimate result of all this very-large-scale integration is at least a 2:1 reduction in the number of ICs used in the current system (see Fig. 5-17). All video and audio intermediate frequency amplifiers and detectors will be collected in a single intermediate processor unit, all video processing and sweep in another signal processing unit, audio digitizing and processing in a third IC and central control a fourth. This leaves only the audio and video power outputs, vertical and horizontal outputs, the surface-wave acoustical filter (SAW) and the tuners as independent units. Along with these will come wideband comb filtering for luminance and chroma, RGB inputs at no additional cost, a phase-locked loop countdown deflection V/H system, stable under all signal conditions, automatic gray scale tracking, digital stereo, and digital factory alignment stored in nonvolatile memory.

Mr. DeVoe also touched on the MAC (European) system of wide bandwidth companding which will require RGB inputs for full benefit. This, of course, we'll discuss by the end of the chapter in hopes there may be something reasonably new to offer as a result of a visit to RCA's David Sarnoff research laboratories in Princeton, New Jersey.

Right now, however, we do have some information (including our block diagram) on Toshiba's CZ2094—the first digital receiver based on the ITT series of ICs available commercially on the U.S. market. We were genuinely surprised at its flexibility and quality. This is a nice receiver and better than good for a first-run product.

Toshiba's CZ2094

The first digital television receiver of any

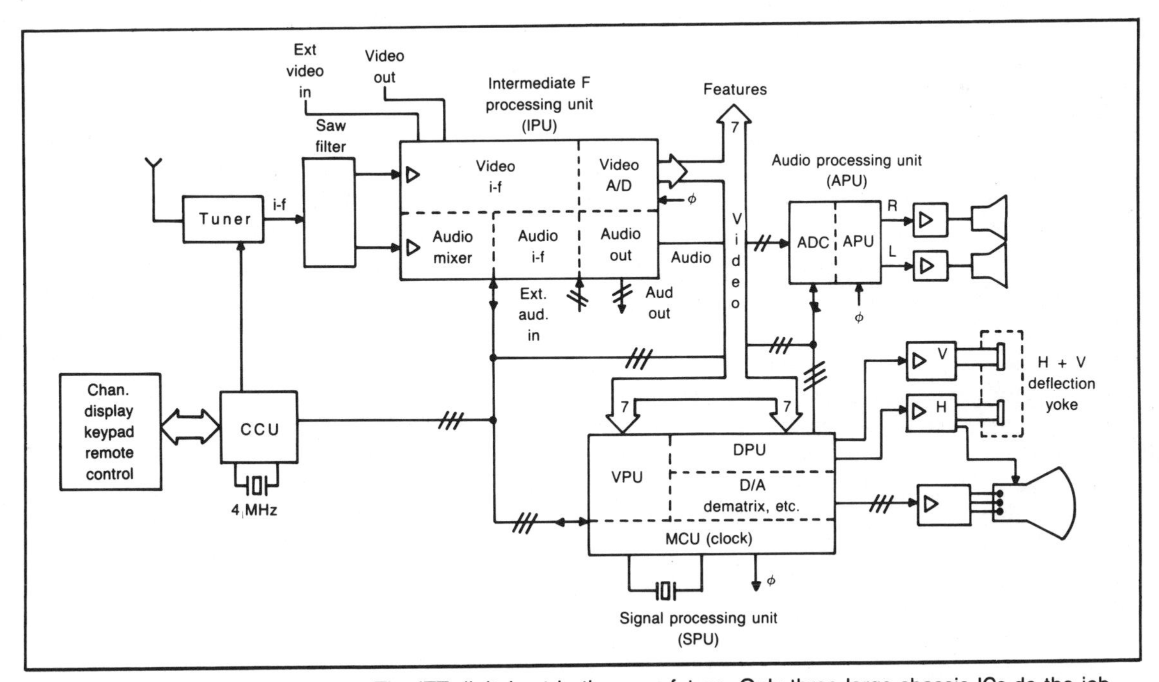

Fig. 5-17. Shape of things to come. The ITT digital set in the near future. Only three large chassis ICs do the job.

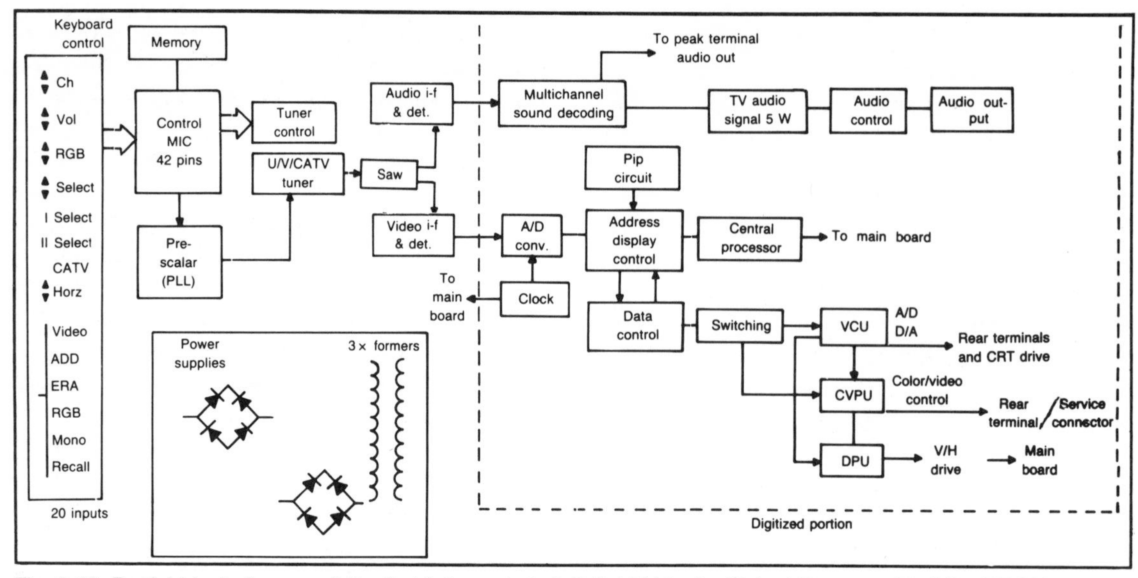

Fig. 5-18. Partial block diagram of the first fully-marketed digital TV in the United States—a Toshiba CZ 2094.

description to be actively marketed in the United States, this 20-inch, super-rectangular tubed receiver/monitor features picture-in-picture (PIP), comb filter, and BTSC-dbx multichannel sound, including both TV stereo and SAP (second audio program). With full U/V/CATV tuning and power consumption of 99 watts, a block diagram of the digital portion of the set is shown in Fig. 5-18, following the video detector. But the real story of this TV is not what is obviously apparent, but what it actually does with these digital circuits.

For starters, all controls are digital and most, such as Video/RGB switches, remote sensor, recall and volume/channel up/down, power, programming, and stereo right/left, up/down, are on the front panel. On the right side, however, are contrast, brightness, tint and color potentiometers for the PIP feature. On the rear are a Teletext receptacle and separate RGB computer inputs, three sets of video/audio inputs and separate video/audio outputs, external 8-ohm speaker drives with 5-watt outputs, supplementing composite video inputs and outputs on the front panel. Remote and front chassis controls deliver superior digital tuning and command functions, with on-screen readouts and bar graphs for channel, mono/stereo, and a second dual-sized picture moveable about the four quadrants of the face of the cathode-ray tube.

Does all this mean that the CZ2094 has superior pictures? Better than most would be a truthful answer since most of the features are digital controls rather than outstanding picture improvement. However, statistics such as 43 dB S/N ratios for luminance and chroma, 4.2 MHz rf input high-frequency response, and better than 40 dB L/R stereo separation certainly does mean a great deal. Power supply regulation also approaches 100 percent, along with 90 percent dc restoration. There's *still* picture for freeze frame, along with remote adjustments for base, treble, stereo separation, color, sharpness, brightness, tint, and contrast. If these are confusing to the average layman, a Reset button automatically presets all these parameters for you.

The innovative and extensive remote control has 32 velvety-soft touch keys producing digital command signals via an infrared (IR) carrier effective up to 16 feet and ± 30 degrees on either side of the receiver's remote sensor. Using two series AAA 1.5 V batteries, the unit lights a red LED as each command is entered for such entries as PIP, Quad, Source, Swap, Still, RGB, Cont-1, -2, STR/L2, Timer, and the usual direct and up/down audio and video programming. Translated, PIP stands for picture-in-picture (two sizes), STR/L2 means stereo or SAP, QUAD means PIP placement among the four quadrants of the picture tube, SOURCE signifies air or baseband programming, and SWAP will reverse the positions of the main and secondary images on the screen when using PIP. STILL, of course, is freeze frame, and there's also a TIMER button that will turn the receiver off automatically for periods up to 180 minutes. Cont-1 controls bass, treble, and stereo separation, and shows a movable bargraph with adjustments, while Cont-2 does the same for color, sharpness, brightness, tint and contrast. For those who can't handle all this there's a RESET button that automatically presents or resets these various settings for you.

We'd call this a *now* rather than an interim digital receiver since Toshiba has obviously decided to devote considerable engineering talent to the project, as well as entering into an agreement with the Westinghouse Corp. to manufacture cathode-ray tubes in New York State. It's highly possible America will be hearing a great deal from this well-known Japanese company during the next several years as digital receiver techniques continue to mature and grow.

You might also like to see laboratory test results of the CZ2094 done previously for *Modern Electronics* magazine earlier in 1985. Unfortunately, most relate to the analog portions rather than digital since there was a total shield about the 1s and 0s department that wouldn't permit safe logic or oscilloscope probing. The lab analysis is shown in Table 5-1.

When components connected to certain ICs, or the integrated circuits, themselves, are changed, Toshiba also has a special digital interrogator and programmer called a JIG. Having 64 keys, this test instrument, with its own central processor (CPU) and ROM, will sample and extract factory-generated information from the receiver's 1024-bit EEPROM, and then deliver this information to the DPU 2500 deflection processor and the CVPU 2100 NTSC comb-filter video processor (Fig. 5-19). In this way, picture sharpness, contrast, brightness, tint, and color are set for "standard" factory conditions, while the color subcarrier frequency, picture symmetry, height, and brightness may also be JIGG'd for programmed operating response. Since you may be interested in the extent of the JIGG's programming scope, Toshiba has given us permission to publish these instructions just as they appear in Toshiba's service manual recently issued. See Table 5-2.

MAC

MAC stands for multiplexed analog component signals, and is a proposed system apparently originated by K. Lucas and M.D. Windram of the Independent Broadcasting Authority (Great Britain) in the beginning of the 1980's (or at least they were the first to offer a well-documented report). We in the United States are paying considerable attention to many of these main proposals, although some modifications must be made to adjust to the NTSC

Ac operating range			90 to 130 Vac
Power drain (w/signal input) at 117 Vac			81 to 85 W rms
Voltage regulation (100 to 130 Vac)		Low voltage	134.1 to 134.4 99.8%
			14.7 to 14.8 99.3%
		High voltage	28.4 kV 100%
Tuner/system sensitivity	Vhf	Ch. 4/12	−4/−9 dBmV
	Uhf	Ch. 15/60	−8/−8 dBmV
Agc swing (before distortion)			59 dB
Dc restoration			90.8%
CRT color temperature			9000°K
S/N luminance, chroma			43 dB
Convergence			99.9%
Barreling/pincushioning/flagwaving			none
Staircase linearity			Slight black compression
Horizontal resolution through rf			4.2 MHz
Horizontal resolution through baseband (video)			5.13 MHz
Vertical resolution (pix. center)			450 lines
Stereo *channel* separation			>40 dB
Adjacent channel separation			22 dB
Audio response at baseband			>100 kHz
Audio response through 6″ oval speakers			variable and 12 dB down at 10 kHz

Table 5-1. Toshiba Model CZ2094 Digital Television Receiver/Monitor Laboratory Analysis.

Table 5-2. Toshiba's Digital Interrogator and Programmer Instructions.

■ CONTROL KEY FUNCTION

(1) CONTROL KEY

Key Name	Function
WRT STR	Writes the adjusting data into EAROM. CPU in digital TV begins operation. LED indicates [|] [| .] only.
STR	CPU in digital TV begins operation. LED indicates [|] [| .] only.
MOD CLR	Resets the adjusting item. LED indicates [○|○] [△|△] only. (○ ○ : Adjusting item No. △ △ : Data)
RED RTO	Stops CPU operation in digital TV and reads out data of EAROM. LED indicates [E|n] [d|] only.
WR STO	Stops CPU operation in digital TV, writes in the factory-preset data into EAROM and reads out them. LED indicates [E|n] [d|] only.
SBRT △ (Subbright)	Makes the adjusting data UP (+1). LED indicates [○|○] [△|△] only. (○○ : Adjusting item No. △△: Data)
P-ACL 1 ▽	Makes the adjusting data DOWN (−1). LED indicates [○|○] [△|△] only. (○○ : Adjusting item No. △△: Data)
CONT SET	Acquires the adjusting data when one of keys among O to F is selected. LED indicates [○|○] [△|△] only. (○○ : Adjusting item No. △△: Data)
WDR 0	Means "0" key in selecting numerals of the adjusting data. LED indicates [○|○] [△|△] only. (○○ : Adjusting item No. △△: Data)
WDG 1	Means "1" key in selecting numerals of the adjusting data. LED indicates [○|○] [△|△] only. (○○ : Adjusting item No. △△: Data)
WDB 2	Means "2" key in selecting numerals of the adjusting data. LED indicates [○|○] [△|△] only. (○○ : Adjusting item No. △△: Data)
COR 3	Means "3" key in selecting numerals of the adjusting data. LED indicates [○|○] [△|△] only. (○○ : Adjusting item No. △△: Data)
COR 4	Means "4" key in selecting numerals of the adjusting data. LED indicates [○|○] [△|△] only. (○○ : Adjusting item No. △△: Data)
COG 5	Means "5" key in selecting numerals of the adjusting data. LED indicates [○|○] [△|△] only. (○○ : Adjusting item No. △△: Data)
COG 6	Means "6" key in selecting numerals of the adjusting data. LED indicates [○|○] [△|△] only. (○○ : Adjusting item No. △△: Data)
COB 7	Means "7" key in selecting numerals of the adjusting data. LED indicates [○|○] [△|△] only. (○○ : Adjusting item No. △△: Data)
COB 8	Means "8" key in selecting numerals of the adjusting data. LED indicates [○|○] [△|△] only. (○○ : Adjusting item No. △△: Data)
SUBC 9 (Subcarrier)	Means "9" key in selecting numerals of the adjusting data. LED indicates [○|○] [△|△] only. (○○ : Adjusting item No. △△: Data)
MODE A	Means "A" key in selecting characters of the adjusting data. LED indicates [○|○] [△|△] only. (○○ : Adjusting item No. △△: Data)

Key Name	Function
BUST B (Burst Level)	Means "B" key in selecting characters of the adjusting data. LED indicates ○○ △△ only. (○○ : Adjusting item No. △△: Data)
Fsc Asw C	Means "C" key in selecting characters of the adjusting data. LED indicates ○○ △△ only. (○○ : Adjusting item No. △△: Data)
KIL 1 D (Killer 1)	Means "D" key in selecting characters of the adjusting data. LED indicates ○○ △△ only. (○○ : Adjusting item No. △△: Data)
KIL 2 E (Killer 2)	Means "E" key in selecting characters of the adjusting data. LED indicates ○○ △△ only. (○○ : Adjusting item No. △△: Data)
WDM F	Means "F" key in selecting characters of the adjusting data. LED indicates ○○ △△ only. (○○ : Adjusting item No. △△: Data)

(2) ADJUSTING ITEM SELECTING KEY

Key Name	Item No.	Data	Function
WDR 0	00	Adjust	Reads output data of white drive, and is automatically set with Red Axis AKB turned on.
WDG 1	01	Adjust	Reads output data of white drive, and is automatically set with Green Axis AKB turned on.
WDG 2	02	Adjust	Reads output data of white drive, and is automatically set with Blue Axis AKB turned on.
COR 3	03	Adjust	Reads output data of cut-off, and is automatically set with Red Axis AKB turned on.
COR 4	04	00	Regulates noise converter (A/D converter) into the ON or OFF condition.
COG 5	05	Adjust	Reads output data of cut-off, and is automatically set with Green Axis turned on.
COG 6	06	01	Sets ABCL1, and regulates ACL effect.
COB 7	07	Adjust	Reads output data of cut-off, and is automatically set with Blue Axis AKB turned on.
COB 8	08	01	Sets ABCL 0, and regulates ACL effect.
SUBC 9 (Subcarrier)	09	Adjust	Adjusts color sync. Turns color sync. ON-OFF switch off. **Keep in mind to press MODCLR after adjustment.**
MOD A	0A	52	Selects mode of video processor. MAA2200(VPU):5A MAA2210(CVPU):52
BUST B (Burst Level)	0B	38	Adjusts ACC gain.

ABCL	ABCL	ACL
1	0	effect
0	0	−20 dB
0	1	−10 dB
1	0	−6 dB
1	1	−3 dB

Table 5-2. (Continued from page 113).

Key Name	Item No.	Data	Function
Fsc Asw C	0C	00	Turns color sync. on or off. Checks adjustment of color sync.
KIL 1 D	0D	5B	Sets upper limit of color killer.
KIL 2 E	0E	5B	Sets lower limit of color killer.
WDM F	0F	Adjust	Corrects output level (red, green and blue axes at the same time) of detection pulse at white drive.
SBRT (Subbright)	10	34	Changes the AKB cut-off level for three axes at the same time.
P. ACL 1	11	10	Sets contrast correction data (F). Contrast gain = $S(F_0-F)$ F_0: Constant (31)
P. ACL 2	12	07	Sets contrast correction data (S).
UNI COLR (Unicolor)	13	20	Sets color reduction level comparing to contrast. (00 = Same level ~ FF = Reduction zero.)
MIN CONT (Minimum Contrast)	14	0A	Sets contrast level with contrast turned to minimum. (Adjusting range of user is fixed.)
COLR SATU (Color Saturation)	16	00	Sets color remainder level with color gain turned to minimum.
CONT SET (Contrast)	17	1A	Sets contrast standard level. (Cursor position) Cursor position also changes with value.
BRT (Brightness)	18	Adjust	Sets brightness standard level. (Cursor position) Cursor position changes with value.
PIX QUAT (Picture Sharpness)	19	05	Sets picture sharpness standard level. (Cursor position) Cursor position changes with value.
COLR SATU (Color Gain)	1A	11	Sets color gain standard level. (Cursor position) Cursor position changes with value.
TINT	1B	10	Sets tint standard level. (Cursor position) Cursor position changes with value.
RGB CONT	1C	07	Sets RGB contrast. (Cursor position)
DRIV R (Drive R)	1D	6E	Sets AKB drive standard level. On red axis.
DRIV G (Drive G)	5E	63	Sets AKB drive standard level. On green axis.

Key Name	Item No.	Data	Function
DRIV B (Drive B)	1F	63	Sets AKB drive standard level. On blue axis.
COFF R (Cut-off R)	20	42	Sets AKB cut-off standard level. On red axis.
COFF G (Cut-off G)	21	34	Sets AKB cut-off standard level. On green axis.
COFF B (Cut-off B)	22	30	Sets AKB cut-off standard level. On blue axis.
DPC L (S-correction)	23	7A	S-correction in vertical linearity.
DPC C (Symmetry Correction)	24	20	Symmetry correction in vertical linearity.
VERT HEIT (Vertical Height)	25	Adjust	Sets vertical height.
VERT DC (Vertical DC Level)	26	50	Sets DC level of vertical saw-tooth wave.
VERT MOD 1 (Vertical Mode 1)	27	03	Decides whether vertical is locked to horizontal or not.
VERT MOD 2 (Vertical Mode 2)	28	8B	Sets vertical count down mode.
HORZ MOD 1 (Horizontal Mode 1)	29	07	Sets horizontal system.
HORZ PWID (Horizontal Pulse Value)	2A	42	Sets horizontal drive pulse width. Caution: Do not change without due cause to prevent damage of horizontal output transistor.
HORZ PHAS (Horizontal Phase)	2B	0F	Regulates horizontal position.
HORZ MOD 2 (Horizontal Mode 2)	2C	28	Decides whether horizontal is locked to fsc or not.
AFC MOD (AFC Mode)	2D	90	Sets characteristics of AFC filter.

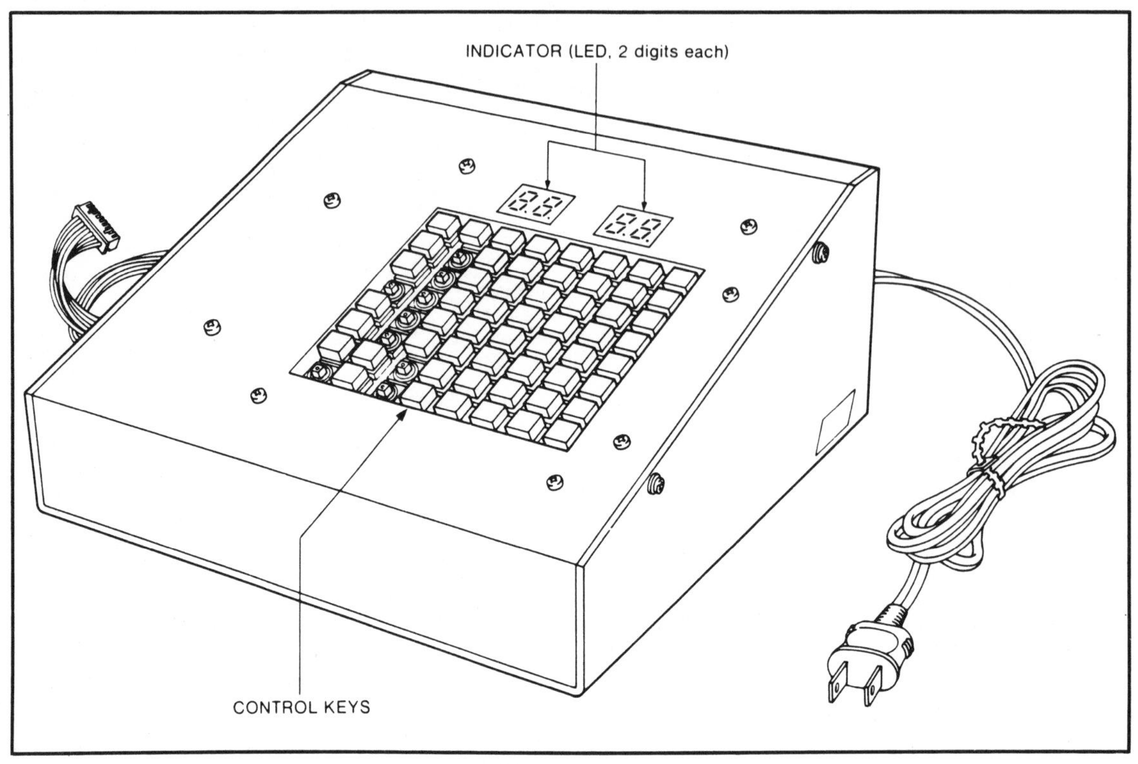

Fig. 5-19. Toshiba's 64-key IC alignment servicing JIG.

and system. Apparently, the idea is receiving considerable attention by several laboratories and manufacturers and may well be the answer to a number of problems dealing especially with satellite (and possibly fiber optics cable) transmissions to transmit more accurate data and also broaden bandwidths for higher definition and resolution.

Due to the noise-voltage response in FM video systems there's an imbalance of signal-to-noise ratios between chroma and luminance, with the color subcarrier and its sidebands affected most. Only with preemphasis filters to increase high frequencies is this imbalance partially equalized, but not without additional problems. There are also nonlinear distortions present due to restricted bandwidths and the usual FM discriminator nonlinearities, in many instances resulting from high-amplitude, high-frequency signals that overdeviate. In addition, there's an FM threshold problem in satellite receivers that often appears in carrier-to-noise ratios of less than 10 dB. High subcarrier levels and FM deemphasis only increase this difficulty, making objectional interference apparent to TV viewers.

In an effort to counter all this, the MAC

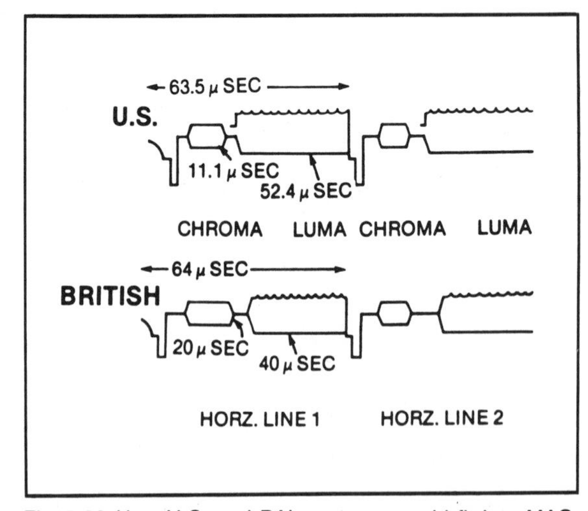

Fig. 5-20. How U.S. and PAL systems would fit into MAC.

system—as originally applied to German PAL (phase alternating line)—would time-compress the usual 52 μsec luminance signal to 40 μsec, increasing the overall bandwidth proportionally. Then 20 μsec of each line would be available for chroma before filtering. Calculations predict weighted S/N in the chroma channel could be improved by as much as 5 dB (6 dB being 100 percent improvement in terms of voltage), with luminance S/N unchanged. Now, crosscolor, crossluminance, and narrow color bandwidths are eliminated and, with no high-frequency subcarrier, signals would be considerably less sensitive to other usual distortions. A diagram of British-U.S. line segments is illustrated in Fig. 5-20. Both versions are already proven and awaiting the right moment for introduction and probably commercial use. Sound, by the way, in the B-MAC system is digitally multiplexed into the TV line-blanking period, and the composite signal then FM modulates the carrier.

Chapter 6

Cathode-Ray Tubes

A GREAT DEAL OF WORK AND STEADY PROGress has been and is taking place among the relatively few television picture tube manufacturers remaining in the United States and several in Japan. With TV receivers ever widening video bandpasses and the advent of TV/monitor combinations, CRT requirements are becoming much more demanding as full 2 MHz color and 4.2 MHz luminance video combine with multichannel sound to bring America *the* complete NTSC receiving instrument.

Thirty years ago just achieving some sort of color and projecting it through the air was considered astounding, even for a paltry one thousand dollars. Today, for that price, you have an incomparably more reliable, full picture and sound equipment that will operate for years with few if any service repairs, and delivering *very* superior visual and audible signals which, before 1980, for instance, were hardly imagined. The year 1984, then, should be remembered not solely as a 30-year anniversary of color but as the advent of complete NTSC receiver video and multichannel sound—with special thanks to RCA for full spectrum color and Zenith for multichannel sound.

Not only are we now enjoying vastly improved conventional type picture tubes with in-line guns and internal magnetic shielding, but experiencing considerable progress on flat, high-voltage-less picture tubes which will shortly become our coveted and long- sought picture on the wall. The technology is virtually there, and only the price keeps some very new designs from the market. Meanwhile, digital TV enters the picture, bringing with it further video and sound improvements and another step along the way to the ultimate in complete home entertainment. These are, indeed, good and prosperous years for consumer products video, whether it be computers, cameras, VCRs, certain games, or just plain, old fashioned television receivers with lots of new innovations.

To understand and appreciate some of the dynamic improvements that have developed during the past 30 years, however, you should have a little history background to preserve continuity. The evolution hasn't been easy, and a great deal of time and money have gone into developments which will be described in more or less detail, depending on their relative importance and current usage. As usual, we consider the present of more specific import than the past, and the future a significant goal if realizable.

DELTA GUNS AND TRI-DOT PHOSPHORS

Since we still have some of the old almost "original" details of several of these first cathode-ray tubes, we'll print a few drawings and include a brief description covering the tricolor guns, grids, and various beam-shaping magnets. All this should serve as solid basis for the newer material that will bring you up to speed in today's rapidly evolving world. Back in 1956, the cutaway view of a color cathode-ray tube appeared as shown in Fig. 6-1. It has a very simple tri-gun structure for red, blue, and green electron beams, several sets of magnets to influence the direction of these beams, the usual electromagnetic deflection yoke, a glass or metal tube envelope, a finely constructed metal shadow mask, and approximately one million red, green, and blue phosphors deposited on a glass faceplate which, in this instance was said to be 21 inches in diameter. Approximately one-half inch behind the phosphors is the aperture (shadow) mask having a small opening for each group of RGB phosphors. The mask is aligned so that each beam may only excite one color at a time, producing three individual primary color images, which the eye perceives as a full color picture because of the variety of colors and their placement on the face of the tube. The three beams, of course, scan in unison but are carefully controlled in intensity to produce the necessary saturations.

All three colors with identical intensities produce white, and in order to create acceptable color pictures, the three guns must be accurately aimed. Each beam is actually larger than the shadow mask aperture, but since they collectively reach the mask from three different angles, only a specific dot is excited. When unintended dots are illuminated, undesirable reds, greens, or blues show up where they shouldn't, causing what are known as impurities. And since the eye doesn't like unreal colorations, this problem needs to be overcome.

Like any other tube type, the gun(s) of this CRT has a coated cathode, a control grid, screen grid, focus, ultor collector grid, and a group of beam converging pole pieces (Fig. 6-2). However, we still have to converge all these electron streams into their respective channels, and so the need for a purity magnet, blue lateral beam positioning magnet, as well as other static and dynamic con-

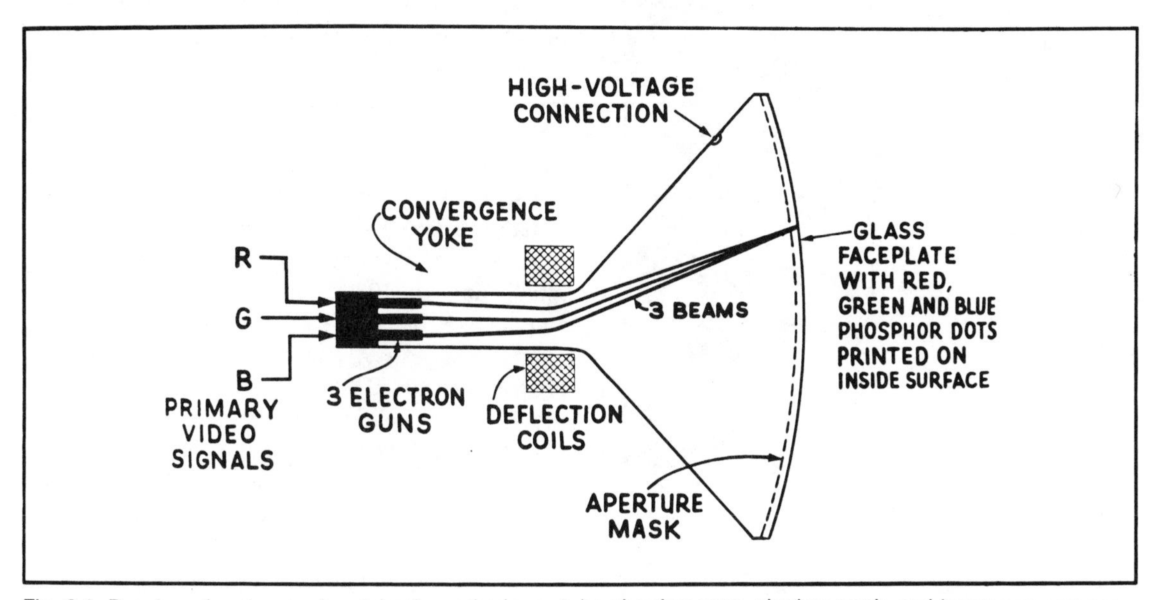

Fig. 6-1. Drawing of early round metal color cathode-ray tube showing guns, shadow mask, and beam convergence.

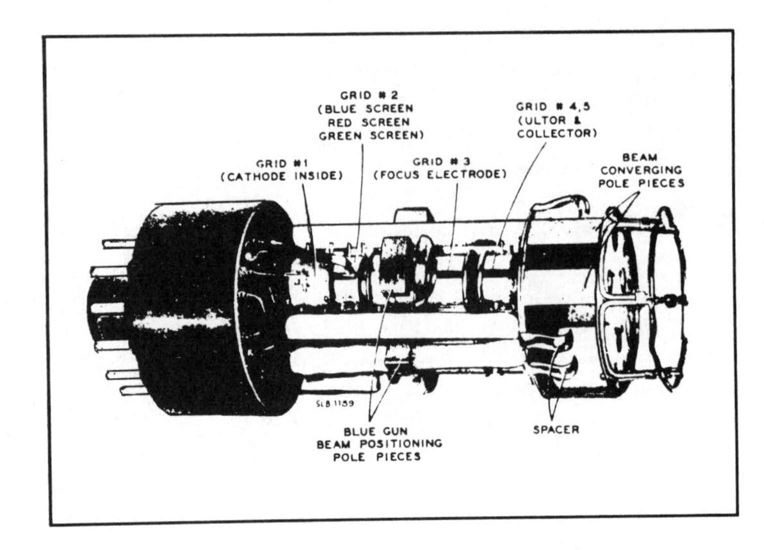

Fig. 6-2. Electron gun structure of an early 21AXP22A color picture tube.

vergence correctors that not only align all beams at the tube's center, but correct magnetic convergence sweeps at tube edges in sync with horizontal scan.

To do this in the early sets, dynamic convergence coils and static convergence magnets had to be placed forward of the purity magnet and close to the large deflection yoke (Fig. 6-3). And the means to control these dynamic sweep coils required special RL tilt and amplitude circuits usually positioned elsewhere in the cabinet. With such coils one adjusted the magnitude and tilt of parabolic correction current waveforms to the three beams with a series of nine controls, several of which were interdependent. Most static (center) convergence magnets were screw-adjustable but remained part of the convergence assembly on the CRT neck. In those days, you were wise to check first for pure red, blue, and green fields, remove any contamination with an ac-operated inductive degaussing ring, examine center convergence for accurate beam registry, then begin to work with sweep dynamics. Usually an 85% overall convergence was by far the best that could be hoped for. It's only been in the middle and late 1980s that superior convergences and life-like colors have been available. In succeeding receiver designs, a pincushion magnetic assembly was also added either in series or parallel with the deflection yoke to straighten as much as possible the V/H sweeps. Today, of course, the process has become immeasurably simpler, including a great deal of self convergence along with computer- matched deflection yokes and picture tubes, virtually eliminating static convergence as it was originally known.

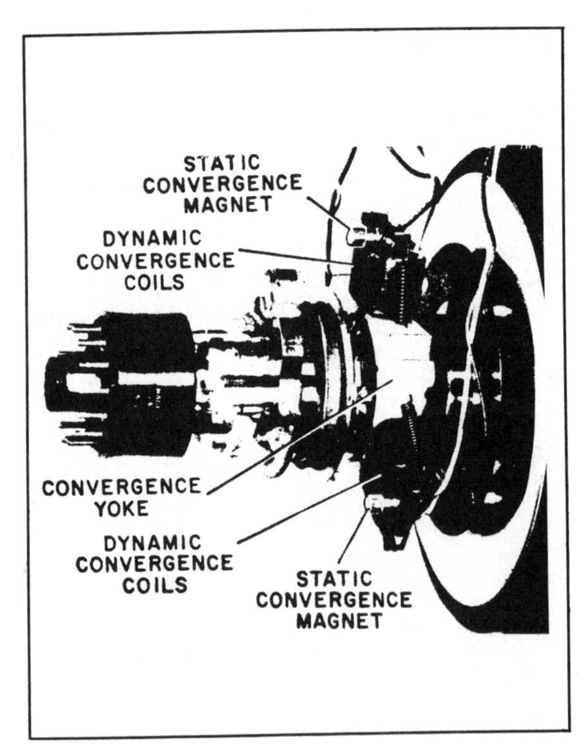

Fig. 6-3. Static and dynamic convergence assembly for color CRTs in the late 1950s.

Shortly after the 21AXP22A was released by RCA, a kinescope "boot" and polyethylene plastic cover were added on the outside of the tube to protect against high voltage and form a 2200 pf high-voltage filter capacitor. Contact with a "live" metal tube and its 20 kV accelerating potential was, indeed, a shocking experience that sat many of *us* down for a few moments to think it over. A rectangular version of this tube was also manufactured, but shortly thereafter only glass envelope tubes became available—thank goodness!

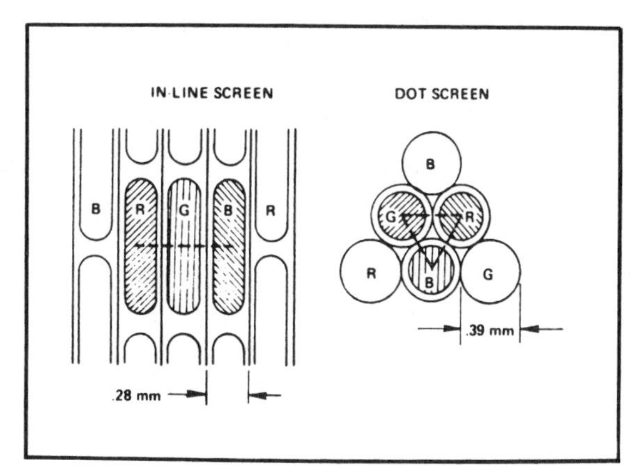

Fig. 6-5. Delta dots and in-line slot apertures compared.

RCA Develops In-Line CRT

Still, the delta gun—120° cathode separation (Fig. 6-4)—continued among both domestic and import receivers until 1973 when RCA introduced the first American unitized, closely spaced in-line gun, combined with a precision-wound toroidal deflection yoke. An inherently self-converging color system, this tube used "slot" mask apertures and vertical-striped phosphors instead of the usual three RGB dots. (Fig. 6-5). According to RCA, the precision static toroid (PST) yoke supplies a stronger magnetic field for CRT line focus improvement without trapezoidal misconvergence "blue" droop distortion and elimination of internal convergence pole pieces. In addition, there is said to be a longer magnetic field for fine-tuning convergence, a unitized gun assembly, and a new method of aligning both beams and the deflecting field. A drawing of the new gun is shown in Fig. 6-6.

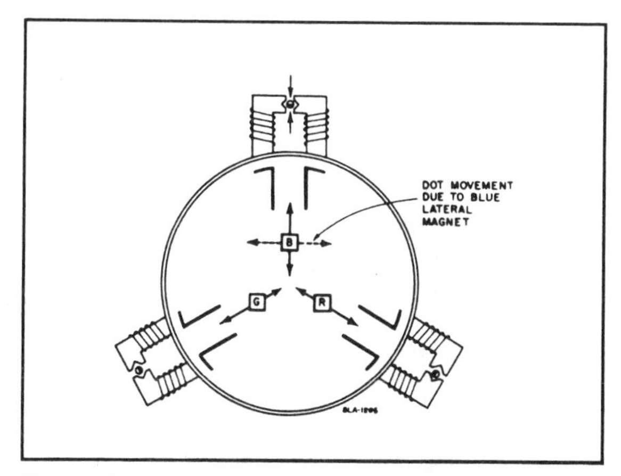

Fig. 6-4. The 120° delta deflection beam guide assembly.

In alignment, horizontal and vertical yoke movements aligns the deflecting field's center with the three beams. Having the middle gun fixed in position, horizontal yoke adjustment increases height and width on one of the outside beams while decreasing the size of the other, while vertical movement rotates one beam clockwise and the other counterclockwise. Final convergence occurs when the two outside electron streams are bent via a special angle in the main focus lens. During the time focus voltage assumes a constant relationship with anode voltage, overall convergence is generally independent of receiver electronics. Four magnetic rings circling the electron guns do the trick, along with the torodially-wound yoke where all wires are placed in dedicated grooves of molded plastic cemented to the core. RCA says the added uniformity and low impedances, especially designed for solid-state electronics, induce "less than half" the convergence errors detected in saddle-wound yokes and their accompanying color systems.

Shortly thereafter, General Electric introduced its own version of an in-line CRT system, but kept four static and four dynamic static convergence adjustments, which did not need matching CRTs and yokes, but did require a ring modualtor plus horizontal and vertical parabolic convergence currents—sort of an old-new compromise. Industry, however, adopted the RCA design and today almost all in-line tubes follow this general pattern with additional refinements that we'll describe as the chapter continues.

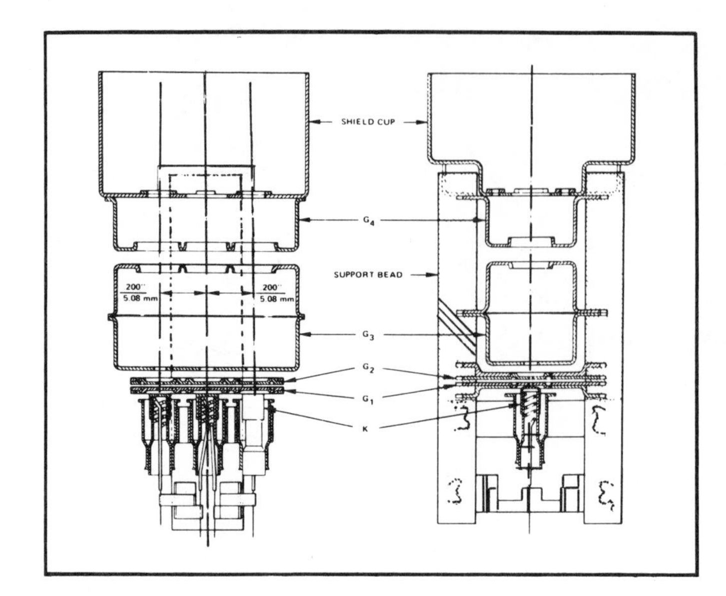

Fig. 6-6. The newly-developed in-line gun (courtesy RCA Consumer Electronics Div.).

Zenith's EFL Gun

Another version of an in-line gun with extended field lens (EFL) was also introduced in 1979 by Zenith Electronics Corp. (formerly Zenith Radio) and its tube subsidiary, the Rauland Corp.

With tube neck measuring 29 mm (now the accepted size), this was the first 25-inch slot mask CRT made for the American market and included, rare-earth red phosphors, high light transmission faceplate, negative guardband chromacolor screen, and a hybrid self-converging deflection yoke. Unlike some other tube-deflection systems, yokes and CRTs are entirely interchangeable. The electron gun has five grids, with 7 kV applied to grid 4, and other focus potentials of 12 kV for grids 3 and 5, extending the field of the prime focus lens, thereby minimizing any aberrations of the beam. (Figure 6-7.)

This new lens is said to eliminate many bipotential and einzel lens problems, at the same time improving resolution. Its deflection yoke is hybrid: toroidally wound for vertical and saddle wound for horizontal, with maximum deflection current of 1.35a vertical and 5.6a horizontal.

On the tube's neck are three magnetized rings used for purity plus 4-pole and 6-pole adjustments, the latter two being static convergence adjustments to aim the red and blue beams alongside the fixed center beam. Dynamic convergence consists of tipping the deflection yoke first vertically and then horizontally to converge any errant lines at 6 and 12 o'clock and then 3 and 9 o'clock. Because of cost, unfortunately, this very effective tube is no longer being used.

Sylvania's Dusting Process

This manufacturer—who has recently been acquired by North American Philips—has a straightforward process of putting one of these color tubes together using special screen rooms,

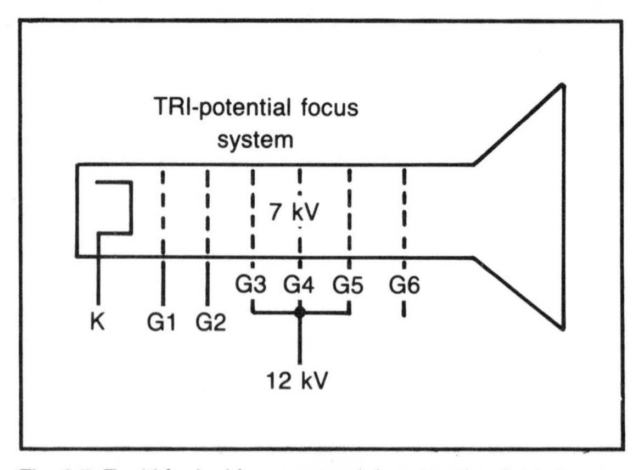

Fig. 6-7. Zenith's dual focus potentials extending field of main focus lens for high-resolution image at both high and low brightness levels.

precise optics, and a unique dusting means of applying the phosphors rather than pouring traditional liquid slurry.

Blue, red, and green phosphors are applied to the various sized faceplates in the screen rooms, a computer system handles the exposure table for the dot and phosphor-stripe process. Black material is deposited on the faceplate with windows and the tube is then lacquered for a smooth surface before being aluminized. After several special coatings, the bell and faceplate are sealed together at an oven temperature of some 450 °. Following these applications, the panels are vacuum sealed, the gun assembly is joined and sealed, and all internal gas evacuated. Afterwards, an external aquadag coating is added, the tubes are tested automatically and then shipped to the user.

In applying phosphors, a photolithographic emulsion is deposited on the faceplate and, then, while tacky, said glass receives a "cloud" of phosphor dust which adheres uniformly. Sylvania (NAPA) says "this goes a long way toward optimizing brightness and minimizing non-uniformity." The company has been producing color picture tubes in New York state and Ohio since 1953.

NEW CRT DEVELOPMENTS SOLVE PROBLEMS

Now that you've had the benefit of sufficient background to understand somewhat more intense detail, we'll proceed in that direction with the latest developments of the mid-1980s, including a special look at Mitsubishi's fine projection tubes, recognized in this category as a perennial leader. Mitsubishi will also be showing, for the first time at 1985 Los Vegas winter Consumer Electronics show, a 35-inch direct-view glass tube which is the largest sold in the United States to date, and Panasonic has announced a 45-incher but primarily for commercial applications rather than the consumer markets. It may be that glass envelope weight and the higher vertical-horizontal-high voltage potentials may not turn out to be the limiting factors after all—possibly just a shortage of glass. It is also understood that the 40-inch projection sets of 1984 will be reduced to 37-inch receivers of 1985. So what's the real difference between direct view and projection? Probably resolution and money, because the projection units are decidedly more expensive, and forever more difficult to maintain. But they are becoming sharper all the time.

Possible Problems

This is not to say there aren't problems with direct view tubes. There continue to be registration errors due to shadow mask thermal motion and vibration. Precise registrations of electron beams on associated phosphors exist only if such beams pass cleanly through shadow mask apertures and strike the exact phosphor for which they're intended. Naturally, shadow mask and screen alignment are critical and any spacing change between the two will result in misregistration. Usually, the mask-frame assembly is supported by only three or four bimetal springs welded to the frame, and under some circumstances this assembly is likely to oscillate, especially with 25-inch picture tube.

Studies have shown that a 3-stud mask-frame assembly has a resonance between 58 and 63 Hz. Four-stud assemblies, however, indicate no significant resonance below 500 Hz. Of course, ac line (mains) frequency is 60 Hz, and vertical color scan frequencies are 59.94 Hz—so a condition of oscillation is very possible in any of the 3-stud units. Apparently, this sets up a beat frequency that's equal to the difference between the scanning frequency and oscillation frequency and is seen in the picture as occasional "winks" or "blinks," and is commonly known as shadow mask flutter.

Although the 4-stud suspension doesn't resonate about 60 Hz, there may be problems in both mask retainers with the advent of "boom boxes"—hi-fi or multichannel stereo sound accompanied by 15- to 30-watt amplifiers. Such a dynamic force could easily find its way through the tube mounting to the shadow mask assembly, causing extra vibrations that result in additional blinks. If audio repetition rates are slow enough, vibrations may appear as a free oscillator, while faster ones would result in a "driven" oscillator or oscillation.

In top quality tubes you might not expect this to happen because frictional steel dampers can overcome much of these problems and basic design the rest. It is something, however, of which the industry and servicer should be aware when there are customer complaints about picture problems for which there is no electrical explanation. Further, some of the less expensive replacement CRTs may retain such problems from past designs that never have been corrected. It is reassuring to know, however, that steady progress in both thermal expansion and vibration problems has taken place and, with the newer faceplates and closer tolerances, you should find considerably fewer problems than even a few years past.

Moire, another "problem" that hasn't forcefully surfaced in some years, may again make an appearance as picture tube manufacturers and chassis designers decrease both "pitch" for shadow mask apertures and phosphor widths and increase (double or triple) vertical-horizontal sweep speeds for smoother, fine-line pictures.

The term *moire* originated with a mohair ribbed fabric that appeared in a wavy design. It may appear on picture tubes as lines-to-line shadings or variations as electron beams pass through their shadow mask openings. As phosphor efficiencies and electron gun optics improve, smaller beam spot sizes result, forming sharper raster lines. If beam diameters are small compared to distances between shadow mask apertures, it's possible beams will entirely miss holes on successive lines. Consequently, hole modulation of the RGB beams can generate a variable frequency appearing to the eye as areas of maximum and minimum brightness that form a swirling pattern. Usually more noticeable *without* video transmission, moire is considered a real problem only if it surfaces on program material at the usual 6- to 10-foot viewing distances.

Temperature effects, too, may also plague CRT makers with any change to faster sweeps. Shadow masks often absorb as much as 75 to 85 percent of the accelerated electron beam energy, and their temperatures naturally increase with additional excitation. Expansions, therefore move the expanding mask outward, causing misregistration on CRT phosphors. Impurities and poorer convergences result in less resolution and definition, nullifying excellent sweep characteristics and superior luma-chroma chassis design.

In the 1970s, RCA developed a Perma-Chrome system which moved the shadow mask somewhat closer to the CRT's faceplate, permitting more desirable expansion and allowing good registry between beams and phosphors. With the introduction of 110° tubes, however, as well as increased anode voltages, mask and frame temperatures did *not* increase proportionally during the first few minutes of turn-on. To overcome this, not only was the mask blackened to increase thermal emissivity, but the back of the phosphor screen was also blackened, permitting greater heat sinking. Now, the misregister difficulty became greatest between the center and right and left edges of the mask. This necessitated a screen printing lens development to adjust beam landing. To further complicate the picture, typical phosphor line widths on in-line tubes are .28 millimeters compared to a .39 mm tri-dot phosphor tube. Just a 100 μm misregistration, according to RCA, results in 30 percent of the beams to miss its phosphor compared with 14 percent for the tri-dot CRT.

A *Super Arch mask* was therefore engineered to minimize any change in spacing between mask and faceplate so any center "doming" was virtually eliminated and a computer check of contouring showed as much as a 20 percent increase in anti-doming efficiency at tube center and no doming at the edges. (See Fig. 6-8). Also, both phosphor lines and the shadow mask apertures were also bowed, permitting line curvatures from edges to tube center and substantially improving both definition and convergence, in addition to greater and more uniform faceplate illumination.

RCA's Coty-29 System

Now very much an industry standard, with many manufacturers licensed to produce it, the Coty-29 *system* for RCA represents a further development in small-necked, in-line picture tubes that has taken place in the early and middle 1980s. Instead of simple tube and shadow mask improvements, RCA has now combined the deflection yoke,

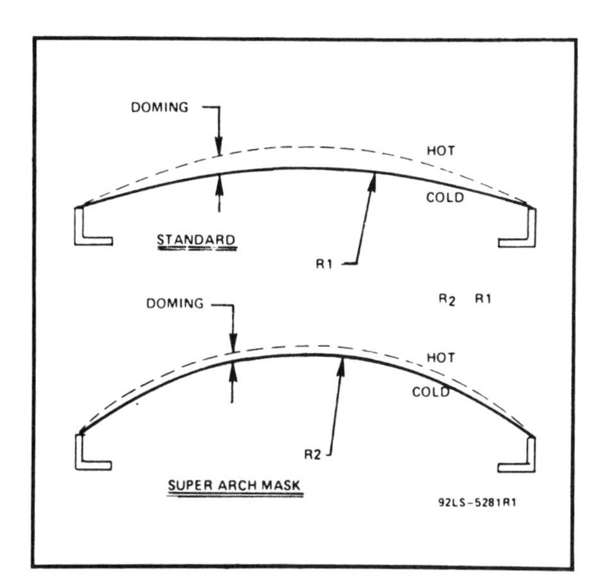

Fig. 6-8. The Super Arch mask substantially reduces doming (courtesy RCA Tube Div.).

receiver electronics, and picture display glass bulb into a complete system "optimized for . . . cost and performance."

As A.M. Morrell of RCA's Picture Tube Division of Lancaster, Pa. described it, the Coty-29 (for 29 mm diameter neck) system is the outgrowth of the following:

- RCA introduced in-line guns with self-converging 90 degree deflection yokes in 1972, featuring a unitized in-line gun and bipotential lens with focus voltage 20 percent of the anode voltage. Yoke windings were torodial for both horizontal deflection and beam-to-beam gun spacing was 5.1 mm.
- In 1975, a 110 degree deflection version of the tube was introduced to Europe, but had "slot" optics to minimize deflection defocusing. The two torodial windings remained, but a quad winding was also included to help with convergence.
- Now a hybrid yoke was developed retaining torodial vertical windings, but replacing torodial with a saddle horizontal winding to increase inductance, offering a better impedance match for transistor drivers. Beam-to-beam spacing was also increased to approximately 6.6 millimeters to permit a larger lens diameter for and improved focus. Simultaneously, focus voltage went to 28 percent of the anode voltage to further accommodate the larger lens diameter.
- Two years later multi-element focus designs with increased focus voltages emerged, along with improved gun designs and ST yoke corrections for North-South pincushion distortion, but not East-West.
- Shortly thereafter, field gradients in the yoke were altered with biased windings and flux shaping to solve E/W pin correction. But this not only increased yoke costs but introduced greater beam distortion.
- In 1980, according to Mr. Morrell, tube neck diameters were reduced from 29 to 22.5 mm to reduce both yoke costs and required deflection power. However, this change necessitated beam-to-beam spacing reductions to 4.8 mm and gave gun designers problems with good focus.

All this, of course, led directly to the evolution of Coty- 29 which remains in 1985 as the cathode ray tube mainstay in consumer-type receivers, although the Square-Planar tube (to be described later) should largely supplant it in time.

Coty-29 goals were to develop a system compatible with all tube sizes, 90- and 110-degree deflection angles, and of modest cost with good performance. Initial development was the XL gun which improved focus, reduced yoke size, narrowed beam-to-beam spacing, and permitted better convergence, whose errors are usually proportional to beam spacing. Although beam spacing reductions do increase screen tolerances, internal magnetic shields in these tubes substantially minimize earth magnetic fields effects.

Actually, two XL guns were developed for the Coty system (Fig. 6- 9), one with a HiPI gun for bipotential focusing, and the other having a double bipotential DB gun with two additional grids and a focus voltage of 31 percent of the anode voltage. This was done to develop the smallest beam diameter possible because all self- converging yokes cause "deflection defocusing."

Both guns have slot optics for beam forming

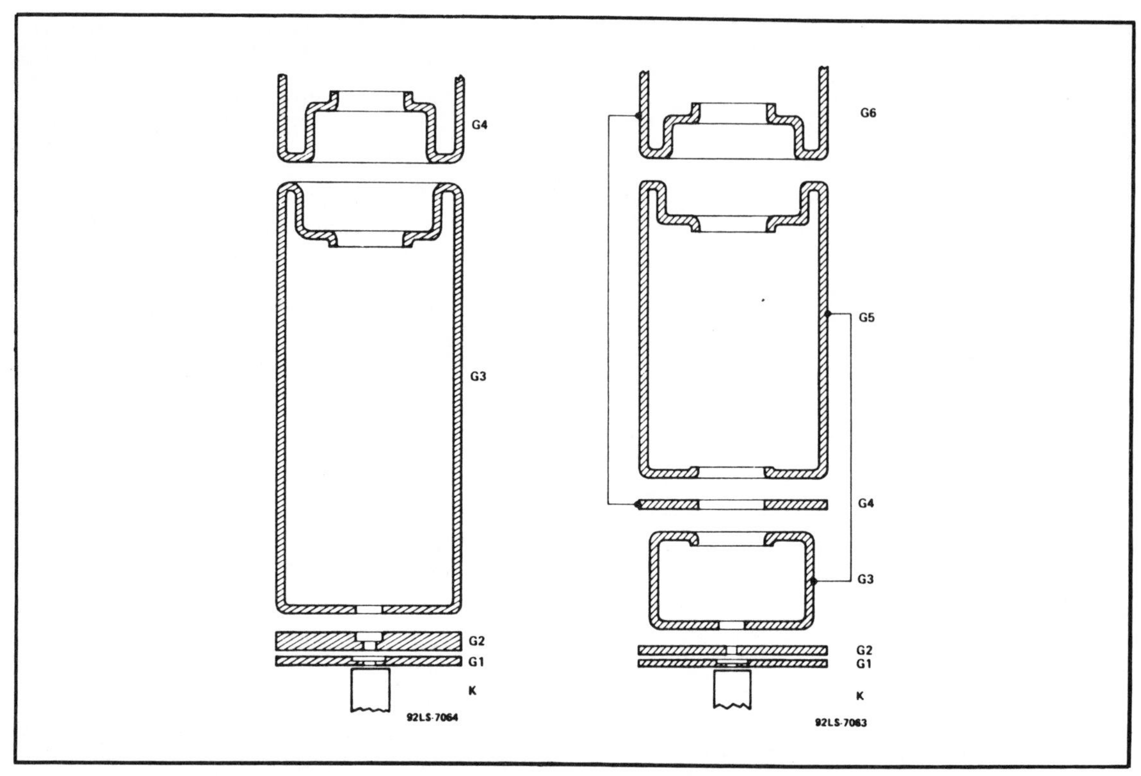

Fig. 6-9. Gun outlines of the XL Hil and XL DG guns (courtesy RCA Tube Div.).

to generate different focal lengths for horizontal versus vertical deflection, compensating somewhat for pincushion-shaped horizontal and barrel-shaped vertical deflection fields.

The main lens is larger and reduces "aberrations," and tube neck diameter remains at 29 millimeters. Both highlight spot sizes and flare are reduced in the design, and optional parameters for both the XL lens and slot optics provide best center and edge focus, significantly improving any yoke defocusing action.

The yoke was both computer and experimentally designed to "closely" match the electron beam path, has lesser material content, reduced power requirements, and is comparable to previous yokes designed for 22.5 mm neck diameter tubes. The same is true for the 110° version because of lesser beam-to-beam spacing and the new funnel contour, resulting in a 10 mm overall tube length reduction for both types. Deflection power requirements have, as a result, also been reduced as well as a 3 KV saving in high voltage in certain circumstances.

Next, the Square-Planar Tube

Not to be outdone by on-shore or off-shore competition, RCA in 1984-1985 has introduced its Square-Planar color picture tube to take advantage of psychophysical eye properties to create the illusion of a flat picture. In so doing, engineers at the Video Component and Display Division, Lancaster, Pennsylvania have developed a new face panel and a shadow mask to match. Prime problem was faceplate thickness which would result from attempting to manufacture a flat front surface. According to F.R. Ragland, Jr., R.J. D'Amato, and R.H. Godfrey, if screen edges were planar, the center portion of the faceplate could "deviate substantially" from the horizontal plane without changing the flat screen perception.

We won't go into all the details, but the end result produced a "perfectly" rectangular screen along with conventional faceplate curvature. For a

27-inch tube, for example, they found that the screen may have a curvature radius as small as 4 meters and still seem to be planar. Further, a 26-inch SP design with a 110-degree deflection angle can have an approximate radius of 1 meter along both minor and major axes and along the sides. In addition, a center panel thickness of 15.3 millimeters would satisfy center stress requirements, or just over 1/2 inch.

Increased front panel radius also affected shadow mask performance when receiving high beam current. Radius curvature was varied from center to edge for flatter contour towards the center and sharp curvature near the edge. There was also a variable horizontal slit pitch introduced for additional major axis curvature. This new mask's performance is said to approach that of the standard spherical mask and, with an anti-doming coating, reportedly is superior to the standard product. RCA Lancaster further says the new design offers a "standard for the development of new families of color displays." Certainly it presents a very square, full picture approaching scenes you would ordinarily appreciate with the naked eye. A picture of the new tube appears in Fig. 6-10.

Fig. 6-10. New planar, highly rectangular 26″ picture tube and accompanying receiver announced by RCA Group Vice President Jack Sauter (courtesy RCA Consumer Products).

The Coty-29 Yoke Design

Having already referred to the hybrid yoke arrangements used with modern color picture tubes, there's no better place than this to describe the Coty-29 product as it currently exists, with most credit going to J. Gross and W.H. Barkow of RCA's Princeton Labs.

Picture tube neck diameters were the initial problem—principally to reduce the large 36 millimeter color tube necks to 29 millimeters so that bulky, long deflection yokes might not only be shortened but contents reduced considerably. Then with the XL gun, its 5.1 mm beam spacing and full ultor (accelerating) voltage, considerably better resolution resulted.

The large-necked tubes come as part of the 120-degree delta gun arrangement and tri-dot phosphor patterns for both round and early rectangular faceplate CRTs. Here, average misconvergence *decreased* with *increased* yoke lengths. Torodially or saddle-wound yokes predominated in early line color tubes, also, and created trap or astigmatic errors in their four corners. It was also customary to reduced horizontal deflection stored energy by using long horizontal windings about ferrite cores. Once more, length and not efficiency predominated.

Smaller beam spacings in the Coty-29 tubes allow beam predeflections in their necks where the highest stored energy is located. A tradeoff was then made between elongating horizontal coils in the direction of the gun and extending the flared portion towards the screen. Four permanent magnets for the screen corners helped take out the negative residual traps and also by moving the horizontal deflection center towards the screen and the vertical deflection center towards the gun. Higher harmonics of coil windings was then reduced so that residual convergence errors became less than 1 millimeter, but also allowed use of minimum magnitude higher harmonics, lessening high-order raster and convergence errors further. Ferrite core length was further optimized for best electro-optical performance and smallest vertical power consumption.

The finalized Coty-29 system now offers the same convergence numbers as its open-throat forerunners but without either internal or corner magnets and comparative statistics as shown in Table 6-1.

Actually, this was a two-system development, with the open throat design leading the way to the ultimate Coty-29 version that's now so popular throughout the country. Figure 6-11 tells the story.

PHILIPS ECG CFF LENS

Another 1984 innovation and advance was also supplied by North American Philips (heretofore Sylvania) of Seneca Falls, New York by releasing its Conical Field Focus (CFF) Electron Lens. This, along with velocity scan modulation, has given brother company North American Philips Consumer Products (Magnavox, Sylvania, Philco) a worthwhile competitive position. In addition to those three brands, it is also understood that Sharp, Sanyo, and Toshiba have ordered tubes with this feature.

According to optics and engineering personnel at ECG, reduced spacing in the new, narrower-necked tubes required better performance electron lenses, especially nonoverlapping lenses in the smaller 22 mm guns. This led to the investigation and ultimate development of one-piece lens ele-

Fig. 6-11. 110° Standard and 110° Coty-29 deflection yokes photographed side by side (courtesy RCA Tube Division).

Table 6-1.

	36 mm Saddle/Saddle	29 mm Semi/Torodial Open Throad	29 mm Semi/Torodial Coty-29
Horizontal coils	230	197	157
Vertical coils	350	170	76
Internal field formers	13	-	-
Astigmatism magnets	-	1.8	-
External field formers	(corner magnets)	1.3	-
NS pincushion magnets	31.2	6	2.3
Ferrite cores	514	308	220

ments in a deep-drawn design that turned out to be not only inexpensive but highly repeatable in production, ensuring uniformity in the finished product. Engineers E.I. Zmuda, D.L. Say, and B.F. Lucchesi claim that the new lens has an improved lens diameter between 20 and 30 percent over regular lenses.

Called a CFF (or Conical Field Focus) arrangement, the one-piece assembly (Fig. 6-12) derives from a solid piece of metal with overlapping cones. When curve depth is maintained at no more than 33 percent of cone diameter, "freedom from astigmatism and similarity of the three spots" can be maintained in a one-piece lens element. There are no demanding preassembly steps and lens integrity is maintained throughout.

CFF design and tube-gun assembly operations. Figure 6-13 shows plots of the electron beam in horizontal and vertical planes between screen and lens. Although lens "regions" differ considerably in the two planes because of overlap at the screen, both beams converge to nearly equal focus and spot sizes without astigmatism.

The CFF system is offered by Philips in both standard 29 mm and 22 mm sizes in 19- and 25-inch and 13- to 19-inch diagonal sizes, respectively. Later, the same CFF technology is to be extended to the squarer and flatter 26-inch tubes scheduled for production during late 1984 and 1985.

Both color picture tubes in the 22.8 and 29 mm neck series have in-line guns with bipotential focus, internal magnetic shields, and usually a choice of 52- or 85-percent light transmittance, slotted shadow masks and Chromatrix II lined matrix screens for best contrast and reduced sensitivity to vertical misregistry, in addition to a variable pitch mask. All are matched to either T-Type or H-type hybrid deflection yokes for smaller spot sizes with improved picture resolution. Static convergence and purity devices are mounted on the necks of these tubes with permanent magnets to converge outer beams on the center beam. The purity magnet compensates for extraneous magnetic field effects and generates a magnetic field perpendicular to the tube axis to move beams horizontally. The yoke, itself, may be adjusted by tilting in vertical or horizontal planes for best convergence, as well as a small rotational adjustment. Picture centering develops from passing direct current through each pair of deflecting coils as required and specified. Beam registry on phosphor lines results from adjustment of the purity magnet, and static conver-

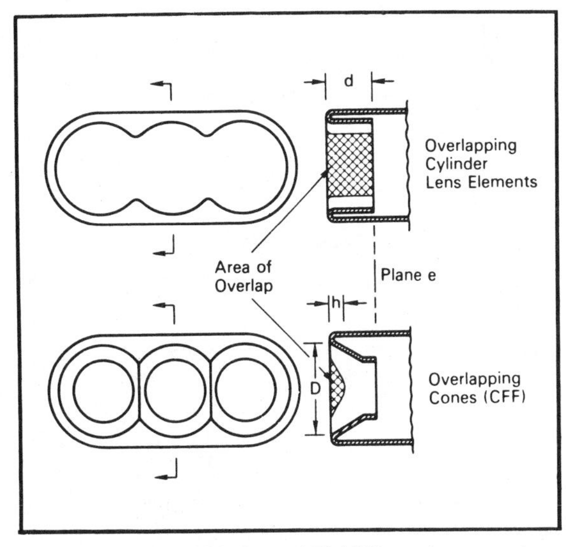

Fig. 6-12. Philips ECG Conical Field Focus in a one-piece assembly. Cones and cylinder overlap, with area compared (courtesy Philips ECG).

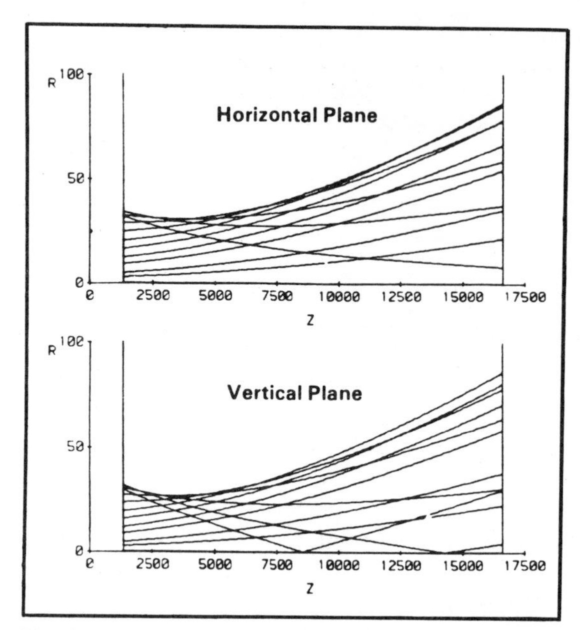

Fig. 6-13. Electron beam plots in dual planes (courtesy Philips ECG).

gence takes place as permanent magnets on the static converger and purity device are adjusted.

As do most, if not all picture tubes today, require that video must be applied to the three cathodes as RGB information, with brightness levels included. Automatic degaussing takes place as a gradually diminishing alternating current passes through two series coils symmetrically placed, on top the glass funnel and produce a vertical cross-axial field.

The tri-color, phosphor-lined screen consists of an array of narrow, closely-spaced phosphor lines arranged in groups, with each line separated by a thin, black opaque coating. Groups consist of red, blue, and green phosphors aligned with appropriate slots in the shadow mask. The opaque coating in high transmission face panels delivers higher brightness than non- matrix tube types with similar contrast ratios. With lower light transmission panels, the opaque material results in greater contrast than equivalences without it.

Pigment in the phosphor system consists of Europium-activated Yttrium oxide for bright reds and certain alternations in sulfide phosphor compositions for blues and greens. By tinting red and blue phosphors, ambient light is selectivity filtered and only the phosphor's characteristic color is reflected. A measurement diagram for one of the newer sized 22.8 mm neck, 19-inch tubes is shown in Fig. 6-14, just to give you an idea of the way the glass bulbs and mounting brackets are specified. Inches are above the lines and millimeters below. You will note this is a 90-degree deflection tube with the customary long neck and larger bulb than the 100 and 110-degree versions. To give you an idea of what a 100-degree, 25-inch tube looks like and how it, too, is specified, one is shown in Fig. 6-15. This one has a 29 millimeter neck with either 52 or 85-percent light transmittance, according to user preference. The E_{A2} anode accelerating voltage is specified in this tube as 30 kV.

JAPANESE CONTRIBUTIONS

Probably the prime contributors in offshore cathode-ray tubes in recent years have been Mitsubishi and Sony. While only Sony uses its so-called one gun color tube exclusively in consumer TV, the company has made system innovations that are important, especially beam modulation. Therefore, fairly concise reports should be forthcoming for each, explaining at the least the better points of selective efforts. As for Mitsubishi, this very large corporate entity has been a specialist for years in picture tube innovations, particularly projection tubes, and is also showing innovations in direct view CRTs as well. Both, therefore, deserve space in any worthwhile picture tube review, even though full details may not be available because of shortfall technical literature distribution either here or abroad. We'll glean what we can, however, from the IEEE Transactions on Consumer Electronics—and that should offer a reasonable modicum of enlightenment. First, Mitsubishi.

Mitsubishi

A giant manufacturer of automobiles, airplanes, and sophisticated electronics has recently signed with Westinghouse in a joint venture to build cathode-ray tubes in New York state. At this writing, the portent of this action is not known and proba-

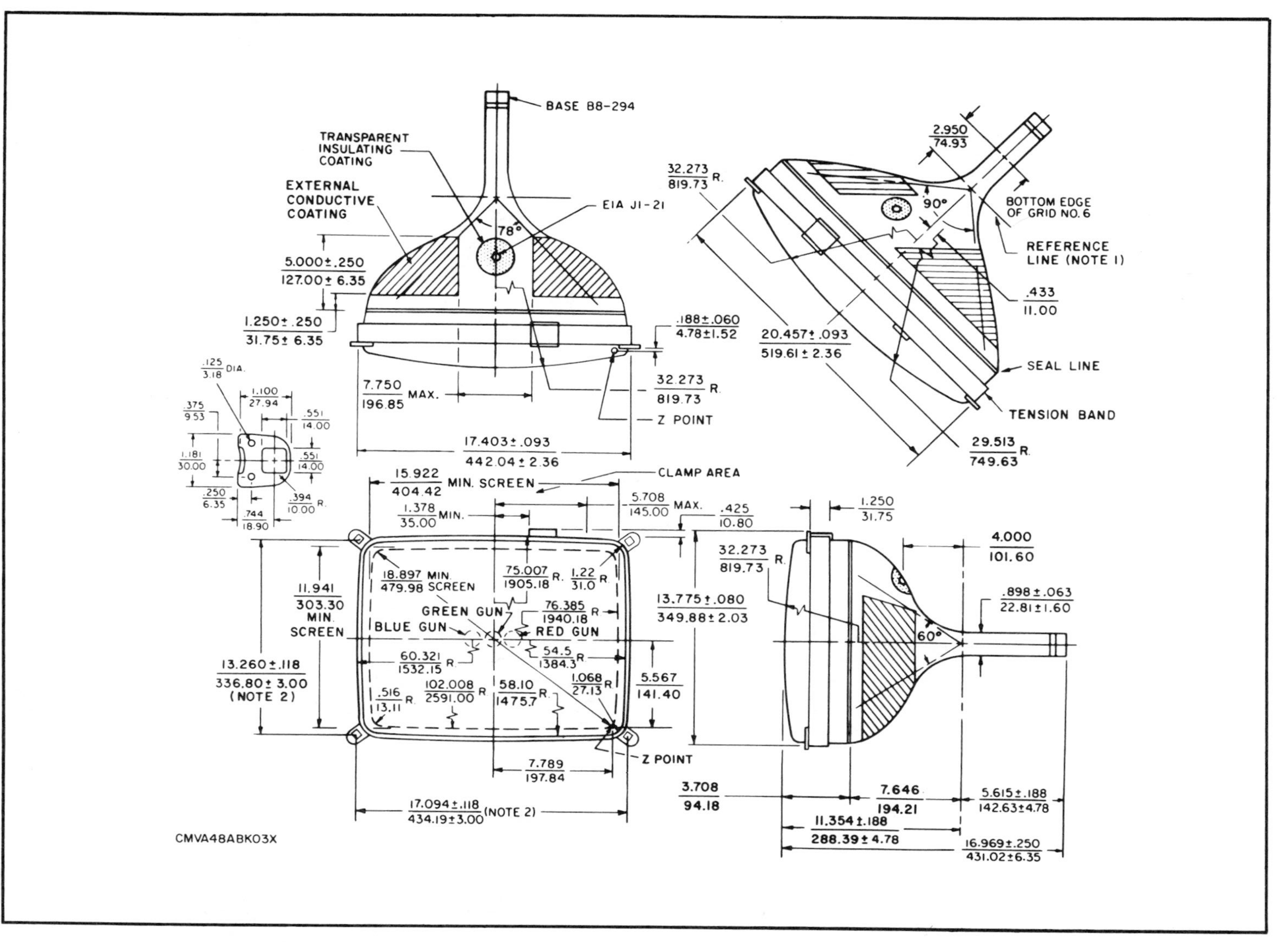

Fig. 6-14. A 22.8 mm neck 90° deflection CFF tube (courtesy Philips ECG).

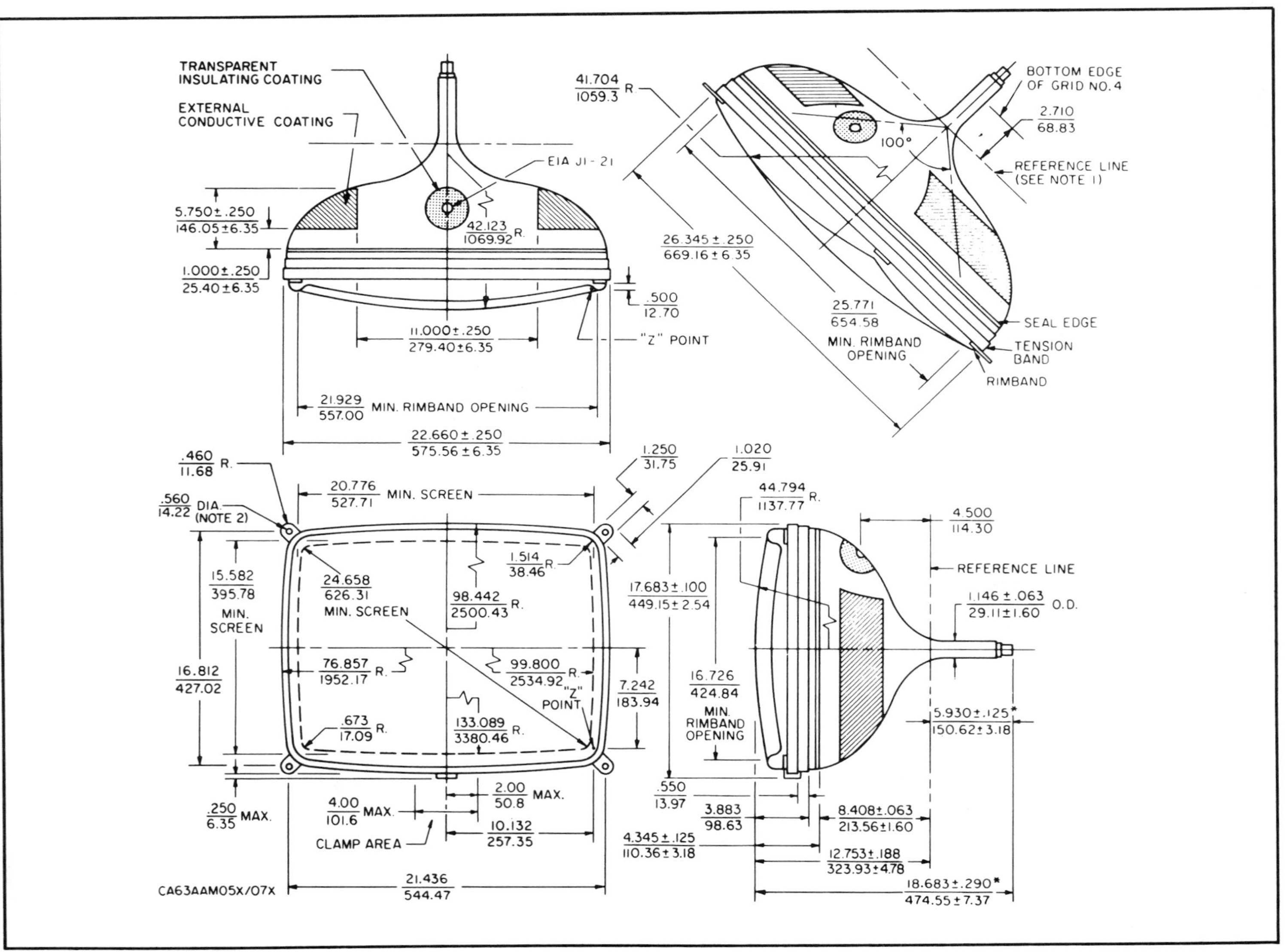

Fig. 6-15. A CFF 25-inch tube with 29-mm neck and 100° deflection (courtesy Philips ECG).

bly won't be until the factory (already in being) begins production on whatever products the two corporations have in mind.

In the meantime, Mitsubishi has introduced a new Diamond Vision™ tube with 0.4 × 0.4 millimeter apertures (slots) instead of the conventional 0.63 mm shadow mask openings, in addition to further refined faceplates and phosphors.

Leading up to this announcement, Mssrs. Shimizu, Kobayashi, and Iwasaki of Mitsubishi Electric Corp. have published an IEEE paper largely describing the phosphor and color fidelity research that preceded it.

To begin with they describe the brightness-reflectivity ratio of any television screen as:

$$BRR = Br/R$$

where Br is the tube light output, and R the tube face reflectivity.

Following introduction of the black matrix in the 1970s and also that of pigmented phosphor, further improvement in color fidelity was needed because of imperfect color reproduction, especially in the cyans, greens, magentas, and deep reds.

Initial effort was exerted to develop a new faceplate that showed selective light transmissions in the visible light spectrum. Afterwards, phosphors with compatible and complimentary characteristics could be matched to it.

The new faceplate contains a "small amount" of neodymium oxide (Nd_2O_3) and other "colorant ". The percentage being one percent, and it was also noted that neodymium doped glass varies in apparent color with various types of illumination. Such dichroism had to be reduced and also a modification of the faceplate color. This was done by the addition of chromium oxide (Cr_2O_3) that would absorb blue.

Now the problem was to attain light output at high efficiency, and at the same time improve phosphor colors. The blue phosphor matched well, the red phosphor also had its peak intensity coinciding with that of the faceplate and became even a deeper red. Green, however, which is often derived from red and blue in the television receiver's demodulator, was another matter. Activators had to be added to better control these emissions; therefore, green's peak intensity was established at 545 nanometers and its color gamut at 530 nanometers. With a larger color gamut (range) the tube offered better color fidelity, and so the change was considered having (Zn,Cd)S:Cu, Al phosphor for light output and a ZnS:Cu, Al phosphor for color gamut, with (Zn,Dc)S: Cu, Al being eventually used as the green phosphor.

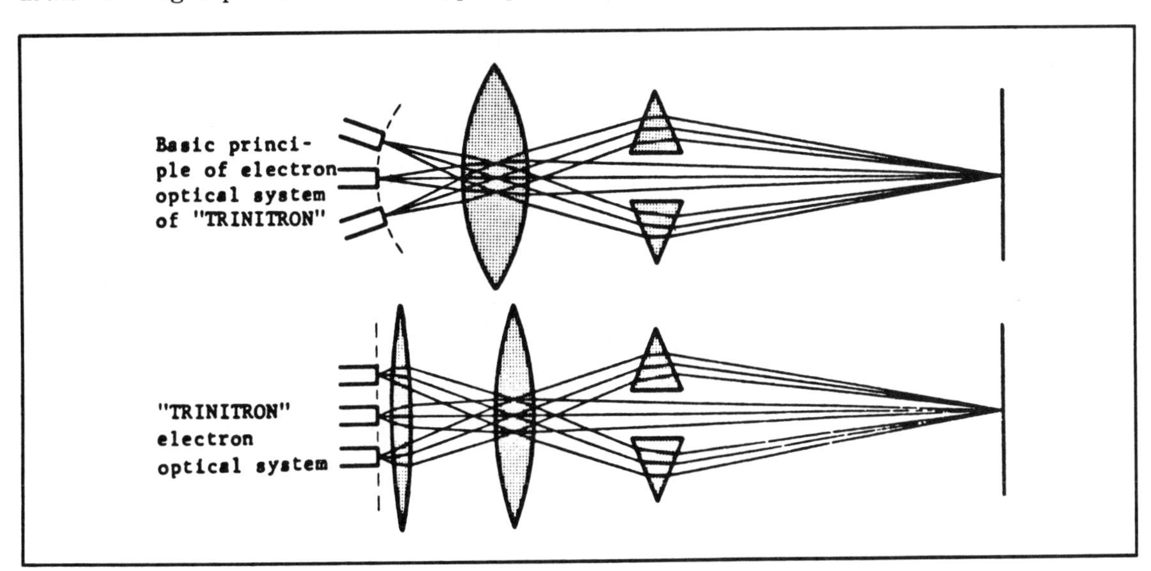

Fig. 6-16. Sony's Trinitron lens system (courtesy Sony).

Ambient light for the final measurement was a florescent lamp having a color temperature of 4200 °K which showed an increase of 40 percent in the brightness-reflectivity ratio (BRR) and much deeper tones for the three primary colors. The blue field, especially, was significantly improved, along with blacker blacks and considerably increased contrast. The authors further state that a multi-flower picture even offered a three dimensional effect as the various shades were displayed.

Sony's Trinitron

Although this tube has three conventional cathodes like any other color picture tube, it has a common lens system for the three beams that actually consists of two electron lenses with wide apertures and a pair of electron prisms for color convergence (Fig. 6-16). According to Mssrs. Yoshida, Ohkoshi, and Miyaoka of the Sony Corp. wide aperture lenses furnish extra brightness and picture sharpness with less aberrations.

A new aperture grill has also been developed for better beam transparency that's as easy to manufacture as a conventional shadow mask (Fig. 6-17). In operation, the three RGB beams cross one another at main lens center, with side beams diverging from their crossing points deflected by two electron prisms by a simple change of direction rather than additional focusing. In the illustration, three cathodes occupy their conventional side- by-side plane and the outer pair are deflected towards the center by a weak electron lens positioned forward of the cathodes. All three beams then cross one another in the main focus lens, called an Einzel lens. All three then pass through the electrostatic deflector and the aperture grill to the striped phosphor screen, striking it at desirable incident angles. The electrostatic deflector consists of four parallel plates, with both inner and outer plates connected to each other electrically, and inner plates "given the same potential" as the phosphor screen. The outer plates are actually at negative potential with respect to the inner plates and a difference of several hundred volts separates the two pairs.

Tube convergence consists of shifting one or both outer beams sideways for proper registration.

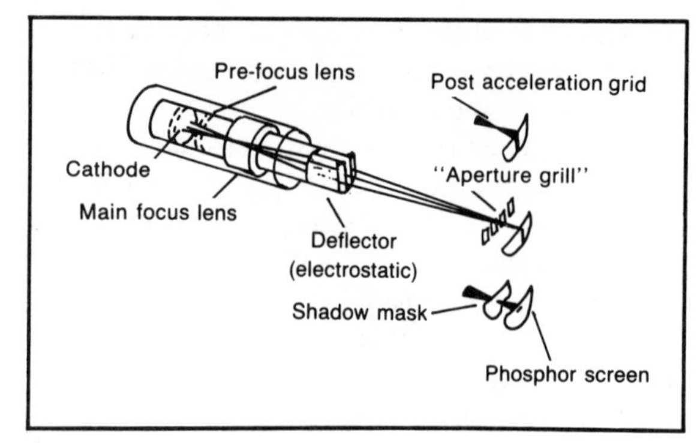

Fig. 6-17. Cathodes, lenses, and aperture grill of Sony's Trinitron tubes (courtesy Sony).

To do this, adjustment of the right and left electron deflector plates occurs by applying a synchronized parabolic deflection wave. As the authors explain, since the ratio of convergence voltage (difference) and deflection plate peak voltage is always constant, separate adjustment of the two voltages is unnecessary. The three Sony engineers claim 20 percent additional electron beam transparency at grill center, and about 15 percent at tube corners and is free from Moire ripple effects.

VELOCITY MODULATION

Better known as beam scan velocity modulation (VM), this principle (as does so many) goes back to 1951—and probably beyond—but is now receiving due recognition by both Sony and Magnavox (North American Philips). While not strictly a cathode- ray tube subject exclusively, its description in this chapter will probably invite more attention than in CRT driver and color demodulator portion of the book because of its electronic effects on the resultant picture.

Sony's Fuse, Yamanaka, and Saito did a computer simulation of the Trinitron system and its beam spot shape to begin the analytical approach using two axially symmetric functions. The general idea was to develop a sharper picture by modulating the CRT's beam scanning velocity with brightness variations of the original picture, at the same time modulating beam current intensity with the incoming television signal.

After a fair amount of calculus and fast Fourier transforms, the three analyzers reached the conclusion that output picture amplitudes of at least one frequency component of the final picture increases as frequency increases. There were, however, some serious disadvantages and so a difference instead of a differentiation signal was used and the original video signal passed through two identical delay lines. One signal was then used to modulate beam current intensity, with the difference signal modulating beam scanning velocity. This aids spatial frequency response of the human eye since it peaks at middle frequencies. At the same time, large amplitude variations also exist in television pictures and blooming is often visible. With VM, however, beam spot size does not increase substantially and blooming is controlled satisfactorily.

Magnavox, meanwhile, in its 19C4 and 25C4 Phoenix III receivers claims sharper pictures, better small area contrast, and improved (apparent) focus.

The Magnavox version actually changes the horizontal scanning velocity during picture transitions by speeding up and then slowing down electron beam velocities so that black/white transitions occur more quickly, and vice versa for white/black transitions, which would then develop more slowly, giving the appearance of additional picture sharpness. It's all done with phase advance.

Scanning velocity changes in this system result from an auxiliary deflection coil mounted on the neck of the picture tube aft of the deflection yoke. A scan velocity modulation circuit drives this coil from a signal whose strength varies as picture information changes. There's also a noise coring arrangement preventing undesirable amplitude noise effects in the picture.

The circuit for this scanning velocity modulation (SVM) consists of an AGC blanker and detector, differential amplifier, buffer, a second differential amplifier, an SVM driver and a pair of push-pull outputs to the SVM coil, identified here

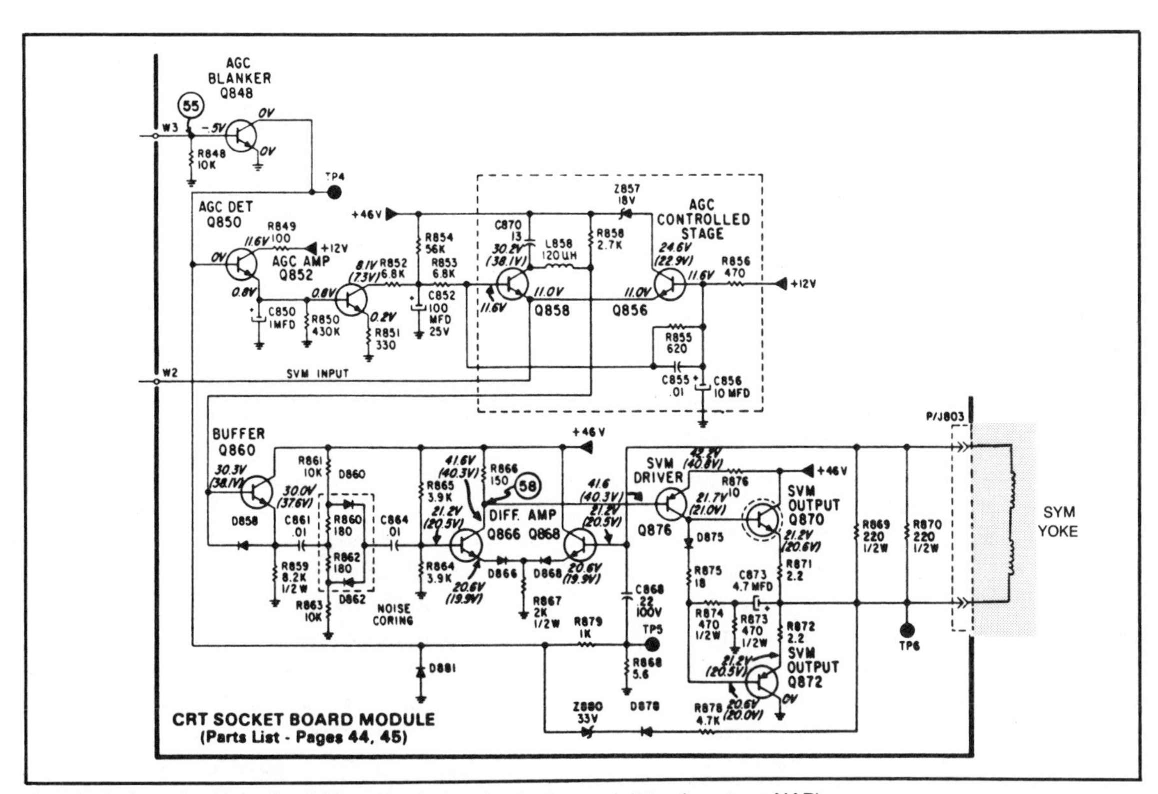

Fig. 6-18. Actual schematic of Magnavox' signal velocity modulator (courtesy NAP).

as a yoke (Fig. 6-18). Inputs to the 5-transistor AGC blanker and processors comes from a vertical pulse that is used as a bias source for the main differential amplifier (described later) and also as a heavily filtered dc source for the first differential amplifier via the detector and buffer transistors preceding it.

Main signal input for SVM originates from the Peak SVM Driver and the comb filter/auto sharp board on its way to the main video amplifier. Due to biasing, filtering, and differential action, highlights of this video reach the input buffer below and are routed to the main differential amplifier in either positive or negative-going information through the two equally-biased back-to-back diodes.

The main differential amplifier reacts to this more negative or more positive signal voltage by conducting, selectively, varying drive to the SVM driver and its push-pull outputs. The waveform at TP6 shows a small negative-going blanking signal at the vertical rate that would be used along with video levels to establish black/white/black transition times for variable velocity scanning as well as AGC-controlled amplitudes.

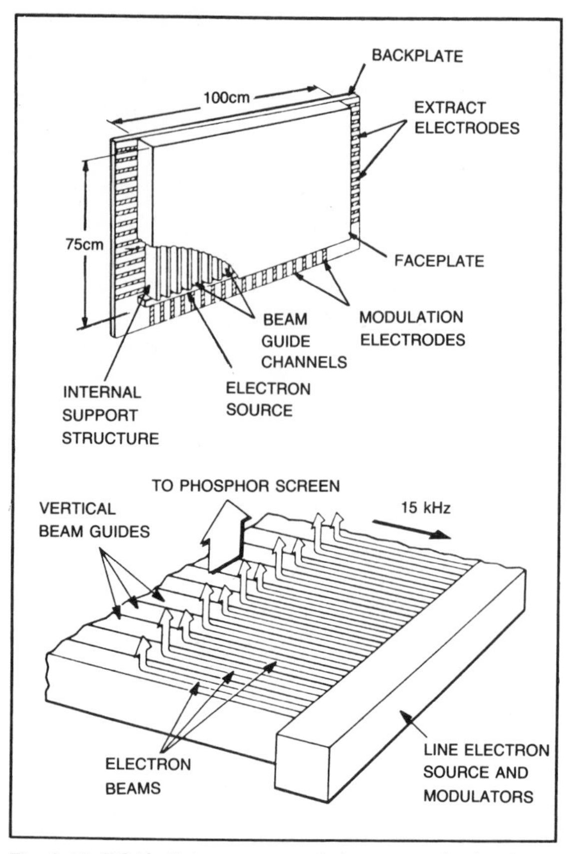

Fig. 6-19. RCA's flat screen panel shows promise but is on back burner for now while liquid crystal research proceeds (courtesy RCA Laboratories).

PICTURE ON THE WALL

This advance is still somewhere down the road just waiting for a whole boat load of venture design bucks, a few technical breaks, and the usual vagaries of the marketplace, however and whatever they may turn out to be.

RCA's much heralded flat screen panel measures 75 × 100 centimeters in full color "matching or exceeding those of today's best receivers." Less than 4-inches thick, this unit was to have required almost 300 W and weighed less than 110 lbs, at least these were the design goals. The unit (Fig. 6-19) consisted of two flat panes of glass and four side walls. Inside are electron sources and beam guides, replete with internal support ribs capable of handling atmospheric pressures of more than eight tons per side. The back of the unit has electrodes and the faceplate coated with color phosphors—all part of the brainchild of Dr. Thomas L. Credelle, Head, Advanced Display Systems Research, David Sarnoff Research Center, Princeton, N J. He used multiple electron sources with a single modulation point for each beam guide. Information may then be shown one line at a time. These beams flow down a guide until they find an energized "extract" electrode; and then a negative voltage at this point deflects the beam 90 degrees directly toward the phosphor screen.

Operationally, for each color a single line of video is stored and then divided into 40 sections, while the following line clocks out in parallel but more slowly. Beam currents are distributed within the various modules by scanning the extracted beams via digital processing and 6-bit A/D converters. Vertical lines are selected by the 15 kHz commutator, a group of high-voltage switches. Dr. Credelle says that although this system still shows promise it is going on the "back burner" for a while and RCA is now diligently investigating liquid crystal displays for which there may be considerable and favorable developments in the forseeable future, including low power.

The first miniature flat-screen color television receiver was introduced at the 1984 Chicago Consumer Electronics Show by Hattori Seiko Co. of Japan. Measuring some 6 × 3 × 1.2 inches, it has a 2-inch diagonal screen and retails for around $500. Unlike slow watch and calculator crystal readouts, however, Seiko deposited thin film transistors at each of the 52,800 pixels on a single glass substrate so they can be individually switched on and off. Because each pixel answers to a single electrical impulse, response times are better and contrasts more acceptable. The usual red, blue, and green filters are used for the colors and cover each pixel electrode. For a blue field, for instance, red and green transistors are turned on and only blue excitation may pass. Obviously, a 2-inch screen won't result in a demand product, nor will some of the overall performance—just now—but, there is more than novelty promise apparent, and there may be considerable effort on the part of a number of manufacturers and their R & D establishments to further refine and develop such principles. Apparently RCA thinks so.

Toshiba Has a Larger B/W LCD Display

This Japanese company has also announced a 12-inch diagonal liquid-crystal device display that can be used for message presentations, computers, business machine terminals, etc. With a viewing area of 253 × 194.6 millimeters, the LC display has been manufactured to replace CRT monitors and will show graphic patterns as well as numerals, alphabets and symbols up to 25 lines × 80 characters in monochrome. Claimed are fast response times, good resolution, light weight, and low power, with a single operating voltage of +5 volts. Weight 800 grams.

Divided into two electrically isolated blocks, the LCD consists of 640 horizontal × 200 vertical dots, for a total of 1280 data lines. Black figures on a green background offer a satisfactory 7:1 contrast ratio, and the LCD module is driven by a 1/100 duty ratio multiplex technique.

A system block diagram appears in Fig. 6-20, showing video, clock, and horizontal/vertical sync inputs into the timing logic, A dc/dc power supply and 64K RAMS. Outputs from timing derive as serial data, latch and clock pulses, a framing signal, the LCD display panel and its LSI upper and lower drivers.

According to the Kawasaki, Ihara, Kuramochi, Sakayori and Fujii paper published in the Nov. 1984 Consumer Electronics proceedings, newly-developed CMOS X-Y LSIs drive the LCD panel with a 1/100 duty ratio and an 18-V swing. Since the LCD is driven by a "two-line-at-a-time" method, video must be converted into an 8-bit parallel data format, transmitted and then stored in two 64K static memory RAMs. The readout, then, is into the upper and lower blocks and onto the nonglare front faceplate to increase contrast and reduce ambient glare.

Although this display is not in "living color" yet, it's not wild imagination to suggest that diligent work in the not too distant future will make it so. Hence, at least one type of picture on the wall seems plainly visible already.

PROJECTION RECEIVER CATHODE-RAY TUBES

Except for a predictable expansion in certain high-voltage and video-output electronics, projection and direct view receivers are very similar, with prime differences being only the optics. Improvements in both brightness and luma-chroma

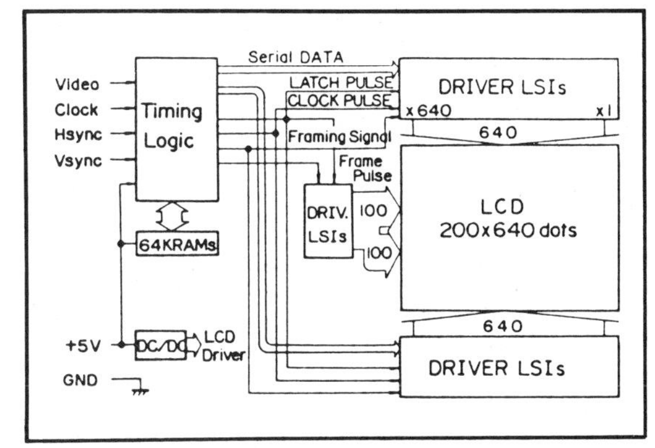

Fig. 6-20. Toshiba's 12" B/W LCD display (courtesy Toshiba).

resolution are now, however, the rule rather than the exception, and projection sets are competing closer and closer on an equal basis with their counterparts, electrically, but not in price. Also, 35- and 40-inch direct view tubes have been recently announced by several Japanese manufacturers, and these will certainly have an effect on the projection receiver market where approximate screen sizes are in direct competition. Unfortunately, however, the larger the projection front and rear screens become, the dimmer the image; therefore, except for monitor projectors that can use the surface of a smooth wall for their pictures, the outlook for projection TV at this time isn't substantial except in certain expensive commercial applications where giant screens are a necessity.

Regardless, you should know at least some of the latest developments, and these we'll report upon in available detail.

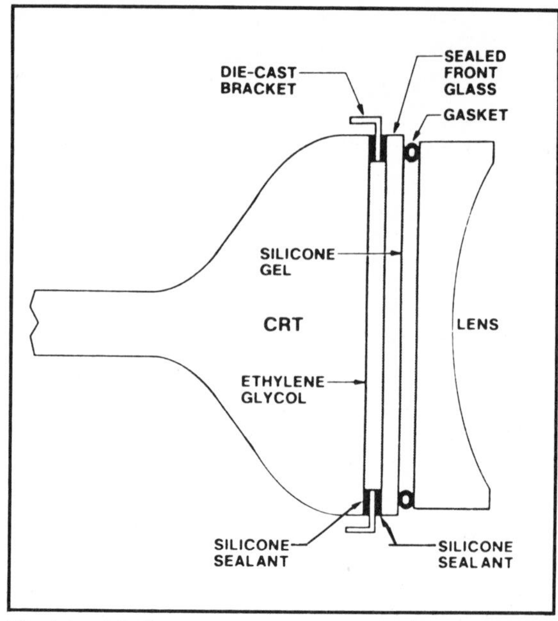

Fig. 6-21. RCA's 7-inch projection tubes with liquid cooling and silicone gel coupling (courtesy RCA Consumer Electronics).

Projection Optics

With the recent introduction of liquid-cooled cathode- ray tubes and liquid-cooled optics, 165 ft. lamberts of picture brightness (nominal peak white) and a contrast ratio of 60:1 has already been achieved, and there's more to come as even better optics and receiver chassis are developed.

According to RCA, contributing factors are a better lens coupling system, greater high voltage, and an improved black-striped 45- inch screen. In projection TV, of course, there are no shadow masks. To learn more about these developments, let's take them in order, beginning with lens construction and cooling. A drawing of the new tube and its contents are shown in Fig. 6-21.

These are 7-inch cathode-ray tubes receiving 32,000 volts accelerating at zero beam current that normally operate at 1.1 mA. Because of additional high voltage "drive," anode leads are attached to the CRTs with a silicone sealant to reduce chances of arching. With such higher accelerating voltage and accompanying electron flow, these tubes develop considerable heat. So to improve working life, liquid cooling in the form of ethylene glycol fills space between a front glass panel and the CRT to heatsink developed heat to a surrounding metal bracket. The result: better brightness, contrast, and cooler tube operations.

Then, to reduce light losses between the CRT's faceplate and the first lens, a silicone gel solution which eventually turns into a paste, bonds the lens to the CRT assembly. Lens and CRT are separated by a rubber gasket before being attached to an aluminum mounting bracket which aligns the tube for maximum dynamic focus of the picture. Replacement tubes are supplied complete with bracket and gel so there'll be no problem in field assembly. Any loss of fluid may be replenished through access holes in the assembly or by changing the entire unit if there are aggravated problems. Low fluid levels are usually visible as off-color line and poor focus in the lower parts of the screen. Improvements are especially listed for RCA-s J-line of receivers.

We also understand from Japanese Hitachi/Panasonic sources there have been certain phosphor improvements that now make liquid cooling unnecessary, in addition to a light output of 200 ft. lamberts. This information, however, seems proprietary at this time.

Chapter 7

Power Supplies

OFTEN REGARDED AS ONE OF THE MORE MUNdane topics in color television, it isn't so anymore. In the dark ages of the 1960s and 1970s, nice, hefty, expensive ac line transformers used to isolate receiver, test equipment, and maintenance personnel from hazards of the "hot" chassis, at least among the better sets. To avoid half-wave and full-wave ripple filtering, added expense, and unnecessary bulk, the day of the switch-mode power regulator and supply dawned and has been with us ever since. This is not to say that there aren't still half- and full-wave rectifiers floating around with one side of the ac line to hot chassis— there are— but not in the better receivers, and especially those that are TV/monitor combinations requiring maximum isolation from the ac mains, as our British cousins are wont to declaim. For in the switch and regulator mode systems, power generation often derives from the 15,734-Hz horizontal frequency, followed by a transformer, then half-wave rectification and filtering, which makes this a much easier task at that frequency. There are also such items as optical couplers (really diodes) that, in time, will probably be able to handle sufficient current to do a good job of complete line isolation at relatively reasonable cost. Eventually, or so it seems, this critical problem of receiver-power company isolation won't be a big deal at all but, for the time being, it will be a worthwhile subject to investigate. In this study, of course, we'll begin with a small "re-hash" of what used to be, just to bring you up to speed. After that, we'll use recent receivers with "live" schematics to illustrate the different systems in current vogue; and, if possible, pick up any new trends that may be appearing shortly in the market.

Short dissertations on voltage doublers, active filters, SCRs in current supply and power shutoff modes may be interesting also— and they will be covered in detail.

Not precisely a "fun" chapter for either reader or author, it is, unfortunately, a necessity because the entire power supply structure is somewhat empirical along with a few substantial developments, especially in integrated circuits and precision regulation—which could be one and the same. In consequence, we might as well start with half-wave rectifiers and work up so that you will have a solid foundation for understanding some of the more complex electronics and circuits that are yet to come. Even a little knowledge is not necessarily a "dangerous thing" as some erstwhile academic wag pontificated in the past. But certainly substantial knowledge can become both a joy and a pot of gold as well. Fortunately, advanced electronics neatly fills the bill and the till. Let's make the most of it!

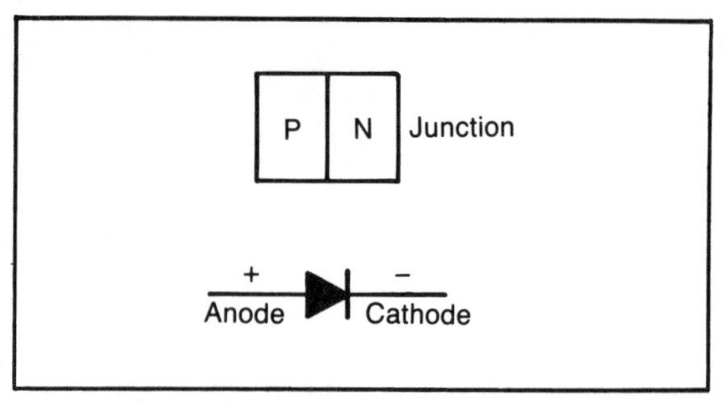

Fig. 7-1. Universal diagram of a simple P/N junction diode.

RECTIFIERS

Primarily, now, we're speaking of removing power line alternations and allowing smoothed dc current to flow into the various systems and circuits of receptive equipment as a full-fledged power supply. Power supplies exist wherever there are passive or active devices that need voltage and current for their operating source. Most equipment has a common negative to "ground" and therefore needs a positive operating voltage that will flow through the gear with sufficient current to return to the source and complete the circuit without undue loss.

By undue loss, we mean what is called a "stiff" current, not a weak, lossy one that will not sustain equipment operational requirements. Therefore, any power supply needs to be capable of sufficient *watts* $P = E \times I$ (dc or rms) to furnish good operating potential plus something extra to withstand the rigors of future leakage, shorts, and opens that are the inevitable heritage of active electronics.

Because television receivers no longer have any tubes but the CRT, we'll spare you the agony of large glass bulb rectifiers, which have burned many of the unwary hands during the '50s and '60s, and concentrate on semiconductors which are far more efficient, longer lasting, need very little space, and require no high-current heater voltage.

The half-wave rectifier simply passes one half of an ac sine wave's alternation and bucks the other half (Fig. 7-1). If its forward current-handling ability (I_f) and peak inverse voltage (E_{peak}) characteristics are adequate, this diode will develop a pulsating dc current output that can be smoothed (filtered) and used as a positive or negative operating potential, depending on its direction and forward or reverse biasing. Such biasing may also be used to block an ac voltage or signal input when convenient, permit waveform clipping, clamping, or bypassing—all by the simple addition of a resistor or two and some available dc bias source. For utility in the diode family, a single PN junction silicon diode can't be beat (germanium diodes are hardly used any more except in unusual circumstances where low-voltage on-potentials require only a few tenths of a volt). Further, germanium is forever temperature sensitive whereas silicon is considerably less so, but does require between 0.6 V and 0.7 V turn-on even in the small signal category.

In discussing diode applications, we won't differentiate between standard power and small signal rectifiers other than to comment that power diodes are not especially fast reacting devices, do handle up to many amperes of current, while small signal diodes are usually used as either signal blocks or signal couplers, occasional level clamps, and video/audio detectors of one description or another. If you think that diode use has diminished since the advent of the integrated circuit, just look at any IC schematically and you'll be amazed at what you see. This is the prime reason we've taken the time to explain the uses and operations of the simple cathode/anode semiconductor without which solid-state electronics could scarcely survive.

All materials (in general) may be classified as *conductors* (such as wire, many metals, etc.), *semiconductors* (diodes and transistors), and *insulators* (porcelain, wood, plastics), depending on whether they can or cannot conduct electric current. This relates directly to the number of relatively free electrons in the material. Good conductors have many free electrons while insulators have few and their resistances are very high. In between, of course, are the semiconductors, which may be doped with impurities to reduce their resistance and make them considerably more useful.

When N and P-type semiconductor materials are joined, some of the free electrons cross the P/N junction and recombine with holes and vice versa, resulting in a space charge region called a depletion layer. The P-type then acquires a slight negative charge and the N-type has a small positive charge. This flow of current is called a *diffusion current* and results in equilibrium on either side of the junction (Fig. 7-2).

To induce further (and controlled) current flow in the diode, a biasing device such as a battery or pulse voltage must be applied across the diode to make it conduct. N-type material free electrons are attracted to the plus (positive) battery terminal, and P-type to the negative terminal. As either or both of these potentials increase in value, the diode's space charge region is narrowed until the energy barrier depletes and excess electrons from the N-type material can flow across the junction, continuing as long as an external potential is applied. Depending on the applied voltage, this may result in either high or low current through the semiconductor and it must be contained within its specified parameters. In the reverse bias condition, the space charge region becomes wider and no more than *leakage* current flows between the two types of material.

As the battery or pulse potential increases, however, electrons from the bias source will overcome the space charge region and flow freely creating what we know as conventional current flow, which is *always* in the direction of the arrow. You'll find this rule even more appropriate when analyzing transistors, for there the collector is reverse biased and the emitter current is the *sum* of base and collector currents.

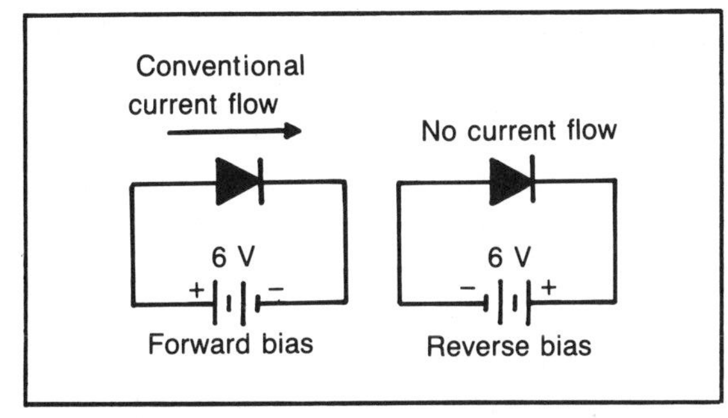

Fig. 7-2. Diodes forward and reverse biased do and do not conduct respectively.

Diode rectifiers may appear in half wave, full wave and bridge, or four diode configurations. There are also voltage doublers, triplers, and quadruplers, if you're interested in these also. Explanations follow.

Half wave rectifiers have really already been explained except to show subsequent filtering (Fig. 7-3). In this example, let's suppose we're generating a small radio or TV biasing voltage, with R_{in} the current surge limiting resistor, diode D, its filter capacitor C, "acting" choke R and the RL load. Here positive ac is blocked, negative ac passed and RC filtered, and delivered to the load as slightly rippling dc at 60 Hz. The larger C, the less ripple through R and R_L. Were another capacitor C2 added on the other side of R choke, there would be better filtering and less ripple. With a conventional inductor in place of R, the ripple factor would be even more improved. At a 60 Hz rate, however, this is the *least* desirable of all rectifier power supplies.

BRIDGES, DOUBLERS

The half wave *voltage doubler*, used largely in tube-type black and white TVs is another "hot" transformerless power source, but this time with better filtering and considerably higher output voltage (Fig. 7-4). On the negative alternation C_{in} charges through D1 to 150 Vac, on the positive alternation C1 discharges through D2, charging C1 and C2 by way of the L_{choke}. With all three capacitors fully charged, the output amounts to approximately 300 volts P-P, accounting for filter and choke losses as the power alternations are processed and smoothed, but still at 60 Hz output.

A basic *full wave* (120 Hz) rectifier circuit complete, with power isolating and partial transient suppressing transformer used for so many years in entertainment products power supplies—and is still

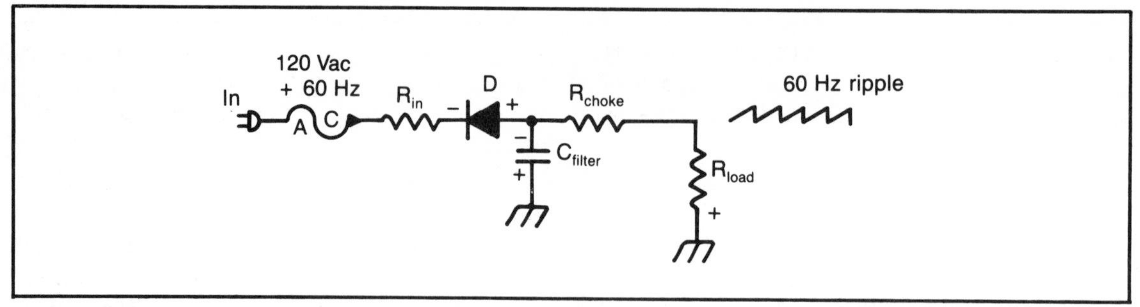

Fig. 7-3. A negative output half-wave power supply.

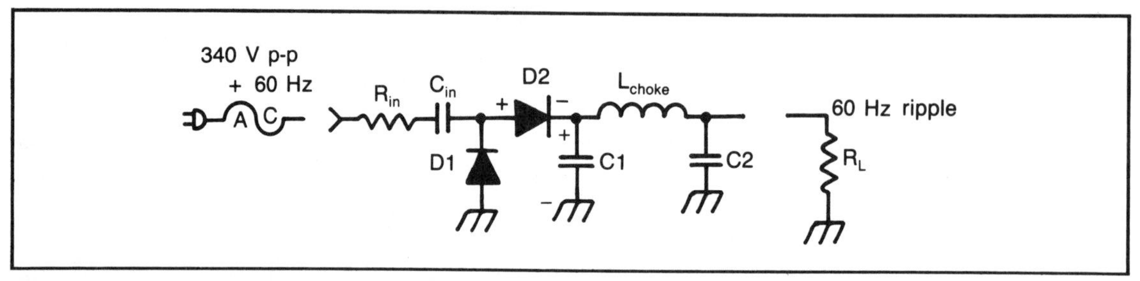

Fig. 7-4. A half-wave doubler fully filtered power supply.

found in VCRs, some portable TVs, and better radios, except that the bridge rectifier (four diodes) is preferred (Fig. 7-5). Such a transformer will step up or down the incoming voltage and completely isolate any chassis from the line. Center-tapped, D1 and then D2 conduct on positive half cycles, furnishing a 120 Hz full-wave rectified current at their cathode junctions. This frequency is considerably easier to filter than 60 Hz, and offers a more desirable supply. Seldom are there transformer failures in such a full wave circuit and when diodes require replacing, both should have substitutes of equal PIV and I_f abilities. Sometimes, also, a capacitor is shunted across the secondary, forming a 60 Hz resonant circuit and producing a square wave ripple output. With the secondary saturated, primaries may handle larger or smaller currents but the secondary will regulate rather well from 90 V to 130 V at about ± 5 percent swing.

The *full wave voltage doubler*—not often seen in consumer electronics—appears in Fig. 7-6. It consists of parallel diodes D1, D2, stacked capacitors C1, C2 and an output resistor that's quite important to circuit operation. On the positive half cycle

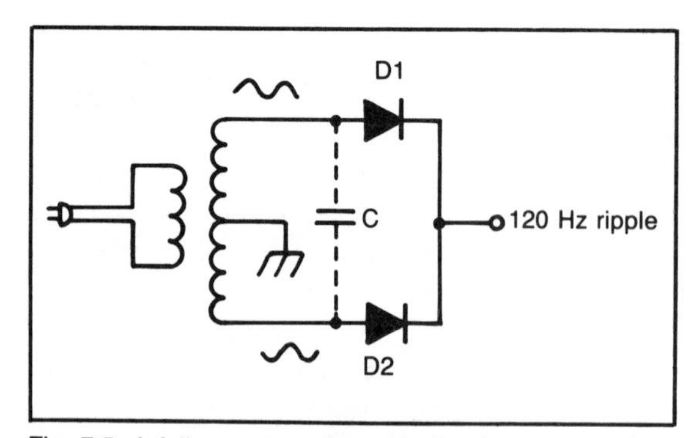

Fig. 7-5. A full-wave transformer-isolated power supply.

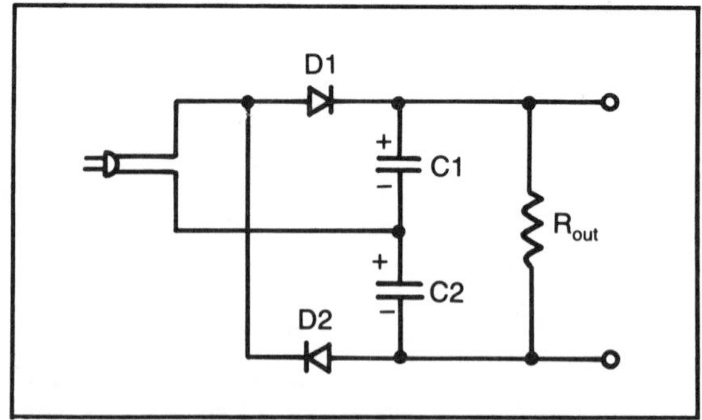

Fig. 7-6. Full-wave transformerless voltage doubler.

D1 charges C1, and on the negative half cycle D2 charges C2. Currents in the two capacitors then add and a resultant voltage develops across R_{out}. Should R_0 not have been included, both capacitors would have immediately charged to their peak potential. With R_0 however, one capacitor will discharge while the other is charging, producing a double-value output.

Full wave bridge rectifiers are now the most common power supply circuits and may be either transformer-isolated or connected directly to the ac line. Protected or "hot," they still deliver excellent power, are efficient, and produce a full-wave 120 Hz output that is easily filtered by a simple L/C choke-capacitor arrangement that's inexpensive. The transformerless version is illustrated in Fig. 7-7 since it is the more prevalent. If a transformer were used, the hookup would be precisely the same as the full-wave supply previously illustrated in Fig. 7-5.

The way this one works is a little more complex than the others because of the four diodes and their various polarities; so correlate the two straight and dotted arrows and we'll explain. On the initial positive alternation diode D1 conducts charging C1, and current returns to the line through D2. On the next positive alternation D3 conducts charging C2 and current returns to the top of the line through D4. Of course, as each set conducts, the other is biased off, delivering only positive polarity current to each of the two filter capacitors on either side of choke L.

Here you have a very efficient power supply that's hardy and relatively easy to troubleshoot

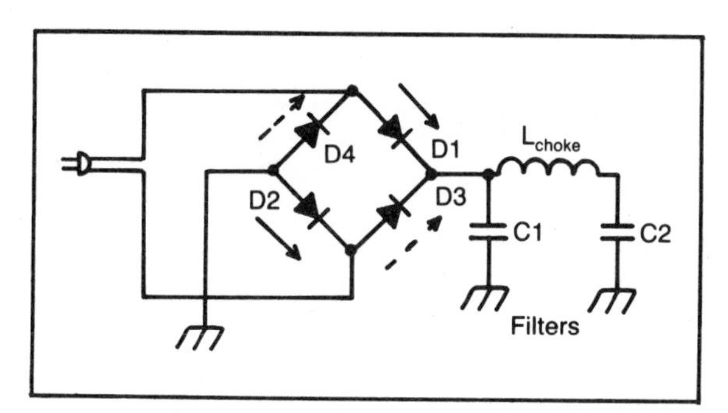

Fig. 7-7. Transformerless full-wave bridge rectifier—very popular.

(with an oscilloscope) since each diode is either conducting or not. But, if filters, choke, and the rectifiers are all working according to design, the results are gratifying. The one problem in all these transformerless power supplies is the lack of chassis isolation endangering both test equipment and operator. You'll need an isolation transformer every time the back comes off one of these "hot" receivers or you're taking the chance of a lifetime and, unlike craps, you can only lose once.

CLIPPERS AND CLAMPS

This portion of the chapter isn't profound, but in dealing with diodes you should have a working knowledge of what these bits of silicon will do when energized in certain circuits. They're also useful for small voltage drops—0.7 V per diode singly or in series and will abort any unwary waveform if biased and circuit-injected properly. In conjunction with zeners, some of the results are almost remarkable. To lead off, let's do a bit of positive and negative single diode clamping to let you experience "the feel" of what these little PN junctions have in store.

Take a simple RC circuit as in Fig. 7-8 and hang a diode across the output in the polarity shown, remembering the input is ac coupled and half the signal is above and half below dc reference. During the negative portion of the input, the diode conducts and charges capacitor C. With polarity wave reversal, the diode is back biased, capacitor C tries to discharge through R, but the time constant is long compared to the input wave cycle and so not much charge is lost. With additional positive half cycles, capacitance adds the input voltage to double the input magnitude, clamping the output at a positive level above zero or dc. Reverse the diode so that its cathode is to ground and you have a negatively-clamped waveform of the same proportions but firmly held *below* reference ground.

Diode limiting is equally simple, so suppose we show double diode limiting as a practical circuit (Fig. 7-9). With two different polarity diodes in the circuit and R1 the current limiting resistor, D1 or D2 will conduct more or less in response to the negative or positive polarity of the incoming waveform producing a limited (and clipped) response on the

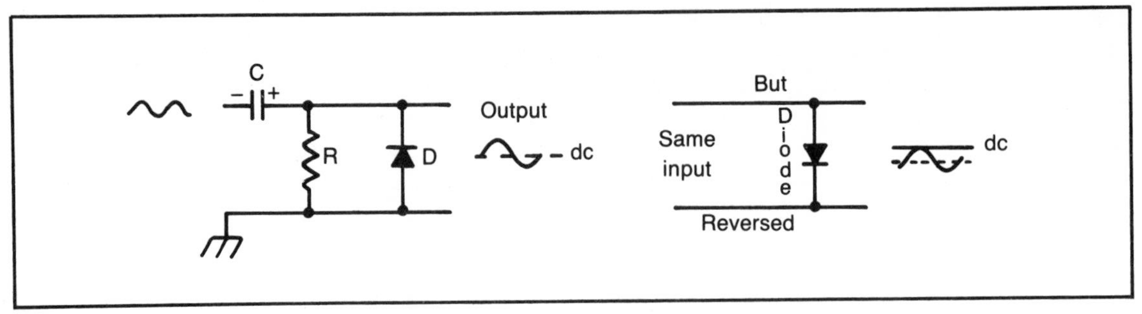

Fig. 7-8. Positive and negative clamping circuits.

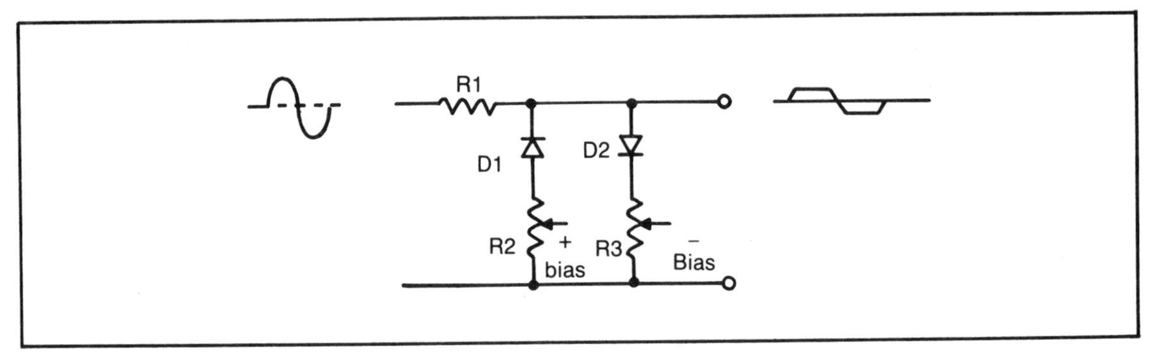

Fig. 7-9. A double shunt limiter controlled by ± dc applied through R2-R3.

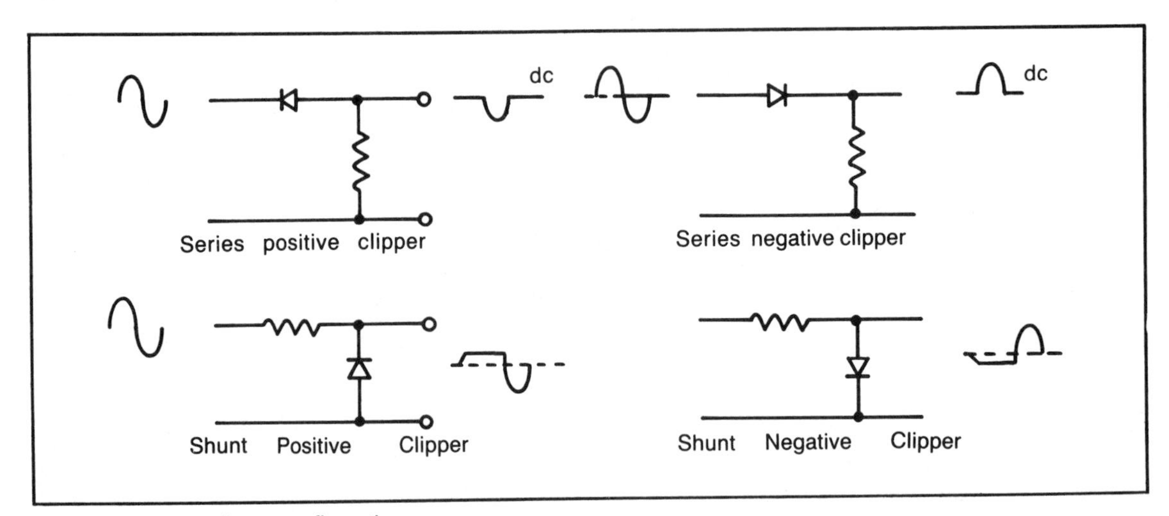

Fig. 7-10. Various clipper configurations.

output. Were these diodes shunted directly to ground instead of separate bias voltages, you would have an output only 1.4 V in amplitude because of the usual 0.7 V drop across each on either side of zero (dc) reference. This latter (no bias) dual diode circuit is often used in analog applications where voltage swings are deliberately limited for one reason or another.

In the clipping category, you may have either positive, negative, or combinations of clipping done very simply with just a resistor and a diode. Four of these methods are illustrated in Fig. 7-10. If they look like ordinary rectifiers, that's exactly what they are except for a small residual portion of the clipped waveform appearing in the shunt example. This is because of the usual 0.7 V diode drop in shunt configuration, permitting just a small amount of the clipped cycle to pass. We won't bore you with obvious details other than to caution that diode outputs are current, not voltage, and whatever develops through or across them is only a voltage-developed display if some "load" resistor has done its job. Otherwise, simple rectification produces the other half of the sine or square wave input virtually intact since on alternate half cycles it conducts little or not at all.

Obviously there are all sorts of clipping and clamping circuits that can be designed to do all sorts of wondrous things. With a working knowledge of the basics you shouldn't feel lost with any of them.

ZENER DIODES

No power supply chapter can really be begun without some worthwhile discussion of zener diodes. Their world belongs to all of us, especially bias circuits and voltage regulators, without which there probably wouldn't have been any. Zeners, although sometimes inclined to be noisy, may also be used as coupling devices, especially in integrated circuits. The prime use for zeners, as you should be well aware, is either as a reliable reference or the actual voltage regulators themselves. Today, 5-percent zeners are fairly inexpensive, are easy to insert, and may also aid poorly regulated circuits in *any* electronics that can supply just a little additional current to run the zener. To reinforce this small argument, we'll resurrect a few simple number plug-in type equations that will help put any and all zeners to work.

Such diodes operate in reverse bias, avalanche breakdown modes from a simple rectifying PN junction. When reverse voltage increases beyond the voltage breakdown point, current avalanche across the junction increases rapidly and heavy conduction occurs. Between conduction and nonconduction, the break should become quite sharp (if it's a good diode), and is known as the zener knee.

In this avalanche region the voltage remains essentially constant at whatever value its manufacturer has specified, but current continually varies. There are, however, temperature characteristics that must be taken into consideration as well as zener biasing and filtering, which are highly important to satisfactory performance. We might also add there are two separate voltage breakdown mechanisms in zeners: In zener breakdowns, voltages *decrease* as PN junction temperatures increase; but in avalanche breakdowns, voltages *increase* as junction temperatures increase. Zener breakdown usually occurs at less then 5 volts, while avalanche is evident above 6 volts. Therefore, designers must offer at least adequate current to ensure satisfac-

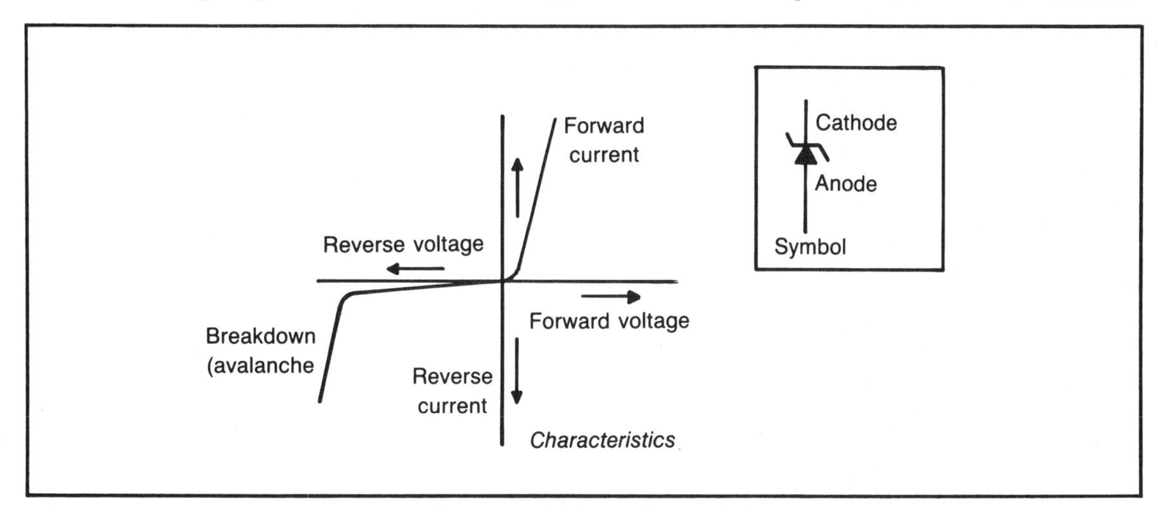

Fig. 7-11. Zener diode symbol and forward/reverse breakdown characteristics.

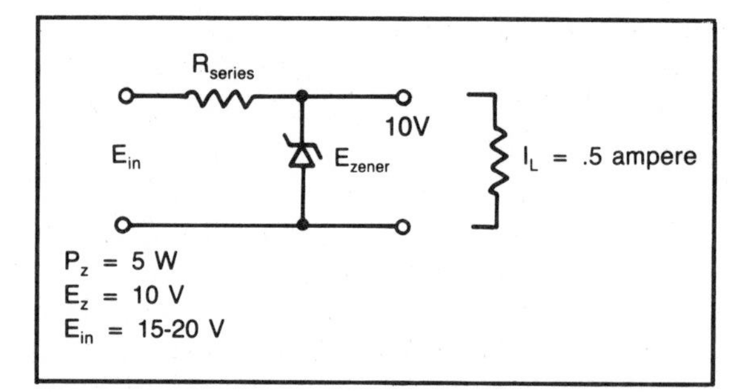

Fig. 7-12. A simple zener diode and shunt regulator.

tory regulation, and pay strict attention to thermal temperature characteristics and environments. Ordinary forward characteristics of zeners are virtually identical to normal rectifiers. (See Fig. 7-11 for details.) You must also be aware that zener impedances usually decrease with current increases, and between 3 and 8 volts there can be considerable temperature coefficient problems. But below 3 and above 8 volts, temperature is almost independent of current if there is no appreciable internal heating.

For practice, let's take an example of a simple shunt regulator illustrated in Fig. 7-12 and determine the series resistance needed to keep it regulating as well as the actual power dissipated. Given are the zener power dissipation, input voltage variation, zener regulating voltage, and output current required.

$$R_{series} = \frac{E_{in}\text{ (min.)} - E_{zener}}{1.1\ I_{Load}} = \frac{15 - 10}{1.1 \times .5A} = 5/.55$$

$$= 9 \text{ ohms}$$

$$P_{zener} = \frac{E_{in}\text{ (max.)} - E_z - .5}{R_s} \times 10 = 9.6 \text{ W}$$

Therefore, to supply maximum load current of 500 mA, you would need a 10 W instead of a 5 W zener and the series resistor would be 9 ohms, with a conservative power rating of 10 W (use P = E × I with I being 1460 mW from the sum of 960 mW and 500 mW). Don't try to get 1 ampere load current out of this same setup, however, or you're in trouble! With this sort of power loss, it's no wonder that zeners in any sort of current applications at all are used only as reference diodes, usually with additional electronics and pass (power) transistors carrying the load.

To give you an idea of what we're discussing, the example in Fig. 7-13 typifies a good regulator circuit with the major expense being only the power transistor. Biased on by the R_B base resistor, the zener is avalanched (or zenered) and supplies relatively constant voltage for the base of the transistor. Output voltage amounts to the diode reference minus the base-to-emitter drop, which is small. Small current drain changes appear as a voltage drop across R_B turning this transistor on more or less and maintaining an equilibrium E_O output and supplying further regulation to the circuit. R_S now becomes the protective series resistor for the entire stage, while R_B furnishes current for the zener in addition to extra bias for the pass transistor.

When repairing circuits of this type, always replace defective components with precise substitutes, including the diode and transistor.

SCRS

How important silicon controlled rectifiers have become, especially in consumer products! Whether as regulators, power diodes, or security devices, SCRs have earned a well-deserved niche

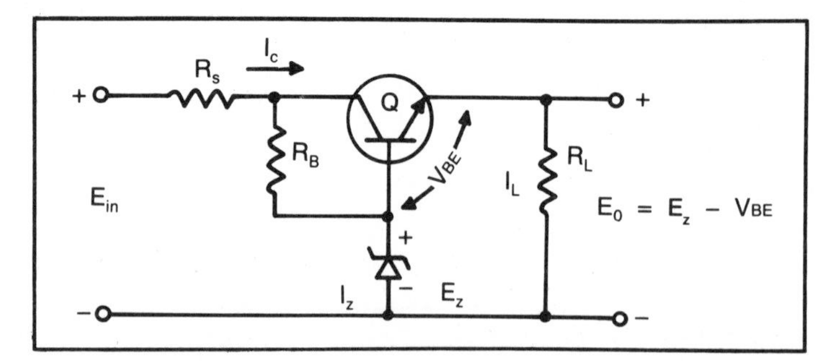

Fig. 7-13. A reasonable two-semiconductor series regulator in common usage.

in semiconductor technology. They're relatively inexpensive, easy to use, and highly effective in current transfers, conducting many amperes when properly gated with alacrity and durability. Nor is their circuit operation difficult to explain—actually it's quite simple and very straightforward.

The SCR has only three electrodes; a gate, cathode, and anode. Schematically it becomes a PNPN structure best explained as a pair of P- and N-transistors in regenerative configuration; that is, the N-channel collector goes to the P-channel base, and the P-channel collector goes to the N-channel base.

A unidirectional device—current flows only from anode to cathode—it is symbolized by the three-in-one diagram shown in Fig. 7-14. The SCR, also called a reverse-blocking triode thyristor, can be shown in three "states" as in the figure: ON, OFF, and Reverse Blocking. In the ON state, SCR current is limited primarily by the impedance of the external circuit; in the OFF state, only a small reverse blocking current flows and the device offers a high impedance. Should the reverse breakover voltage exceed the reverse breakdown voltage, the SCR is subject to thermal runaway, burns red and disappears in an acrid cloud of smoke.

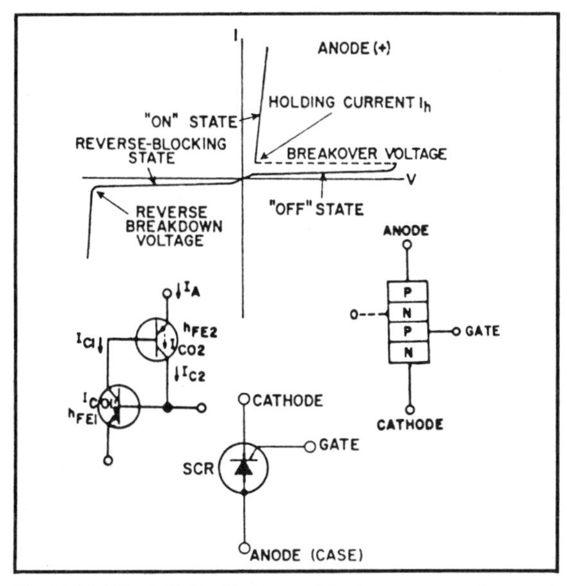

Fig. 7-14. The SCR in theory and fact—a very useful semiconductor for many applications.

For consumer applications, the SCR usually has some pulsating or alternating variety of anode potential and either a receiving network or just plain common at the anode. When a pulse or dc energy charge sufficient to trigger reaches the gate, the SCR turns on more or less depending on the amount of gate potential and the rapidity of its arrival. The SCR then stays in conduction until the anode potential either subsides or goes away completely. In other words, once triggered the SCR remains in that state until the positive potential on its anode quits either by alternation or outright cutoff. SCR turn-off times depend on reverse recovery and gate recovery times; similarly, turn on times are governed by delay and rise times and are defined as the time between gate initial signal and when current through the thyristor arrives at 90 percent maximum value via a resistive load.

When these devices are used in forthcoming circuits such as regulators and diode bridges, you may then understand their complete operation quite easily. They're remarkable bits of silicon, designed specifically for low frequency operations, particularly power supplies. Naturally there are many more sophisticated regulators than this one, and we'll be coming to them as we begin to look at those in existence today. As you will see, there's not a great deal left to chance in these more advanced units because better electronics, superior integrated circuits, and more complex and sensitive equipment requires, first of all, superior power supplies and virtually 100 percent dc voltage regulation.

BEGINNING FROM SCRATCH

It's probably instructive to go back a few years and look at what was then a "modern" color television power supply, delivering 80 volts at 1.5 amperes. Although it uses a programmable unijunction transistor (PUT) that's not popular any more, the principle can be adapted with other, newer solid state devices and there are switching actions that lead up to what have become very popular and cost effective switch-mode power supplies almost universally in favor during the middle 1980s.

As illustrated in Fig. 7-15, this one has the usual

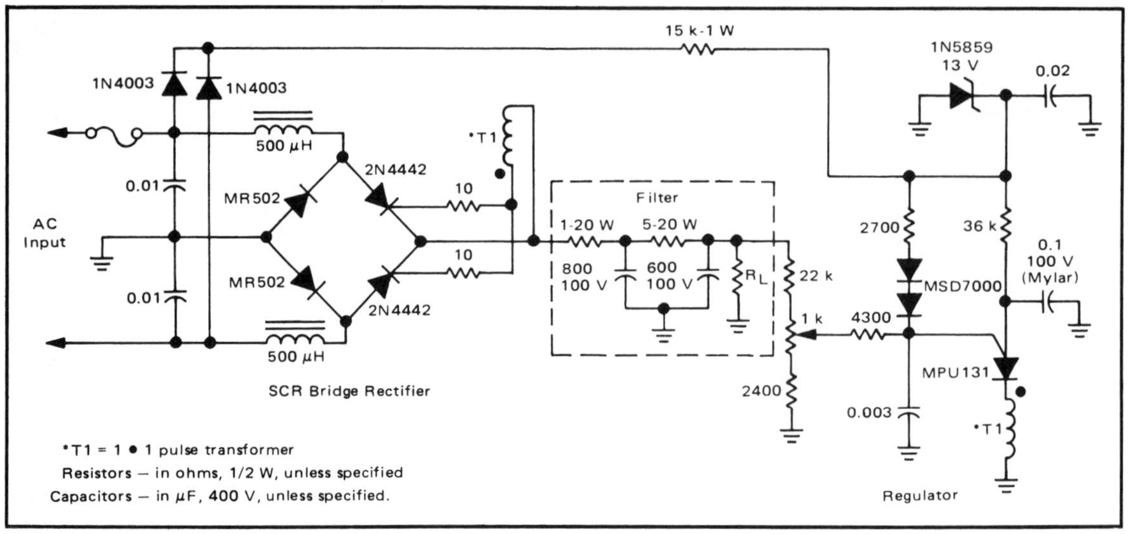

Fig. 7-15. An older example of a 1.5 A 80 V color TV switching type power supply. T1 is actually a pulse-coupling transformer (courtesy Motorola Semiconductor).

bridge rectifier input, with a twist, a 13 V zener clamp and reference, along with the usual RLC filtering, and an MPU regulator, which is the PUT.

At the ac input you first see a pair of IN4003 diodes which supply rippling dc to the anode of one MSD diode, the PUT, and also the cathode of a 13 V zener regulator and clamp and its 0.02 μF capacitor. The two 500 μH chokes on either side of the bridge rectifier and 0.01 μF capacitors are the line filters, while the two MR502 diodes are the common returns to the ac line. Like all silicon controlled rectifiers, the two SCRs conduct when gated on and both rectify and control the amount of output voltage, depending on the on-gated time delivered by T1 across the two 10-ohm current limiting resistors. Of course, as soon as the power line's positive alternation goes negative these SCRs turn off anyway and wait for the next cycle to turn them on again. When line voltages are high, the SCR(s) turn on later in the cycle supplying less current output and, therefore, less developed voltage.

The gating device is the MPU131 PUT which fires when its gate voltage reaches a certain level. By increasing this gate voltage, the PUT fires later, or by decreasing this voltage through the 1 k potentiometer, it fires earlier permitting a longer duty cycle and more output current, developing higher voltage. Current through the primary of T1 develops voltage in the secondary of T1 which then enables the two SCRs in the bridge rectifier and they fire with incoming positive half cycles of ac line voltage.

The filter portion of this switching supply is nothing more than RC conventional, and the 80 V output is delivered across R_{load} in parallel with divider resistors and the PUT bias potentiometer.

You can probably identify this as at least one of the early switch-mode power supplies which have been considerably refined since this one came to life. Certainly PUTs are out and SCRs are decidedly "in" as everyone in the TV design business looks for more efficient and cheaper ways to deliver voltages with adequate current and little ripple. Our thanks to Motorola Semiconductor for this one. Too bad Motorola Chicago decided to leave the television business when it did. Motorola Semiconductor, however, is still plugging and will be contributing a great deal in the chapter on chroma.

DELCO'S EARLY SWITCH-MODE CONTRIBUTION

Yes, the same company that makes radios for General Motors in Kokomo, Indiana developed this early switcher, and it's a *good* example of those that succeeded it even though not precisely in the same form and fashion, as you'll shortly see (Fig. 7-16). Along with pulse shaping circuits and a number of primary and secondary transformer windings, the Delco version also has its own 15 kHz LC oscillator that can be synchronized with any television's own horizontal sweep, a stage which has been removed in later models and driven directly from the set's own 15,734 Hz, usually digitally controlled, horizontal countdown arrangement. (See chapter on horizontal/vertical sync.)

At any rate, this switch-mode low voltage supply is coupled directly to the ac line through F1, F2 fuses, the D1, D2 diodes and their stacked capacitors are connected as a full wave transformerless voltage doubler, and there's a fair amount of filtering before powering the primary of the power transformer (upper right).

Transistor Q1 and its assorted RLC components constitute a variable frequency (tunable) excitation Colpitts oscillator with L1 and the two 0.015 capacitors making up the tank circuit, partial load and feedback. Output is across the collector to the emitter of Q2, whose base is biased by dc through the 2.5 k potentiometer above. Q2 controls the bias level of Q3, limiting the maximum pulse width, as transistors Q3 through Q5 form amplifier pulse shaping circuits that control the timing for gating transistor Q6 through coupling diode D5 and the RC input network following.

When the supply is operating, current for Q5 through Q7 originates from the upper primary of the power transformer, ensuring proper startup and continued operation. Q6 and Q7 are a stacked pair, delivering large output pulses to the lower portion of the primary which, in turn, excites the secondary at a horizontal rate. The value of the four windings, their half-wave diodes, filter capacitors, and loads then produce well-filtered receiver supply voltages of 350, 75, 30, and 100 Vdc for operation of the entire receiver. Power supply regulation depends

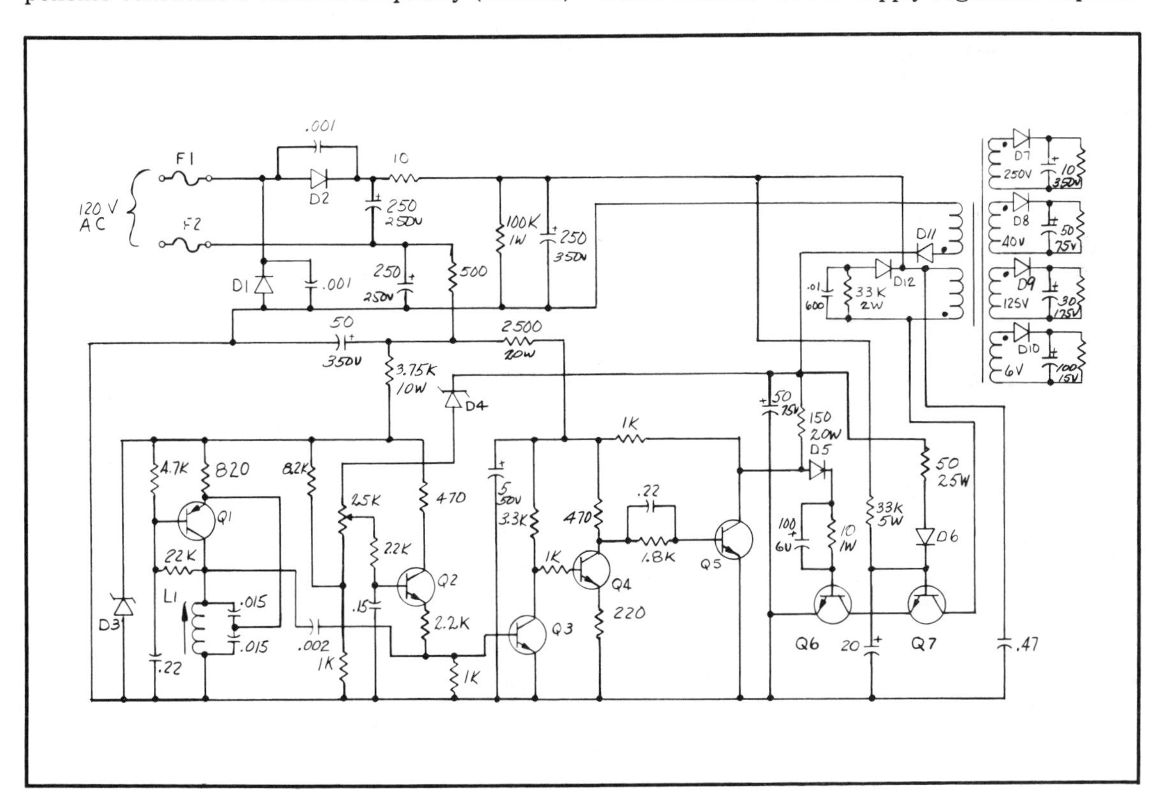

Fig. 7-16. An early 15-kHz switch-mode Delco power supply.

on pulse-width on and off times controlling Q6-Q7 conduction resulting from load demands coupled by varying dc levels to the base of Q2 via zener D4 and the upper transformer primary which responds with more or less current to secondary load requirements.

Having the secondary of this supply fully transformer-isolated, the primary circuits are "hot" while those in the secondary are "cold" and *safe* for man and his equipment. Suffice to say, this is at least one method of designing a hot-cold power supply that does regulate a number of output voltages and requires very little half-wave filtering to reduce ripple for touchy transistor and IC operating circuits. You can do equivalent regulation also from the flyback (high voltage) transformer if its primary B+ is highly regulated, permitting not only fairly stiff and reliable secondary voltages but good high-voltage regulation, which is *essential* for adequate performance in the newer monitor/TV and digital receivers. And as high definition and fully digitized receivers make their appearance, the days of sloppy operating voltages will have disappeared altogether, particularly in the better chassis. It may be, also, that the IC manufacturers will concentrate on packaged regulators shortly with even more effective and simplified results than we know now. Certainly there's a need if sales can reach an acceptable volume.

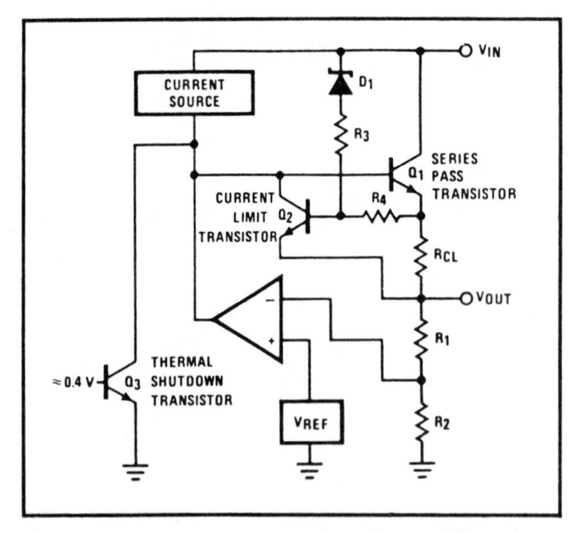

Fig. 7-17. Basic block diagram of an effective IC regulator.

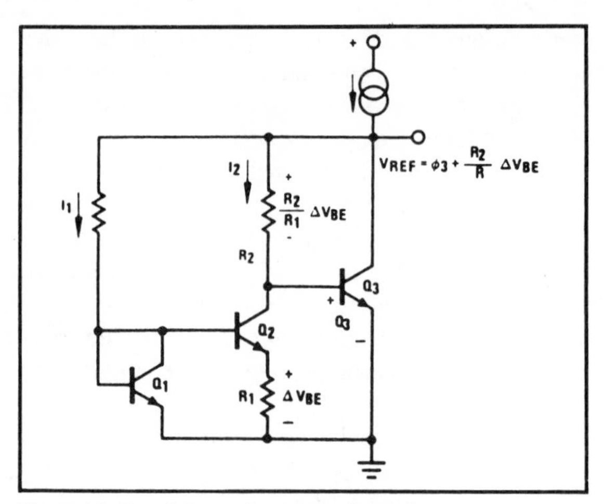

Fig. 7-18. Band gap reference schematic.

IC VOLTAGE REGULATORS

Such IC regulators may well take the form of some National Semiconductor units that have always been highly successful. One such example is illustrated in Fig. 7-17, which includes a reference, regulating comparator amplifier, current limiter, and thermal or possibly overvoltage shutdown arrangement for additional protection. As is customary, series-pass transistor, Q1, furnishes the regulated voltage and fluctuating current out, the large differential amplifier senses voltage load changes and delivers more or less regulating current to Q1, and Q2 serves as the current-limiting semiconductor that "fires" when excess loads become apparent.

Here's what happens. When the breakdown potential of zener D1 is exceeded between V_{in} and V_{out}, current flows through D1, R3, and R4, developing a greater voltage at the base of Q2 than across RCL. Q2 then fires and diverts current from the base of Q1, reducing its output and preserving the regulator.

National also offers a reminder that high-power regulators are endangered more by thermal characteristics than electrical ones, and fluctuations in *either* load current or input voltage result in thermal changes. Therefore, shutdown transistor Q3 was also added to the circuit to prevent possible burnout. On the circuit, Q3 is placed next to pass transistor Q1 and its base biased to approximately

0.4 V, just below turn on at room temperature. If Q1's temperature rises to an unsafe level, Q3 turns on and once again diverts current from the base of the Q1 pass transistor, reducing its output to a safe level.

The V_{ref} for this unit may either be a close tolerance zener or *band gap* device such as shown in Fig. 7-18. National states that in a band gap arrangement, two monolithic transistors "operating at different current densities develop a predictable voltage" offering lower noise and better long term stability than zeners. However, there are controllable problems with initial voltage control, temperature drift, and some thermal gradient effects are slightly worse among band gap circuits if there isn't considerable design care.

In the simplified schematic, if the emitter and collector resistors, R1 and R2, of Q2 are "properly chosen," the positive temperature coefficient of delta (change) V_{BE} can be cancelled, resulting in virtually zero TC drift, making this a very stable reference indeed. As you can see, all currents are shown in the directions of collector-to-emitter and Q1, with its base and collector ties, is no more than a temperature compensating diode.

TODAY'S POWER SUPPLIES

The final portion of this chapter will include both switch mode and other common varieties of contemporary television power supplies taken directly from schematics of the latest receivers, accompanied by operational descriptions. By now you should have assimilated enough background to be able to judge their advantages and deficiencies without further editorial comment since some are outstanding and others are not, depending on both design and application. With the newer receivers ordinarily requiring much less power than the old vacuum tubers, regulation has become simpler but, at the same time, considerably more stringent for both temperature considerations and accuracy. Therefore, you may like what you see or have other ideas. Rest assured, however, that with video system sales soaring every year in a healthy and ever-growing industry, things to fill the various needs will be forthcoming as demand requires. In the developing three-continent race, Europeans, Asians, or Americans are bound to do it—and let's hope success belongs to the United States.

The RCA Version

To a good deal of the industry belongs this credit, so we can't award RCA an exclusive again since they've had several power and regulatory systems in recent years. However, this one is very straightforward, delivering well-regulated 115 Vdc to the horizontal output and flyback, horizontal driver, and vertical deflection stacked amplifier. Without a traditional oscillator, RCA describes its CTC 126 regulator as a pulse-width modulator (PMM) with two reference inputs, and only a single common (ground) for the entire chassis.

Beginning with the usual hot chassis bridge rectifier input and fair filtering, this voltage supplies 150 V potential for the flyback and dc bias for regulator SCR 101, noted on the simplified diagram as an arrow "to SCR gate" (Fig. 7-19). It also reaches the base of Q101 error amplifier through a 1- and 2-percent precision resistor divider bias array. No. 2 is a 20 V pulse from the flyback's secondary (that also serves as the CRT heater pulse) to the base of buffer (called an oscillator) Q104 which is used as a timing waveform generator for regulator predriver Q102. When turned on it discharges C109 and turns off Q102. Between flyback pulses C109 charges positively through R108/R109 and fires Q102, whose emitter is clamped by 6.2 V CR107, and whose base is biased by CR105 and R109. When Q102 conducts, it removes positive dc bias from the base of the Q103 PNP regulator driver and this transistor then fires the SCR101 gate, causing this semiconductor to conduct, delivering current for the 115 V regulated supply.

The regulation portion of the circuit develops at the base of Q102 which receives a reference voltage from the screen/focus resistor assembly originating with the flyback. Should B+ try to decrease, voltage at the base of Q101 begins to fall, its collector supply increases and timing capacitor C109 charges more rapidly firing Q102 sooner, Q103 does the same, and there's a longer gate pulse for the

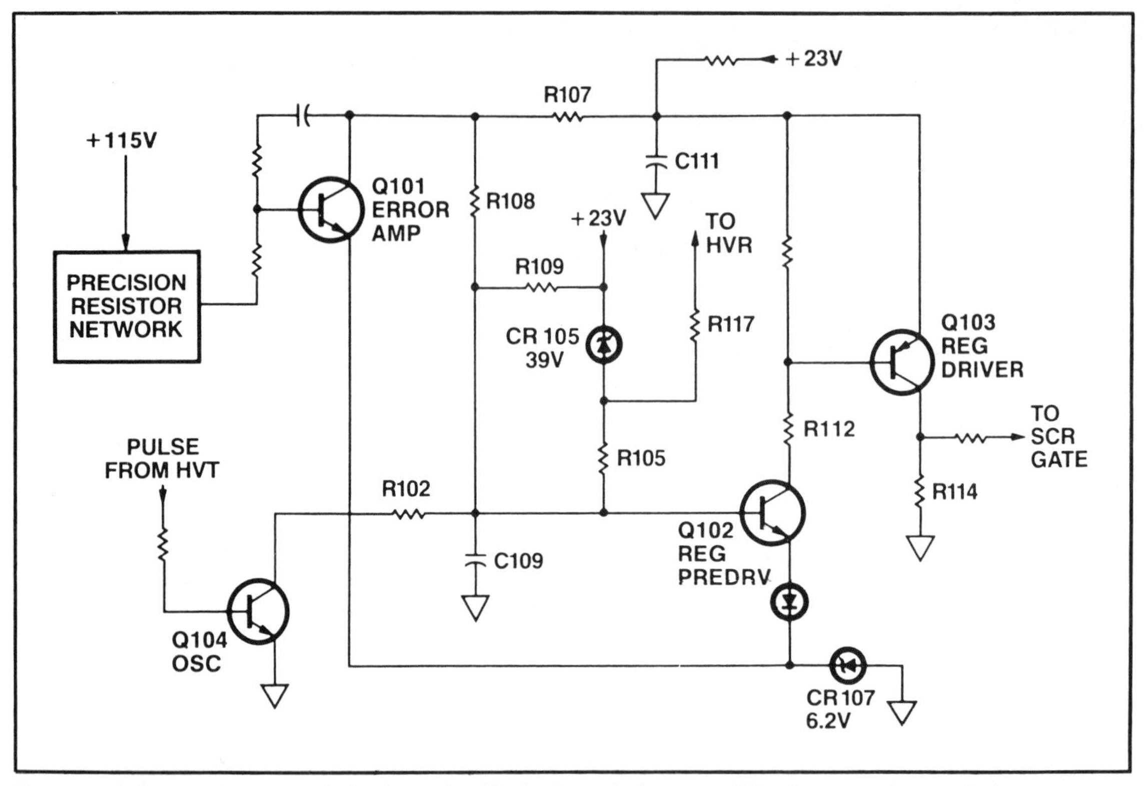

Fig. 7-19. A B+ regulator control circuit—a simplified schematic (courtesy RCA Consumer Electronics).

SCR generating additional current and, therefore, additional developed voltage. The reverse is true if B+ rises so that the SCR will conduct for a shorter interval delivering less current. Now, should beam current through the screen/focus assembly change rapidly, it can reach a high point where CR105 avalanches and prevents Q102 from "drastically" disturbing pulse-width output from Q103 to SCR101.

Zenith's Pulse Width Modulator

Zenith, in its 9-160 supply and sweep module is involved with just enough circuitry and a proprietary pulse-width modulator to make supply electronics rather interesting. Actually, this one amounts to a pretty good challenge for anyone who is not one of the designers both as to parts and complexity. We hope there's no confusion following the description in which we'll include the degaussing function (for both the record and good measure) even though that operation is so routine today that it's taken entirely for granted, being a relatively uniform circuit with completely predictable results. Its purpose, of course, is to wipe out any stray fields collected by the shadow mask and avoid unsightly splotches of red, blue, green or purple that always seem to lurk at the CRT's corners. If you're anxious to duplicate this condition or give your dinner host acute indigestion, just hold a magnet near the tube's center and wave it around a bit to thoroughly foul up all colors. But afterwards, be sure to have a degaussing coil handy to quickly right your wrong, then apologize on bended knee or protect your jaw.

As usual (for the 1980s) action begins with hot and return sides of the ac line connected to a bridge rectifier. However, before full power is applied to the bridge (Fig. 7-20), a current must pass for a few milliseconds through the degaussing coils, clearing the shadow mask of any incorrectly aligned magnetic field impurities. Then, just as soon as thermistor becomes hot in a few milliseconds, it increases in resistance and full current flows into the large and transformer-isolated small bridge rectifiers to begin the unregulated dc supply voltage flow. Through the LX3351 flows 150 Vdc for the horizontal output transistor collector supply and primary of the HOT transformer. The 15 Vdc voltage at (1) counts only when the receiver is initially energized and kick-starts the 25 V supply through CR3376 and the 18 V supply through CR3352, the latter being filtered by 100 μF CX3330. The air spark-gap protects against lightning and the 2 ampere fuse affords some insurance against short circuits. When all systems are started CR3376 and CR3352 are reverse biased and the two voltages are shut down, but the 15 V source continues to be used as a 60 Hz reference by the pulse-width modulator whose schematic appears in Fig. 7-21. As you can see, this special integrated circuit receives a filtered and zener-regulated reference voltage through pin 3 from two sources, a horizontal countdown pulse from the horizontal driver via the high-voltage shutdown circuit, develops a sawtooth-triangle at pin 4, and receives a flyback pulse along with the rest. A voltage divider with potentiometer controls the dc level at the same pin 4, and operating voltage derives from the 12 V supply through a filter coil and 10 μF filter. Its output at pin 1 goes directly to the base of the Q3301 horizontal driver. With all remaining voltages now generated by primary windings and rectifier-filters on the flyback, control of horizontal driver conduction is very important.

At such time as the sawtooth-type (triangular) voltage exceeds the value the reference voltage, Q9, Q10 draw more current from Q13 than do Q11, Q12, this turns on positive feedback switch Q6, Q7, and Q5 which delivers considerable current to the base of Q3 overdriving Q3 and also output Q1. This results in a square wave being developed which starts and ends as the sawtooth intercepts the reference level at pin 3.

A horizontal pulse from the countdown circuit also energizes Q14, amplifies and clamps these pulses between ground and B+ and producing another square wave which is then shaped into a triangular (or sawtooth) wave via RC devices between pins 4 and 5, and completing at pin 4. At pin 3, a sample voltage from the 15 V supply and a sample of 18 V sweep-derived power are mixed. Should either supply vary from its nominal set, the square wave output at pin 1 will vary in duty cycle, changing the on and off times of the Q3301 horizontal driver, thereby supplying pulse-width modulation.

Should high voltage increase between 4 kV and 6 kV over its normal value, the shutdown circuit receives a flyback pulse of about 60 volts. Whenever this pulse exceeds some set value, rectification takes place through CR3353, charging C3351 which discharges through a filtered and carefully selected resistor network. Should this filtered dc voltage exceed the breakdown value of CRX3355, QX3351 turns on and so does QX3338. They constitute a feedback switch which locks, once triggered, driving pin 7 on the pulse-width modulator positive turning on Q2 which shunts the collector of Q5 to ground and kills both Q3301 drive and high voltage. The 15 V isolated supply maintains the shut-down condition until the receiver is either restarted or the condition repaired. Transistors Q8, Q13, and Q4 are loads and current supplies for Q10, Q11 and Q5, respectively, on the PWM module IC.

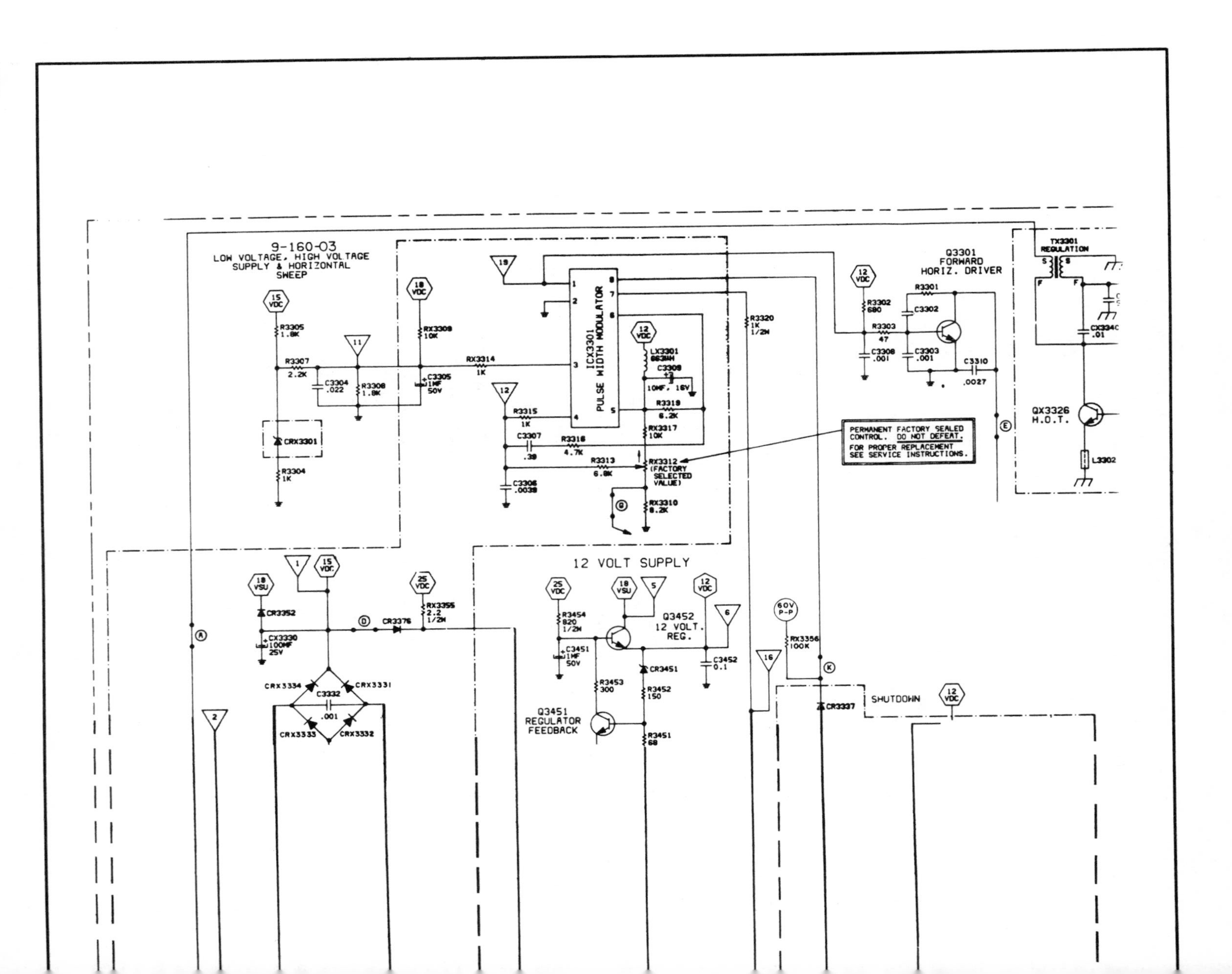
9-160-03
LOW VOLTAGE, HIGH VOLTAGE
SUPPLY & HORIZONTAL
SWEEP
ICX3301
PULSE WIDTH MODULATOR
Q3301
FORWARD
HORIZ. DRIVER
TX3301
REGULATION
QX3326
H.O.T.
PERMANENT FACTORY SEALED
CONTROL. DO NOT DEFEAT.
FOR PROPER REPLACEMENT
SEE SERVICE INSTRUCTIONS.
RX3312
(FACTORY
SELECTED
VALUE)
12 VOLT SUPPLY
Q3452
12 VOLT.
REG.
Q3451
REGULATOR
FEEDBACK
SHUTDOWN

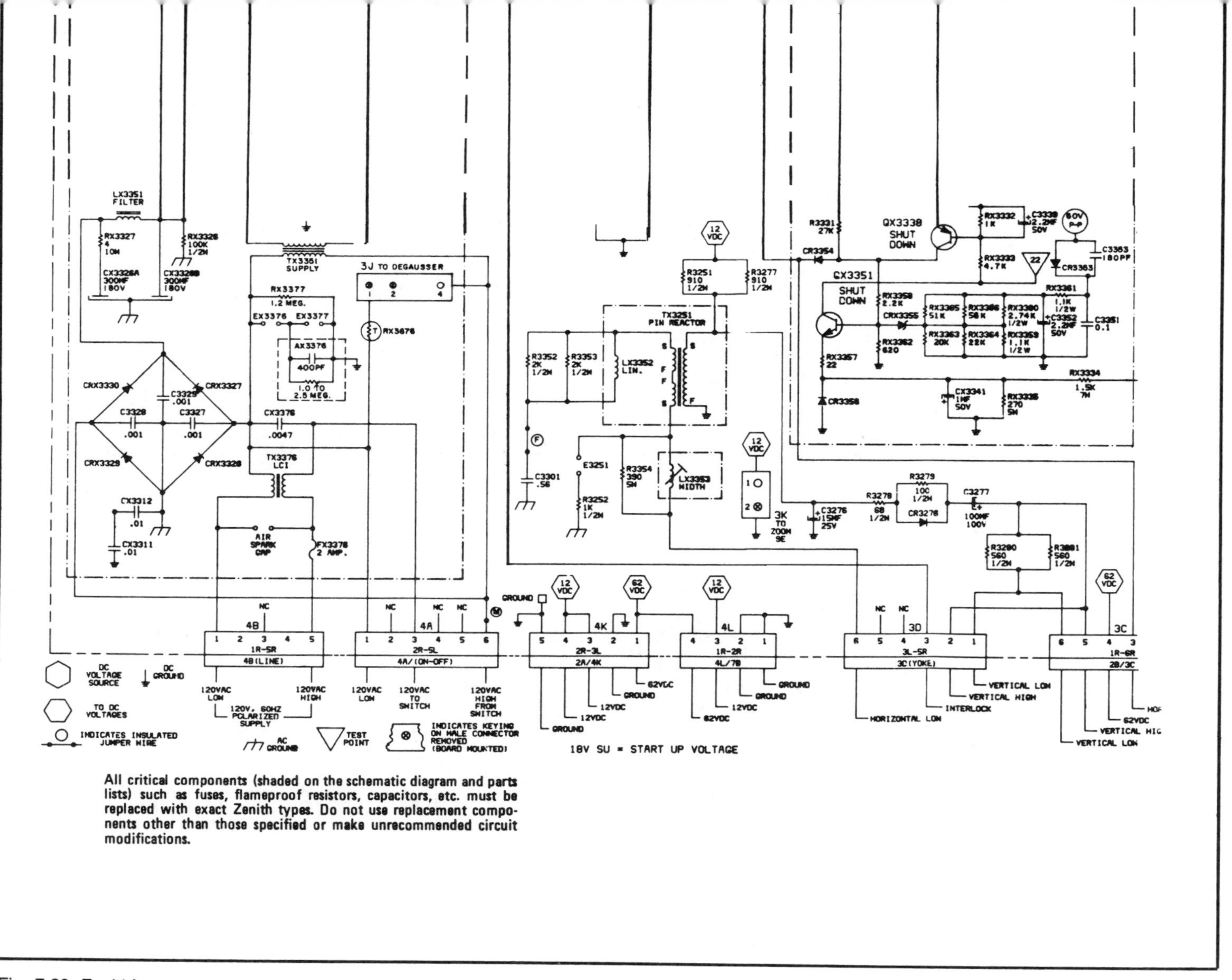

Fig. 7-20. Zenith's power supply and modulator schematic (courtesy Zenith Electronics).

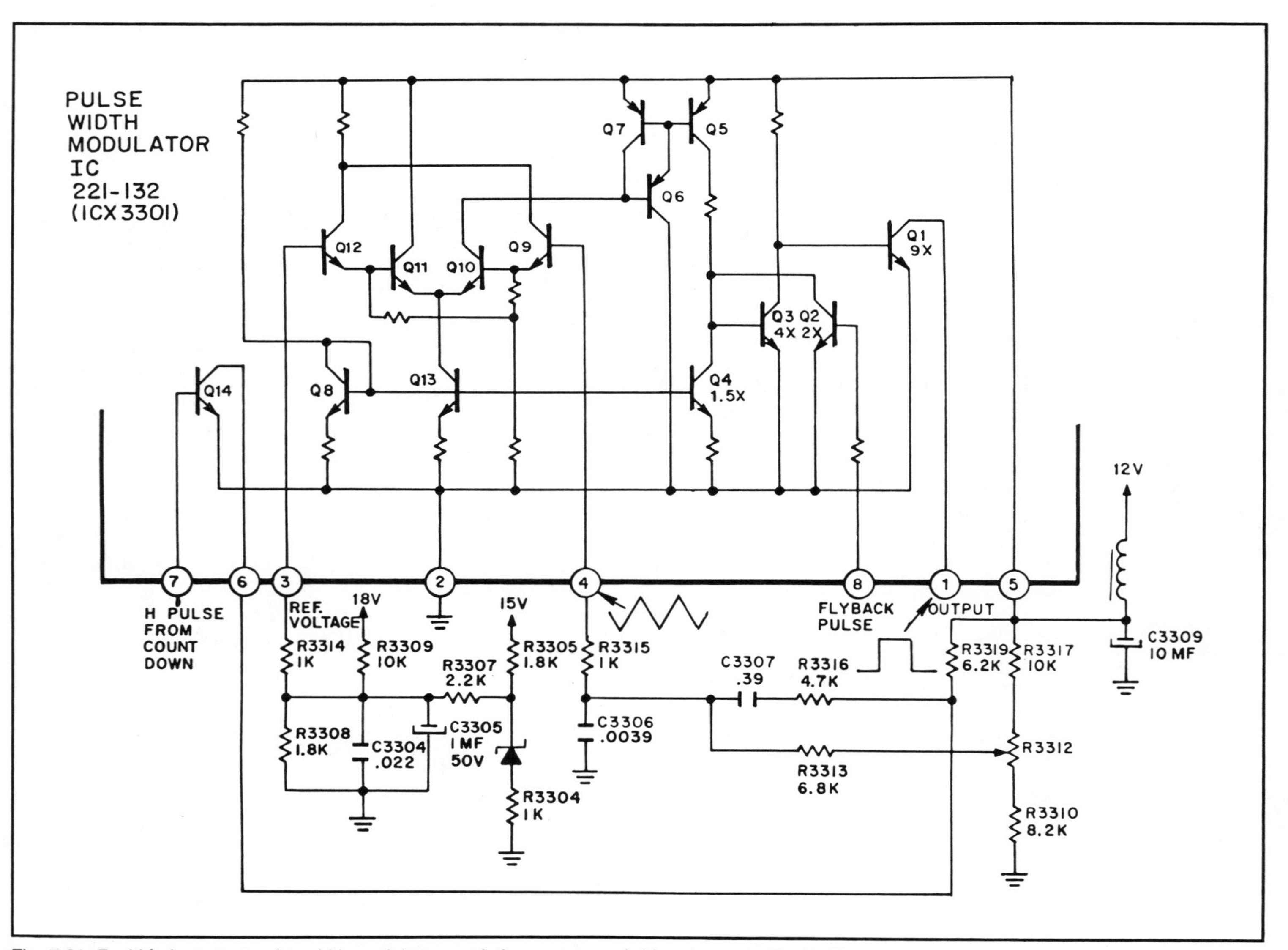

Fig. 7-21. Zenith's important pulse-width modulator regulating system and driver (courtesy Zenith Electronics).

Chapter 8

Tuners and I-fs

THESE ARE A PAIR OF INTERTWINED TOPICS that are difficult in one sense and reasonably simple in another. Unfortunately, we can't cover their complete spectrum, mainly the major philosophies outstanding in the 1970s and mid 1980s. Some of this, however, may be somewhat sketchy because significant bits and pieces are now Japanese, particularly video/rf mode on-screen identifications, and the sons of Nippon don't often supply minute details to go along with their evolutionary, fine-tuned developments. Nonetheless, we'll try our best from the fonts of Sony and Panasonic and see what a little cultivation can germinate. As for the U.S. manufacturers, there's no problem since a great deal of what's available is already published material, and a few extras make excellent explanations and appropriate detail.

Video i-fs are a totally different matter since our own semiconductor houses are turning these out by the millions, and all sorts are accessible for handy reference, complete with excellent engineering details. So between or among several diverse sources and levels of descriptive applications, tuner-i-f documentation should be reasonably complete and most certainly interesting and up-to-date. We would also expect that these contemporary philosophies should continue for some time since microprocessors and phase-locked tuning controls have only really blossomed in the past several years. This especially is true for the Japanese who depended on discrete voltage-varactor channel tuning for years in both television receivers and videocassette recorders. As you will see, however, the picture is changing very rapidly as enormous efforts are made to become competitive in today's highly developed and rapidly moving video market, and useful sophistication has become the watchword.

TUNERS

While we could cheerfully digress to the ancient past on turret and switch-type tuners, the video world is turning away from mechanics as rapidly as it can, and electronics overwhelmingly predominates. Therefore, we'll offer a brief description of one or two manual tuners, but major emphasis has to be on the highly-controlled electronic "marvels" pouring out of today's factories both here and abroad. With their remote controls and multi-functions, especially the monitor/TV variety, there seems to be no end to useful innovations produced each twelve months among May-June introductions throughout the country.

Tuners, however, are still very difficult to design, especially the rf amplifier, whose gain, noise, SWR, and cross modulation largely determine how the overall tuner will perform, whether such amplifiers are MOSFETs, bipolars, or whatnots. Cross modulation—where extra signals generate positive and negative byproducts—is especially critical relative to signal gain and spurious beats, and must be carefully designed and shielded to avoid chassis and outside rf pickup. Nowhere is this more true than in digital receivers or computer areas where 1s and 0s are flying constantly at varying or fixed rates with amplitudes (in some cases) large enough to modulate the main signal and cause problems. Whether cheap or expensive, tuners still must offer an adequate 6 MHz bandpass, good adjacent channel rejection, a broad channel capture ratio, negligible drift, mechanical ruggedness, relative ease of servicing, and acceptable low-oscillator radiation. The electronic varieties, of course, have singular advantages over the mechanical group, including voltage regulation, better agc control, automatic fine tuning control (AFT) and other electronic devices to keep them channel-locked when once tuned. Further, there's not the mechanical wear associated with switches, knobs, contacts, and tuner wafers which inevitably clog with dust, oxidization, and wear. If there's a choice, go electronic every time, especially where improved MOSFETS are the rfs for both U/V tuners since cross modulation will be improved from 10 to 15 dB and rf agc is considerably more effective since one gate of the MOS is under direct control at all times.

There's more to the tuner than just shielding and distortionless amplification, however, don't forget the local oscillator (1st detector), and mixer which must heterodyne any and all incoming channel frequencies between 54 and 806 MHz (channels 2 through 69) down to a mean 44 MHz center frequency in order to properly position both video and audio carriers at specified points on the i-f response curve. In the old tuners, there were all sorts of traps and tuned circuits between the tuner and i-f circuits, but that has now largely been superceded by surface wave acoustical filters (SAW) which do the job much more easily and accurately than RLC tuned circuits ever could. After the rf, however, the tuned circuits return, including oscillator and mixer inductances which must be frequency-accurate for satisfactory i-f generation results.

In switch-action manual tuners, series switched inductances in shorting actions select to 6 MHz bandpass for the channel tuned, then, through mutual coupling, pass the received information through to the mixer. The tuned oscillator, below, operates at 45.75 MHz above the video carrier, and also connects to the mixer. The down beat difference, then, becomes the 44 MHz i-f signal which is then i-f-amplified and audio-video detected removing its 41.25 and 45.75 MHz respective carriers, leaving an intercarrier differential of 4.5 MHz for the resulting sound.

The VHF tuner illustrated in Fig. 8-1 is a very simple device with only a 300-ohm impedance input, a safety RC coupling network and a 300/75 ohm balun introducing 54 to 216 MHz hi-lo VHF to the i-f high-pass filter and the 12 channel tuner coil positions that follow. This manual channel selector is gang-tuned so that the primary and secondary of the rf amplifier as well (naturally) as the oscillator are tuned simultaneously for required interaction. The mixer, directly above the oscillator, then combines rf and oscillator frequencies in its base circuit, and the transformer in its collector passes only the 44 MHz frequencies needed by the i- f. Eight and six picofard capacitors furnish loop feedback for the oscillator, with B+ being supplied through "B" and the 220-ohm resistor and series coils for all three tuner stages.

The bipolar rf amplifier receives agc through the indicated lower terminal and, when the tuner is switched to UHF, these frequencies are introduced through the UHF input (lower left), with the rf and mixer now acting as two additional amplifiers for further i-f amplification. At this time, of course, the local VHF oscillator (LO) is disabled. Note that

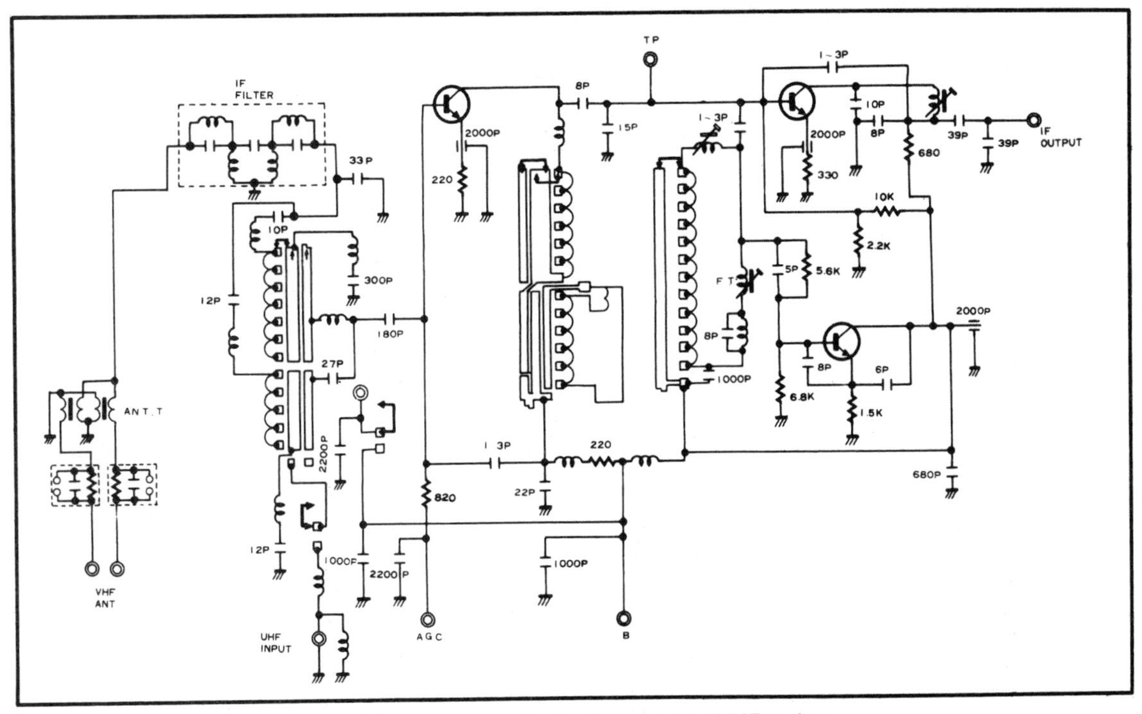

Fig. 8-1. A very simple switch-type rf tuner is shown to illustrate manual VHF tuning.

only the local oscillator and mixer outputs may be inductively tuned, with all else being fixed.

The UHF tuner is totally different from VHF in that there is no FM (88-108 MHz) break between channels and no separation such as that from 216 MHz (the end of VHF) to 470 MHz, the beginning of UHF and channel 14. But these UHF channels are at considerably higher frequencies, therefore wavelengths are shorter, line-of-sight more acute, and signal pickup becomes even more difficult. But there are usually fewer UHF channels in any one area than VHF stations, and they often broadcast at considerably higher power (Fig. 8-2).

Consequently, TV manufacturers decided that an inexpensive UHF tuner shouldn't DX anything over a few miles and, therefore, included no rf amplifier, only a diode mixer, and the usual transistor oscillator. Add three tuning capacitances in a ganged tuner, attach these to oscillator and mixer lines, and you have a pretty skimpy UHF receptor.

In urban and close suburban areas, these rotary but now detented tuners do work to a certain extent, although signal-to-noise ratios aren't superlative. Considering the number of components, you might wonder how it operates at all—and sometimes it doesn't, at least very well. But then, again, frugal people do purchase spare receivers. This one is a *prime* example. Presumably 1S1926 isn't the mixer date of manufacture but an appropriate part number. The diagonally striped portions are actually large inductors called *lines*. Output, of course, is directly into the VHF tuner for additional amplification, which is also a good test point to introduce a sweep frequency when aligning i-f amplifiers—usually a *necessity* when swapping tuners or making i-f substitutions or repairs. The factories do it, why not you? We'll demonstrate in the troubleshooting chapter.

MOSFET Front Ends

It wasn't until complementary-metal-oxide field-effect transistors (MOSFETS) with dual gate inputs became available that tuner gains *and* noise was satisfactorily controlled. High gain-capacitance

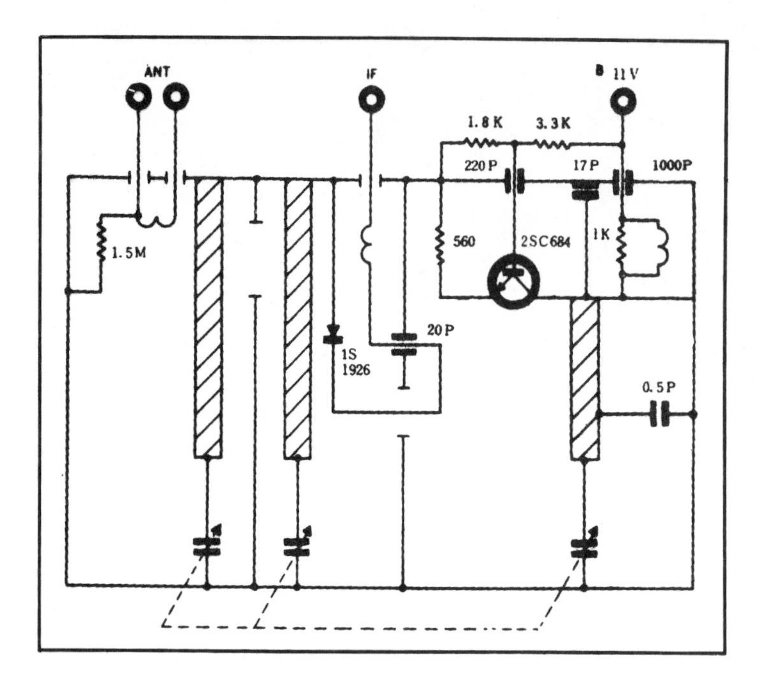

Fig. 8-2. Some manual tuning remains at UHF, too, but only in the cheapest receivers.

ratios at higher frequencies are MOSFET characteristics, with power gains and noise figures inversely proportional. Consequently, MOS transistors have become universally popular in U/V tuner front ends, along with additionally advanced and sophisticated electronics resulting in tuners with no moving parts, improved rf and mixer cross modulation, high input impedances, better gains, and excellent shielding and reliability.

Then, with the introduction of MOSFETS came high-low channel bandswitching and the use of voltage-driven varactors to select the individual VHF channels, or UHF channels, as it turns out, although the two tuners were never then in the same package. Such transistors are depletion-mode devices; that is, they're wide open until pinched off. Typical power gains for early rf and mixer stages amounted to between 30 and 40 dB, with good impedance matching and less than 1 percent cross-modulation characteristics. A typical tuner of this 1977 era is illustrated in Fig. 8-3.

As you will note, only the local oscillator is bipolar, with both rf amplifier and mixer being N-channel MOSFETS. In this tuner, switching diodes are identified by symbolic switches, while varactors have capacitances drawn on their right side. All told, there are six switching diodes and four varactors doing the electronic jobs of high-low VHF change and individual channel selection formerly reserved for a box full of coils and dozens of metal-tipped contacts scraping along together. As most already know, varactors are special back biased diodes operating as capacitors, but totally dc-controlled, often from 0.5 to +28 V. Source potential is a regulated supply delivering carefully controlled voltage drops across precision resistors. In older TV tuners and cheap video helical-scan recorders, such voltages tuned individual channels, which were usually limited in number to a total of approximately fourteen.

In the 1985-1986 TV and VTR lines, all the better receivers will retain varactors for channel selection, but instead of adjustable resistors, microprocessors with phase-locked loop oscillator tuning will replace the little thumbwheel precision resistors, and tuning accuracy, memory storage for channel programming, and digitized remote functions will offer receiver/recorder owners a vastly improved means of broadcast station selection and accuracy than in the past. Usually, too, CATV channels are included also.

In this particular tuner, it's already obvious

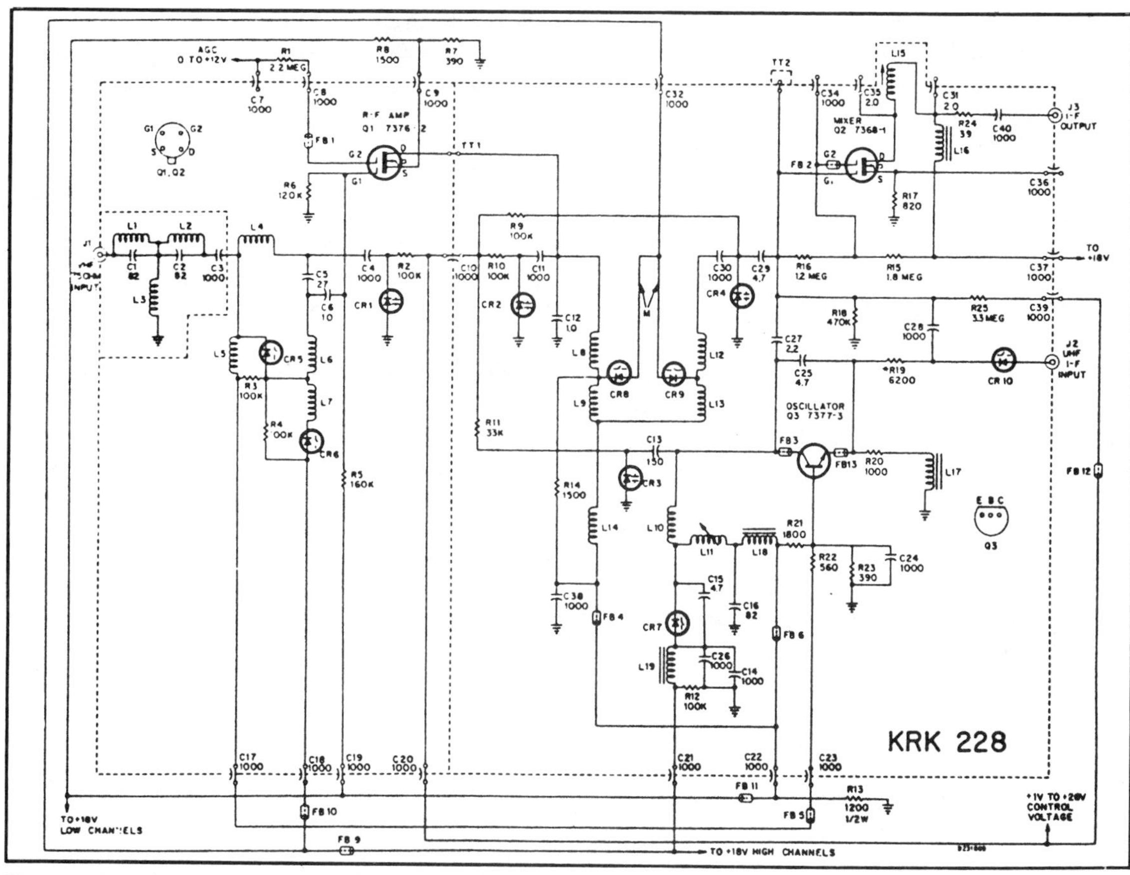

Fig. 8-3. VHF varactor tuner with MOSFET rf amplifier and mixer (courtesy RCA Consumer Products).

there are two sets of voltages, one for hi-low VHF at fixed potentials and another stepped variable for the usual 13 channels. In the simplified diagram of Fig. 8-4, switching diodes CR7-CR9 are in parallel, with CR5-6 in series. During high-band switching, approximately + 18 V is applied to the anodes of the three parallel diodes supplying operating energy to the rf amplifier and local oscillator collector through L8, L10, and L19, as well as a local oscillator base voltage via CR5-6 through R22-23. L5 and L7 are now in parallel, and inductors L9, L13, and L14 are effectively shorted out, permitting the tuner to operate at higher frequencies. With low-band switching, these and other allied inductors are returned to the circuit as diodes CR5, 6, 8, and 9 are back biased and CR7 cut off also. The tuner now operates at lower frequencies, permitting variable discrete voltages to tune the CR1-4

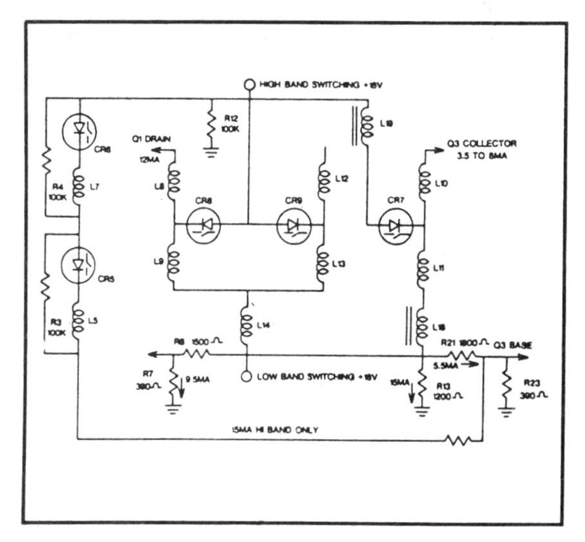

Fig. 8-4. High-low channel bandswitching (courtesy RCA Consumer Products).

varactors, selecting individual channels between 2 and 6. With UHF tuner input, switcher CR10 disables the local oscillator with back bias and the mixer becomes another i-f amplifier stage.

UHF tuners are not nearly so complex since varactors can easily handle frequencies between 470 and 806 MHz (Chs. 14-69) and therefore require no hi-lo switching. Here varactors are soldered to both tuner lines and disc capacitors, minimizing lead inductance (Fig. 8-5). There are a few more peaking coils, the usual diode mixer, and a bipolar oscillator; but the prime differences are an rf amplifier and separate common base bipolar i-f output, compensating for gain loss through the rest of the tuner.

The rf MOSFET readily responds to the usual i-f delayed agc (automatic gain control), offering better cross-modulation characteristics, lower noise, and extra gain. The mixer is known as a hot carrier diode having a constant voltage bias to improve cross modulation. Drift occasioned by CR5 is compensated by C16 with its negative temperature coefficient. I-f amplifier Q3 supplies load for the mixer equalizing impedance variations, with parameters adjusted for equivalent noise factor-gain to be compatible with VHF and agc controls. The three initial lines are adjusted in frequency by inductors L2, L6, L7, and C15 so rf and oscillator responses will track.

The foregoing should serve well as a detailed introduction to varactor and switching diode-controlled solid-state tuners with MOSFET rf amplifiers and their advantages. These general design guidelines are carried forward at least into the late 1980s, and will probably continue until front end integrated circuits take over in digital receivers by

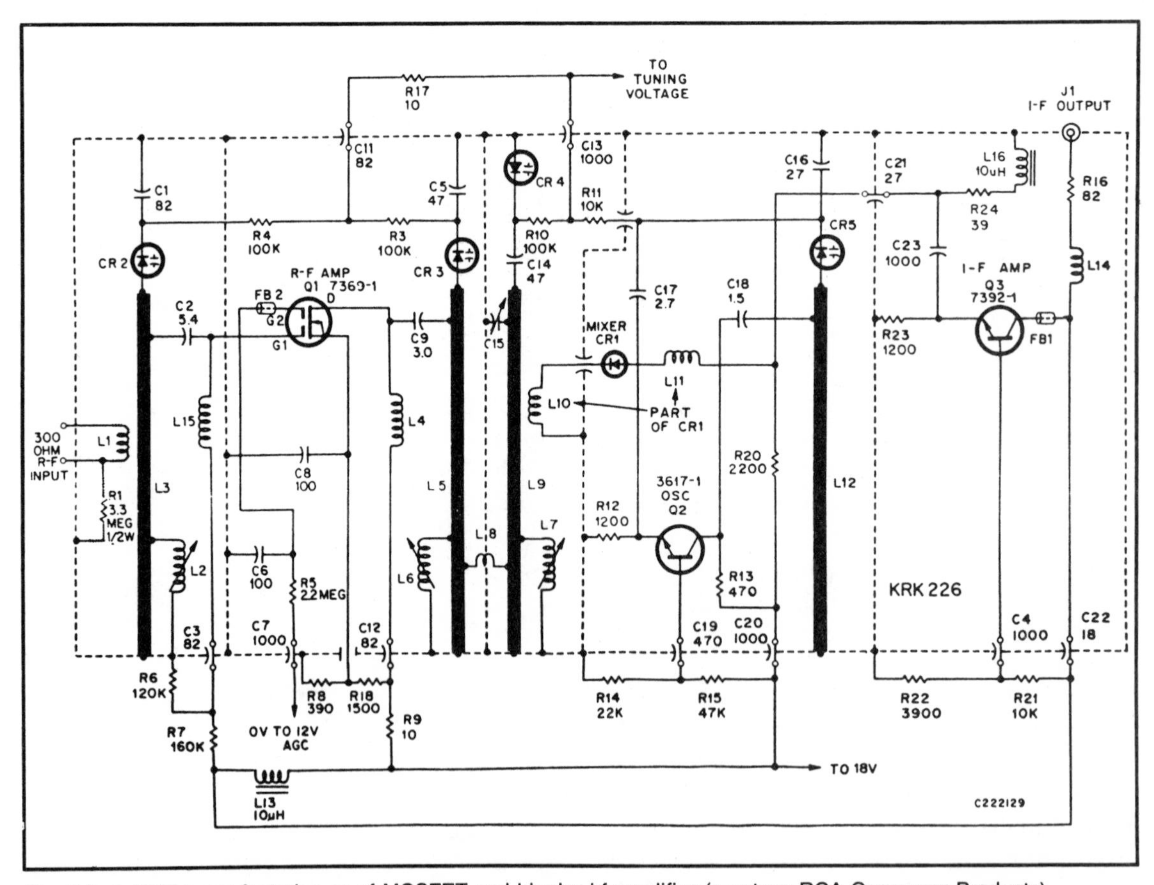

Fig. 8-5. A UHF tuner featuring an rf MOSFET and bipolar i-f amplifier (courtesy RCA Consumer Products).

or before 1990. By then there may be two-stage amplifiers and mixers, further removing undesirable rf pickup and even image frequencies generated within the tuner itself. On this, however, the crystal ball is unclear since a great deal of work still remains to be done to substantially prove this theoretical concept.

Meanwhile, the great advance in engineering between the 1970s and now for tuners, at least, centers on superlatively designed digital controls for all sorts of display, frequency lock, and command deliveries that have all been made possible by highly flexible programmable digital logic and microprocessors. Drawing relatively little current and working on low or medium voltages, many of the newest receivers have forsaken analog controls altogether and even display relative volume, picture, brightness, sharpness, etc. settings quantitatively on the screen so an operator can judge almost any control condition or situation that's programmable. In addition, remote programming now will do almost anything manual programming can, so convenience and reliability factors have become astonishing.

Tuner Programming

As usual, to begin with, let's go back a few years and look at some initial tuning systems to "get the feel" of digital control as it applies to both channel selection and visual displays. Then we'll expand to include current practices of putting most or all tuning and control settings as visible readouts on the receiver's cathode-ray tube. In this way, understanding won't be as difficult as though we immediately described a highly complex system such as some of the TV/monitor sets currently being released to the market. You may find this approach helpful and, as usual, consistent with as much tutorial help as possible.

Probably the best and most useful explanation of a system buildup we've come across is one appearing in the IEEE Consumer Electronics Transactions of some years past by Gregory Ledenback of National Semiconductor. However, unlike the original explanation, we'll start with the phase-locked loop (PLL) portion, then work through the various system phases in an attempt to make electronics life bearable for even those not passably familiar with such logic hardware.

The PLL begins with a fixed-frequency oscillator which is divided down to some reference frequency and compared to that of a varactor-controlled tuner (Fig. 8-6). A 1200 bit read-only-memory (ROM) determines this division from the particular U/V channel selected and delivers appropriate instructions to the phase comparator receiving constant outputs from the oscillator. The phase comparator, in turn, detects any difference between the crystal and tuner selected frequencies and returns a dc correction voltage to the tuner to lock in the desired signal. Also connected to the program counter is a 4 MHz clock, divider, and stall detector output that aids in the programmed countdowns.

Other outputs from the ROM include outputs for U/V band selection (which we've just discussed) as well as hi-lo VHF switching. There's also a mode control to switch from PLL to automatic fine tuning (AFT), with AFT coming from the receiver and the slope position of the video carrier. You also see control logic and up/down converter in the upper left portion that's subject to PLL operations. Outputs here are both system time and 4-line binary coded decimal (BCD) commands into the 1200-bit ROM.

The PLL controller must then interface with a keyboard and display circuit having the same

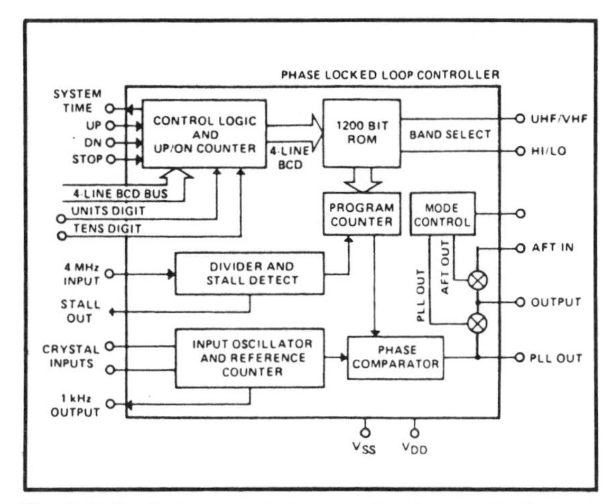

Fig. 8-6. Block diagram of phase-locked loop controller and its operations (courtesy National Semiconductor).

BCD multiplexed output which connects both to the PLL and a display bus (Fig 8-7). In this way, the interface circuit controls the bus and the PLL assumes a *slave* state with keyboard control of the channel selecting ROM. The PLL reference counter frequency of 1 kHz inputs the keyboard decoding and control logic block, which also accepts keyboard scan and memory stop, in addition to interfacing with an up/down counter, clock, and video display controller. The keyboard decoding portion contains scanning and other logic and also outputs memory sync, while the clock and remote interface responds to up/down signals as well the 4-line BCD output it furnishes to the control logic and up/down counter on the PLL.

Keyboard or remote input format may either be a number and "channel select" or a series of two numbers for each channel with no channel select, with an up/down system having two scan rates depending on whether the button is pushed and held or simply engaged individually. On screen readouts originate from a BCD 7-segment character generator with placement controlled by horizontal and vertical sync inputs. Clock times are also displayed the same way after being digitally programmed via the video display controller and output decoder. An external RC oscillator, apparently, provides the clocked output.

Both up/down and direct access channel tuning are combined in Fig. 8-8 for convenient comparison. The entire controller uses P-channel, metal gate,

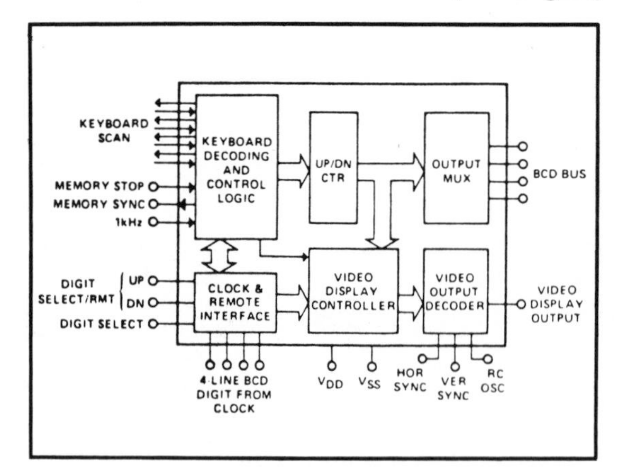

Fig. 8-7. The keyboard and display interface logic (courtesy National Semiconductor).

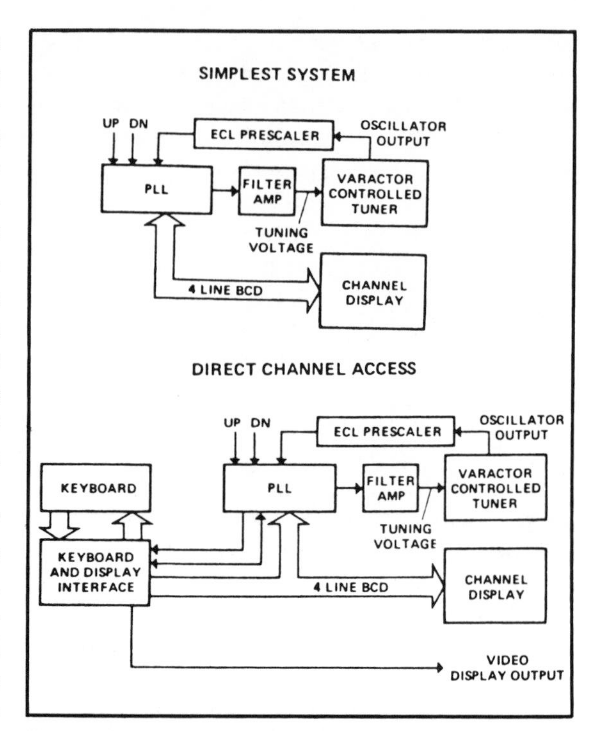

Fig. 8-8. Both up/down and direct tuning systems are block-diagrammed by National Semiconductor.

ion-implant technology in the form of building blocks to accommodate either low cost or more feature-inclusive accommodations for the better receivers. In those days, control keyboards had but 16 functions including direct address, up/down, channel select, display, something called instant tune, and time. Today many advanced systems have 32 or more commands and considerably more flexibility to control video, VCR, and audio, with some determining ranges of color, tint, and sharpness, as well. A great deal has taken place in both convenience and technology in the past decade—and that's just a comfortable beginning. Another 10 years may see many of us working from our homes in a multitude of occupations that normally require an office or other enclosure far from the immediate dwelling.

The foregoing should have prepared you, at least in block diagram form, to accept receiver keyboard and remote tuning for precisely what it is: logical access and precise channel selection for all U/V/CATV and satellite transmissions (second hand) that are available in the middle and late 1980s. What we will now consider are some of the actual logic operations that make all this possible. Therefore, be prepared for some rather stiff electronics and sophisticated programming as we work with microprocessors, ROMS, RAMS, gates, flip-flops, and nonvolatile memories.

G.E.'s Electronic Tuning

General Electric's tuner system is a very good example of advanced sophistication to begin with, having some very interesting features. The remote function for this particular system is especially noteworthy for its 12-hour channel Lockout and chairside channel programming. In fact, G.E. is still the only receiver we know that can deliver full broadcast station select programming in memory from its remote.

The remote, itself, has no memory but is a relatively conventional IR (infrared) transmitter operating on two AA 1.5 batteries, clocked with a 455 kHz ceramic resonator and divided down to a carrier frequency of 37.9 kHz. This carrier is keyed by a pulse position code developed by IC901 with mute, power, volume, up/down channel, direct address, enter, and add/clear channel programming. Each keyboard entry creates a different 8-bit data code, acceptable to the receiver's microprocessor *only* if the correct code is entered, thereby preventing undesirable access from outside sources such as other transmitters.

The PIN diode detector accepts the infrared carrier transmissions and connects the signal to the input of a linear amplifier IC with a voltage gain of 1,000. With further amplification and limiting, peak detection delivers a logic output to the receiver's internal IC600 logic processor which carries out instructions from microprocessor IC601.

Microprocessor IC601 receives either remote or keyboard command signals, initially turning on ac power, supplying audio control voltages, channel tuning energy, LED channel indications, PLL data, and storage of favorite and last channel on nonvolatile memory IC800 located nearby (Fig. 8-9). There's also comparator AFC monitoring for an automatic fine tuning search called channel *block-out.* LED display immediately responds with key entry for units and then tens digits, or reverts to a prior channel if *enter* doesn't follow in five seconds.

The microprocessor has an oscillator operating at about 400 kHz. When powered initially, all counters are reset and I/O chip IC600 is also energized. The microprocessor will then read the Electrically-Alterable Read-Only-Memory (EAROM) when the receiver is initially switched on following microprocessor reset, after which *all* channels may be programmed for U/V/CATV frequencies at the operator's pleasure. Also in the microprocessor is a program counter and subroutine register, a 1024 ROM with 8-bit words, an R output latch, a 64 RAM with 4-bit words, decoder instructions, X, Y, and accumulator registers, and output latches.

IC600 is an I/O 44-pin flat pack containing a shift register, pulse-width D/A converter, keyboard data multiplexer, LED display driver, reset counter, bandswitch driver, and an internal RC clock that pumps along at some 350 kHz (Fig. 8-10). Information entering pin 27 goes to a 16-bit serial-in, parallel-out shift register which is clocked at pin 30. The clock also operates the microprocessor's reset counter and the D/A volume control converter with all else in the I/O slaved to the microprocessor data and clock outputs.

A data enable pulse into pin 29 results in the contents of the 16-bit shift register being loaded into a 16-bit memory, where it responds to a 16-bit code. Outputs control keyboard or remote inputs, LED display drivers, bandswitch, and audio volume outputs.

Data input to IC500 arrives in 4-bit words and written to six internal latches identified as L1 through L6. Load Address Enables are sent when pin 5 is pulsed, with frequency data loaded as pin 4 is pulsed (Fig. 8-11). The initial three 4-bit words are VCO channel frequencies in binary, and the final three 4-bit words are 31.25 kHz offset steps in binary.

For tuner control, phase-locked loop action from IC500 can synthesize local oscillator frequencies to 1023 MHz, with +2 MHz being the

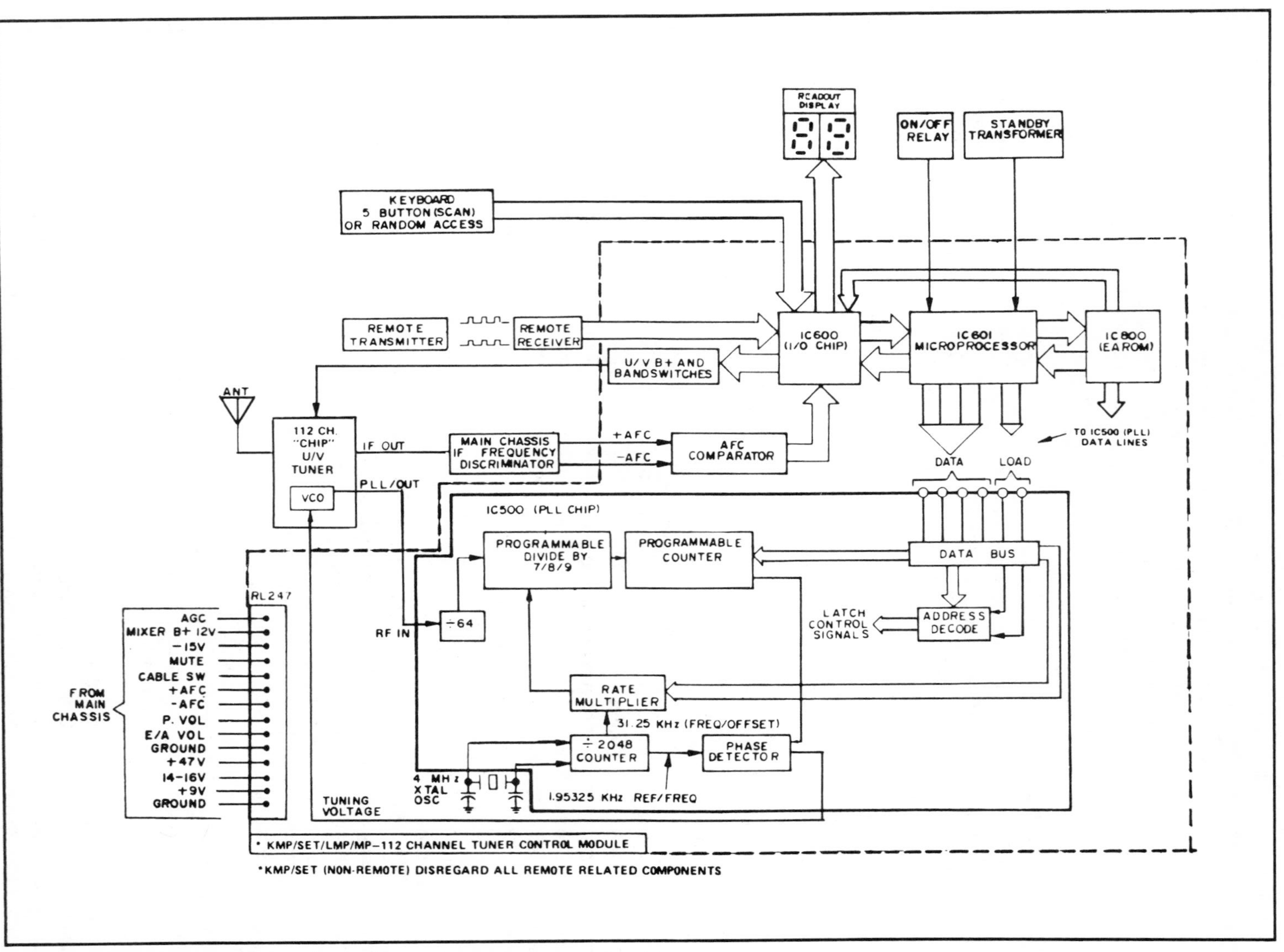

Fig. 8-9. 112 channel tuning system block diagram (courtesy General Electric).

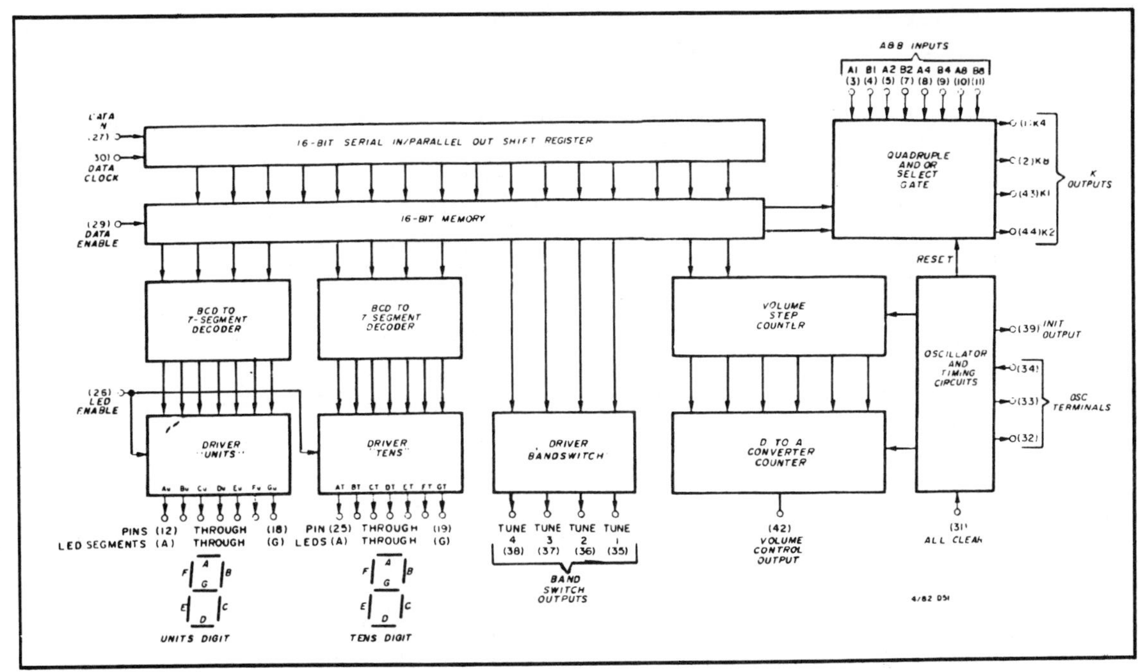

Fig. 8-10. An internal block diagram of the IC600 I/O integrated circuit (courtesy General Electric).

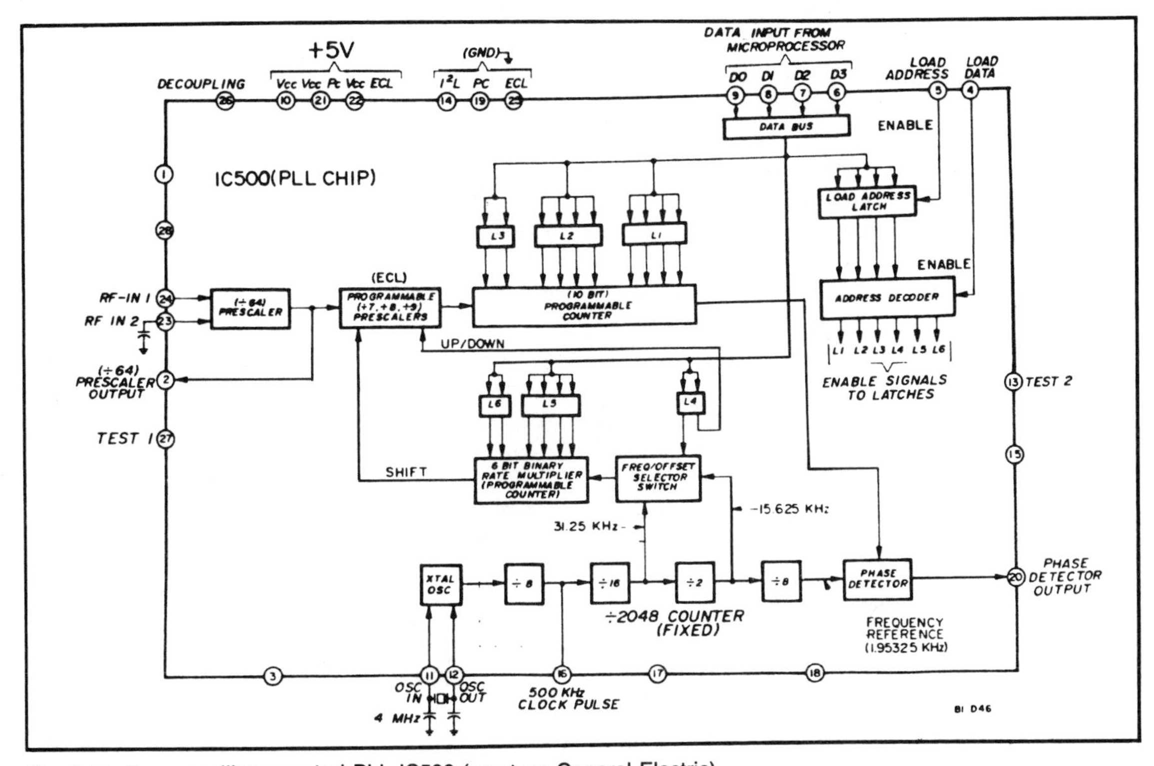

Fig. 8-11. Tuner oscillator control PLL IC500 (courtesy General Electric).

maximum deviation permitted in frequency offset. This data and channel divider ratios are loaded sequentially by the microprocessor and transferred to the PLL programmable divider prescaler and rate multiplier. Incoming rf is first divided by 64 and then divide-by- 7/8/9 prescalers so that samples of the tuner's local oscillator (LO) can be compared with synthesized frequencies generated by the 4-MHz crystal oscillator at the bottom of the diagram.

LO outputs are further divided by a 10-bit programmable counter whose outputs continue to a phase detector to be compared with the various divisions of the 4-MHz oscillator and clocking by a 500-kHz clock pulse, amounting to a divide-by 2048 fixed counter. The two 1.953 kHz frequencies are then compared, with any differences going to an adaptive loop filter especially designed to handle deviations of less than 2 MHz, then on to the tuner's LO. There's also an AFC comparator (not shown) operating on IC600 which will also increase or decrease frequency as the occasion demands.

Audio circuits receive their volume controls from IC600 via pulse-width modulation ranging between zero and 50 percent of the duty cycle in a total of 64 steps. Such steps are at the rate of 10 per second.

Two additional operations in these control circuits should also be noted because they are important in channel capture and exclusion: AFT search and General Electric's special Blockout.

On either remote or receiver keyboard whenever the Enter button is engaged, the AFC detector is sampled following a 160 msec delay. Should the response be positive, a +250 kHz offset is loaded into IC500; if negative, a −250 kHz offset is loaded instead. Sampling continues at 30 msec intervals until a change takes place in AFC detector output. After that, the step size reduces to 31.25 kHz, checking in both directions for any subsequent AFC change. Four consecutive 31.25 kHz crossovers with only one same-direction step completes the action and the PLL locks at the correct channel frequency. AFT search limits are +2 MHz and −1.3 MHz away from channel frequency and will not continue after initial lock without another Enter button command.

Blockout commands are processed through the IC800 EAROM (not individually illustrated) which operates with an address accept code that is clocked into IC600, and then a read code. A 5-button scan keyboard and random transmitter access allows the *blockout* to be programmed for a full 12 hours, cancelling any channel in the tuning range that might contain undesirable information. A 9,9 entry results in display blinking for five seconds, channel numbers to be blanked are next entered preceded by a minus sign, display continues to blink for another five seconds and then changes to 99. A 4-digit security code of any random numbers (but you better remember them) is next entered, and then the Enter button seals the exclusion. To unblock these channels before the 12-hour period terminates, 9,9 must be re-entered, followed by the proper 4-digit code. Otherwise, these channels will continue to be blacked out for a full 12 hours ± 12 percent.

Without trying to analyze virtually impossible details, the preceding should suffice to at least acquaint you with the highly complex operations of contemporary tuner controls. Some are less severe, others somewhat more so, especially where video 1,2,3, several sets of rf inputs, and video/audio outputs are included also. Perhaps these can be covered in the receiver section since such latest additions apply equally to both digital and analog receivers because all the tuning controls are digital anyway.

Tuner Timing

Since most of this is done by computers these days, program charts are more in order to bring you up to date than basic timing diagrams, but such charts don't mean a great deal unless you have the entire system or a large subsystem laid out before you. Since we've just done that, a repeat wouldn't add that much new information and would probably go unread. So we'll stick to the original schedule and uncover a fundamental timing diagram just to illustrate what the programmer and/or designer has to do initially to get his "show" on the road. A somewhat antiquated one but entirely applicable illustration is available from Motorola which includes both a system flowchart as well as the *basic* timing diagram rather than a series of extensive 1s and 0s which might be even more confusing than helpful; therefore, we'll go with this approach and keep everything simple. By tuner timing, of course, we mean overall equipment timing also—the various controllable functions.

The software flowchart is a good start for any programming venture because it includes all the basic operations using an 8-bit NMOS microprocessor and a CMOS synthesizer for tuning and bandswitching. What the designer had to consider when making his determinations and the various 1-9 button sequences used for the purpose precedes since you must outline objectives first before actually entering the program sequence. Since both graphic explanations are appropriate and virtually self-explanatory we'll include them "as is" with an accompanying word or two of description.

Numbers and circles are the keyboard buttons to be used in channel designations, up/down scans, set on/off, clear, audio up/down, etc., followed by time set, programming, mute, channel recall. The Description column should tell the rest. The entire diagram outlines what is intended so that programming of that intent may follow. See Table 8-1.

The software routine is very direct and should not be at all confusing. After Start there's Power Up, a No-Yes 60N loop test, a remote code operation, clear key and timer sequence, a code check, and D/A data refresh. There are some more steps but they're not wholly essential to your becoming familiar with what the programmer has to consider. And at least, here, you have system startup with the tuner system ready to program, including D/A converters for such items as volume, picture, brightness, tint controls, etc. See Table 8-2.

Implementing the foregoing from the remote control, Motorola selected frequency-shift keying

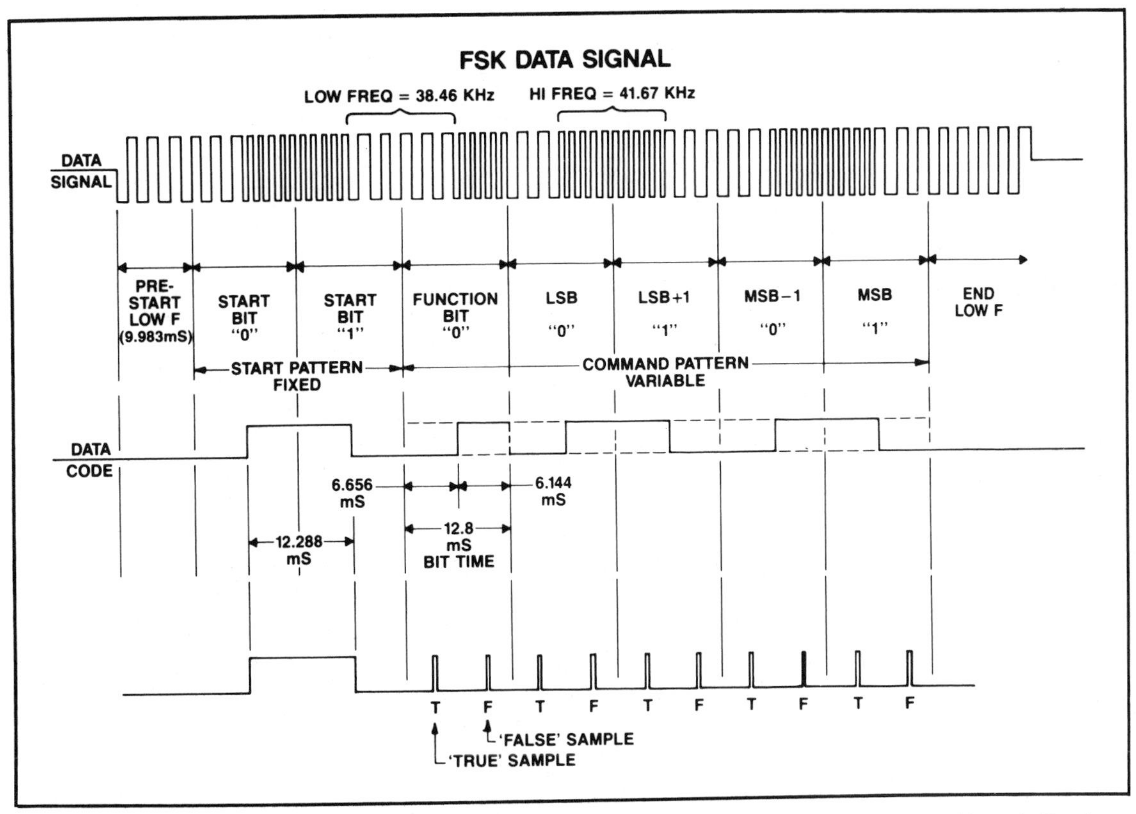

Fig. 8-12. Remote frequency-shift keying transmits all receiver commands in this application (courtesy Motorola Semiconductor).

Table 8-1. Keyboard Functions.

FUNCTION	0	1	2	3	4	5	6	7	8	9 (SET)	AU	PX	HU	SC	TM	MU	CL	RE	↑ (ON)	↓ (OFF)	DISPLAY	DESCRIPTION
KEYBOARD BUTTONS — ○ INDICATES KEY IN SEQUENCE																						
SET ON / CHANGE CHANNEL	①							②													0 7	LEADING ZERO SUPPRESSED ACCEPT FOR INITIAL INPUT
CHANNEL SCAN UP / 'DOWN'														①						②	LAST CHANNEL	STOPS ON NEXT "GOOD" CHANNEL
SET OFF																	① ②				——	TWO KEY PRESSES TURNS SET OFF
CLEAR																	①				LAST CHANNEL	SINGLE KEY PRESS RETURNS KEYBOARD TO STANDBY MODE
AUDIO UP / 'DOWN'											①								②		LAST CHANNEL	
PIX 'UP' / DOWN												①								②	LAST CHANNEL	STANDARD CUSTOMER CONTROLS
TINT 'RED' / GREEN													①						②		LAST CHANNEL	
TIME DISPLAY														①							12 45 36	HOURS THEN MINUTES THEN SECONDS } ONE CYCLE THEN RETURN TO CHANNEL DISPLAY
TIME SETTING	③		⑤	⑦		④		⑧	⑥	①					②						(blank) 05 28 37	• DISPLAY BLANKS WAITING FOR HOURS INPUT • HOURS DISPLAY THEN BLANKS WAITING FOR MINUTES INPUT • MINUTES DISPLAY THEN BLANKS WAITING FOR SECONDS INPUT • RETURNS TO CHANNEL DISPLAY
PROGRAMMING SET 'ON'	⑦ ⑧	④		⑥					⑨	① ⑤					②				③		(blank) 19 (blank) 30 (blank) 88	• DISPLAY BLANKS WAITING FOR HOURS INPUT • HOURS DISPLAY • THEN BLANKS WAITING FOR MINUTES INPUT • MINUTES DISPLAY • THEN BLANKS WAITING FOR CHANNEL INPUT • DISPLAYS CHANNEL THEN RETURNS TO LAST CHANNEL
PROGRAMMING SET 'OFF'	⑤ ⑥ ⑦		④							①					②					③	(blank) 20 (blank) 00	• DISPLAYS BLANKS WAITING FOR HOURS INPUT • HOURS DISPLAY • THEN BLANKS WAITING FOR MINUTES INPUT • MINUTES DISPLAY THEN RETURNS TO LAST CHANNEL
SOUND MUTE																①					LAST CHANNEL	INTERCHANGES TWO ADJUSTABLE AUDIO LEVELS
RECALL CHANNEL																		①			NEXT CHANNEL	INTERCHANGES TWO OR THREE "FAVORITE" CHANNELS FOR CHANNEL HOPPING

Table 8-2. Software Routine.

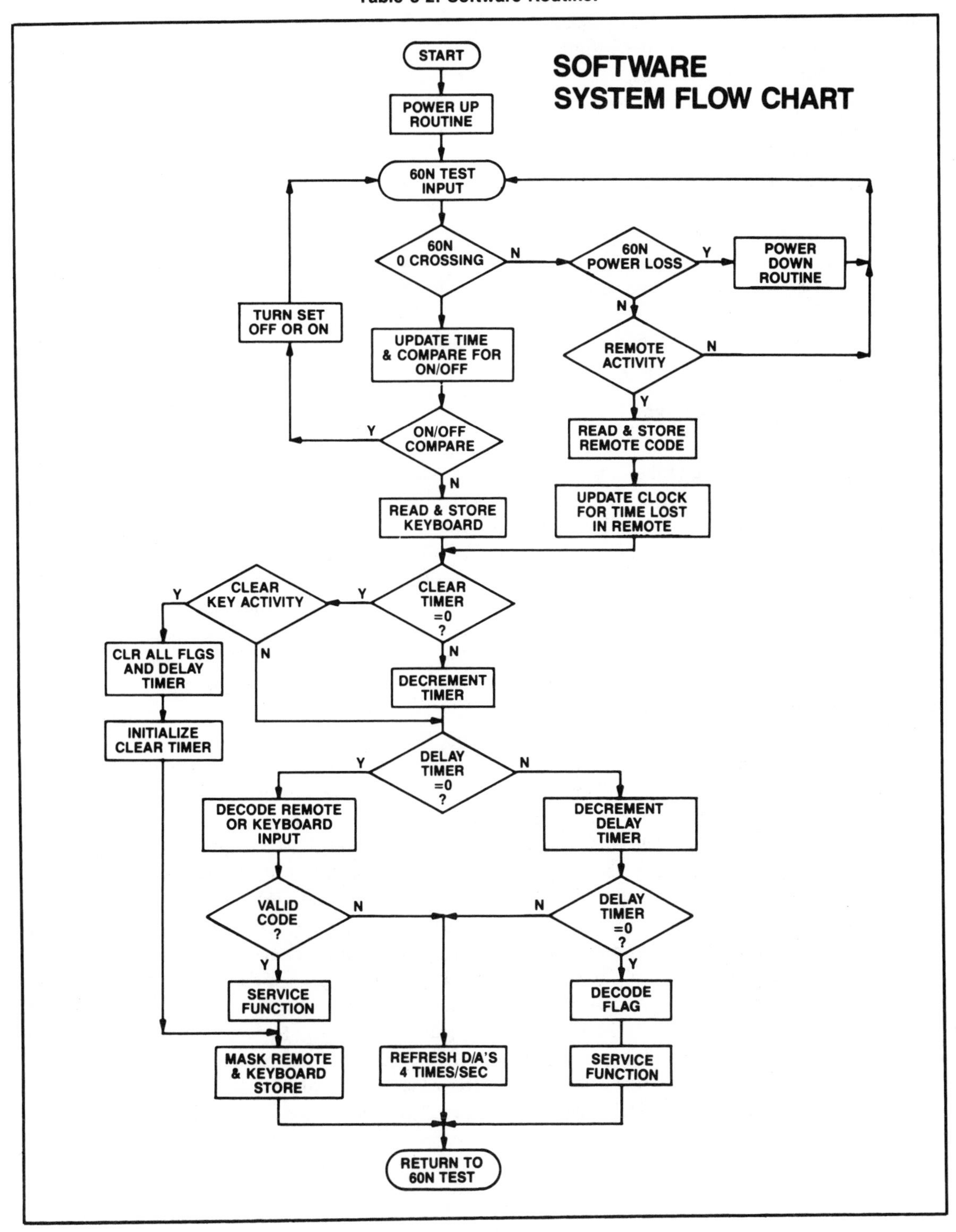

(FSK) and a CMOS IC as the hardware. It can generate a 5-bit, 32-data word code, decoded easily with phase-locked loop. The data signal is illustrated in Fig. 8-12 in both low- and high-frequency data conditions, along with function bit signals and data code timing. Both start fixed patterns are identified, with variable patterns following to the end of the data signal. Extraneous interference or incorrect signals are ignored by the receiver as it awaits corrected data before executing the remote commands. A block diagram of the entire system is much like the one just explained for General Electric so, in the interest of space and redundancy, it won't be repeated.

CRT Alpha/Numeric Displays

Translated, this means read/write numbers and letters on the cathode-ray tube. It's a very convenient technology instituted several years ago by both ourselves and the Japanese and elevated to prime status in 1985-1986. Not only are channels, time, video and rf indicated, but the latest wrinkle involves pickets of chroma, tint, sharpness, luminance, volume, picture, bass, and treble spread across the lower half of the screen—selectable as to varying degrees of lows or highs symbolized by the length of the respective pickets. All this, of course, is in color, with one prime color identifying each major function. Rf inputs, for instance are in green, and video in blue. Stored recall might have a red border, while something else might be in between. If you were persuaded that character generators weren't a part of television, then you have another think coming. They're all sorts of characters involved with TV today.

For our example, we've selected a Panasonic NMX-L4 chassis and display system to work with since it may have some worthwhile variations from run-of-the-mill receivers already in the field (Fig. 8-13). This one has microcomputer control, a phase-locked loop (PLL) frequency synthesizer, character generator, and nonvolatile MNOS memory that stores all sorts of information.

Keyboard or remote control furnishes microcomputer inputs that then converts them to 8-bit channel information. From this, 40-bit PLL division and tuning-band selections are generated, including the CATV switch position command. Band and divider data continue to the PLL and band-switching blocks, which control the tuner's local oscillator, along with feedback to the PLL that locks the oscillator to its proper frequency.

From the data address, information goes to both the MN1228Q and 82-bit favorite channel data. With programming via the microcomputer, when a channel is selected, a 1-bit is chosen, and when a channel is erased, the computer reads a 0 bit. With programming, up/down channel search finds the nearest 1 bit and stops in whatever direction the operator commands. So that off-frequency channels from video games and computers don't confuse the PLL, automatic fine tuning immediately operates as soon as any channel is tuned, pursuing the offending frequency. The microcomputer also changes its tuner counter setting and full frequency lock is established. Both main and swallow counters are used in this operation.

Having the tuning cycle established and favorite channel data set in nonvolatile MNOS memory according to specific address, specific channel information, time, or whatever else may be programmed can appear on the CRT via on-screen display MN1227A (Fig 8-14). This is an interesting IC and considerably complex since it must be clocked, have horizontal and vertical position controls, a timing generator, data memory, and character size control. The final output goes through a series of NAND gates, reaching the RGB final CRT amplifiers and its three cathodes.

Horizontal and vertical positioners determine the display positions of the various characters, relying on horizontal (HD) and vertical deflection (VD) pulses, counting from the 5.8 MHz clock. Four characters sizes are available in this particular application, according to computer selection, as well as color display. Character data memory has 60 RAM storage locations, each addressed by a 7-bit code, while letters and numbers have 6-bit address codes. When characters are displayed, that part of the picture used is erased for the second or two any character remains on the screen.

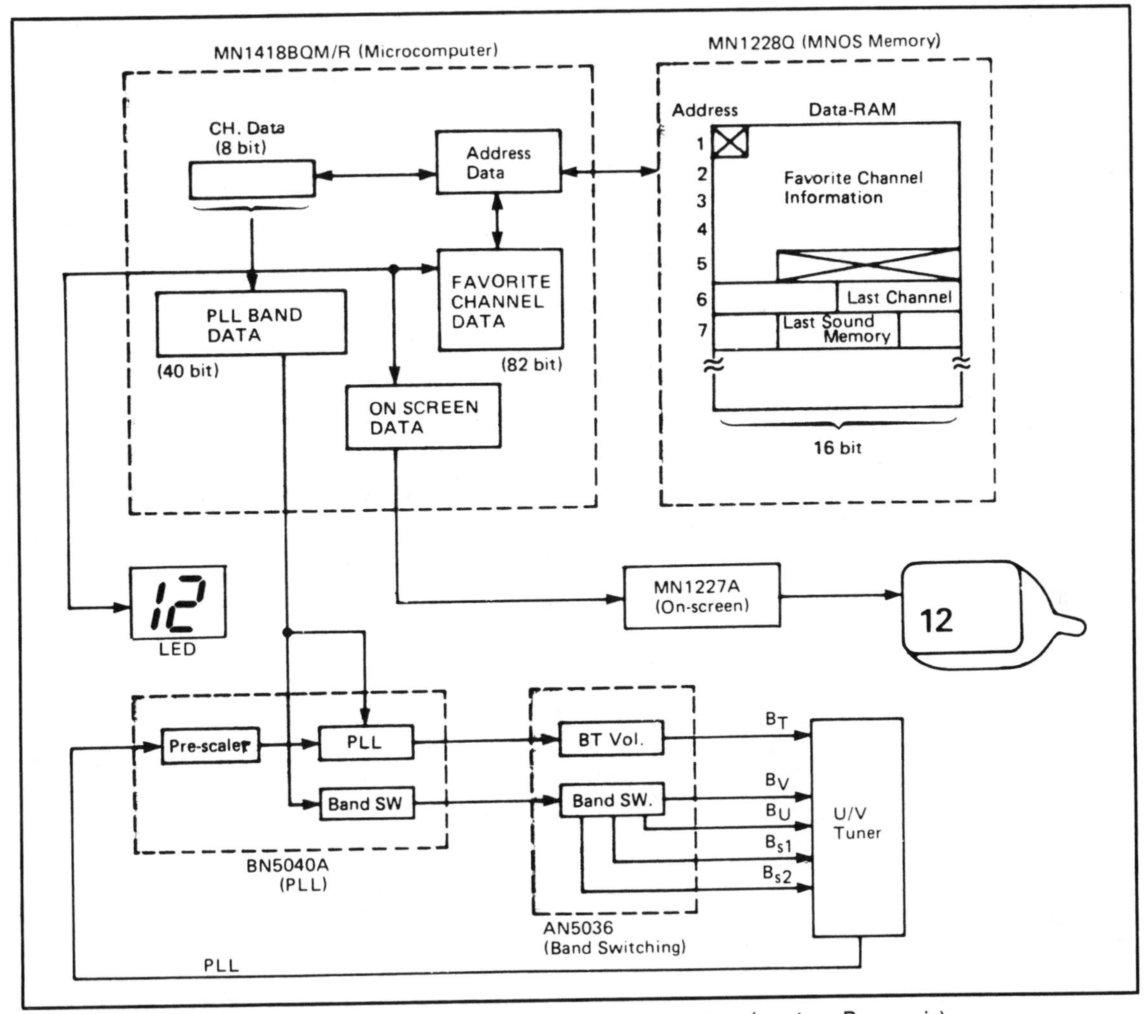

Fig. 8-13. NMX-L4 chassis bandswitching, data display, and tuning system (courtesy Panasonic).

VIDEO I-FS

Intermediate frequency amplifiers are not exclusives of television or radio receivers, but are common to almost all communications equipment to boost selectively tuned front end signals into usable information, which may be shaped, detected, and audibly or visibly displayed. In the early days of television a 3-stage video i-f amplifier was a noteworthy achievement, and four stages were found only in the *most* deluxe sets on the market. Then there were some early color receivers that tried to get away with two, but were highly unsuccessful and extremely difficult to align. In those days, of course, every stage was either RC or transformer-coupled, usually the latter, and either stagger (offset frequencies) tuned or overcoupled, producing i-f responses such as those illustrated in Fig. 8-15.

Having a great deal to do with efficient i-f amplifiers, was the coupling and filter network between the tuner and i-fs. Prime objective here was to "anchor" the various frequencies emanating from the tuner so that sound, video, and chroma appeared where they should on the intermediate frequency response curve. If they didn't, then the video passband would be narrowed, sound might

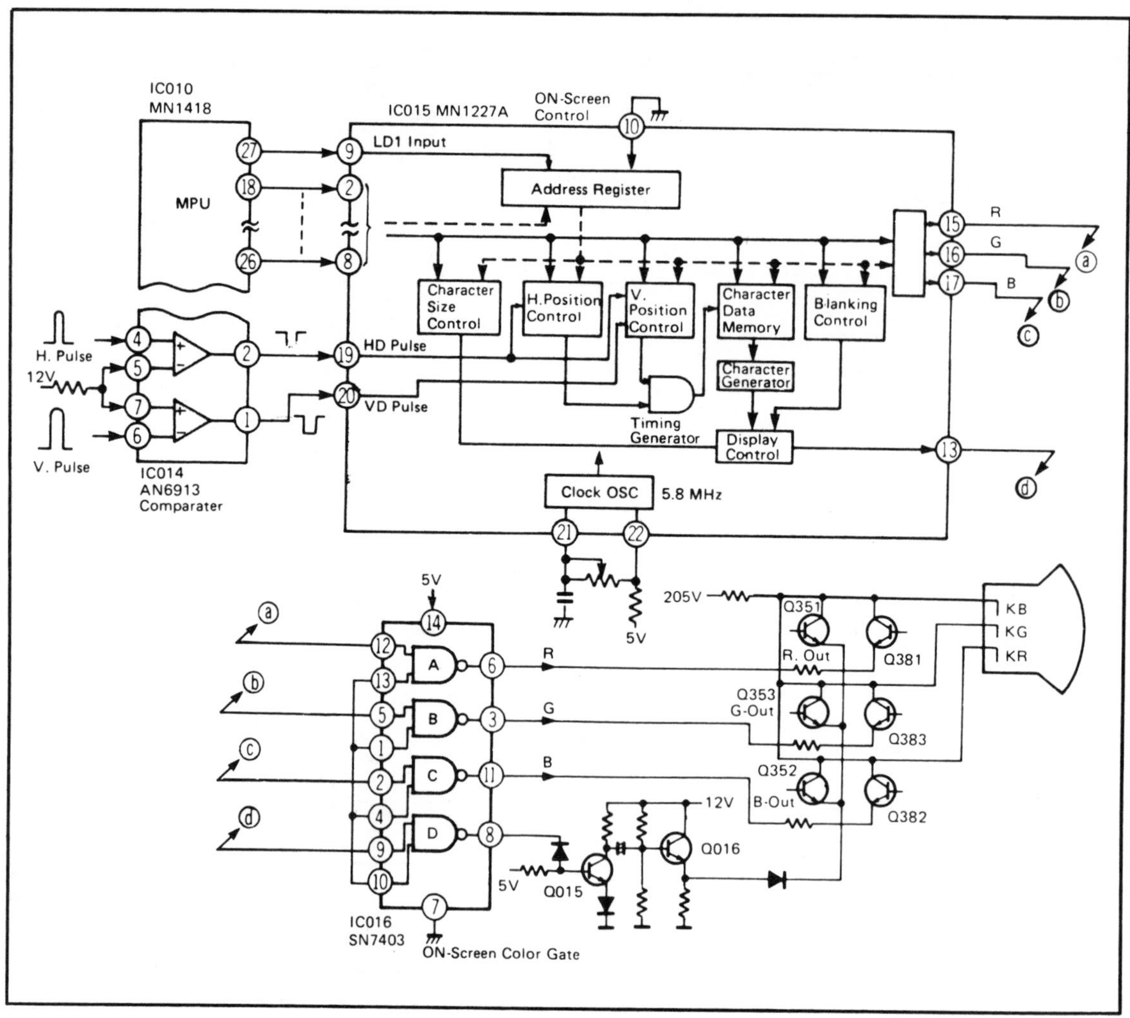

Fig. 8-14. The MN1227A on-screen alpha-numeric display IC and outboard logic gating (courtesy Panasonic).

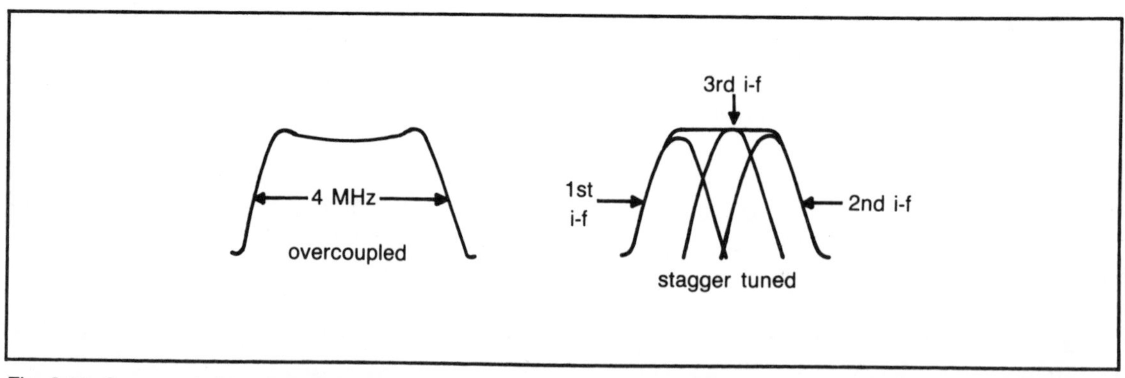

Fig. 8-15. Overcoupled (or doublecoupled) single i-f response versus stagger-tuned version. Both can deliver adequate bandpass, but stagger tuning is less expensive.

get into video, and chroma could well become nonexistent. Nowhere is this more evident than on a sweep generator's response curve during i-f alignment—a point that will bear considerably more discussion in the final troubleshooting chapter. Older receivers, whether vacuum tube or transistorized, are especially vulnerable to i-f problems or drift as passive components and active devices age under the influence of heat and time.

Fortunately, however, with the advent of excellent integrated circuits, alignment is but a negligible factor anymore, especially if there are no inductors to peak or capacitors to tweak. As long as there's substantial impedance matching between a broadband tuner and broadband i-fs, alignments are seldom attempted or required. If they are, there's usually the synchronous video detector to contend with that will destroy most i-f waveforms without a very special demodulator probe. We will, however, investigate this further in Chapter 15. For now, let's return to the i-fs proper, look at a pair of the oldies, and then analyze one of the newest we can find for your anticipated edification. As you will discover, i-fs and their allied circuits are some of the least difficult in the receiver—at least to explain, but not to design. For engineers they are tough! We might also say that in the chapter on multichannel sound, we'll go through the i-fs again, but this time emphasizing *both* sound and picture, since for adequate stereo-SAP audio-video separation, these new i-fs have to be excellent or there are *serious* problems.

Tuner-I-f Coupling Networks

We'll exhume a couple of antique service manuals as the first of these for tubes and transistors, then segue as gracefully as possible into surface-wave acoustical (SAW) filters, hopefully imparting some worthwhile information along the way. Fortunately by 1973 most color receivers were either partially or wholly transistorized, so it's doubtful if even history is worth that much regression. But, as promised, we will select a couple of oldies in the solid-state category and explain the coupling networks of the '70s era.

The first appears in a modular M20 chassis by Admiral, which used to be a big name in the business, especially black and whites (Fig. 8-16). Tuner information enters the i-f input network across bandshaping coil L200, then into a bridged T 47.25 MHz and parallel resonant shunt 41.25 MHz traps that became familiar circuits with solid-state receiv-

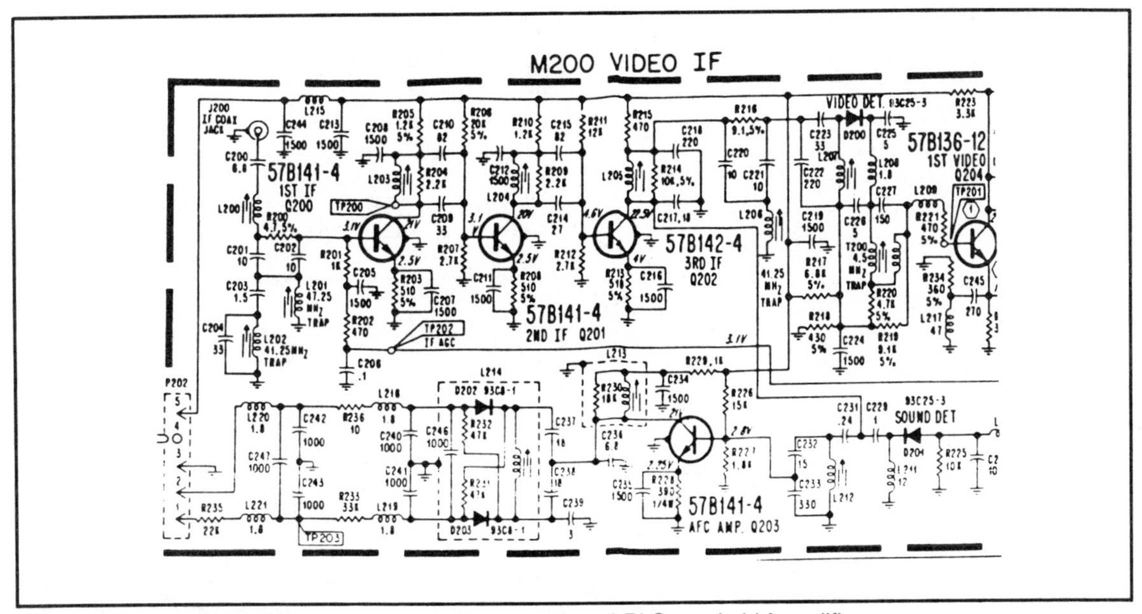

Fig. 8-16. An early Admiral M20 chassis coupling network and RLC-coupled i-f amplifiers.

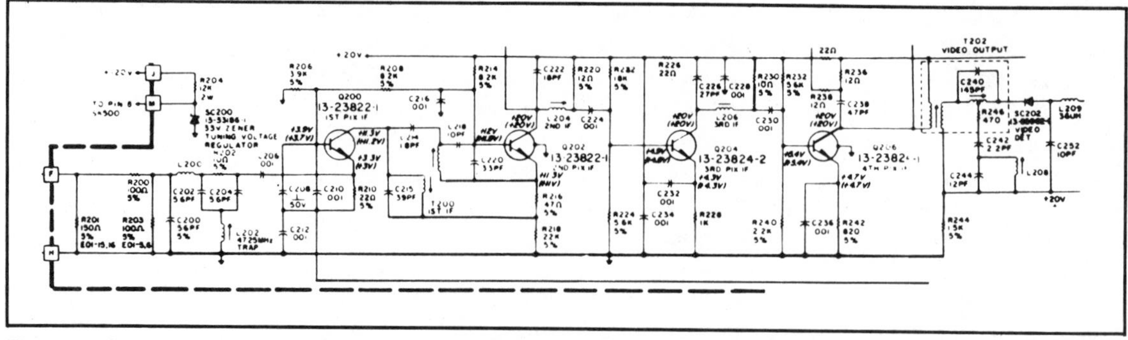

Fig. 8-17. Sylvania's EO1 four-stage video i-f with weak/strong signal bandpass response (courtesy NAP Electronics).

ers having discrete i-fs. The 47.25 MHz trap is the lower adjacent sound channel trap, while 41.25 MHz audio carrier must be kept at virtual ground reference on the response curve to offer all necessary response waveshaping and to keep audio well separated from video. In this coupling network, there are but three series and shunt inductors to tune, while some others in those early days had more. Note that i-f agc is applied directly to the base of the first i-f amplifier, which was not especially good design engineering because of obvious loading. There is further and final 41.25 MHz sound trapping following the i-f amplifiers to prevent all sound from entering the succeeding video baseband amplifiers.

The second example (Fig. 8-17) comes from a Sylvania and an EO1 solid-state receiver then manufactured in Batavia, New York, but now part of the North American Philips (NAP Consumer Electronics) organization in Tennessee.

This was a 1971 set, and Sylvania's first all solid-state effort. Coil L200 impedance matches between tuner mixer output and i-f amplifier input, with C202, C204 and L202 forming the 47.25 MHz lower adjacent channel sound trap in a high pass constant K configuration. T200 in the collector of Q200 is adjusted both for good passbanding and also to position the 41.25 MHz sound carrier close to the baseline overall response. L208, just before the video detector, is the final sound carrier trap. You will note this receiver has four video i-f stages, but this time with agc being applied to the emitter of first i-f amplifier Q200 and also by dc coupling to the emitter of 2nd i-f amplifier Q202. Therefore, as the agc transistor conducts less, positive dc bias on these i-f transistors increases and they both conduct less, responding inversely to larger input signals. At *minimum* gain, T200 positions the picture carrier at maximum, drops chroma down, and narrows the overall passband so that maximum luminance is passed until stronger overall signals are once more received, returning i-f responses to normal. Clever, eh?

That should be enough of a convincer on these two types of coupling devices to understand there must be precise alignment of both their responses and those of the video i-f amplifiers to pass maximum signals through to the video detector (then a half- wave diode) and amplifiers. We're now ready to proceed directly to SAW filters and see what they do and why.

Surface Wave Filters

Surface-wave acoustical filters (SAWs) were introduced to television receivers in the later 1970s

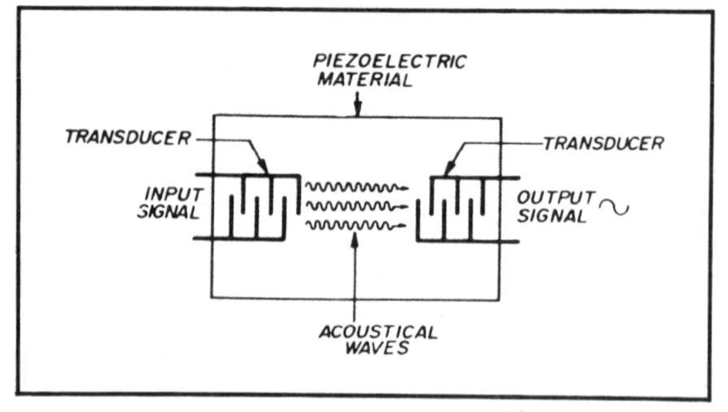

Fig. 8-18. A surface wave acoustical filter (SAW) that bandshapes and replaces most front-end traps.

(Fig. 8-18). Between 1977 and 1982, Toshiba alone produced 40 million of them extracted from Lithium Niobium ($LiNbO_3$) and Lithium Tantalum ($LiTaO_3$), the latter being the i-f filter for television. With transducer electrodes at either end of the piezoeelectric material, a voltage impressed across the crystal distorts it and results in waves being transmitted the length of the crystal at acoustic speed. Here the output transducer develops a proportional voltage for the i-f input. However, since reflected waves off the substrate's edge result in distortions, barriers at substrate ends absorb most or all these undesirable effects. If the input transducer's electrodes are weighted and each pair of fingers responds to different input frequencies, the filter's response looks like that of an i-f curve with very sharp skirts and highly desirable bandpass. Usually a SAW preamplifier is added to overcome insertion losses which are a characteristic of the filter. In most receivers only a 47.25 MHz adjacent sound trap remains to be tuned, with all others now supplanted by this effective device. Failure rates are in terms of 10^{-9}.

Integrated Circuit I-f Amplifiers

Hardly a new development but certainly a welcome one, integrated circuits have now taken over the i-f, agc, sound detection, and video demodulation in all modern television receivers, with additional developments on the way because of multichannel sound. Here we'll discuss the ordinary variety that will be around for some time, leaving further sophistication and explanation to the chapter on TV stereo and SAP, the second audio program that's usually included.

A good example without frills is RCA's CA3153G "gold-chip" dual in-line 16 pin pack that features fast response agc, sample and hold, wideband i-fs, linear video detector, low noise, and an internal shunt regulator. Block diagram and schematic are shown in full in Fig. 8-19 for your convenience.

Signal input from the tuner enters at pin 3 after passing an i-f filter between pins 3 and 5, which has a capacitor i-f clamp. Transistors Q2 and Q4 provide amplification, with Q3 also delivering some energy to Q11 in the agc delay. Transformer coupling between pins 9 and 11 then supply this signal to the detector and video amplifier.

As it is processed, negative-going sync from the video output at pin 16 is applied through Q41 and diodes D8 and 9 to the base of Q35. With inversion, and appropriate biasing from R59, video is clipped and the remaining sync becomes coincident with the keying pulse through pin 1. Q42 turns off because of the keying pulse, C13 charges through the positive sync pulse through D9 with an amplitude proportional to video. This voltage then passes through transistors Q35, Q36, Q38, and Q40 for R57 connected to terminal 2, and becomes the charge current for this external agc filter. During keying, D7 and Q37 offer a constant current discharge path for this capacitor at pin 2. Therefore, the capacitor only retains the *difference* in charge and discharge currents. R57 also connects to Q7, and agc bias to the i-f amplifiers is applied through Q5 and Q6 at pin 8. Tuner delayed agc develops at pin 10 from Darlington-connected transistors Q12 and Q13, whose conduction is partially governed by the manual delay bias at pin 7.

There is also a noise gate in the circuit resulting from C11, R54, Q32, Q33 and Q43. When high spike noise appears in the agc feedback loop from the video output, these transistors conduct and reduce loop gain, dimimishing said interference if not eradicating it completely. The shunt regulator consists of zeners Z3 and Z4, and Q31-Q34, responding to changes in any B+ voltage variations, while C6 and Q24 act as the video peak detector for Q26, Q28, and output Q29.

An Updated I-f

Without going into minute detail, a later in-circuit video i-f, complete with SAW filter appears in Fig. 8-20. The preamplifier delivers adjacent channel sound-trapped 47.25 MHz i- f signals to the SAW filter which bandshapes the response with inductive aid and is differentially applied to the U300 i-f processor. The intermediate frequency amplifiers supply substantial dB amplitude increases while being controlled by internally developed agc voltages. Two tank circuits are used, one for

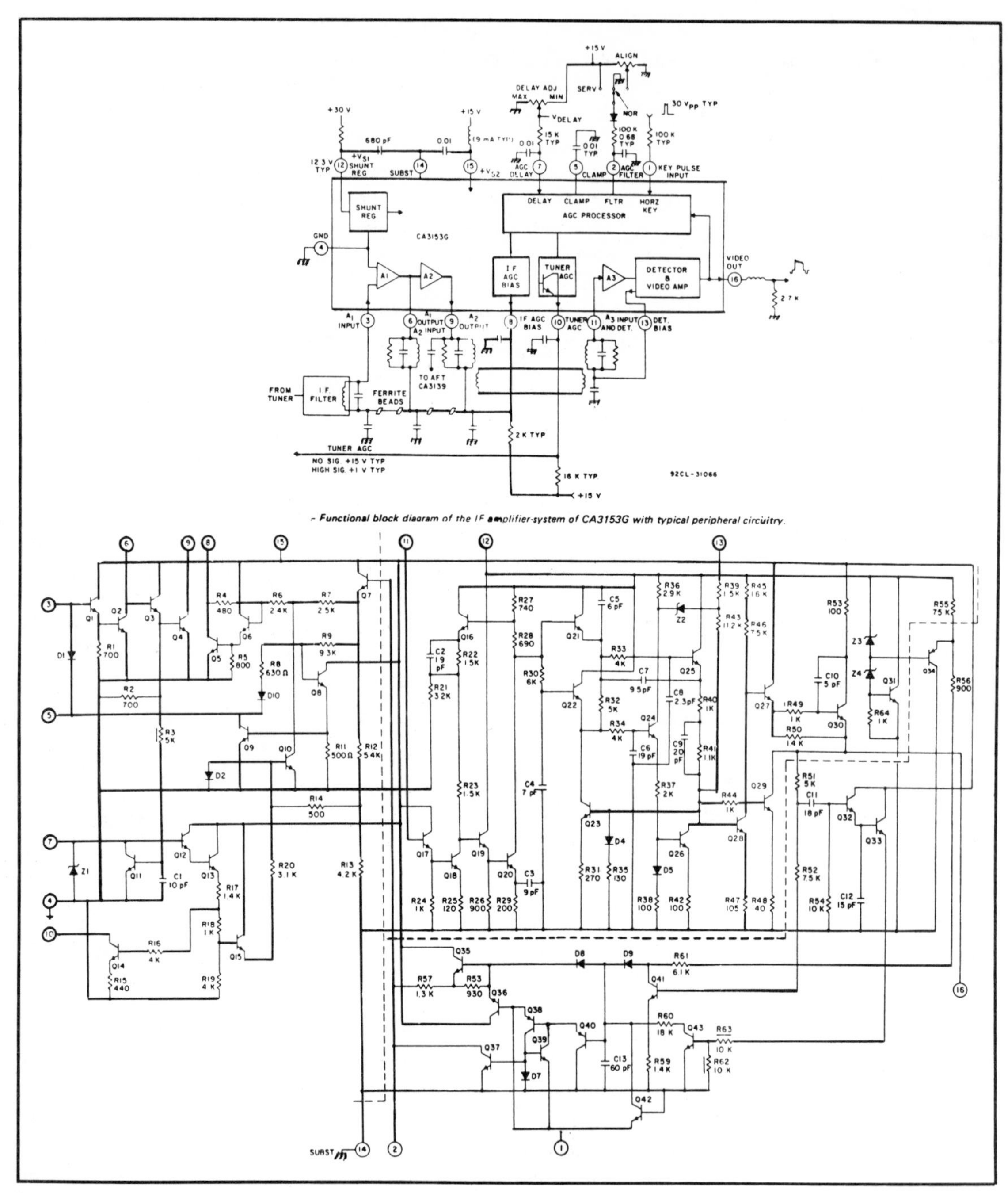

Fig. 8-19. A good working video, i-f, agc and detector IC (courtesy RCA Solid State Div.).

automatic fine tuning and the other for the synchronous video detector. As the video response moves away from its nearly center position on the response curve, automatic fine tuning is applied to the tuner's local oscillator to correct any drift and return it to center frequency. Being synchronously

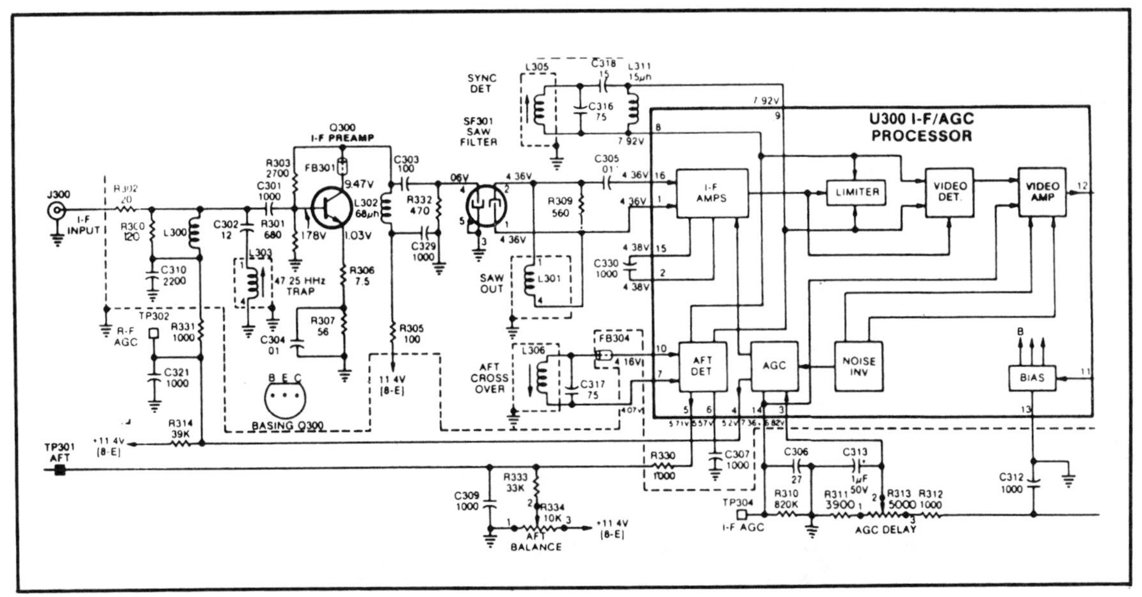

Fig. 8-20. A complete i-f, SAW, and i-f input on a newer CTC121 RCA (courtesy RCA Consumer Electronics).

detected, video amplifier output is now linear and proceeds on toward video amplifiers and also into a centertapped sound coil for audio processing. A 4.5 MHz sound trap later removes any residual sound from video.

Apparently this arrangement of not removing audio until *after* video demodulation has worked quite well for RCA, and a similar system is continued in their later multichannel sound receivers, but with the addition of a special noise reducing circuit on the output.

Chapter 9
Receiver/Transmitter Synchronization

NORMALLY THESE SUBJECTS ARE NOT treated separately in video texts, but they have become so important in recent years with the many new developments that we thought this approach might be helpful for both transmitter and receiver personnel. The NTSC (National Television Systems Committee) specifications haven't changed, but methods of generating said signals have among both broadcasters and the millions of home receivers occupying most American homes today. Therefore, it is our duty to discuss available information as fully as possible for horizontal, vertical, and chroma sync so that you will have full coverage of the subject without having to consult some other reference.

Any of these three synchronous signals must have an initial standard, and that is usually the broadcaster's 3.57945 MHz TCXO-controlled color subcarrier oscillator. Subsequent countdowns, of course, produce the 15,734 Hz horizontal frequencies and the 59.94 Hz color vertical repetition rate. Black and white rates are just a fraction higher at 15,750 Hz and 60 Hz, respectively, but are probably only used now in amateur and possibly a few low-power stations, where color is either too expensive or impractical. So we'll just concentrate on color since it is close enough to black and white to be compatible and serve both systems comfortably. However, as videocassette recorders, cameras, video discs, video games and computers become even more prevalent, receiver sync needs to be updated and evaluated very seriously because of the effect some of these peripheral devices have on the resulting TV picture. In the cheaper sets without monitor facilities and special filter and shaping networks, results are sometimes disastrous. Any computer or any video game, won't necessarily work with a low quality television set and certain modifications are usually needed at considerable expense, if available at all.

Transmitters, however, by whatever electronic means they choose, keep churning out the same NTSC patterns as before, consequently we might as well work through these FCC-approved specifications, tackle a transmitter sync example, then analyze receiver problems and circuits. At chapter's end you will also be informed of the vertical blanking interval and its importance to both transmitters and receivers alike, plus VITS and VIRS test signals now largely "deregulated," but totally useful for broadcaster and set manufacturer. They're still the standards for satisfactory chroma, luminance, bandwidth, and other measurement parameters, whether politicians concur or not. Most of these test signals *will* be used in the final chapter devoted to troubleshooting. See if they don't all make sense. We'll even add extra information on color sidelock generators that can help, too, for those of you in receiver repair or evaluation.

THE SYNC SIGNAL

As illustrated in Chapter 1, this is much the same information (Fig. 9-1) given before but with the addition of two fields of line 21 which has been designated by the FCC as captioning for the deaf. Seven cycles at 0.503 MHz are specified for clock run-in at an elapsed time of 12.901 microseconds. The start bit begins after 27.45 μsec from the leading edge of the sync pulse and two 7-bit + parity ASCII characters follow in field one. In field 2, there's a 9-bit pseudo-random framing code, followed by an active video area of 34.9 μsec also in framing code.

As before, all equalizing, vertical, and horizontal sync pulses are given in terms of a horizontal line's duration, and you see the horizontal sync pulse, itself, followed by a minimum of 8 cycles of burst on its back porch. Below are measurements for both the equalizing and vertical sync pulse. You may remember there are six equalizing pulses before and after the six vertical sync pulses, and horizontal pulses occur only once each scanning line, all sync operations occuring during the vertical and horizontal blanking intervals when there is no picture on the screen, only nonviewable retrace.

Further into the chapter we will discuss the vertical blanking interval in detail, illustrating its contents, and also describing VITS and VIRS test signals which the FCC says are *not* now mandatory. We will also have a word or two dealing with Teletext—an alpha/numeric system imported from Great Britain that has some staunch adherents but who are still relatively few in number and will probably remain so until a way can be found to turn commercial-type data transmissions into a pound of fast bucks. At the moment, the quest seems quite dear.

For now, we'll return to the sync topic exclusively, describing the operation of an old sync generator, then that of a new one, just to give you the flavor of progress during the past 30-40 years. Afterwards, selected receiver sync regeneration will be considered and special waveforms.

An Early Transmitter

The subject of transmitted sync signals was not exhaustively treated in the early training manuals, and the best we can do is offer one of RCA's block diagrams that explains what occurs but not how and why. Nonetheless, we'll try and interpolate this one to some extent, and then bring you immediately up to date with a special piece of synchronizing gear that Tektronix, Inc. has been selling to the broadcasters for years and is used in all sorts of transmitters.

The block in Fig. 9-2 fully represents one of the early tube-type broadcast sync generators, including the originating 3.579545 MHz (it used to be megacycles) subcarrier oscillator upon which all transmitter sync is based. Divide this by 455 and you have half the nominal horizontal frequency of 787.1318 Hz, then multiply by 4× for 31.468.5 kHz for twice the 15,734 Hz actual horizontal frequency; divide by two, then, for the desired sweep rate; but also divide the 4× multiplier by 525, which produces the color vertical frequency of 59.9400 Hz. With these exact figures, of course, you don't have to work with the "nominal" frequencies printed since they are simply examples that cover regular color and monochrome transmissions, both of which were on the air at the time. Modern integrated circuits in many receivers have now gone to this identical method of sync generation and station-receiver lock except that they still depend on horizontal, vertical, and equalizing pulses to provide coincident timings rather than actual countdown from the regenerated color subcarrier oscillation in the TV receiver itself. The reason probably is that synchronizing signals have to be used in places such

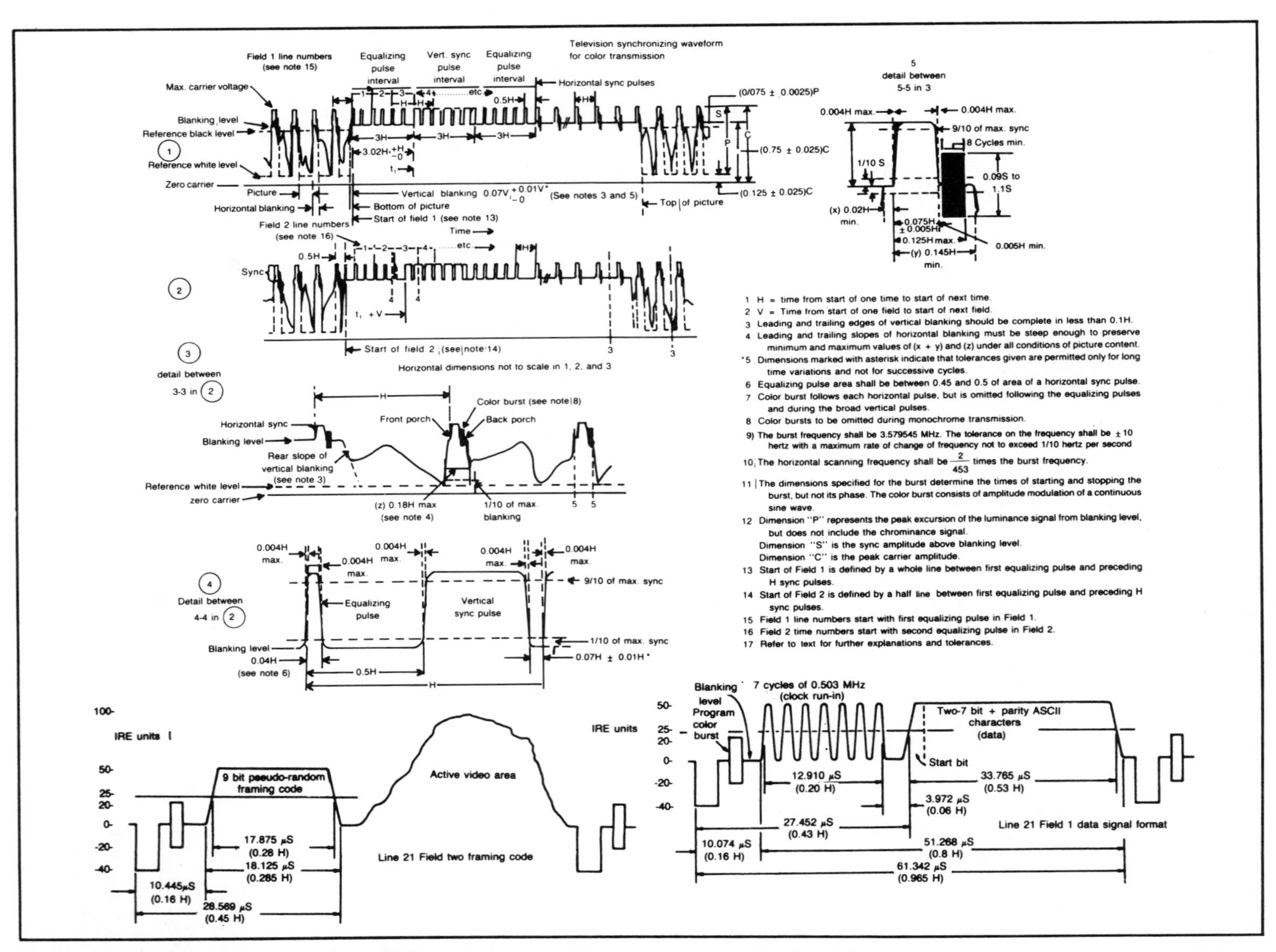

Fig. 9-1. Synchronizing signals used by *all* NTSC television transmitters and receivers in North America.

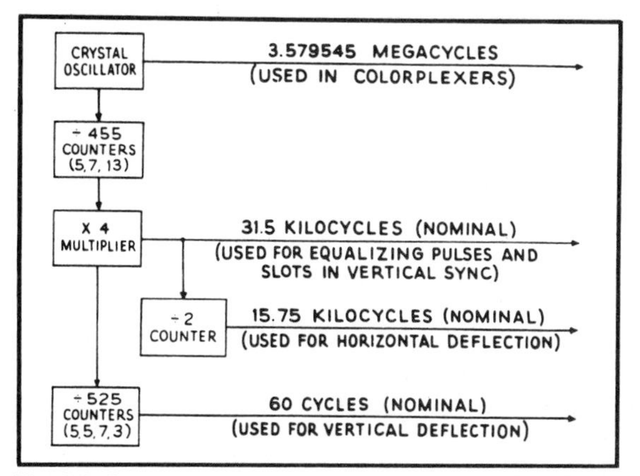

Fig. 9-2. An old transmitter counts down to modern sweep frequencies, i.e. 15,734 Hz for horizontal and 59.94 Hz for vertical.

as AGC, deflection, blanking, etc., so they might as well coordinate the sync sections as well. In addition, the color subcarrier regenerated in the receiver might be off enough to disturb normal sync, therefore the two are separated for now. Later, when all-digital sets appear this may change for a variety of reasons in one or more world systems; but not necessarily ours, although the 50-60 Hz power line difference between ourselves and Europe is always a problem that has to be accounted for. Right now the crystal ball's a bit cloudy, and in the future expediency will probably become king.

Modern Sync Generation by Tektronix

Rather than produce varying types of proprietary transmitter sync generators, the industry seems to have largely left that chore to Tektronix of Beaverton, Oregon, long known for its pre-eminence in oscilloscope and spectrum analyzer instrumentation, not to mention logic analyzers, cable testers, and a variety of other equipment used by engineers, technicians, and scientists the world over (Fig. 9-3).

Largely a modular system, with mainframe and ±5 and ±15 V power supplies operating from an ac power source that is not to exceed 250 volts rms, the 1410 accepts a total of six modules ranging from SPG1/SPG2 NTSC sync generator; TSG3 linearity generator; TSG7 SMPTE color bars; TSG5 NTSC pulse and bar generator; TSG6 NTSC multiburst and video sweep generator; to the TSP1 NTSC switcher/convergence generator.

The sync generator, of course, becomes our primary interest since this is an integral part of the chapter for transmitters of all descriptions. Front panel selectors include internal/external composite sync and subcarrier, LED indicator for shows *sence* of composite external sync, followed by automatic switching to internal; a subcarrier LED does the same for loss of external chroma reference; subcarrier sync incoming or composite to phase-lock timing generator; horizontal delay can delay output sync ± 1/2 μsec; internal-Gen lock internal or external; phase control of subcarrier; on/off VIRS; horizontal unlock; and the BNC panel connectors are for composite blanking, sync, subcarrier and black burst (Fig. 9-4).

According to Tektronix, the heart of this module is a 40-pin MOS LSI sync generator located on the sync timing board which generates all timing signals for the total system. Clocked by a 5 MHz 2-phase pulse device from a 10 MHz oscillator, the U129 device and U152B splits these two pulses and applies them to cascaded push-pull amplifiers, stepping up their amplitudes to almost 30 volts. These are now 320H clock pulses that the U129 counts

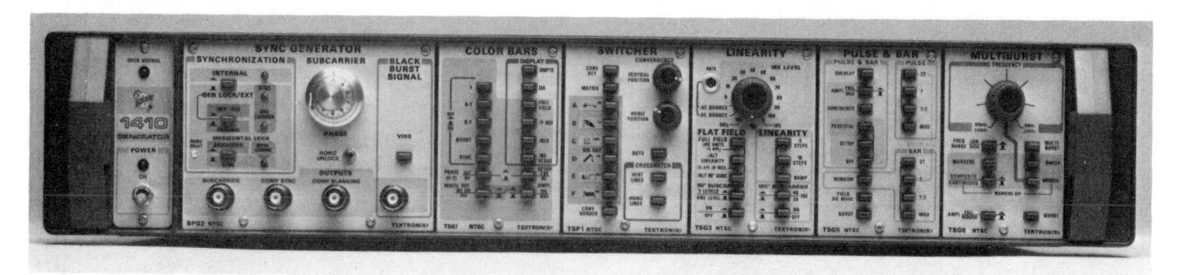

Fig. 9-3. Photo of Tektronix' 1410 transmitter sync generator with all six modules in the mainframe (courtesy Tektronix).

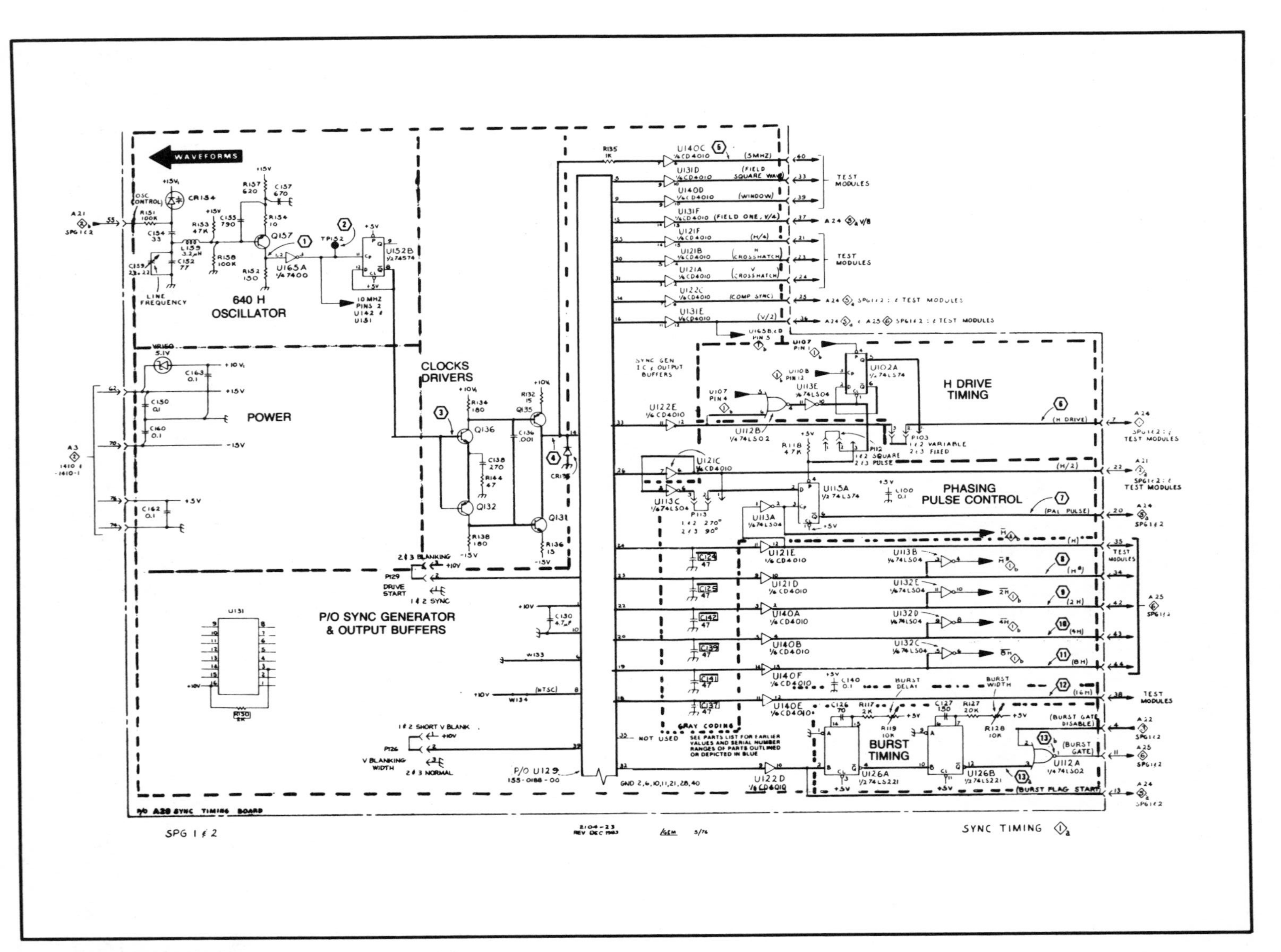

Fig. 9-4. Sync timing schematic I (courtesy Tektronix).

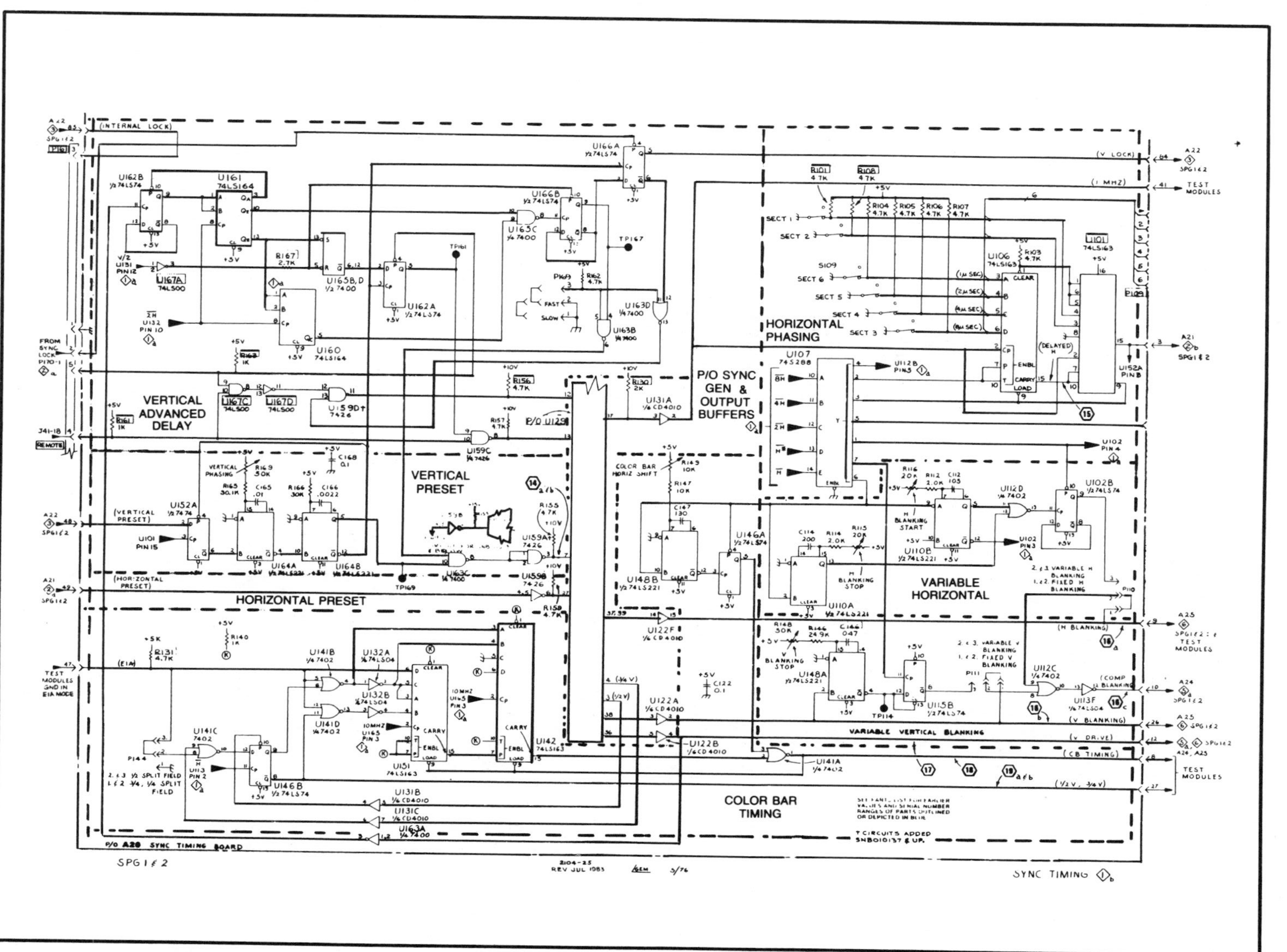

Fig. 9-5. Sync timing schematic II (courtesy Tektronix).

down to 64H, and then to simply H, the 15,734 Hz horizontal frequency. The 64H derivatives from 32H to H then pass into a series of exclusive OR gates and modify sync generator outputs to a Gray code and compatible with T^2L logic to come.

Burst timing takes place in the U126A and B dual monostable multivibrator having external timing components. Horizontal drive signals originate from the sync generator while U102A behaves as an on/off gate controlled by burst disable from the generator logic board. At the end of H drive from U129, its output goes low, stopping the variable H drive.

There's also a special pulse generated by U115A similar to one in PAL systems that regulates subcarrier phase on alternate horizontal lines, whose output may either be a pulse or square wave. Clocking originates from U129 coincident with line sync, with U115A being reset upon completion of H drive.

For color bar timing, U151 and U142 form a two-stage, 6-bit binary counter that counts down from the 640H clock input. In the full field mode, data counts are 63, divided by 65, resulting in an interval between output pulses of 6.5 μsec. In the EIA mode for the first 3/4 field, the counter is loaded for a 53 count, then divides by 75, producing 7.5 μsec bar widths. For the final 1/4 of the field, the data load changes to 34, the counter divides by 94 and −I, W, and Q bars are generated. Vertical advance delay permits slower U129 vertical counter lock to an external reference (Fig. 9-5).

Start and stop pulses for variable horizontal blanking generator U102B originate from dual monostable multivibrator U110. Times between trailing edges of start and stop pulses determine variable blanking pulse durations. There's also variable vertical blanking from U148A and U159B, all originating with U129.

The horizontal counter divides the double subcarrier by 455 for an H rate signal of 2048, there's a sync phase comparator, counter reset logic, and a timing jitter detector. You will also find subcarrier to sync lock timing or simply internal sync, sync and mode switching, AGC operation for chroma gain control, a composite video sync stripper, chroma demodulators, burst detector, a subcarrier output, as well as VIRS and black burst. The modulator is double balanced and forms sidebands proportional to the product of input signals voltages and the carrier signal. Subcarrier automatic gain control makes certain the modulator is always driven with constant subcarrier signal amplitude.

The final portion of the description is rather abbreviated, but you should have gathered the general idea from preceding sync generation formations as they were orginated. With the two diagrams supplied by Tektronix, this brings us to the end of the sync development portion and you now have a pretty good idea of what video transmission is all about.

VERTICAL BLANKING INTERVAL

This has become extremely important to the television industry since Teletext will be carried here, as well as what's left of VITS and VIRS, captioning for the deaf, and network source and time signals. On the basis of 21 lines, times 63.5 microseconds, the vertical blanking interval lasts for 1.33 milliseconds (Fig. 9-6). While not all are now executing specific duties, it is expected that Teletext will largely absorb lines between 14 and 20, excluding line 19 which has been reserved for VIRS, the vertical interval (color) reference signal.

Line 21 has also been set aside for subtitle captioning for the deaf, and line 20 is used by the networks for their program source and time identifications. VITS (vertical interval test signal) used to be required on lines 17 and 18, but has now been "deregulated" so that broadcasters may use any type of signal (including Teletext) on these lines as long as they are within FCC guidelines. As for the others, lines 1-3, 7-9 are equalizing pulses, 4-6 are vertical sync pulses, and the remainder belong to horizontal sync pulses.

At the moment, some other lines will not be immediately available until such receivers which might show vertical retrace lines are off the market. Then, in 1988, lines 10-13 may be used at 50 IRE, and by 1991 at higher levels.

In the meantime, VITS and VIRS still are important to both broadcasters and receivers for test

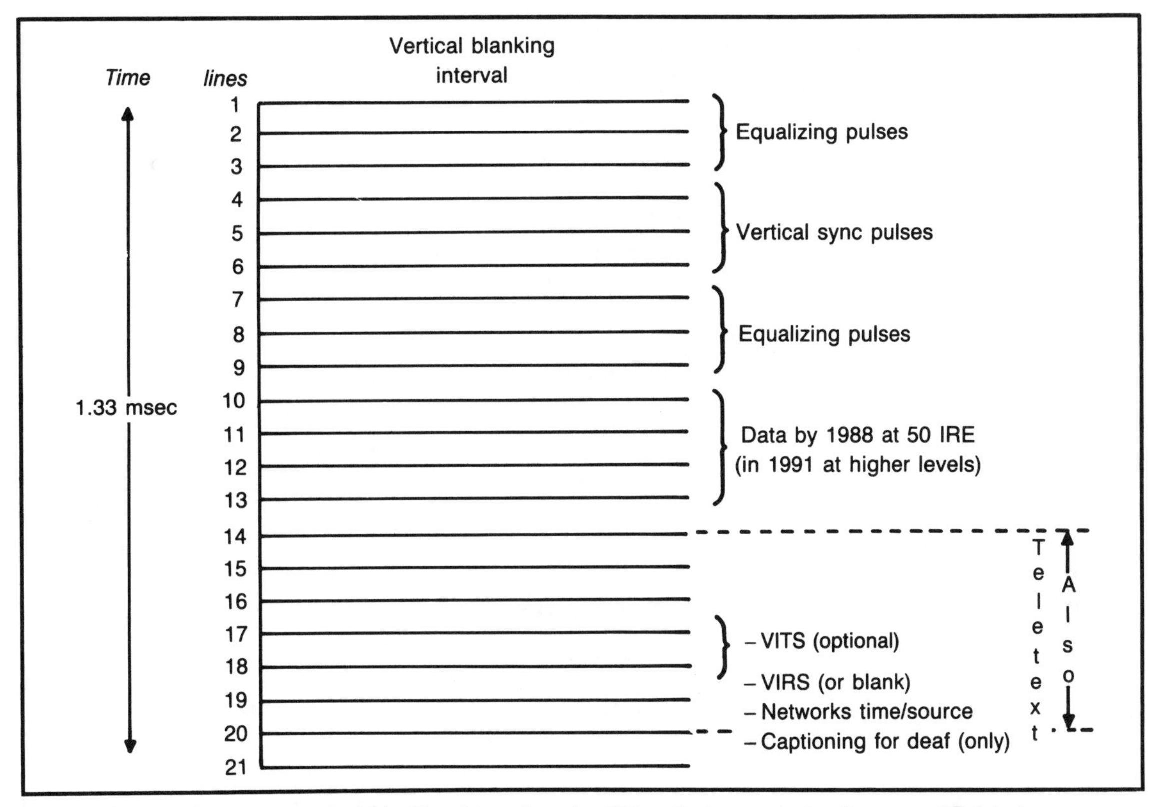

Fig. 9-6. The very important vertical blanking interval carries TV vertical sync, test patterns, and Teletext.

and performance appraisal. Most broadcast-caliber test equipment has long contained such signal generation and we expect they will continue to remain as standards for some considerable time to come. Networks, besides the digitally encoded source and time information, however, may use different signals as various occasions arise, so don't be surprised by changes that may take place continuously or from time-to-time; it's the year of deregulation!

Consequently, a worthwhile description of VITS and VIRS may well be in order for either posterity or continued support of broadcast standards which intelligent maintenance and competition may, hopefully require.

VIRS

Occupying line 19 of the vertical blanking interval, VIRS may or may not be broadcast continually, but nothing else may take its place. Its purpose is to check special program parameters, especially any phase differences between burst and chroma so the transmitter may be adjusted accordingly. In Fig. 9-7 you see the signal waveform extending from −40 IRE at the tip of the horizontal sync pulse to +90 IRE, which signifies a 40 IRE chroma reference amplitude. In between is the program color burst, blanking level, all within 12 microseconds, the chroma reference occupying 24 microseconds, a luminance level of 50 IRE and black reference at 7.5 IRE, both consuming another 24 microseconds. Therefore, from sync tip to sync tip, 63.5 microseconds—the complete time span of a single line—is consumed. Using such information carefully, an entire transmitter chroma signal can be set up accurately, including the necessary luma, black, and blanking levels. Receivers may also use this signal for hue and amplitude control, a topic that will be covered at length in receiver chroma processing discussions.

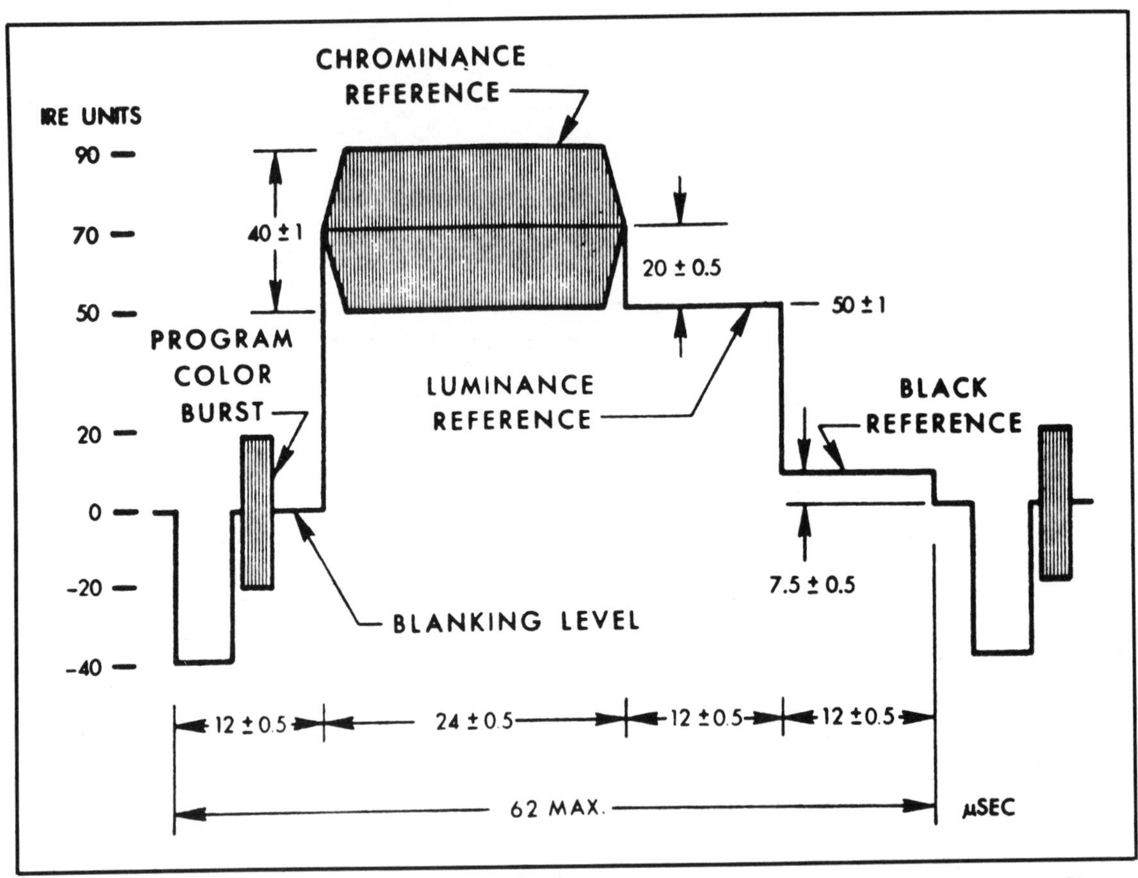

Fig. 9-7. VIRS—the vertical interval reference (color) signal placed on line 19 of the vertical blanking interval. Chroma and burst must be in phase for accurate color.

VITS

A far more extensive test device than VIRS, VITS has been designed as a general diagnostic test signal for high- and low-frequency characteristics, color-luminance lead and lag, amplitudes, color bar placement and linearity, luminance linearity, and ringing. Occupying all of lines 17 and 18 when and if broadcast, VITS was formerly mandatory for all transmitters remote from their studios. But stripping VITS off the broadcast network transmissions and retransmitting it was not to the liking of local stations, so the FCC scrapped it. We'll show you why it's *been* important.

Fields 1 and 2 for VIRS are the same, but they differ for VITS in that multiburst occupies field one of line 17, while color bars are in the same position on line 2. When using an ordinary oscilloscope to separate the two, you'll often sync on either indiscriminately. After line 17, however, line 18 is the same on both fields and, of course, VIRS follows.

Multiburst here has a full IRE amplitude of 140 units beginning with the negative going horizontal sync pulse and extending to white reference (Fig. 9-8). Each burst measures 60 IRE p-p, has at least three cycles of oscillations and extends from 0.5 to 4.1 (or 4.2) MHz, depending on the generator. A breezeway at 50 percent amplitude separates each of the bursts and color burst precedes them. It is used to assess high- and low-frequency response, high-end rolloff, and the percentage of peak carrier level. In television receivers, multiburst is invaluable in determining frequency response in both video i-fs and amplifiers, right back to the picture tube.

Color bars, their positions and levels from −40 to 100 IRE are invaluable for transmitters in placement of the various colors and their relative amplitudes (Fig. 9-9). Consuming a total of 52 microseconds, the pattern begins with the usual negative horizontal sync pulse and finishes with black, each bar 6 μsec in duration. Chroma gain distortion, oscillations, improper amplitudes, glitches, etc. may be gleaned from this precise pattern. Good for evaluating VCRs, which have fixed chroma and luminance levels, this pattern is almost useless in television receivers after the video detector, for only a series of rectangular steps shows at the cathode-ray tube. Here, you're much better off to use a gated rainbow generator which, if clean, cannot only produce a "hound's tooth" pattern of varying but predictable amplitudes, but also a chroma vector permitting total evaluation of chroma processing and demodulation. All TV receivers, of course, are fully luma, chroma, and picture adjustable so the NTSC generator isn't predictable at all once past the fixed video detector.

Composite test signals on line 18, fields 1 and 2 are most valuable in the studio and of only marginal significance in the receiver. Once again extending from −40 to +100 IRE, they feature a modulated or unmodulated staircase, 2T and 12.5T pulses, and an 18-microsecond window (Fig. 9-10). Once again, because of variable picture controls, this or these pattern(s) are pretty useless in a TV *after* the video i-f amplifiers and detector. But for linearity, chroma-luminance lead or lag, and ringing or low-high frequency distortion and rolloff, the staircase, 2T (luma) to 3 MHz and 12.5T (chroma)

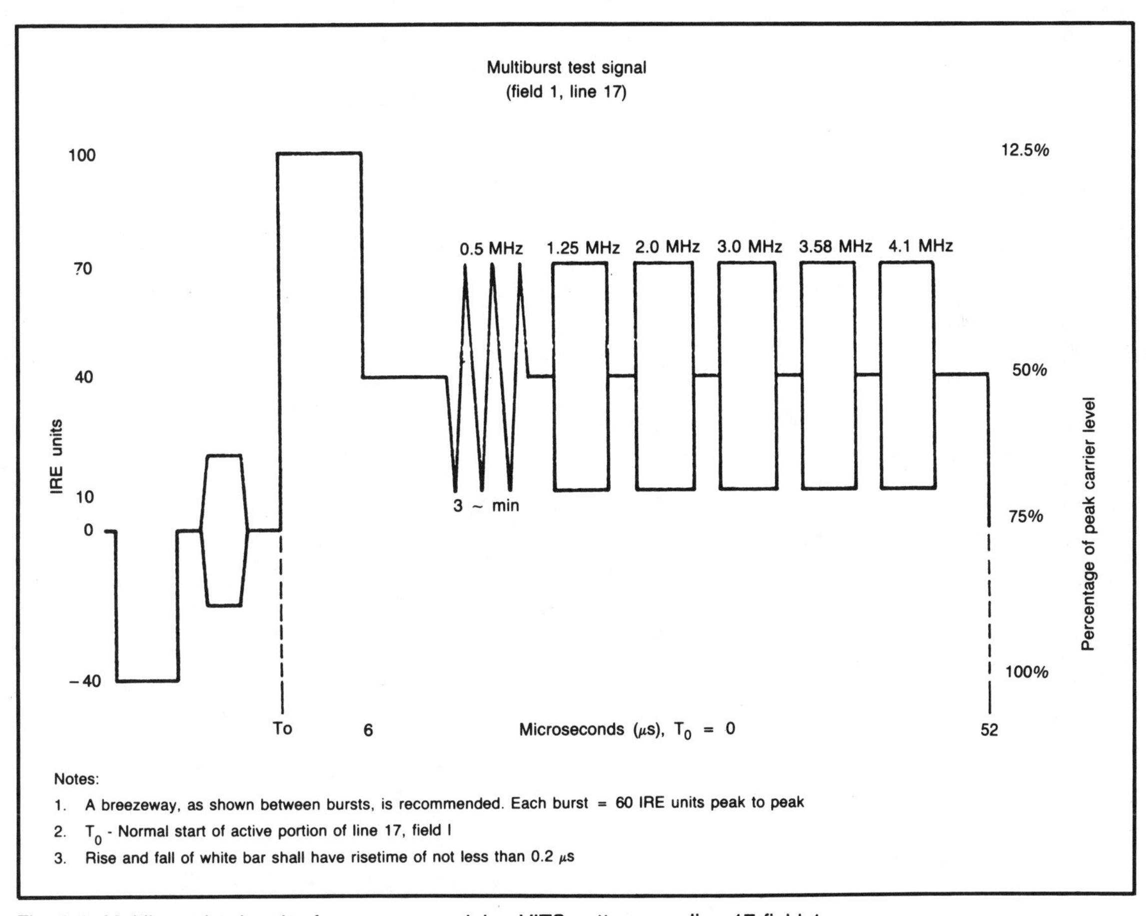

Fig. 9-8. Multiburst begins the frequency-examining VITS patterns on line 17 field 1.

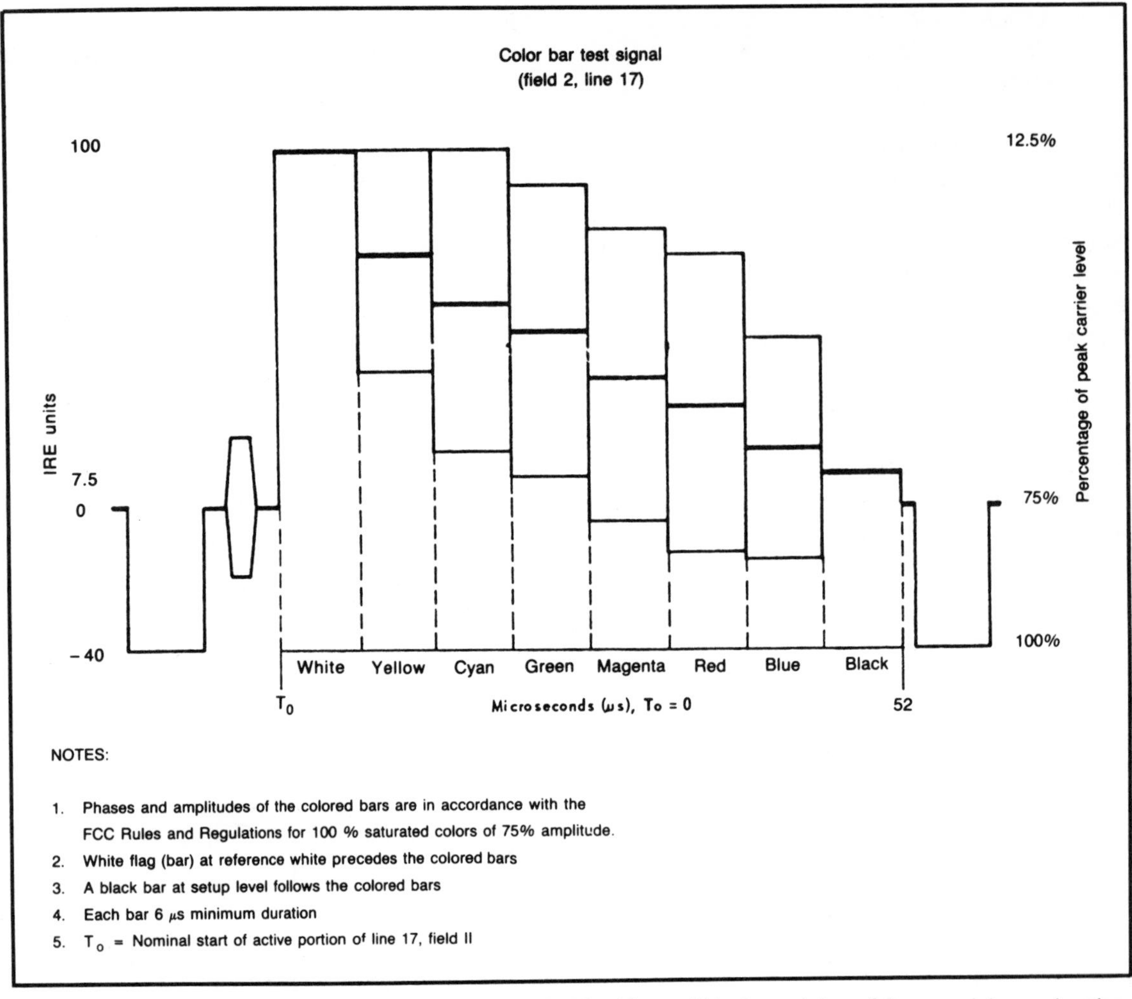

Fig. 9-9. Standard NTSC color bars preceded and ended with white and black, each bar of 6 μsec minimum duration.

beyond 3 MHz are indispensable, while the window does what all fast rise time pulses with long horizontal plateaus do, an that is evaluate gross frequency characteristics. When modulated, of course, the staircase *must* remain in phase with burst or there are evident chroma problems.

From the above, you can readily see why VITS and VIRS have become highly important for transmitters and (often) receivers in evaluating their various parameters. But at any rate, good luck to Teletext and its projected host of users. Let's hope the TV transmitters can handle all these newer intrusions without the necessity of even greater precision test signals. For the time being, at least, the FCC and its Commissioners think they can . . . but then there are those who believe otherwise.

TELETEXT

Teletext is the generic term for several competing technologies which encode digital information during the vertical blanking interval, then decode the same information in a television receiver, displaying it as a full or partial picture on the TV's cathode-ray tube.

An adaption of a British Broadcasting Authority transmission standard originating in 1976, there are now hundreds of thousands of units worldwide decoding telecasts containing alpha-numeric and graphic information of all descriptions. In the United States, it has been thoroughly tested with gratifying results by Field Enterprises and the Chicago Sun-Times as early as 1981 over WFLD TV. Formats of that period were 40 characters by 24 rows and, apparently, continue with 525-line raster compression to 480 scan lines without convergence problems in the vertical striped shadow mask picture tubes. When this is done, the same raster display appears on U.S. tube faces as those of 625-line Europe except the vertical refreshing rate becomes 60 instead of 50 Hz. Compression to 480 lines also reduces flicker and increases vertical resolution. Some receivers now appearing on the market have both raster positioning and height adjustments to accommodate just such inputs.

Another method of raster accommodation, we are told, deletes two chosen lines from the European "font," smoothing diagonals on some letters and rounding others, resulting in fewer scan lines and "superior" results. Cable or broadcast transmissions may be decoded by internal TV decoders or special settop converters, although the latter may be somewhat less effective than the former. Flicker, we are also told, may be reduced in Teletext by using a noninterlaced display, significantly reducing what could become an annoy-

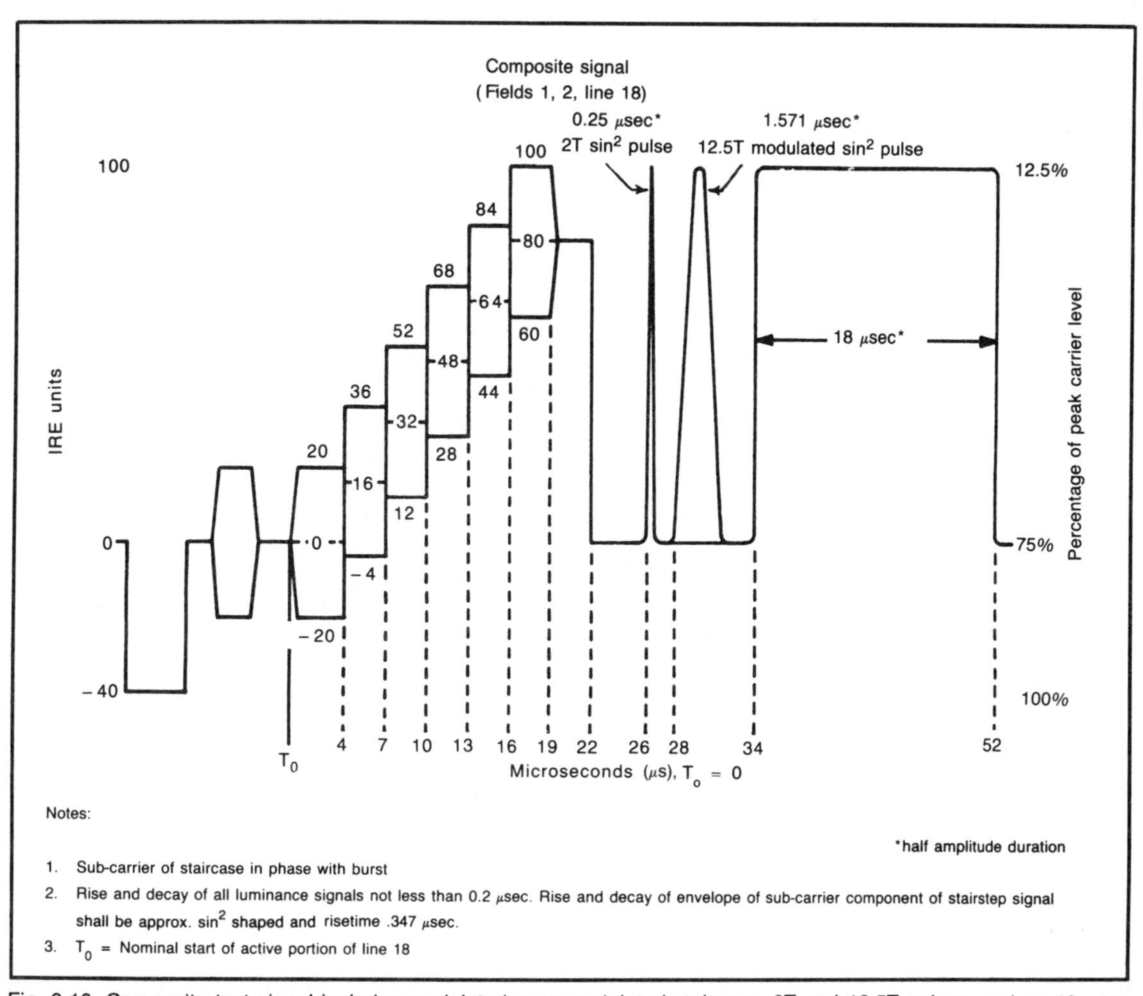

Fig. 9-10. Composite test signal includes modulated or unmodulated staircase, 2T and 12.5T pulses, and an 18 μsec window for low- and high-frequency evaluation.

ing problem under some circumstances. Another matter of concern is amplitude tilt from the tuner resulting from multipath and channel-to-channel variations leading to special design factors in receivers to counteract this problem. Envelope video detectors are most prone to tilt, although quasi-synchronous detectors can be affected also. The FCC-required envelope delay and predistortion may be counteracted in receivers by minimum phase video peaking circuits, 4.5 MHz intercarrier sound traps, and chroma bandpasses. In specially designed Teletext receivers, linear phase to the data slicer is desirable as well as equal balance between high and low video frequencies. A good beginning for consumer-type sets includes surface-wave acoustical (SAW) i-f filters, synchronous video detectors, and well-designed 4.5 MHz sound traps with little loss and negligible group delay, plus a carefully-tuned video-peaking circuit.

Recommended pulse rates for Teletext into the transmitter (at least for testing) are 5.727272 MHz, with pulse shape approximating the Controlled Cosine Rolloff Spectrum with amplitudes around 70 IRE. This varies, of course, in actual transmissions according to ongoing FCC specifications listed for the vertical blanking interval. A kissing cousin, Video-Text, has also been rather extensively tested and has been transmitted on a data channel—with a subcarrier delivering aural information, especially on subscription television services (STV). This is said to be the second largest market for distributing pay TV next to cable television (CATV). The third service in order is microwave multipoint distribution service (MDS). Video-Text, apparently, is specifically directed toward news and information services rather than commercial varieties of enterprise. Library updating for transmissions, as practiced in one instance, took place during the day and was transmitted during the night to the various home decoders receiving an "active" library content of some 65 pages.

In the United Kingdom, the most recent Teletext Decoder advertised—at least the one we know anything about—has three integrated circuits, less than 10 capacitors and a level set potentiometer, making a tidy package that can either be wired or rigged as a plug-in board. Even newer developments should be forthcoming almost immediately because of the 1985 introduction of digital television receivers with this specific capability built in. Videotext decoders using stack coding, on the other hand, are considerably more complex using as many as 20 ICs to fully complete their missions.

A NEW TELETEXT DECODER

As others study Teletext graphics animation and their various parameters, Signetics Corp. and Nabil Damouny have introduced a new Teletext decoder that's user-programmable, I^2C bus-controlled and will integrate with "any" equivalent bus system. Microprocessor managed, it has bidirectional lines for serial data and serial clock, accepts composite baseband and operates either on the vertical blanking interval or full field (over much of the 525 scanning lines).

Designed to operate at either the 6.93 MHz European or 5.72 MHz U.S. data rate, the new decoder contains a super data slicer (video input processor), Teletext controller chip, multi-page memory, and a general purpose microcomputer. It will support up to 8K bytes of memory and may be programmed to receive the normal 7-bit plus one parity bit, or 8-bit byte data. Mr. Damouny says that fixed format World System Teletext is virtually error free because a 1:1 correspondence exists between transmission codes, acquisition and display memory, and the actual screen display position. Because received information is constantly being cycled, any error received in one cycle can be automatically corrected in the next.

A combined block diagram and its TV hookup is shown in Fig. 9-11. Note that Teletext *requires* RGB inputs into the receiver and, therefore, needs a TV/monitor combination for intended design operation. Decoder output has both RGB color signals as well as blanking, and contrast reduction/controls, and therefore "a simple video output circuit might be needed at the output of the Teletext decoder . . . to provide buffer/drive capability and appropriate voltage levels . . ." Settop adapters may also be available, just like CATV cable converters

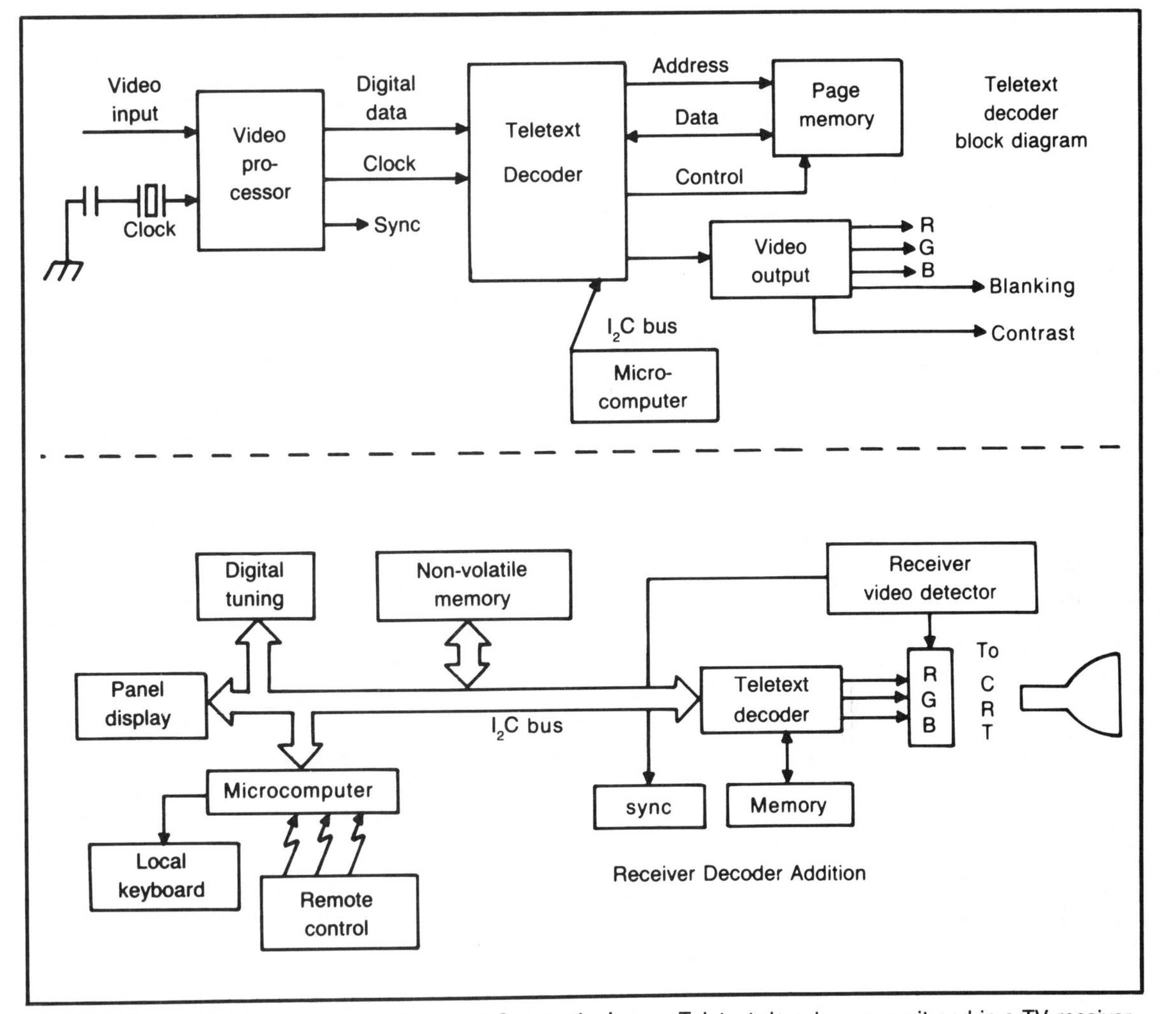

Fig. 9-11. Composite block diagram of Signetics Corporation's new Teletext decoder as a unit and in a TV receiver.

if the TV receiver can't have a built-in decoder.

RECEIVER SYNC

Having seen what Tektronix and the transmitters put together, let's generate a little historical and contemporary perspective of what receivers have done over the past 40 years to become more effective in vertical/horizontal synclock. Digital, by now has certainly taken over, and rightly so since these circuits have to deal not only with synchronisms within the television receiver but also those from outside, as well. This presents a much more difficult situation because there are times when vital equalizing/vertical pulses are compressed or missing, and in order to maintain proper scan frequencies, sync countdown integrated circuits have to pick these up or compensate accordingly. This, of course, was not true in the early days before computers and games surfaced to plague TV sync circuits. It's only been in the '80s that good synclock with most or all inputs has been possible. Now, you may be reasonably well assured that all peripheral devices of worthwhile origin will operate through your color receiver provided it, too, has been adequately designed. Many inexpensive ones still have not since they're far too unsophisticated to handle the changing or ill-formed patterns offered.

In The Beginning

Both horizontal and vertical sync appeared at the grid of the sync amplifier and separator tube since most video was removed (clipped) from the composite signal by a previous video amplifier biased to retain only the sync tips. Additional video is further clipped in the amplifier and separator so that "pure" sync is amplified and inverted to the two RC networks (Fig. 9-12). On the left, the 39 pF capacitor and 10 k resistor form a pulse shaper and filter, sharpening the leading edge of the horizontal sync pulse so it can be compared with flyback transformer pulses and any difference dc-routed to the horizontal oscillator for frequency correction. Such comparison in those days was done either by tube or semiconductor diodes evaluating pulses against sawtooths for high or low crossings.

On the right, you see a series of resistors and capacitors known to anyone in electronics as an integrator. That is, instead of the capacitor going in series as in the horizontal differentiator, these capacitors usually are connected to ground with, perhaps, a dc block toward the output. RC time constants then accumulate at discrete intervals building up a large charging waveform which then serves as the vertical pulse trigger. And that's the way it used to be . . . Problem with this circuit was that all these were discrete components, usually with paper capacitors in the integrator which, over fairly short periods, began to leak and distort the vertical trigger causing indiscriminate rolling. Tubes would also develop interelement leakage and even "slump," failing to execute either clipping or amplification—then everything went awry, as the more polite people would say.

Upon the introduction and fairly widespread use of transistors, a considerably simpler and less current-hogging device came into being that was reasonably effective for the times. Instead of two hot vacuum tubes, this simple circuit used only a single transistor, one diode, three resistors and two capacitors (Fig. 9-13). Here's how it works.

Operating much like a class C amplifier whose on-times are less than half the composite video cycle of 63.5 μsec per line or 11.4 msec per field, the idea is to bias this stage off during video scan times and on when the larger amplitude sync pulses appear at the end of each line or field. In between, TV receiver sync circuits "flywheel" on their own until the next group of sync information arrives to supply any corrections.

In this circuit, horizontal pulses pass easily through the 0.005 μF capacitor into the base of the transistor but do not disturb the cathode back bias of the diode above. When vertical pulses arrive, the

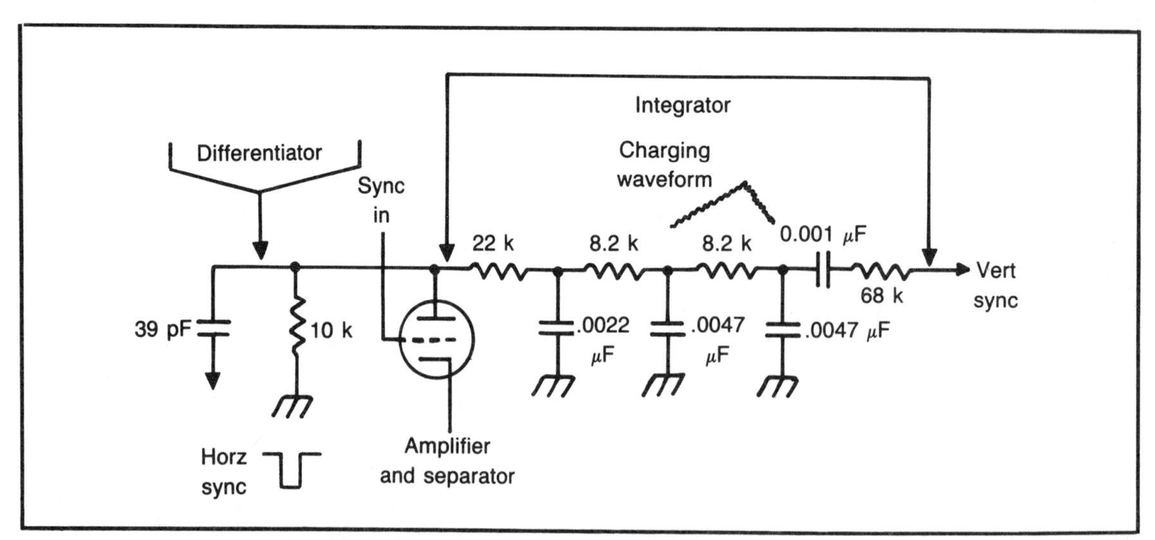

Fig. 9-12. Typical old vacuum tube vertical and horizontal sync separator.

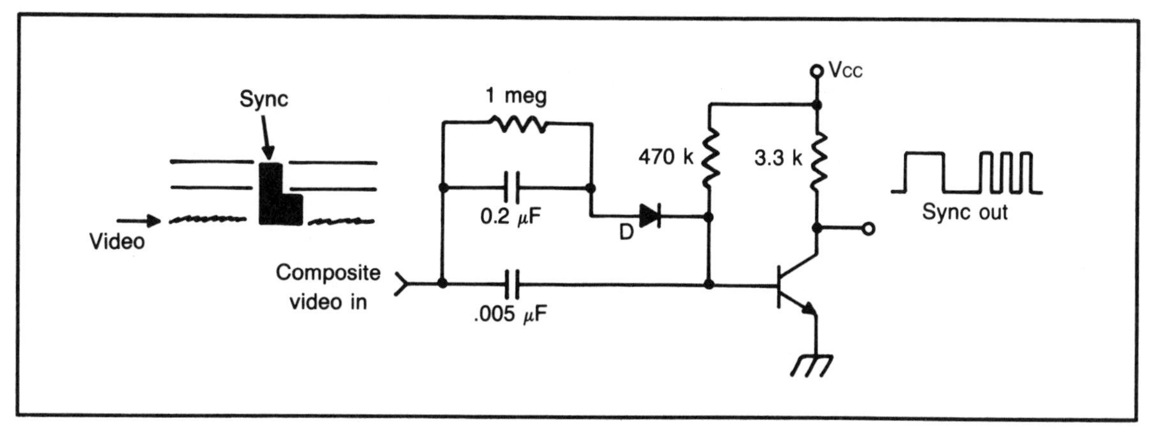

Fig. 9-13. Transistorized separation in a single stage for both horizontal and vertical sync.

0.005 μF capacitor blocks them because of its 530 k-ohms impedance at 60 Hz. However, the charging time constant of the 1 megohm and 0.2 μF capacitor above permits the low frequency pulses to pass, forward biasing the diode and, at the same time, turning on the transistor during the vertical interval blanking time. Leaky diodes or capacitors in this circuit would also cause *both* horizontal and vertical sync problems, but this is much less likely to occur than in tube circuits due to lower voltages and currents and development of mylar and other more durable dielectric capacitors. A good quality silicon diode shouldn't give any particular trouble either.

So in saving many components and doing sync separation in a single stage, designers have increased reliability, instituted considerable cost savings, and lowered power consumption by as much as 50-60 percent. Nonetheness, you must remember this is still a rather inefficient analog approach that contains a number of shortcomings. There is no flexibility, as you can see, and certain peripheral equipment probably won't work with these general time constants, especially computers and some games. Consequently, with the approach of digital receivers, a digital horizontal-vertical countdown circuit was the next logical step. This is what we'll describe next.

Digital Sync Countdown

Let's go back a little ways with this first example just to elaborate on the design philosophy a bit so you'll feel at ease with some of the latest ICs which are based on initial solid-state logic countdowns. It is, of course, with these continually updated ICs that external video sources of all descriptions have been accommodated and served well in our better receivers. With horizontal sync there has been no major problem, but 59.94 Hz vertical sync poses all sorts of difficulties when it and accompanying support waveforms are either distorted, compressed, or interrupted. What we will describe henceforth are all monolithic integrated circuits, normally transistor-transistor T^2L or I^2L integrated injection logic usually contained in many top-of-the-line receivers. Lesser sets frequently take questionable shortcuts but often sell for less.

Texas Instrument's SN76547 horizontal processor and vertical countdown circuit with excellent noise immunity, automatic countdown or sync mode, and adjustable horizontal output pulse duty cycle is a good working example.

Still in use by some older receivers, this I^2L logic is noted for high packing density and low speed-power product, all using standard bipolar structures. The particular IC separates composite sync from composite video and removes noise with comparators; uses phase-locked loop for horizontal sync; has adjustable duty cycle for horizontal drive; detects vertical sync digitally and with special gating for either 525 or 544 lines. A block diagram appears in Fig. 9-14.

Here the sync separator comprises a dual comparator, passing band-restricted information from

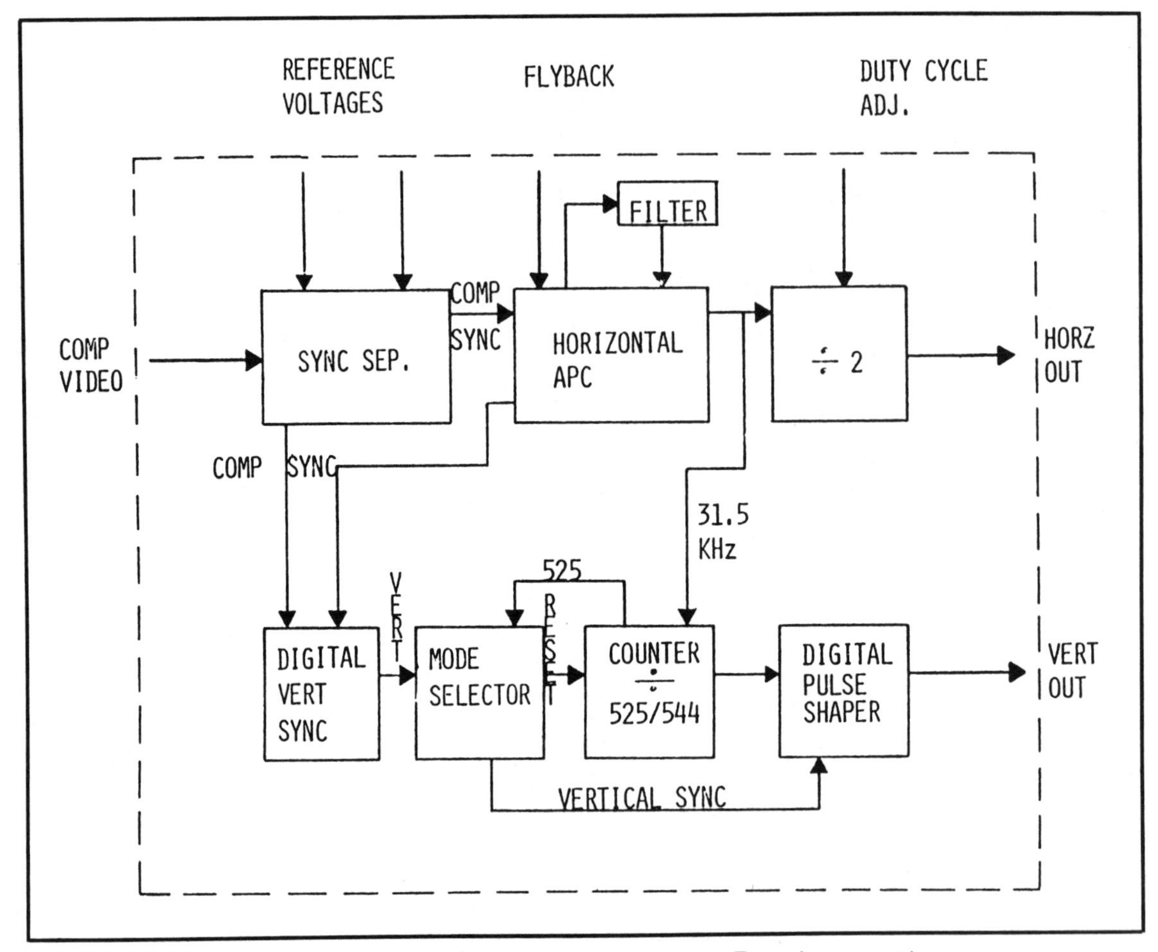

Fig. 9-14. An older vertical countdown and horizontal processor (courtesy Texas Instruments).

a dc-coupled low-level video detector. Video is excluded above a preset dc value, while noise below a predetermined dc value is locked out; therefore, only sync tips and not video and noise proceed as composite sync to succeeding stages. Sync stripping is automatically adjusted by an internal feedback loop taking approximately 30 percent of the composite sync-video signal.

Horizontal phase detector (APC) consists of a phase detector gated externally by flyback pulses and internally by detected composite sync. An RC voltage-controlled oscillator is adjusted to 31.5 kHz, and loop sensitivity of the PLL is also with RC components to select various combinations of pull-in ranges, hold-in ranges, static phase error, and noise bandwidth. The 31.5 kHz frequency is then divided by two for the horizontal output. By applying a voltage from a resistive divider and its dc component, the horizontal duty cycle may be adjusted generally between 50-100 percent.

Vertical sync is detected by a 2-bit shift register clocked by the positive transition of the 31.5 kHz oscillator and into a shift register either as a logic 10 or 01, with any clock and composite sync phase delay greater than 2.5 μsec, equalizing pulses will not be included. Should the phase delay exceed 27 μsec, separation pulses are clocked in and vertical sync would not be retained. Normal delay depends on system delay between clock and flyback of, typically, 8 to 15 μsec.

Vertical countdowns are based on the 262.5 horizontal scanning lines in each vertical field. Doubling the 15,734 Hz rate and then dividing by 525 (lines) resolves the 59.94 Hz, just as it did in the transmitter. Because vertical sync is more touchy than horizontal sync, noise and interference immunity are added for good measure. The SN76547 can also discern between standard and nonstandard vertical sync. It does this in the sync mode as a coincidence gate clocks a 3-bit counter whenever coincidence is detected and clears if no coincidence is detected. But, in switching to the 525 line countdown mode from sync, 8 coincidences in a row are needed, while 8 noncoincidences are required to switch back.

The 525 count window and detected vertical sync are 16 μsec wide. When vertical sync is detected within this window it is considered coincident, and so a field line count of 262.25 would also be coincident. Should countdown be selected immediately, the error between vertical sync and the 525 count would continue for 8 counts or two horizontal lines producing vertical jitter. Therefore, by stipulating 8 coincidences worst case jitter amounts to 1/4 of any horizontal line.

That's about as plain as we can make the vertical countdown story, with Texas Instrument's help. We might add that this is not necessarily a new technology except from some clever twists, but putting it altogether in one IC certainly is a considerable leap forward that's sure to be emulated (nice word) by many others here and abroad. It will be interesting to see how competing solid-state manufacturers accomplish the same result as we proceed. For this one, the schematic indicates 10 stages of flip-flops are needed in this vertical IC countdown. Fortunately, there are few external components or adjustments.

RCA's CA3210E/CA3223E Sync Countdown

Considerably more up-to-date is RCA 24-pin dual in-line medium-scale integrated circuit (MSI) used in both U.S. and European 525 and 625 line versions, respectively. A block diagram of the U.S. CA3210E 525 line system appears in Fig. 9-15, along with the package pinout and functions.

The master oscillator's rate is 16 times the 15,734 Hz scan frequency as set by LC tank components at pins 4 and 5, with AFC control by horizontal sync inputs and the filter loop, and feedback from the 4-stage counter below. From this counter are divided by 4, 8, and 16 outputs, the first two of which are used in the 10-stage counter, and the third enters a ramp generator and also compares its phase with broadcast horizontal sync, timing the entire system. The ramp generator's output, in turn, develops phase control for the horizontal output compared with flyback pulses from the second automatic phase control (APC) loop. Any deviations are then adjusted and continue accurate horizontal drives. The 10-stage counter furnishes timing references for the vertical circuits.

This is done with logic countdown so that the 464th clock pulse sets a sync window generator. Should incoming sync coincide with the 525th clock pulse, a coincidence gate resets a 3-bit counter and this begins vertical blanking and vertical sweep. If incoming sync is either nonexistent or faulty, the 10-stage counter still outputs a pulse at the 525th count, but the 3-bit counter will then count the number of fields without coincidence, and if sync is regained before the 8th field, it will reset and continue normally. If coincidence is not detected, the 3-bit counter pulses a toggle which shifts the countdown mode to synchronization. If sync is still missing, the vertical system free-runs at a frequency set by a 592 count because nonstandard sync occurs regularly between 464 and 592 counts. A vertical ramp drives all vertical deflection. For servicing, a service switch shunts signal through a diode and removes vertical blanking. There's also a skew switch (not shown) that can route dual frequency signals to the AFC where vertical energy may control both the selector and external filter time constants. Such a system permits phase-sync rapidly with nonstandard equipment such as VCRs.

Not shown, either, is an interesting internal shunt regulator consisting of two transistors with common collectors, the first acting as a simple emitter follower, and the second as a regulator. With increasing input, a base reference zener conducts turning on the buffer transistor and then the regulator. This second transistor then regulates the output by varying conduction in parallel with the

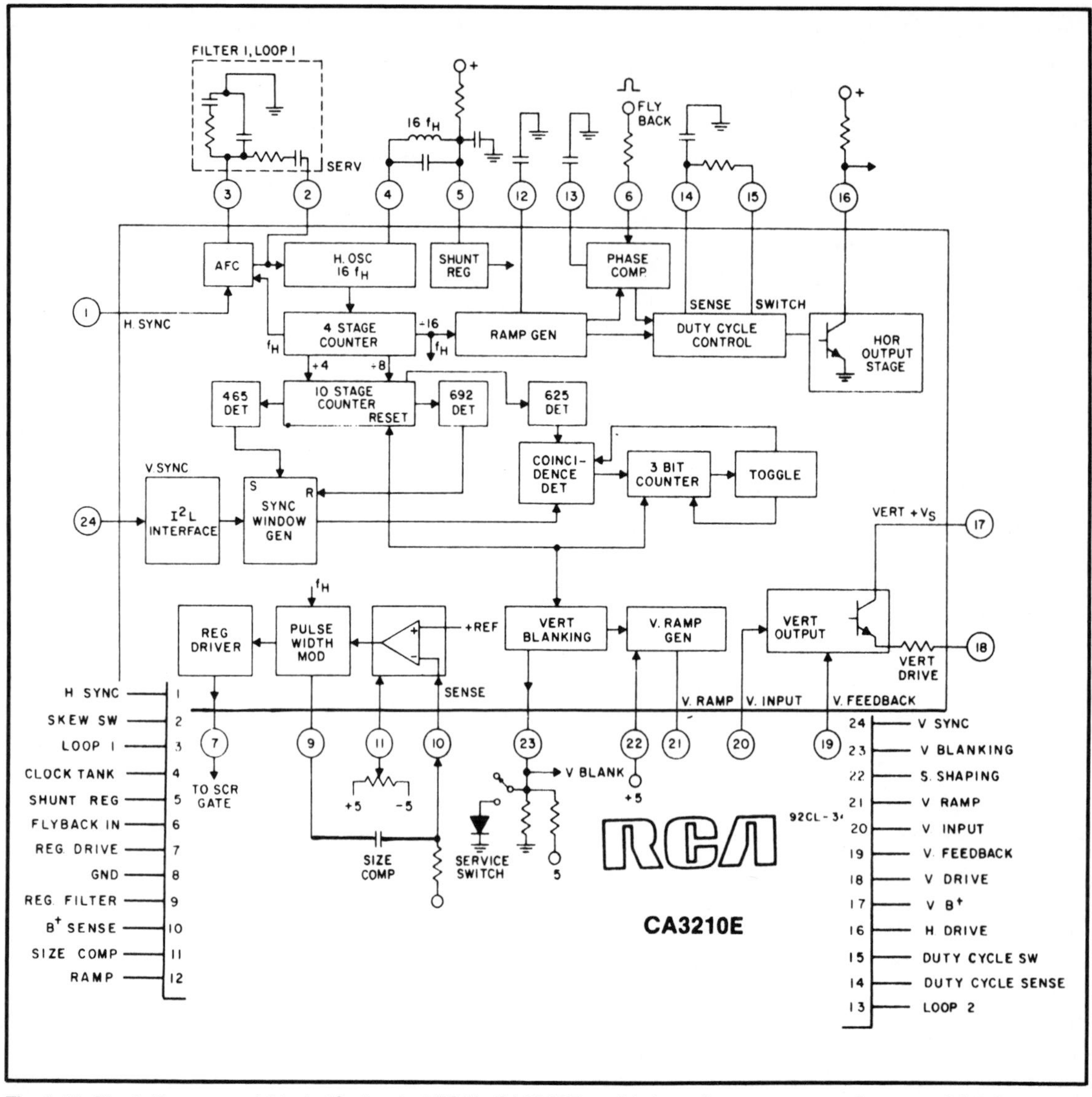

Fig. 9-15. Block diagram and 24-pin IC pinout of RCA's CA3210E sophisticated sync processor (courtesy RCA Solid State Div.).

load: less conduction, a greater voltage output, etc.

The CTC 118 Method

A somewhat similar IC to the foregoing is now used in RCA's CTC 118 chassis, and although some portions and pinouts are identical, the application of this circuit in an actual television receiver may aid your appreciation for increased complexity of ICs with sharply reduced external parts count and considerably more precise execution. You'll note in the IC block a ramp reset switch disable to the vertical blanker, a 4 μsec delay between the 1st loop phase detector and the second and only an 8H counter before the ramp generator. Let's see how this one works:

Called the U401 deflection circuit, the 24-pin IC (Fig. 9-16) combines *all* horizontal and vertical synchronism, including the ability to recognize and

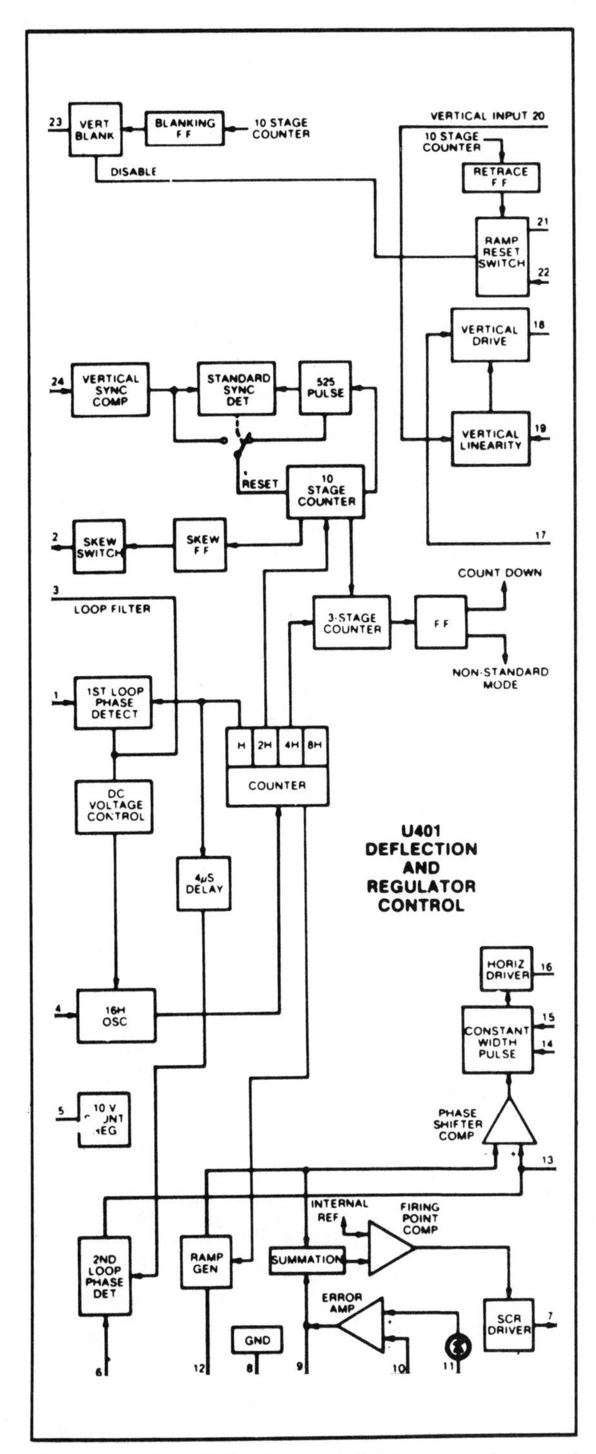

Fig. 9-16. RCA's CTC 118 V/H deflection countdown and processor (courtesy RCA Consumer Products Div.).

compensate for nonsynchronous systems such as certain games and recorders, thus making it virtually indispensable for processing most or all the various video product inputs on today's U.S. market. Using a dual phase-lock loop design with horizontal sync entering through pin 1, the first loop has good noise immunity with longish time constants via a loop filter, supplying dc voltage frequency control to the 16th oscillator below. The second loop compensates for horizontal deflection stage load variable delays with a phase detector receiving 4 μsec delayed signals from the first loop detector as well as inputs from the flyback transformer. And along with a ramp generator voltage delivers its output to a phase shifter comparator and, eventually, the horizontal driver. This second loop also has a filter at pin 13 with faster time constants to filter control signals, maintaining balanced phase between video and horizontal deflection yoke currents.

Master oscillator output goes to a 4-stage counter that has 2, 4, and 8H dividers, and a direct output to both the first loop phase detector and the 4 μsec delay to the second phase loop detector. The 2H output then goes to a 10-stage counter for vertical mode countdown with feedback to a 3-stage counter which also receives digital pulses from the 4H divider and 4-stage counter below. With vertical sync entering from pin 24, standard or nonstandard sync is determined by the 525-pulse count and either accepted as adequate or not. If not, and sync is not apparent by the 8th field, the 3-bit counter toggles and changes countdown to synchronization. Pins 20 and 21 are the vertical input and ramp; 22, the S-curvature yoke shaping (capacitor) network; and vertical blanking may be interrupted by the vertical reset switch.

The ramp generator next to the second phase loop detector receives inputs from the 4-stage counter, comparing this phase with broadcast horizontal timing, sending any correction to the phase-shifter comparator. This stage then maintains a constant pulse width for the horizontal driver. The same ramp generator output also reaches a summation network fed by an error amplifier whose inputs develop from a B+ sensor and a high-voltage "resupply." The summing network established a firing point above an internal reference for the SCR driver which maintains high B+ 150 V system regulation.

You may have to re-read these sync chip IC explanations several times to absorb their full meaning. Granted, such descriptions are neither simple nor simplified. Full details are probably far too intricate and involved to print except in an engineering treatise, along with substantiating equations. But, what we've done here with considerable aid from T.I. and RCA should go a long way towards understanding what the modern countdown circuits do and why. I seriously doubt if these explanations will be found in any similar competing publication. They're tough, but worth the effort!

CHROMA SYNC

Not a subject usually dealt with separately, chroma sync is probably the least understood of all broadcast-receiver mating mechanisms and certainly one of the most important. For without rigid timing between the two, bands of stray colors (or no color at all) grace the picture tube with either standard monochrome or red, green, and blue stripes of varying widths and stabilities. In the days of vacuum-tube receivers, one of the first tasks repairmen had to learn was setting the chroma oscillator with a color bar generator whose timing wasn't always accurate, then touch up phase differences between stations and low/high signal reception with the receiver on broadcast signals. As you might surmise, color receivers in the late 1950s and 1960s weren't all that great, and the combination of slumping vacuum tubes and leaky paper capacitors did make life in the repair lane rather miserable—but profitable.

With the coming of transistors, voltages and currents were reduced, components improved substantially, and the transistor itself usually worked or didn't, providing an output of approximately 90 percent throughout its life span until the day it died by either open or short. Then, with the advent of well-developed and sophisticated integrated circuits, most color problems have disappeared, at least for the average viewer—but not engineers. There are still things to do, among them standardizing full bandwidth chroma processing and establishing positive tint control (a super idiot button) that will accurately reproduce *exactly* what broadcasters are transmitting. Then it will be up to the transmitters to stay on the ball rather than lay total blame on the various grades of market-oriented receivers which may be either great or miserable, depending on their particular origin.

At any rate, let's go through a few of the early and late processes so you'll have an idea of sequential developments. Usually the more complex the system, the better the color control, characteristically among integrated circuits. Interestingly enough, the original ideas on this subject haven't changed a great deal, just applications and methods, as you'll see.

A Chroma Oldie

In this simplified block diagram, composite video and chroma are separated by a tuned chroma takeoff coil, and chroma, along with burst in the horizontal blanking interval, is introduced to the first bandpass amplifier and the color killer (Fig. 9-17). With incoming chroma, the color killer is *not* disabled, burst proceeds through to the primary of T1, reaching the burst amplifier coincidentally with a horizontal sync enabling pulse. The amplitude of this burst governs conduction of the first chroma amplifier and, at the same time, is phase-shifted, detected, and returned to the 3.58 MHz subcarrier regenerator as an oscillator correction voltage to lock timing with the broadcast signal. Such oscillator output is then used as R and B reference phase sinewaves to select proper angles of chroma detection. With no incoming chroma, the first chroma amplifier does not conduct and the color killer shuts down the second chroma amplifier to prevent noise from reaching the demodulators and RGB outputs.

Generally, this is the way chroma and burst operate today even in the most sophisticated arrangements. Of course, you'll see considerably more processing stages and feedbacks in modern ICs, but the original ideas remain and burst still does exactly what it has always done, sync the subcarrier oscillator and maintain amplitude of the

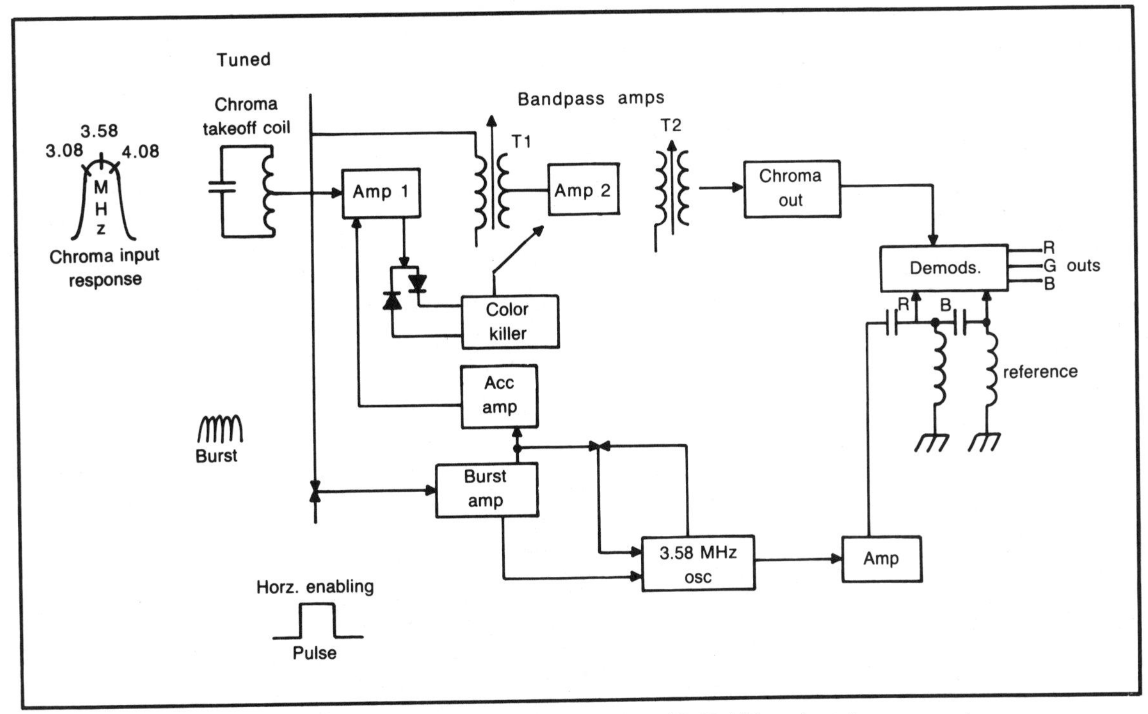

Fig. 9-17. Typical early color system processing burst, chroma, and 3.58 MHz subcarrier regenerator.

color circuits when present. When absent, lack of color amplification turns on the color killer. So with these principles in mind, let's turn to several integrated circuits and see how they do.

The μA780 Subcarrier Regenerator

Built by Fairchild Semiconductor in the early '70s and described by John Chu, this chroma subcarrier regenerator is a classic and decidedly the forerunner of *all* circuits of this type. Although it has many stages and appears quite complex, it has always worked exceptionally well and enjoys respect and understanding from color circuit designers everywhere. Block and schematic illustrations are combined in Fig. 9-18 for your convenience.

Locked in phase and frequency to color burst, the 3.58 MHz regenerated subcarrier derives from a voltage-controlled oscillator phase-locked loop, a phase detector and low-pass filter. Operating in series resonance between pins 7 and 8, the crystal offers positive feedback through pin 6 and Q16 to half the Q17-Q18 differential amplifier below. Via this route, the oscillator output goes to the emitters of Q2-Q3 and Q5-Q6, as well as Q9-Q10. Q19, on the other side of Q18 receives dc bias from the emitter of shunt regulator Q20, which is part of the zener-referenced Q21 through Q24 voltage regulator for the entire IC. Q1, meanwhile, gates color burst on while blanking it in the demodulator output so that Q1 is off while the oscillator output is on, and transistors Q4 through Q11 are also off although the oscillator, of course, is running continually. Any phase change in the oscillator output is supplied by the manual tint control connected to pin 1 which permits a 45-degree phase swing due to an RLC phase lift network connected to the collectors of Q2 and Q3.

To keep the oscillator exactly on frequency, automatic phase corrector compares color burst input phase with that of the voltage controlled oscillator's output and positive or negative corrections are differentially amplified by either Q9 or Q10 to buffers Q12 and Q15. The automatic color control detector performs a similar function for the bandpass amplifier on a succeeding IC, but differ-

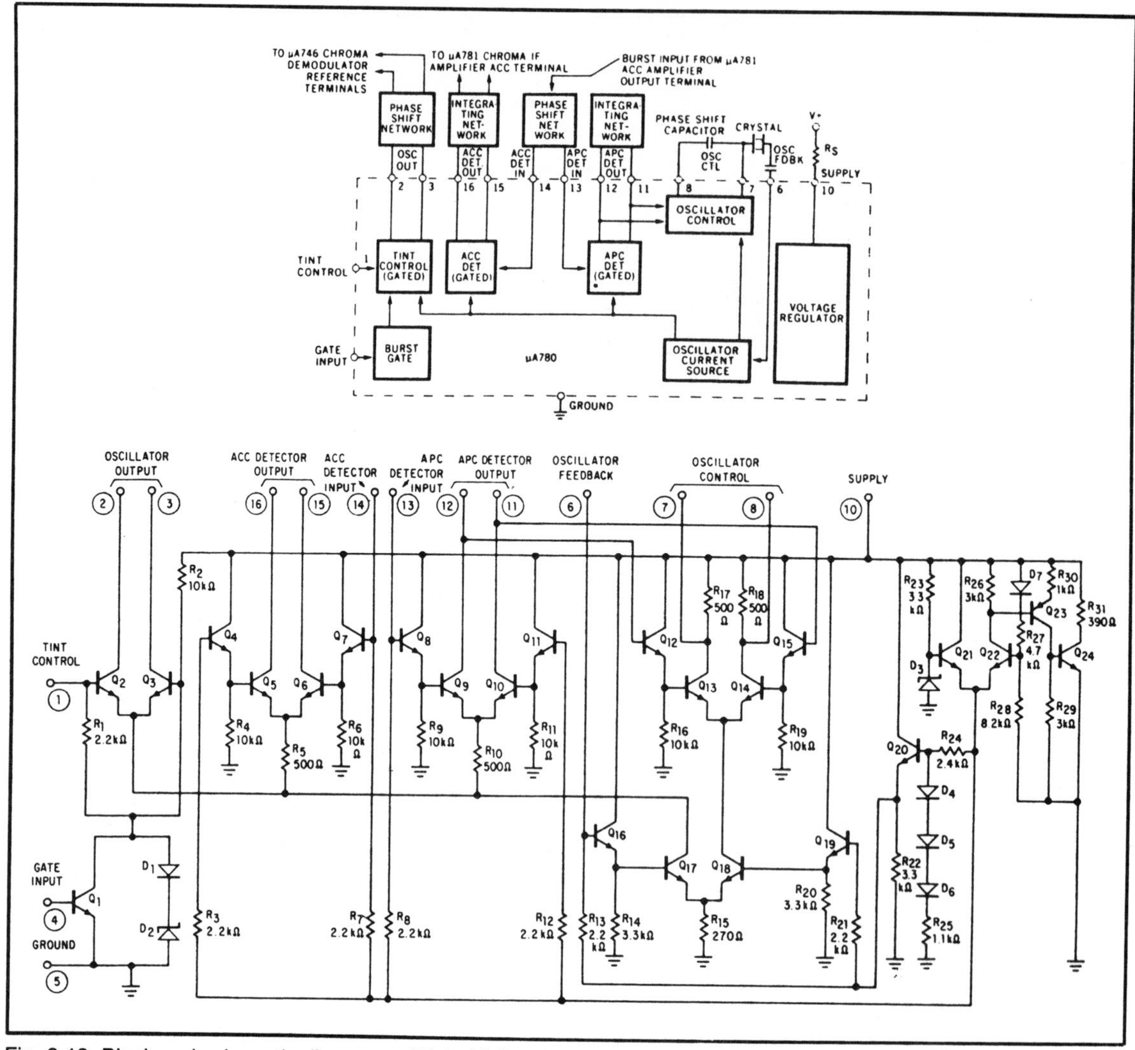

Fig. 9-18. Block and schematic diagrams of Fairchild's μA780 chroma subcarrier regenerator (courtesy Fairchild Semiconductor).

ing in initial phase. Outputs of both are filtered, however, and do their correcting via dc potentials passing through to their regulating stages.

The foregoing should offer just about all the information needed to evaluate other subcarrier regenerators. Additional examples of such oscillator-control circuits are now available since they are included in all-chroma or chroma-luminance ICs that process all these functions on medium (MSI) or large (LSI) scale integrated circuits that are even more utilitarian and reliable than the one just described. In retrospect, all these advances are just a further step toward an entire television receiver on a single chip. When that day will eventually arrive we don't know, but you can bet a lot of people are working on it day and night. Coupled with LCD picture-on-the-wall readouts, a fully developed system should be something to behold. Once manufacturing techniques manage to build such an IC with relative ease, wideband television receivers may even be less expensive than they are today.

Chapter 10
Video Processing

LIKE COLOR PROCESSING, SOME MAY WONDER why an entire chapter needs to be devoted to a topic that seems so straightforward and apparently noncomplex as video processing. From our standpoint, let there be further doubt that *any* of these premises remain true of today's better video products. Until about 1978, we were still working with 2.5 to 3 MHz receivers and wondering why referees striped shirts scintillated with crosscolor. It was only in 1984 that anyone had the temerity to announce a receiver that could reproduce everything a broadcast station could produce. A year or two earlier, the first television receiver/monitors were introduced, and in 1985 many were talking of RGB inputs that had bandpasses up to 15 MHz.

It is these far-reaching developments that have prompted both a general and specific discussion of video processing from the video detector right back to the cathode-ray tube. As input devices continue to increase in complexity and bandwidth, more and more attention has to be devoted to this critical portion of the picture display. We used to think the cathode-ray tube was the prime limiting factor, but not any more. It's the amplifiers, peaking, isolation, grounding, and passbands that will be stretching the designers best efforts to keep up with all the video developments incoming from every imaginable sector. In truth, the television receiver has now become the absolute center of information and entertainment in the home. Computers, video games, VCRs, video discs, and all the rest are just accessories; for without the cathode-ray tube and its picture processors, these ancillary pieces of equipment would remain both blind and dumb. Shortly, many hard copy printers will be available to print out what you may display through these various external devices on the tube. What we're trying to say is there's no end to the video explosion taking place in the late 1980s and we fully expect it to continue on into the 21st century as satellite communications envelop an enormous share of America's and the world's data, voice, and video traffic.

VIDEO DETECTORS

It's not only safe but appropriate to begin with

several different methods of video detection and then bring developments up to date with the latest and best, as we see them, through the various media means and engineering descriptions. For only through linear detectors in both audio and video do we reproduce outstanding sound and pictures. The latest versions of each, of course, use only integrated circuits today, and we'll get to the video segment of these continuing developments shortly. But first, let's look at a nonlinear detector—the humble germanium diode—and see what it had to offer before the era of synchronous video demodulation. Then you should understand why more complex and effective electronics had to eventually reach the market. Nonlinearity in video immediately means distortion and these little semiconductors gave it their all.

The Diode Detector

Simple, inexpensive, and good enough for the early days of television, the germanium diode with its 0.2 V cathode-to-anode drop, has long since departed the path of history in most color receivers and only remains primarily in the $88 black and whites (Fig. 10-1). It doesn't do the job, and here's why.

One reason is that the dc component of any TV signal relates to its mean brightness. If there is no dc coupling, then only picture content produces an average video value—and this isn't good. A dc level must be restored so that black and white information may be positioned around it so that contrast and other controls will have no upsetting effect. An all dc-coupled system, however, will result in the CRT image going white at times to the point of blooming with no incoming signal resulting from negative carrier modulation, which means "whiter than white." But when black level clamping is used, the *absence* of an incoming signal fails to light the CRT and excess cathode current is not drawn.

So here's where the simple diode—series or shunt—becomes involved. Any diode detector will produce a unidirectional output current proportional to its input, less the usual 0.2 V drop. Unfortunately, as the filters charge and discharge, only peak amplitudes of the composite video signal pass across them, along with various spurious and harmonic frequencies generated along the way. Because of the inherent characteristics of the diode itself, such problems are magnified and distortions continue on through succeeding amplifiers directly to the picture tube—nonlinearities, tweets, and all. In this respect, dc restoration has aided to some extent, but even though it's closely controlled, the original problems still remain to a large extent, and half wave rectification—which is forever difficult to control and filter—remains exactly what it's always been: a serious problem.

In order to solve many or most of such difficulties, the IC industry has been hard at work over the years devising synchronous, full wave detection for video signals, and most color receivers today are well represented with first rate video detectors contained in the same plastic in-line packages that hold video i-fs, noise, and agc circuits. Sometimes you'll find audio detectors in the same chip, too. At any rate, there *are* workable solutions in hand, and we should now begin to study their evolution to become familiar with the various manufacturing design techniques.

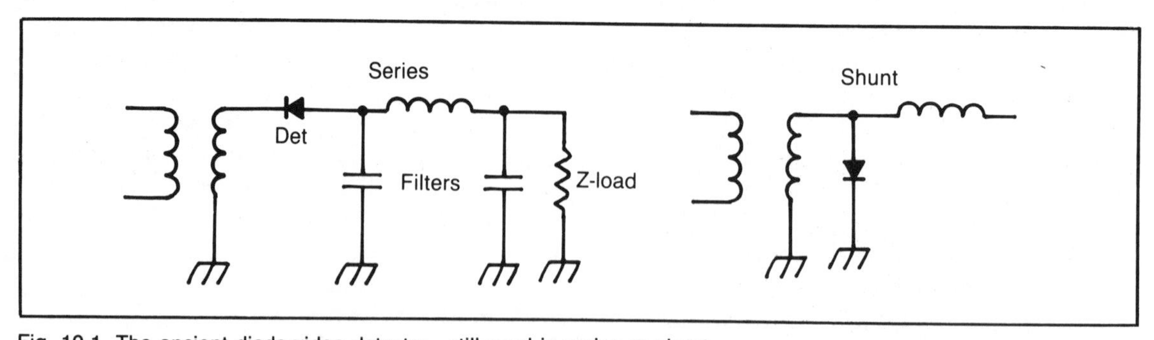

Fig. 10-1. The ancient diode video detector—still used in series or shunt.

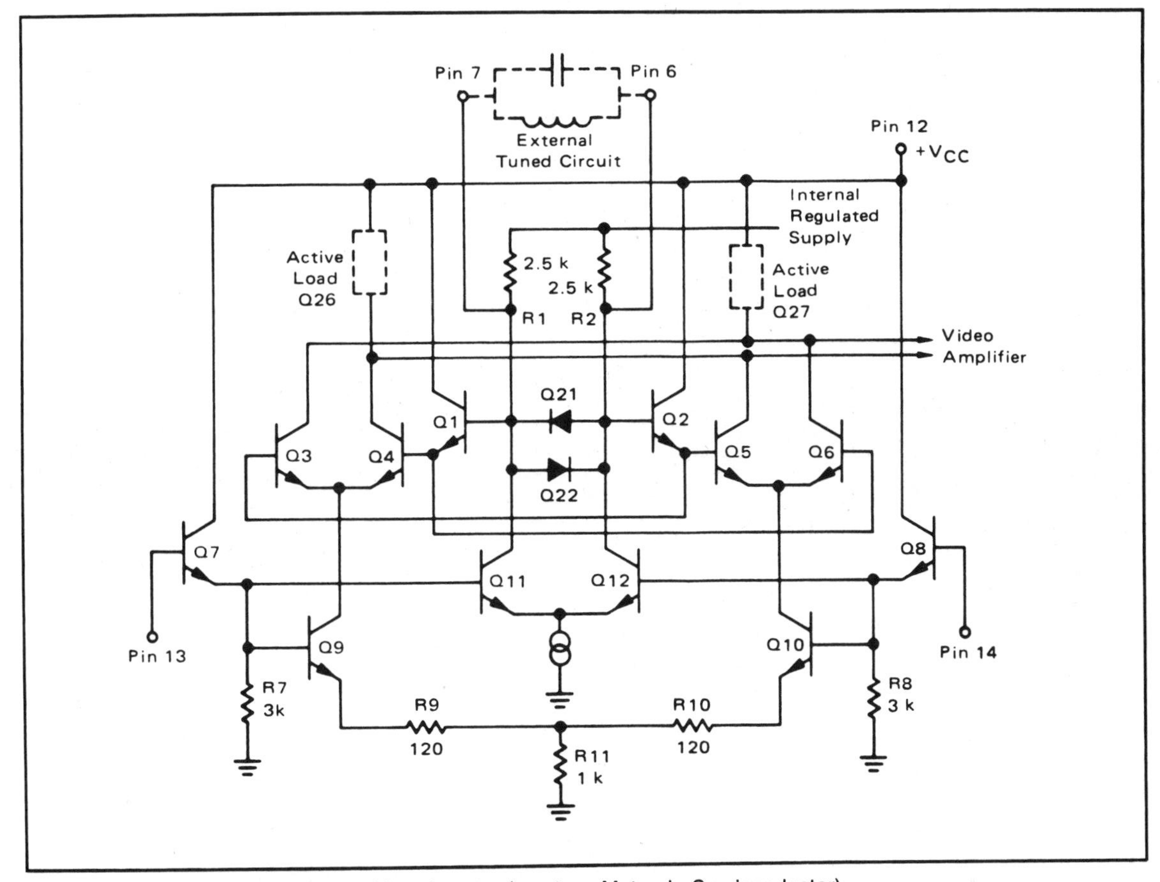

Fig. 10-2. An original synchronous video detector (courtesy Motorola Semiconductor).

A Synchronous Detector

Let's go back all the way to 1975 and a Motorola application note by James Reid on the MC1331. This was a 14-pin plastic IC that mated with a second MC1352 which did all the i-f amplification and agc operations for both rf and i-f. With only one interstage transformer between the two, plus RLC tanks to develop both 4.5 MHz intercarrier and 45.75 MHz video carrier frequency switching, this little IC looked like a winner in those early days of luminance and chroma receiver circuit development, and undoubtedly was.

The "basic" multiplier appears as illustrated in Fig. 10-2, and is described as a balanced multiplier, low-level detector. Transistors Q11 and Q12 are switching differential amplifiers with dual inputs from the preceding MC1352 i-f amplifier via buffers Q7 and Q8, which also serve as current sources Q9 and Q10 for quad detectors Q3 through Q6 above. Additional buffers are Q1 and Q2 for the pairs of quad detectors on either side, and the tuned LC circuit above adjusts phase differences to either 0° or 180° between switching-linear channels for best detection. The two D21-D22 diodes limit phase shifts in the switcher preventing excess differential gain.

In action, synchronous switches Q3 through Q6 are driven at a synchronous rate and alternately conduct during each half cycle, producing full-wave video outputs of incoming i-f information processed through lower differential amplifiers Q9 and Q10. At the same time Q11 and Q12, the limiter amplifier, delivers *video rate* signals to the two switching diodes above to energize the synchronous switches.

The video detector's switching voltage also

drives the separate sound detector, eliminating sound sensitivity compromise and the traditional 920 kHz beat between intercarrier sound at 4.5 MHz and the 3.58 MHz "suppressed" subcarrier chroma.

The sound detector, Fig. 10-3, is also an interesting topic, and although not video, has a tuned 4.5 MHz circuit of its own. The 45.75 MHz video carrier switch voltage inputs through Q17, which is also a current source for differential amplifier-detector Q15 and Q16. A varactor diode regulates the action of buffer Q14, while the 41.25 MHz sound carrier enters the detector through Q13. The collectors of Q13, Q14, and Q16 are all tied together and to the external 4.5 MHz tuned circuit. As Q15 and Q16 alternately conduct at the 45.75 MHz video rate, intercarrier sound is detected at the collector of Q15 and is emitter-followed out at ± 200 kHz through Q19. The remainder of the IC consists of a video amplifier and internal voltage regulator.

Obviously this is no stone age contraption but a reasonably sophisticated audio and video detector combination to overcome many prior faults in discrete circuits that formerly favored simple diodes to do the handiwork. Presumably they were linear,

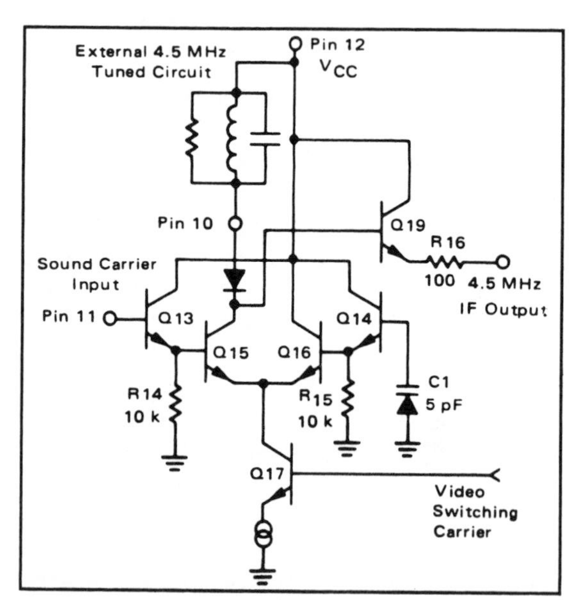

Fig. 10-3. A switching 4.5 MHz sound detector (courtesy Motorola Semiconductor).

but above all, these were very good first tries at putting complex functions in a monolithic integrated circuit and making it work. We'll also see how this squares with some of the later ICs that basically do just about the same things, although somewhat differently and at the discretion of their parent manufacturers.

No, we haven't lost sight of general video processing, but we thought that a good dose of synchronous video detectors might be of considerable help in laying this more difficult portion of the process to rest early in the chapter so we wouldn't have laborious repeats.

Matsushita's Pure Synchronous Detector

Although this particular IC operates at an i-f frequency of 58.75 MHz, there's probably no reason why it couldn't work at our video 45.75 MHz as well, and since it brings much of the subject up to date, we'll describe it with aid from designers Kutsuki, Kubo, and Isobe of Matsushita Electric, Osaka, Japan.

According to Matsushita, diode and quasi synchronous detectors are *not* sufficient for Japanese sound multiplexing or the various Teletext systems; therefore, the development of this new detector, which includes I and Q as well as AFC (automatic frequency control) output. The block diagram and schematic are combined in Fig. 10-4.

The voltage-controlled oscillator (VCO) is the heart of this IC and must have a wide range to follow the LO in the tuner and also deliver a stable signal under normal operating conditions relative to voltage, temperature, etc. The Q detector nongated phase comparator generates VCO control as well as an i-f output, and is separated from the I detector by a 90° internally-generated capacitative phase shift via MOS capacitors to the emitters of Q50 and Q51. The quadrature signal is then amplified by Q48 and Q49 for the bases of the Q119 through Q122, which is the phase detector with its Q114 through Q118 active loads.

Incoming video i-f information reaches differential amplifier Q126 and Q127 bases on its way to common base transistors Q124- Q125, which are part of the phase detector. The i-f input also goes

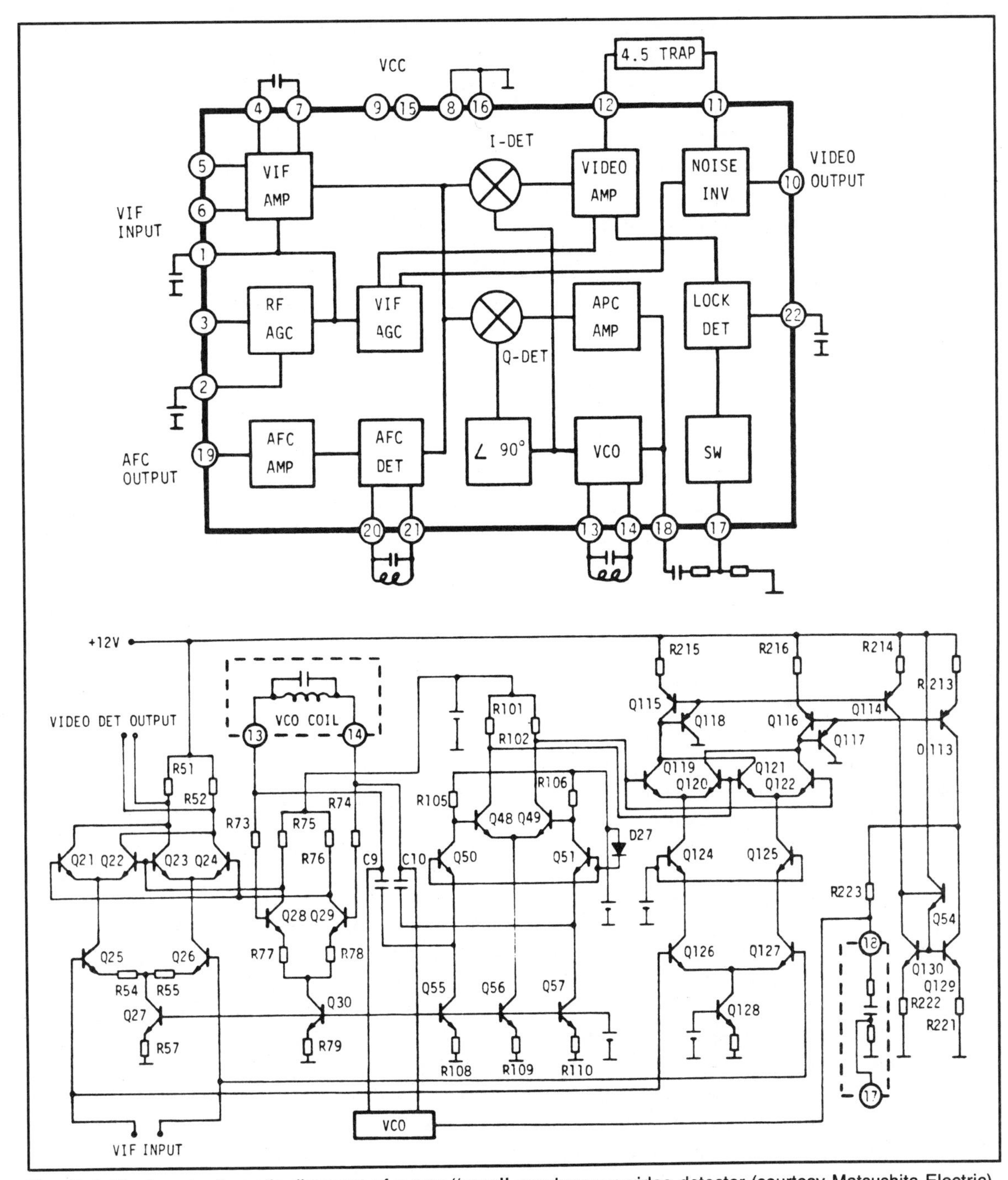

Fig. 10-4. Block and schematic diagrams of a new "pure" synchronous video detector (courtesy Matsushita Electric).

to the bases of Q25 and Q26 to be part of the video output to which the voltage-controlled oscillator supplies switching control to synchronous switches Q21 through Q24 to the left and above, with phase control of the VCO through Q113, R223 and the lock detector. Of course, as i-f currents and voltages are received, the four synchronous switches detect it and deliver a fully demodulated, phase-correct video output across R51 and R52 and pin 10.

Low-pass filter bandwidth is controlled by a "lock-unlock" detector comparing average video detector output level with a reference voltage having a pull-in phase error of 1.58 degrees. In lock, the bandwidth of the low-pass filter narrows, and out of lock it widens, with a pull in range of some 1.3 MHz, rendering automatic phase control continuously to the VCO.

Video Modulator

Watch this one—the heading says *modulator*, not demodulator, and is included here as an example of ICs used by industry to *modulate* composite video information up to channels 3 or 4 rf for nonbaseband (no video/audio) input television receivers. It should further add to your understanding of this general range of modulator/demodulator circuits among consumer products that are in contemporary use.

This time we're describing the outline and actions of Motorola's MC1373, 8-pin rf oscillator and dual-input modulator that generates carrier-loaded information for TV tuners (Fig. 10-5). It may be used for video games, home computers, tape recorders, and test equipment, has good oscillator stability to 100 MHz, and is color and sound compatible, along with overmodulation protection. A single 5 Vdc supply at 12 mA typifies normal operation with an rf output of 1 V at 67.25 MHz.

The rf oscillator is made up of Q13-Q14 in differential configuration with crosscoupling (feedback) through followers Q11-Q12. A tank circuit connects across pins 1 and 2, setting the oscillator frequency. Feedback transistors Q11-Q12 also supply switching excitation for doubly balanced modulators Q7 through Q10 as luminance enters differential amplifier Q4 and Q5 through pin 4 and Q2-Q3. Chroma also has an input from pin 5 into Q4. Q15 and Q16 form stable bias reference for current generator Q6 between the two emitter resistors of Q4-Q5.

Base voltage taking luminance into Q5 may only be offset negatively, so that overmodulation from this source will not occur. Resulting differential information received by Q4-Q5 is then switched at the selected rf frequency and delivered to pin 7 as the modulated video or video/sound combined output and on to an external vestigial-sideband filter (not included).

This is a good and very basic modulator on which to cut your teeth, and most consumer products of similar cost and complexity should operate similarly. There are fancier ones, yes, but that trend will continue and is not overwhelmingly important here, as long as you have a fundamental understanding of the bare-bones product. Are you surprised at the somewhat similar transistor functional arrangement to a *demodulator*? Principles are often equivalent to television antennas which may both radiate or receive—but are really different beasts of burden since receive antennas are usually broadband and subject to little power, while transmitters are single-frequency radiators and must handle thousands of watts through their emitting turnstiles.

With early and late model synchronous detectors described and recorded, you should be ready to proceed with the remainder of the video picture as it enters the various video amplifiers, comb filters, combined luma/chroma processors, and finally to the picture tube. Except for the comb filter and a bit of special video handling, the remainder should not be too difficult or demanding. We'll try and keep it simple—but first, comb filters.

THE ESSENTIAL COMB FILTER

Available in two varieties of very effective circuits with somewhat different approaches, the comb filter was preliminarily discussed in Chapter 11 (Chroma Processing) but without continuing into any expansive detail. The principle of all comb filters remains the same: to fully separate chroma and luminance by delay and line-matching tactics so that each may be introduced into succeeding stages and/or IC amplifiers for final RGB display processing. Along the way, chroma will have to be combed for its 2 to 4 MHz bandwidth (no longer just 3 to 4), amplified, demodulated, and then joined with luminance for red, green, and blue addition

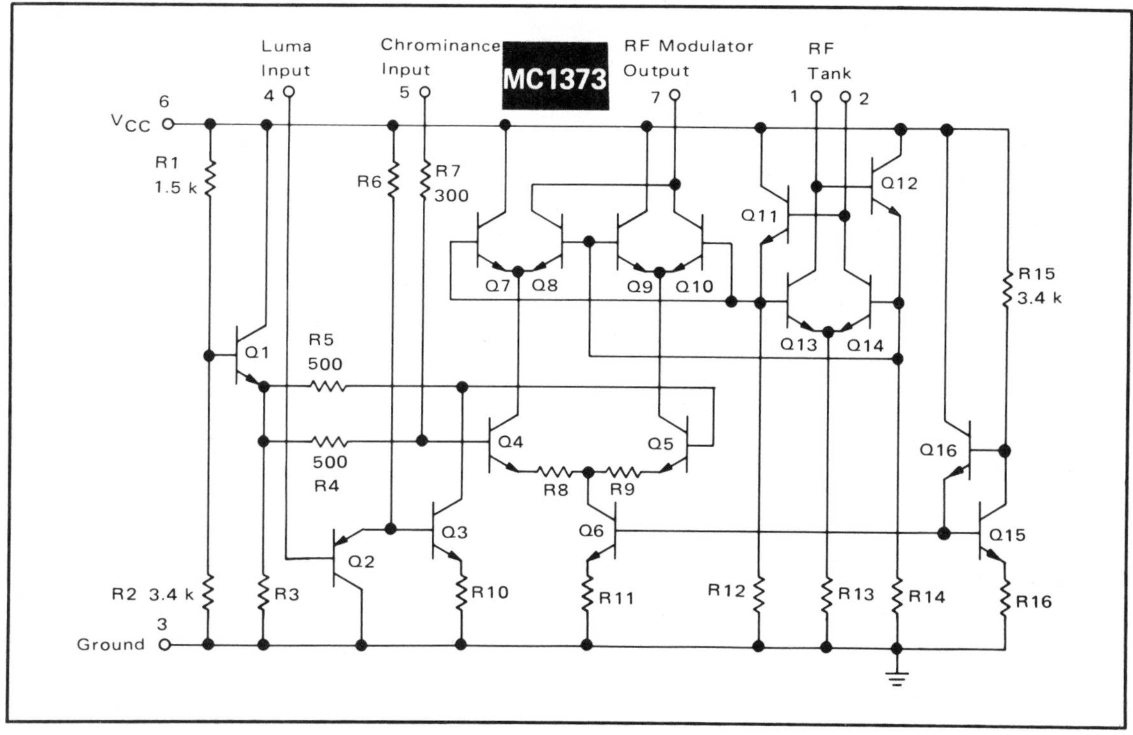

Fig. 10-5. An MC1373 modulator for composite video and sound (courtesy Motorola Semiconductor).

before being applied to the cathode-ray tube. In early color receivers, the CRT, itself, did the luma-chroma grid and cathode combining, but today all tubes accept signals at their cathodes only and grids are at ac ground for reasons such as arching, inaccurate R-Y, B-Y, G-Y color mixes, and just plain economics.

Also, as this color is combed, vertical resolution lost in the delay and separating process will have to be restored to something approaching 450 lines for maximum picture quality. As you will see, the glass block delay line circuits do this rather easily, but the charged-couple device ICs require considerably more complex treatments. For the record, one says the other restores more resolution than the other and also prevents dot crawl. If so, our general measurements have yet verified any great discrepancy, and in the final analysis one seems to work just about the same as the other.

For your edification, we'll take two good examples of these comb filters and work them through so you'll understand their principles and, if there is a preference, make your own selection. A chroma-luminance signal spectrum diagram is included for ready reference (Fig. 10-6).

The CCD Comb Filter

Described initially by Steve Barton of Fairchild

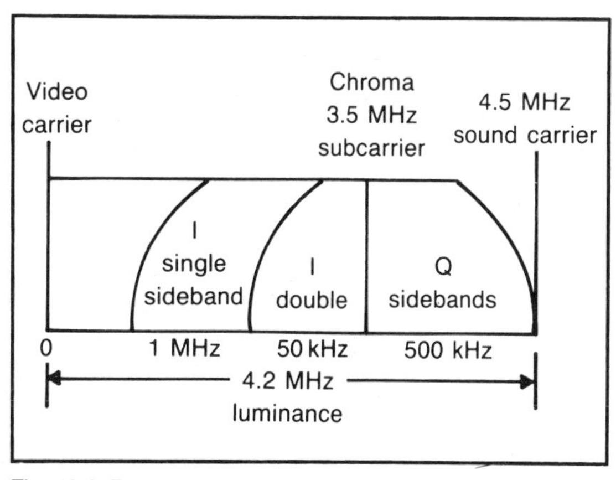

Fig. 10-6. Drawing of standard NTSC luma-chroma spectrum response.

Semiconductor, this is now an RCA-Toshiba device that has been simplified but still retains the basic principles for which it was designed.

As you should already know, older or cheap 3 MHz receivers have a 3.58 MHz LC trap in the luminance channel which round off high-frequency video resolution to 3 MHz and produces ringing, especially on edges of the picture, as well as some dot crawl. When sudden luma-chroma changes occur, they appear to excite the chroma trap and produce these undesirable conditions. By using a comb filter, luminance and chroma are cleanly separated since one is at whole multiples of the NTSC line spectrum and the other at odd multiples, or in half the line scanning frequency slots. Consequently, if one is resolved and then the other, they may be processed individually, chroma demodulated, and then joined in baseband form for CRT display. Luminance, of course, has already been detected, so it is only composite video that must be separated for further chroma detection and luminance delivery.

The basic comb filter and its transfer functions are illustrated in Fig. 10-7, similar to the one in Chapter 11, but this time combined with chroma and luma frequency interleaving characteristics—thanks to Mr. Barton.

The CCD device is actually a 910-bit analog shift register, developing a one horizontal line delay (H1) when clocked at four times the 3.58 MHz color subcarrier frequency, or 14.31818 MHz. Virtually flat to 7 MHz and rolling off at 14.3 MHz, the signal at output 1 is the sum of adjacent or successive scanning lines while color is 180 degrees out-of-phase on these same lines, because it is an odd harmonic and, therefore, cancels.

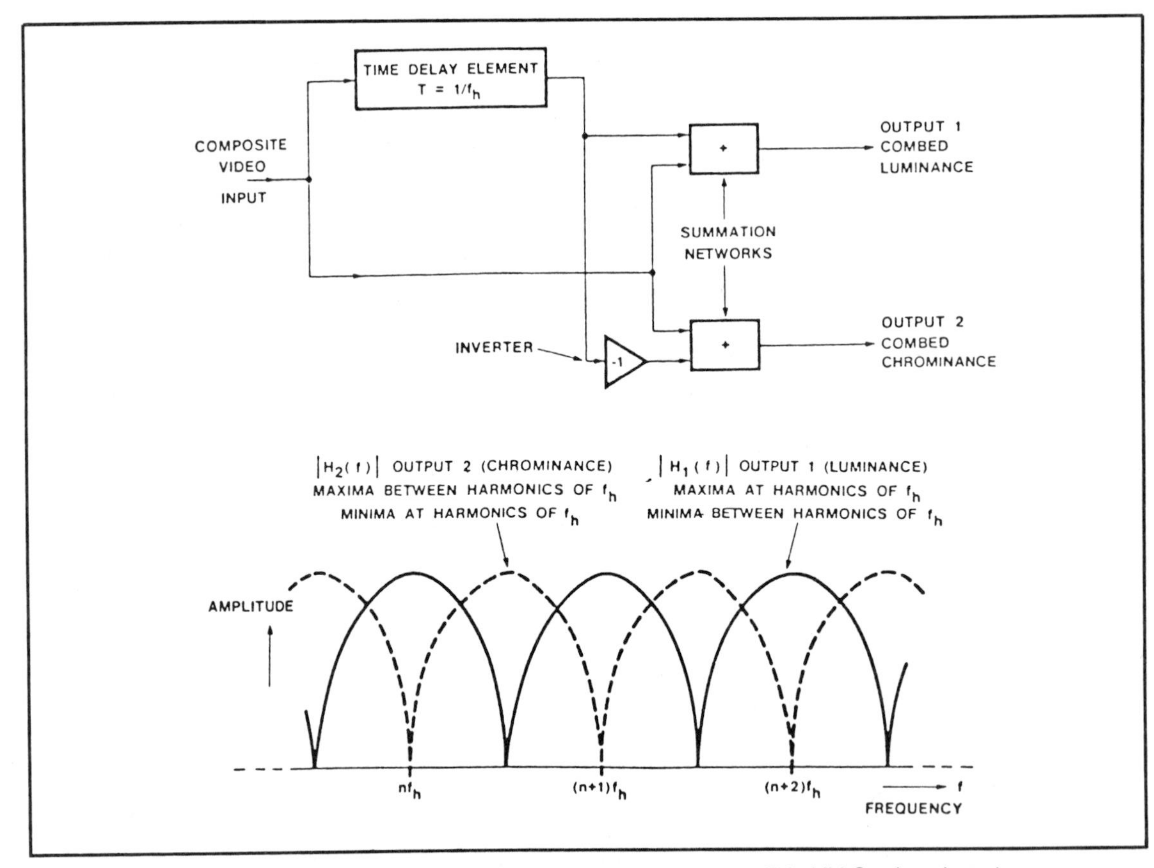

Fig. 10-7. Basic comb filter and its chroma-luma transfer functions (courtesy Fairchild Semiconductor).

At output 2, there is both delay and inversion so that chroma signals are now *in* phase and luminance out-of-phase, therefore only chroma appears this time and not luminance. Such a procedure averages the transmitted lines, but in so doing, introduces a substantial loss in vertical resolution. There may also be errors in stationary pictures on horizontal edges and diagonal lines, but such errors would be evident "only momentarily" when there's motion. The summation networks have both "antiphase" and direct video that act on both outputs and contribute to the chroma combing action.

Vertical resolution loss may be compensated by restricted combing to the interleaving portions of the composite signal or adding to luminance certain information containing the constricted vertical response. The simplest way is to add a bandpass filter between 2.1 and 4.1 MHz or simply a 2 MHz low-pass filter to remove chroma and add this to the luminance signal so that combing only occurs above 2 MHz—a process easier to write than execute with ordinary hardware.

Regardless, RCA has adopted such a comb filter and has used it very successfully since about 1979-1980 in many of its better ColorTrak models. You'll have a description and drawing of the latest version next.

RCA's CCD Comb Filter

Having a CTC 121 book of service data handy, we can work from this since RCA always has a good technical description of its products' theory of operation—and such proclivity is forever an aid to electronic accuracy. As you will discover, the comb-filter circuit here is a rather difficult one, and any and all help is welcome. (See Fig. 10-8). We'll use the full schematic since so much is involved.

As you can see, a 3.58 MHz limiter, 10.7 MHz tripler, a U600 comb processor, inverters, filters, and a nonlinear processor are all involved in the business of separating chroma and luminance, then restoring vertical resolution. Obviously all this effort has paid off handsomely since the CTC 121 and its CTC 131/132 successor have the best pictures probably ever produced in consumer product television. Certainly the recent addition of full 2 MHz bandwidth chroma eclipses all competition in the 1985 model year.

The 16 V regulator has the traditional series-pass transistor with zener base control and very heavy ferrite and capacitative filtering, judging from the number of evident components. In addition, there's also a single 9.1 V zener on the other side of R625 that also clamps another 9 V power source in IC U600, plus the regular 16 V supply.

Baseband video enters the comb filter through a small capacitor at pin 11 while 3.58 MHz subcarrier is introduced to the Q604 limiter via series-resonant circuit L612 and C630. We're now ready to put the process together.

The collector of Q604 is dc coupled to the base of Q603 and its base circuit limited by back-to-back clipping diodes CR602 and CR603. The tuned secondary and primary of L603-4 selects the *third* harmonic of the subcarrier and capacitatively couples a 10.7 MHz frequency into the logic and clock generator in U600. This generator drives the 683.5-element CCD analog storage block so that it couples a charge equivalent to incoming video information with each shift equal to a delay of 9.3 nanoseconds, and with the two multiplied, results in one full line scan delay of 63.5 microseconds—sort of like a bucket brigade, only faster.

Composite video then divides, with some being inverted, then amplified as chroma, and the other amplified without inversion, with both now reaching a summing amplifier. Luminance information, being in-phase from line to line, is then amplified and doubled in value, while chroma is cancelled. The next time around, inverted chroma is added by the summers, doubled, and luminance is cancelled. Luminance is then delayed by 186 nanoseconds before being filtered and peaked for the discrete Q606 luminance inverter.

Chroma, however, is now inverted and has two amplified outputs, one to the chroma buffer amplifier (not shown), and the other to a vertical resolution circuit via pin 13. Chroma is then peaked and filtered by the L600 and associated RC components and added to divided down luminance information at the junction of C637 and C610. The addition of luma has been done to compensate for

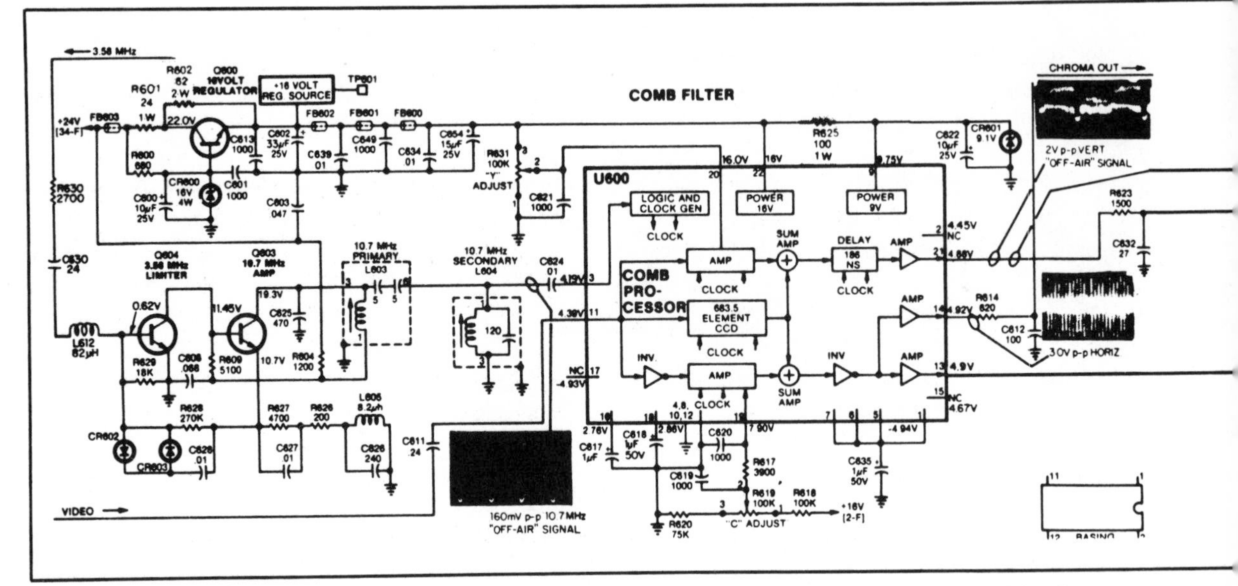

Fig. 10-8. RCA's charge-coupled device (CCD) delay line with intricate vertical restoration (courtesy RCA Consumer Products).

minor video "discrepancies" and line shifts in the video black level. This combined signal now goes to both the Q609 nonlinear processor and also to the base of a transistor in the U701 luminance peaker.

The four CR605-CR608 diodes in Q609's base circuit deliver bias to this transistor in proportion to the ac signal input, which includes feedback from the luminance peaker. The amount of vertical line restoration is then determined by the Q609 gain, selectively, for both large and small signals.

Nonlinear processor signals are now combined with chroma-luma signals that bypass this stage through L611 and enter pin 12 of U701, the luminance peaker. They are amplified and returned the junction of R636 and R750, and combined with the output of Q609, from whence they travel to the luminance buffer (not shown). Luminance just before the DL700 delay line also reaches the U701 luminance peaker through pin 21, finding the base of a differential amplifier. On the other side of DL700 another input enters the other side of this differential amplifier through pin 4. Another voltage from the upper right inverter and the auto peak amplifier also is applied to the emitters of the same differential amplifier. This is actually a feedback loop consisting of combed luminance, delayed luminance, and auto-peaked feedback, and goes directly to the base of luminance buffer Q709 (not shown).

That's the story of this considerably involved chroma/luma/vertical restoration circuit that, we hop, never breaks down, because it may be somewhat difficult to troubleshoot and repair. We should know a great deal more about this breakdown aspect about 1990.

The Glass Delay Line Comb Filter

Used by just about everyone *except* RCA and Toshiba, the glass delay line is considerably simpler to manage and control, with vastly fewer parts and no extended vertical line restoration. Obviously, it's cheaper and seems to do a good job of combing with only several transistors, one or two filters, and usually a plug-in circuit board. Some still criticize it for permitting more than normal dot crawl and not really allowing full high-frequency luminance amplification. But, it's obviously the industry's selection for one reason or another, and our tests show it does a very good luma-chroma separation job, with dot crawl only in very special stationary scenes that are ordinarily very infrequent. In fact,

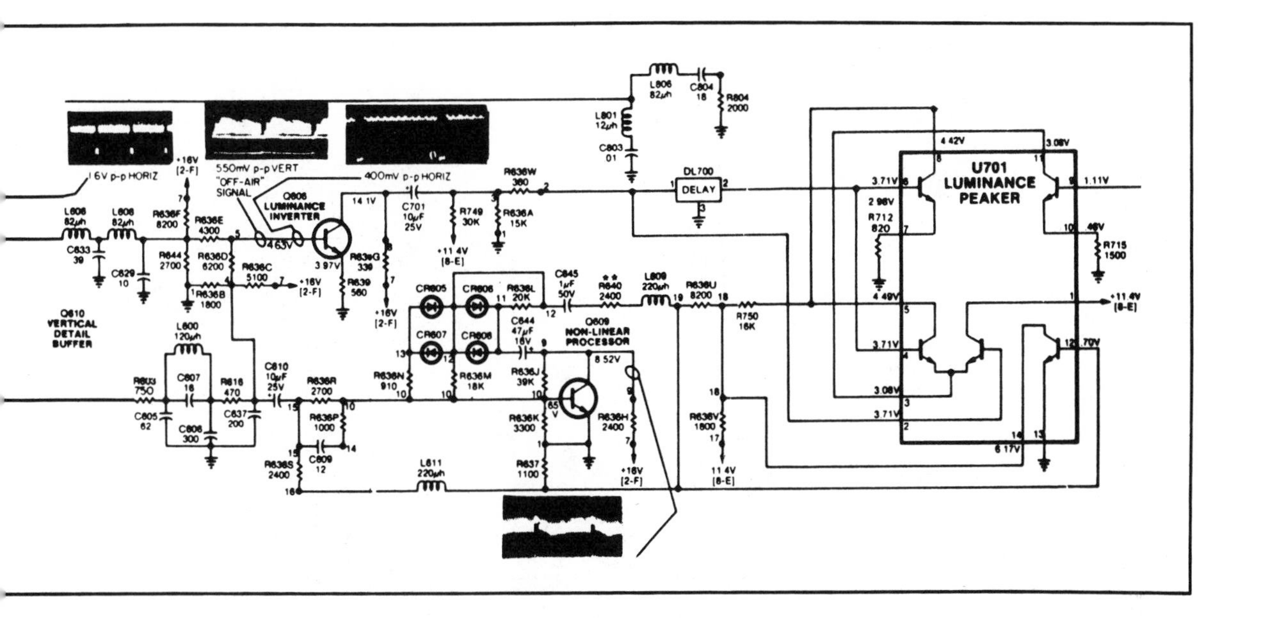

your author who made the final tests for Magnavox, the developer, and certified the results to the broadcast industry as having 25 percent better high frequency resolution than the 3 MHz receivers then in vogue.

Engineered by Bill Miller, these glass delay-line comb filters use a restricted bandwidth variety of comb filtering where double sideband chroma is interleaved with luminance. Since combing is restricted to this area, luma vertical resolution is unaltered with line-by-line changes below the lowest combing frequency of approximately 2-3 MHz. Magnavox maintains that with this type of comb filter, the same fine detail vertical resolution is resolved with any conventional television receiver. Our tests would have to say 'amen' because there have been constant improvements to this day.

Since the same block diagram given earlier in the chapter still applies we won't repeat, but just continue with the schematic description of Maggie's comb filter illustrated in Fig. 10-9. This original filter consisted of eight transistors, the glass delay line, lots of RLC circuits, and some hairy tuning. Today, that's generally changed in every way except the delay line, which is probably much improved also. At any rate here's what it did in the Magnavox T-815 chassis.

Composite video had two paths into the circuit. One directly through to the emitter of Q6 and thence, aided by selective peaking to the luminance output. The other input was inverted by Q1 and applied to the 1H delay line. Chroma, however, had but a luminance comb defeat since it was already contained in composite video, and this input arrived from the color killer.

Phase and amplitude are adjusted by R2-L1 and output balance by R10 into follower Q3, nulling the two out-of-phase signals from about the delay line so that one delayed and one not delayed will null luminance at the R10 wiper producing primarily chroma and only a little low-frequency luminance for the Q5-Q7 output. C8, L5 and R22 introduce appropriate phase and time delays equivalent to the luma channel during signal processing.

Chroma also arrives at the combing portion of the filter through C5, R16, and R17, with L3 assigned as phase adjust and R16 setting amplitude for chroma bandpass. As this combed chroma enters the base of Q6, composite video reaches the emitter of Q6 through R19. With these signals in phase, chroma is cancelled and only luminance travels through Q8 and Q9 to the output. Parallel

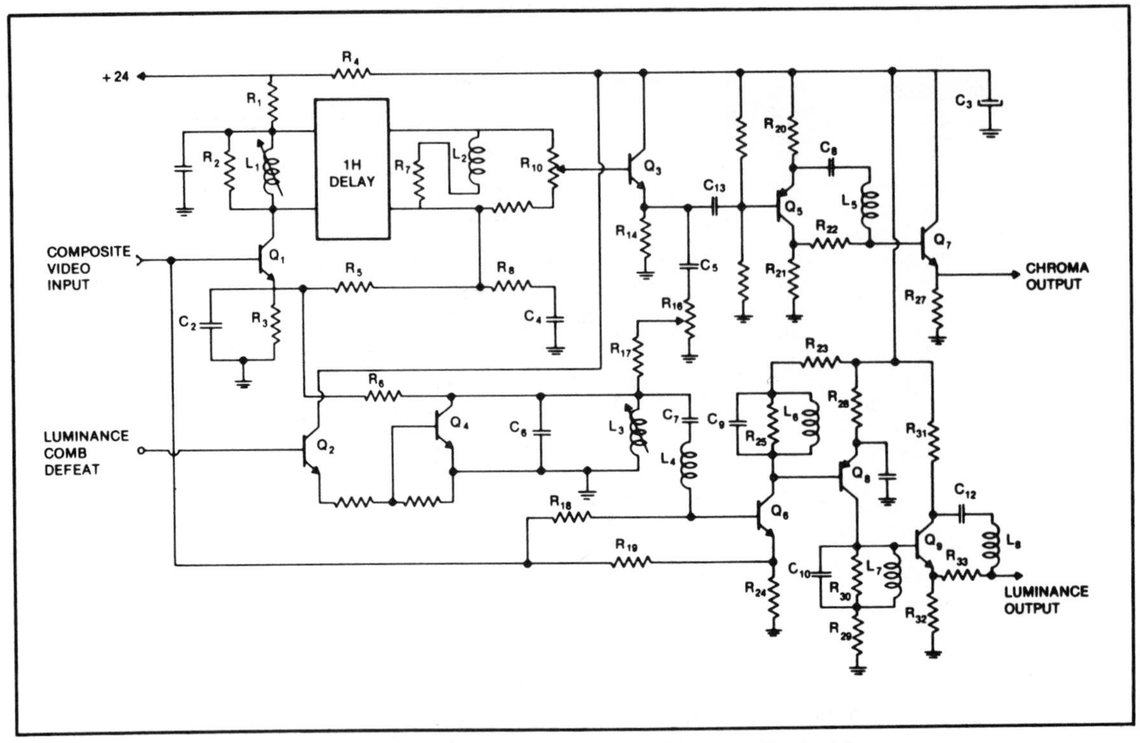

Fig. 10-9. Bill Miller's original Magnavox comb filter (courtesy NAP Consumer Products).

tuned circuit L6, C9, and R25 and companion circuits in the following two stages compensate for low-frequency luminance about 2 MHz filtering through the chroma bandpass.

Quiescent during incoming color, Q2 and Q4 are connected to a separate color killer on an adjacent integrated circuit. With no incoming color, they turn on and shunt any Q1 luminance combing action to ground so the receiver will handle nonstandard sync from such externals as video games and computers without problems.

A Modern Glass Delay Line Comb Filter

While the original comb filter was effective, there was a great deal of impedance and circuit compensation for other signal processing portions of the receiver, and so the parts count was necessarily high. In today's version designed for the 25C4 chassis (Fig. 10-10) there is considerable parts reduction by Magnavox and even electronic simplification, although most of the original principles firmly remain.

Neglecting the four transistors in the sharpness circuit on the same board, this 1984-1985 version has the same 63.5 μsec glass delay line, a delay line driver and only three additional signal processing transistors, not counting the 24 V series regulator—quite a difference from the original of some eight years past. It's easier to describe, too, and we won't indulge in excruciating details.

This comb filter retains the comb filter mixing and null principle as composite video enters the delay line, which you can think of as a transformer, with the delayed signal developed between the upper and lower outputs. There is also a direct video signal from the emitter of paraphase amplifier Q6 which reaches VR18, where mixing and nulling occurs among direct and delayed signal portions. As composite video is delayed, added, or cancelled, chroma passes through the Q9 luma inverter and out through its collector to the chroma output. But also in its collector are the LR circuits that join with luminance coming from the emitter of luma equalizer Q8, proceeding on into the Derive amplifier, blanker and peak detector of the auto sharpness circuit.

Simultaneously, luminance reaches the base of phase splitter Q10, which also produces paraphase outputs, as Luma 1 and 2, one of which is further phase-shifted by a capacitor on the main chassis, with both then proceeding to a peak driver and white peak attenuator. Chroma, at the same time, is additionally filtered by an LC circuit, entering the chroma amplifier on IC300 (not illustrated), along with suitably delayed video into the video amplifier of the same IC.

With that description firmly in hand, we will leave the rather tricky subject of comb filters and proceed on to the remainder of luminance processing, describing both individual integrated circuits and those in immediate receiver applications.

VIDEO PROCESSING

Following the video detector there must be composite video processing, including the usual luminance delay, so that luma and chroma—one wideband and the other narrowband—travel at the same rates through their respective circuits, reaching the ultimate mixing stage at precisely the same instant. In the meantime, certain amplification, clamping, peaking, and controls must be applied to maintain full frequency response and sufficient leveling to carry this video through all final stages to the picture tube. To further aid in understanding such procedures, we'll begin with a relatively simple integrated circuit and work up, leaving as little as possible to the imagination.

A Basic Luminance Processor

The CA3143E, designed and marketed by RCA, is a good example for initial study since it involves many of the principles now appearing in most all of the better receivers. In this one you have black level clamping, linear dc brightness controls, contrast, peaking, and horizontal/vertical blanking. Once again we'll combine the schematic and block illustrations in a single diagram as supplied by RCA Solid State (Fig. 10-11).

Signals from a tapped luminance delay line are applied to video inputs 1, 2, and 3, where all three are collected first in a peaking amplifier, then to the video amplifier along with a nonpeaked video input. Signals at terminals 2 and 3 are inverted and summed while the one at terminal 1 is not. This permits low-frequency signals to remain unattenuated, while those at high frequencies are, relative to the delay line tap points. Therefore, at high frequencies, inputs at terminals 2 and 3 *are* attenuated

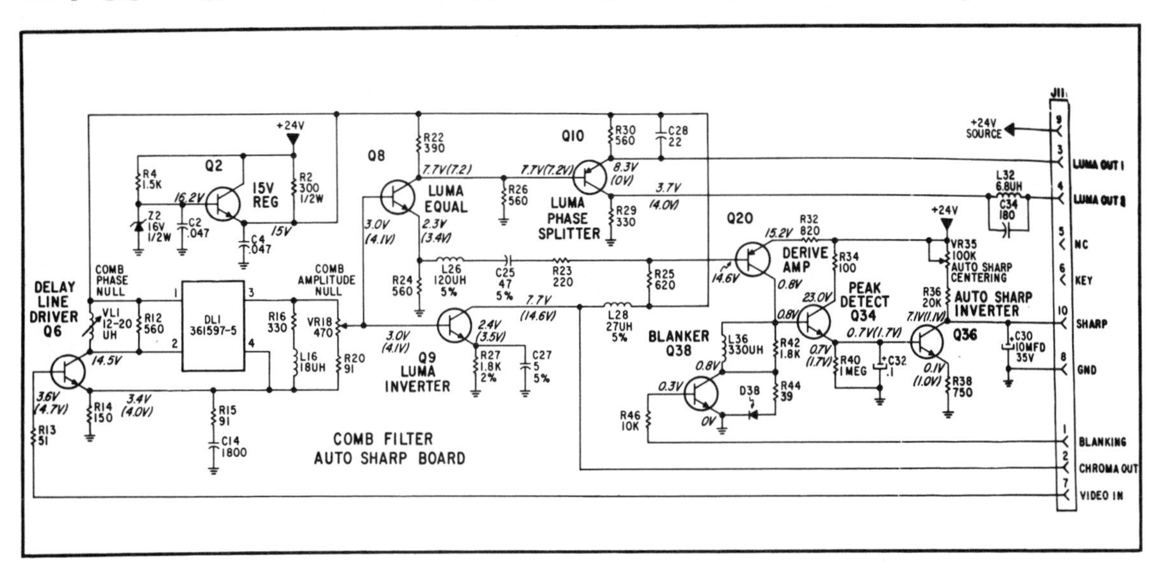

Fig. 10-10. Magnavox' newest glass delay line comb filter and sharpness control circuit combined (courtesy NAP Consumer Products).

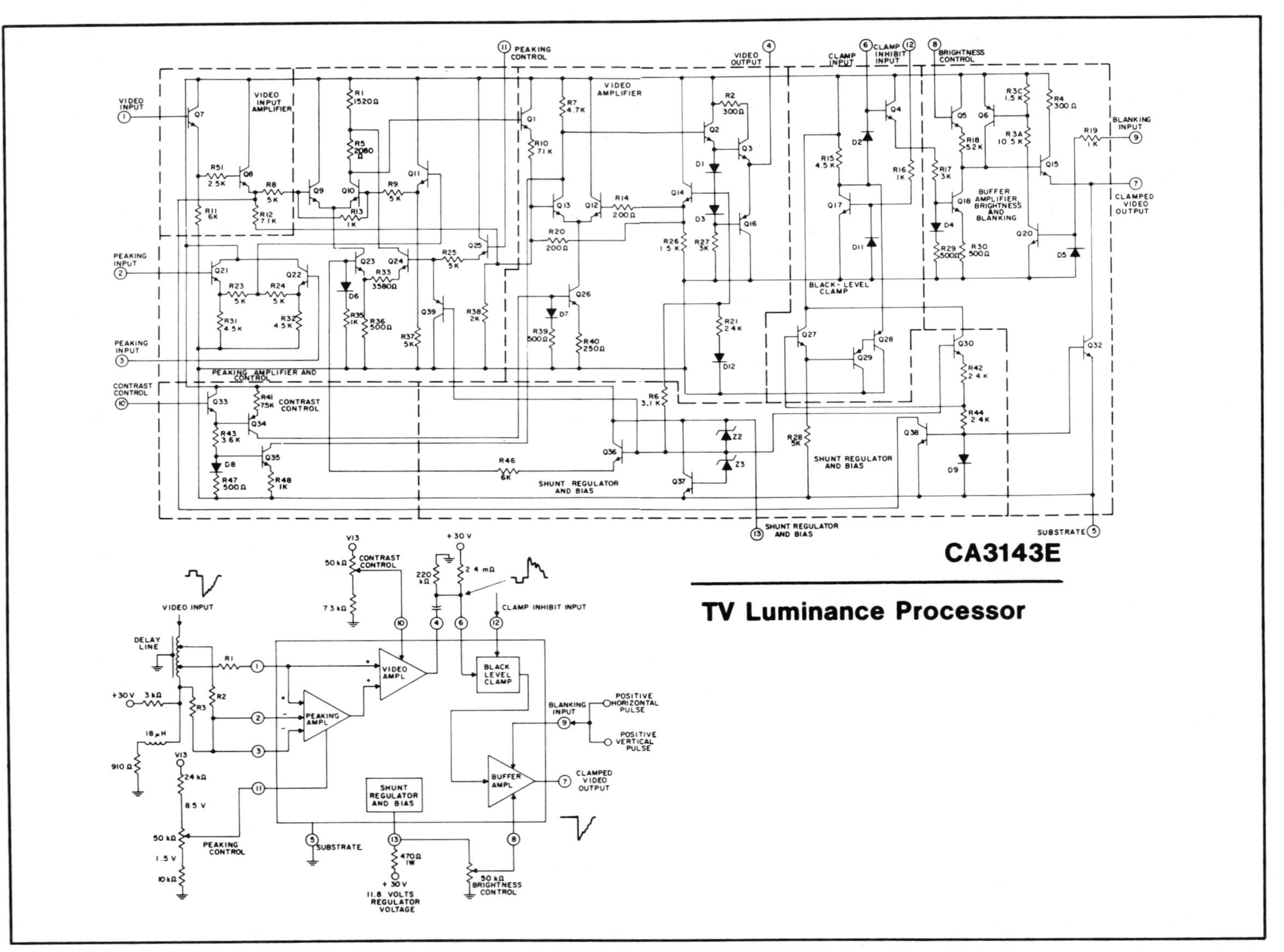

Fig. 10-11. Block and schematic diagrams of the CA3143E luminance processor (courtesy RCA Semiconductor Div.).

and so the peaking amplifier output produces high-frequency video at the video amplifier. An external peaking control sets peaking signal amplitudes according to operator preference.

The differentially amplified and push-pull video amplifier output at pin 4 is ac coupled through to pin 6 into a clamped input consisting of diodes D2, D11, and transistors Q4, Q17. With the signal entering pin 6, D2 clamps to the level of Q28-Q29, and Q4 conducts accordingly. But, during the sync (blanking) period, when a positive sync pulse enters pin 12, Q17 saturates and the anode of D2 goes to ground, disabling the clamp so that D2 and Q4 go to the next level above sync which is the black level, establishing black level clamping.

There's additional signal control as the base of Q18 by D4 as video enters the buffer amplifier and brightness area. You will note there's a blanking input at pin 9 and a dc brightness control at pin 8, with each controlling the gain of amplifiers Q5 and Q6, which are the load for Q18. As blanking enters and Q20 responds, Q6 back biases Q18 and Q15 cutting off these transistors and the video output, while between blanking, voltage from the brightness control varies the gain and, therefore the drive, into follower Q15 and the clamped video output. Contrast is a function of another dc control at pin 10 which regulates conduction of Q33-Q34 and the video amplifier differential pair Q12 and Q13.

For a relatively simple-appearing circuit that turned out to be somewhat complex, perhaps even more than expected. You'll have to become used to these since the more functions designers can lay on an IC chip, the greater operational control is possible with the various TV receivers—a considerable departure from one or two video amplifiers and a couple of capacitors out of the dim past. Video clamping, peaking, and lots of automated controls are now the rule rather than the exception, and modern designers have learned to use more rather than less in the way of strobes and levelers to keep video wideband on track.

The CA3217E Chroma/Luma Processor

So you won't lose interest, we thought a final composite color and brightness "giant" IC with 28 pins to perform *all* operations between the video detector and final RGB amplifiers might be useful here before describing a similar one functionally installed in a television receiver (Fig. 10-12). About the only things it does not include are sync processing, audio, and high voltage—and they aren't exactly luminance and chroma.

This time, however, only the block diagram for the CA3217E will be included since so much chroma is involved and the schematic, itself, is a veritable nightmare of interlocking circuits actually involving more color than luminance. So, we'll mainly talk about the luma functions and leave chroma to Chapter 11 where it belongs.

On the left you see the two inputs from luma-chroma, of course, with the comb filter separating them. Feedback from the blue amplifier and blank/buffer is picked up in the keyed comparator which seems to be controlled both by a synchronizing pulse as well as the brightness control which, of course, furnishes a dc level usually considered bias. This brightness bias, also connected to the beam current CRT limiter, has an input with the picture control potentiometer into the active picture controller. Two outputs from this circuit go to both the chroma 2nd amplifier and also the composite luminance amplifier, automatically limiting their gain with increasing beam current. With luminance fully clamped and regulated, the next step is reunion with demodulated chroma in the luma/chroma matrix amplifier so that RGB-Y chroma now has the Y luminance added and becomes red, blue, and green for final power amplifiers and the picture tube.

You might note in this IC there is also a modulator connected to the 3.58 MHz carrier limiter, a 90-degree bandpass filter and both I and Q chroma demodulators. Modulator inputs are also from the fleshtone phase detector and a chroma limiter, probably a form of feedback for constant fleshtone correction, which may be switched in or out at the operator's preference.

There's also a *sandcastle* circuit input into this IC you should know about. It's a pyramidal type of waveform consisting of horizontal/vertical blank-

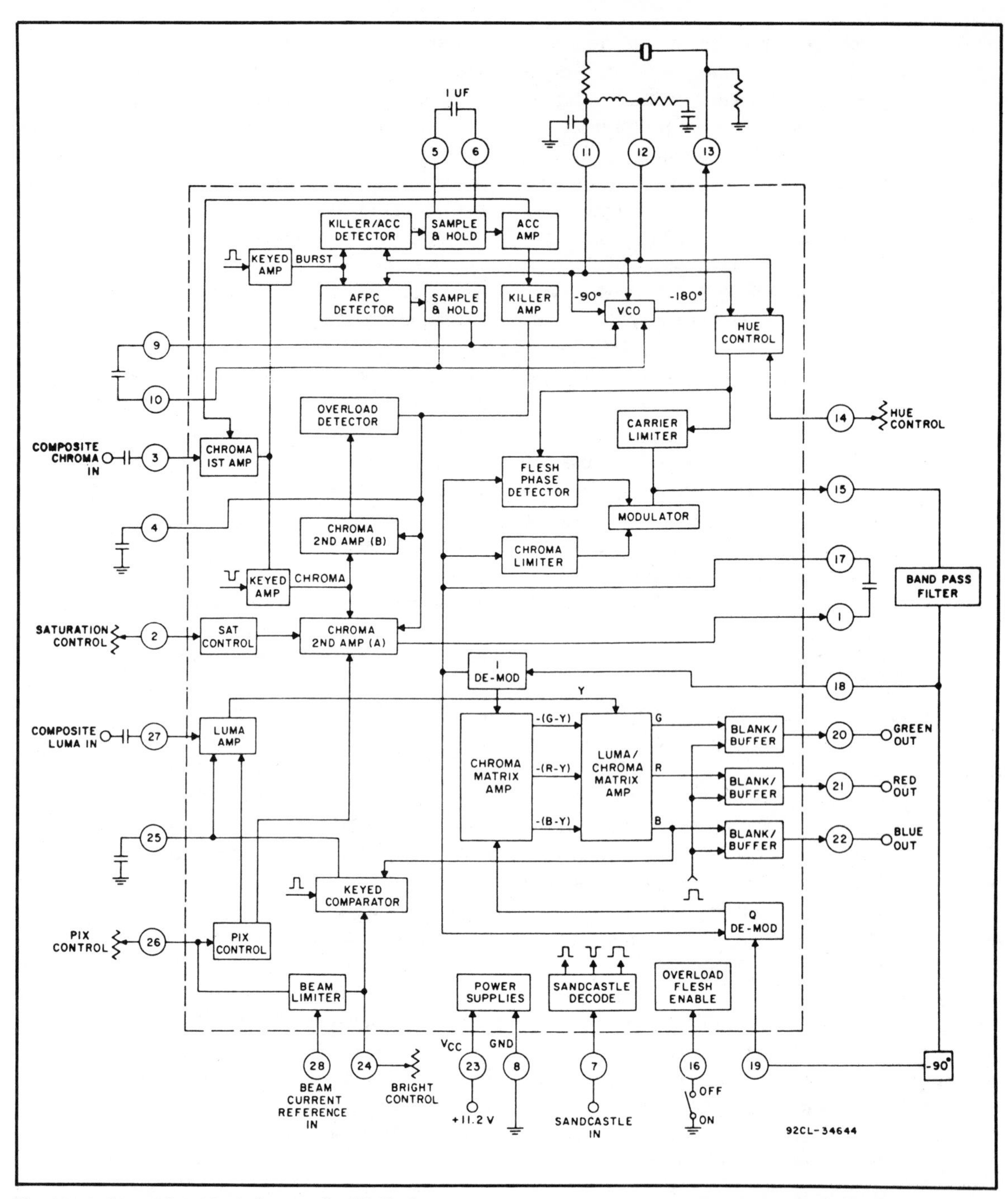

Fig. 10-12. Signal flow block diagram for RCA's CA3217E luma-chroma processor (courtesy RCA Semiconductor Div.).

ing, and color keying. Entry into the IC is at pin 7, and these pulses then deliver critical burst keying, V/H blanking and other timing within the CA3217E. One description of sandcastle inputs also includes black level clamping, but we see no evidence of this among several of the newer RCA receivers. Apparently, sandcastle serves sync and blanking from its composite pulse input and nothing more. A decoder inside the chip separates the three and channels and guides them to their respective waiting circuits.

The foregoing should have served most of the up-to-date requirements through 1985 and beyond relative to individual integrated circuits. Now, for your edification, we'll work with a partial schematic hereafter so you can visualize the entire picture.

General Electric's PM Chassis

For variety, let's look at General Electric's PM-A chassis and its complex chroma-video processor so you'll have a flavoring of the industry for suitable comparison (Fig. 10-13). This schematic portion is shown in its entirety without deletion, although we will not attempt to describe other than the obvious luminance and some chroma functions that are inseparable. As you will observe, it has many of the same operational blocks as several of the previous ICs, but is probably not a proprietary monolithic circuit since G.E. usually designs with off-the-shelf components. Do note the number of test points indicated and also the numerous dc voltages apparent at the various pinouts. This receiver also contains a vertical interval color reference IC that automatically maintains color amplitude and fleshtones without distortion. G.E. calls it VIR and Matsushita (Panasonic) in its receivers refers to the same general circuit as Color Pilot. This IC and remote control TV channel programming are two G.E. exclusives that are marks of excellence.

Video enters IC300 through pin 27 with ac coupling to the "emit peak" block—which probably means high-frequency peaking—and continues on into the contrast control where two brightness controls (one a limiter) and the contrast control all seemed to be ganged together. With heavy filtering, however, this is dc biasing and coupling rather than any involved ac function, so brightness and picture may be considered largely independent and are simply voltage arrangements to induce or reduce selective amplifier action.

With pedestal (black level) clamping and video amplification, luminance continues through the brightness control, is subjected to appropriate blanking, and then continues on to a peaking amplifier and video driver before being reunited with color information in the emitter-base junction of the final RGB amplifiers (not shown).

As you can see, the chroma input, fleshtone control, VCO and I-Q detection are entirely separate from luminance. And even with VIR inputs, color averaging, and an IQ 3.58 MHz gating reference from the voltage-controlled oscillator, this portion of the IC remains totally chroma, even to outputting RGB-Y color difference signals for CRT final amplifier matrixing. You may also see a "range" control for the Q detector which we would surmise is some sort of limiter for this stage since 500 kHz Q transmissions and receptions are thought of as part of the basic I-Q double sideband and do not have the 1.5 MHz range of colors assigned to I modulation, which is usually described as both double and single sideband because of its 1 MHz extended response.

Although a comb filter is not shown installed on this model, the keyways are available for simple plug-ins and it can be attached to the chassis very quickly and effectively. We would assume it has the usual glass delay line and several amplifying and paraphase transistors as do most of these type luma-chroma separators. Not shown, but on the same main signal board, are the i-f/agc/det ic, a sweep-oscillator IC, and a 4.5 MHz sound processor and demodulator. External sections accommodate high voltage, a full-wave transformerless power supply, and the usual RGB cathode-ray tube board mounted final amplifiers. An RLC sharpness control is located in the emitter circuit of the peaking amplifier, followed by inverting amplifiers and a video driver. Focus, screen, and high-voltage outputs are all taken from a series of stacked rectifiers and voltage dividers in the high-voltage transformer's secondary.

Thoughts

We probably could continue on ad infinitum into the mysteries of luminance processing but it would only be a rather stale repeat of what's been chronicled before. As you can see, luminance and

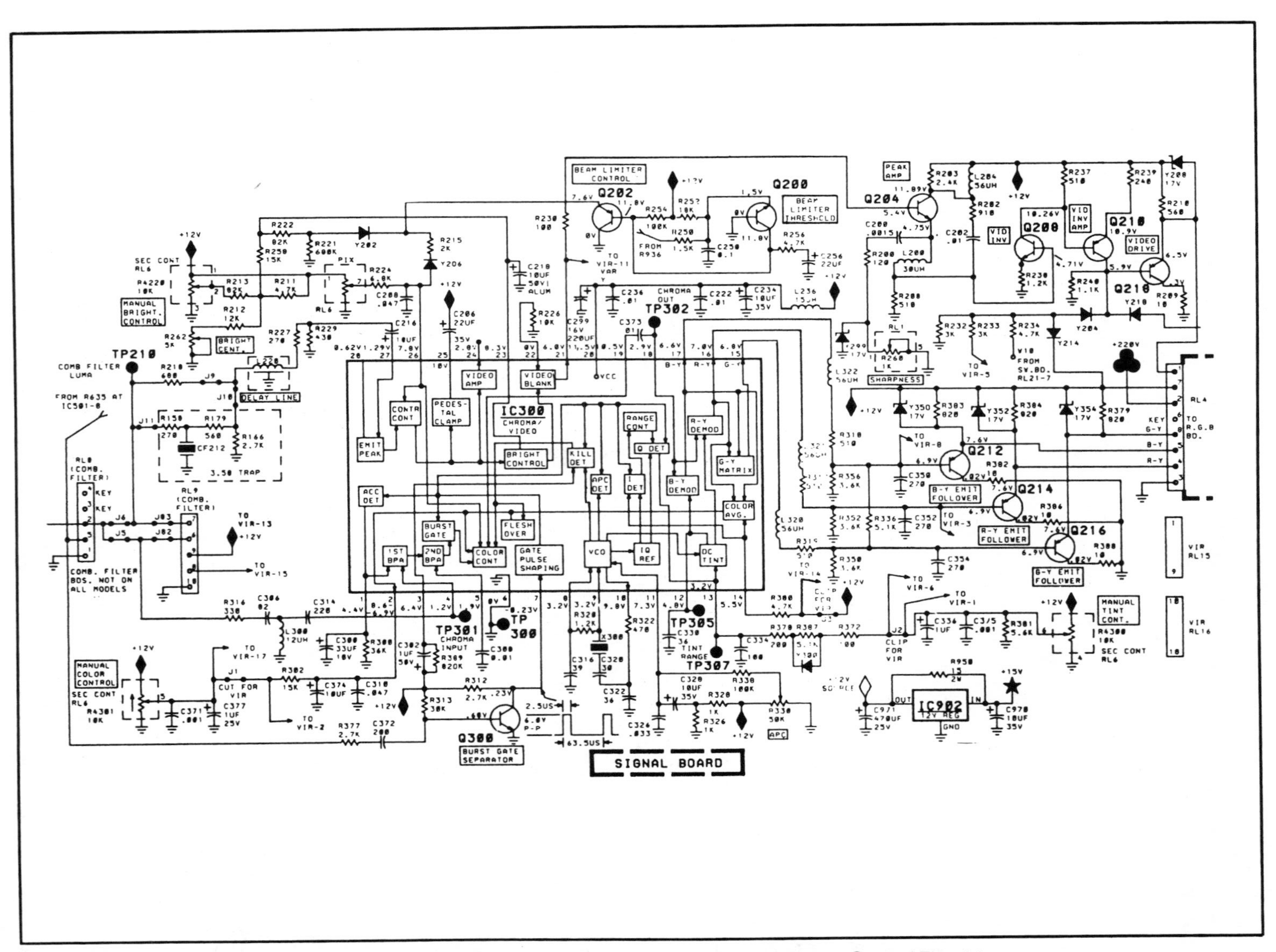

Fig. 10-13. The way General Electric processes its video-color information in the PM-A chassis (courtesy General Electric).

chroma, although now separated for certain electronic treatment and then rejoined for CRT display, may soon be handled differently with the coming of digital multiplexing, resulting in cleaner and more manageable separation-recombination than we've previously known. And, eventually, like the grid-cathode luma-chroma CRT mixing of old, other ways may be found to efficiently put these two essential picture portions together for even brighter and more colorful displays without some of the more laborious processing required in even today's advanced receivers—it's just a thought.

In the meantime, over the past few years U.S. industry has brought luminance from less than 3 MHz response to 4.2 MHz through rf, and out to as much as 8 MHz via direct RGB inputs. And as we know them today, consumer CRTs are about at their limits of good reproduction with baseband approaching 10 MHz. Specialized monitors, on the other hand, can offer even greater extended resolution—as much as double that of the previous figure—but cost is prohibitive and certainly not likely to become part of the consumer market. For even greater consumer video bandwidths, we'll probably have to await further development of liquid crystal displays or some sort of very fast, true-color matrix that will do the trick. Work on these very disciplines is proceeding in several countries right now, but there's no promise of a commercial-consumer product that's salable in the immediate future. Regardless, the video "revolution" continues unabated, and new developments in both luma bandpass and chroma fidelity are occurring every day.

Chapter 11

Chroma Processing

A GREAT DEAL HAS OCCURRED IN THE PAST several years— especially 1984—to warrant a searching look into color processing, its problems and accomplishments. From the days of CBS field-sequential system broadcast in August 1940 to the present, there have been giant strides in the reproduction of "living color" in most marketed receivers, readily acceptable to the viewing public, but not necessarily to television engineers both here and abroad who understood its considerable early limitations.

Original single tube demodulators were almost pathetic in their chroma detection, and it really wasn't until Zenith and Fairchild developed a solid state plug-in detector that chugged along with a flock of vacuum tubes did real chroma, in my humble opinion, begin to flourish. At a later date, RCA announced an IC with automatic color correction, and then the whole process was off to the races.

Presented as a technical paper at the IEEE Chicago Spring Conference on Television and Broadcast Receivers, June 1968, the μA737E finally began to see the light of adequate reds, blues, and greens. The three contributors were John Rennick, Zenith Radio Corp. and Larry Blaser and Derek Bray, Fairchild Semiconductor. They advocated and described a round, large 9-pin IC that could be inserted into a standard vacuum tube socket—and to this day, Zenith has always made most of its ICs socket inserts for easy removal and replacement, even to the large 28- and 40-pin LSI integrated circuits it uses today.

THE μA737E CHROMA DEMODULATOR

Since many of the principles of this original chroma detector exist today in everyone's chroma system in one form or another, we believe that a reasonably thorough description of its operation is more than warranted. Actually it would be a disservice to the industry if not given its rightful place in chroma and television history, for color display and development would not be where it is today had it not been for this *extremely* important 20-year old engineering triumph.

The basis for this entire chroma detection

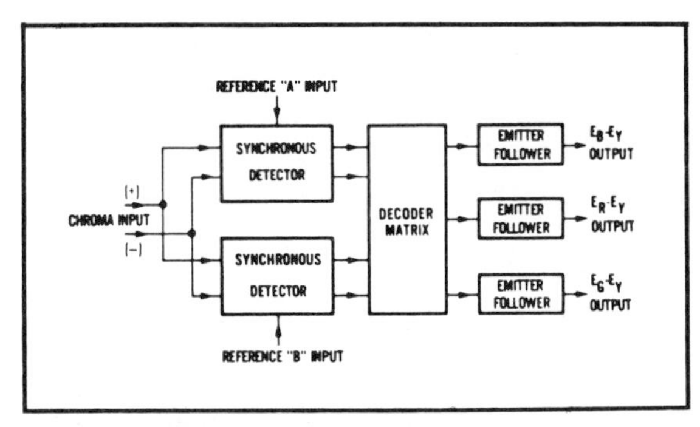

Fig. 11-1. Block diagram of first TV commercial IC synchronous color detector (courtesy Fairchild Semiconductor).

system was that of doubly-balanced *demodulators*—which is the inverse of doubly-balanced modulators (to eliminate carrier) common to all color broadcasting from the beginning of the service. That such a system was not used in vacuum tubes was probably due to tubes themselves as well as RLC passive components which required good-to-exceptional tolerances and temperature stability. Then, of course, there was the added factor of additional costs.

With the development of monolithic, sophisticated integrated circuits that became both stable and highly repeatable, less costly, and considerably more accurate, the time for this specific color detection method had arrived.

In Fig. 11-1 you see the original block diagram as offered by Mssrs. Rennick, Blaser, and Bray. Shown are plus and minus chroma intelligence inputs, a pair of doubly-balanced synchronous detectors supplying a decoder matrix (resistors) with red and blue signals from which R-Y, B-Y, and G-Y signals were fully developed. These, of course, are color difference signals, with the luminance (Y) intelligence (brightness and fine detail) being added later. Reference A and B inputs are the switching voltages supplied from the chroma oscillator.

The synchronous detector portion is shown more graphically in Fig. 11-2. Here "negative" chroma input into Qe and a positive dc biasing voltage into Qf share differential action of the two common emitter-coupled amplifiers with a common current source. As chroma input varies, Qe or Qf conduct supplying the emitters of Qa-Qb or Qc-Qd with chroma information which is separated into hue (phase) and saturation (amplitude). At the same time, chroma oscillator inputs reach the two pairs

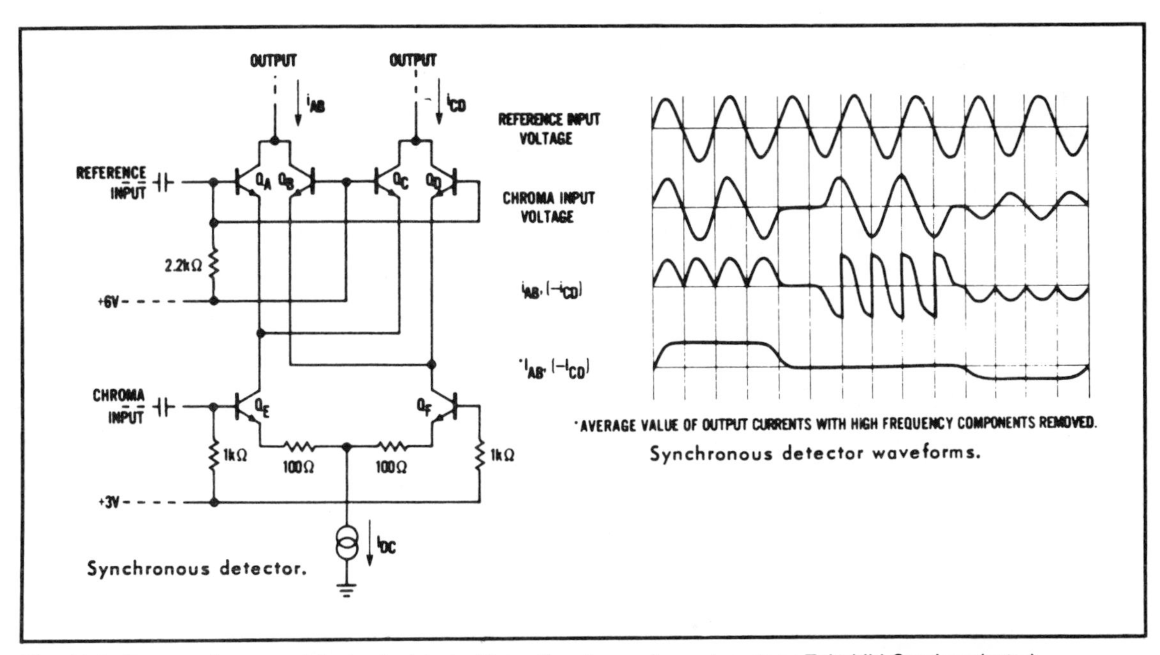

Fig. 11-2. One synchronous detector isolated with pertinent waveforms (courtesy Fairchild Semiconductor).

of switches phase-shifted at approximately 90 degrees, depending on the system. As these two switches pass selected chroma from emitter-to-collector outputs, the doubly-balanced configuration cancels the reference 3.58 MHz and delivers only red and blue information.

Green, although present in a mixture of the other two, is obviously absent and must be recovered. Consequently, a resistive matrix is designed to accomplish this need.

The full schematic of the μA737E (Fig. 11-3) shows how this is done, along with a number of changes which make the operating scheme considerably more practical. Negative-going chroma enters amplifiers Q11 and Q13 via pin 2, while positive-going chroma reaches Q12 and Q14 by way of pin 3. At the same time, Q1 and Q2 supply bias for Q11 and Q14, Q15 through Q18, with Q17 and Q18 being the two constant current generators for the chroma differential amplifiers above.

Now, the doubly-balanced detectors can go to work demodulating the chroma by reference A and B inputs connecting to transistors Q3 and Q6, Q7 and Q10, respectively, at the same time as dc bias from Q1-Q2 effectively switch Q4-Q5 and/or Q8-Q9 when the other switch halves are not conducting. Final results are fully detected chroma in red and blue hues developing across their respective collectors and green appearing at the junction of the 10 k and 22 k resistors, being emitter followed through driver Q19. At the same time, blue emerges through Q21 at pin 9, red through Q20 at pin 8, and green, of course, at pin 7.

This one 9-pin device actually supplanted three vacuum tubes, lots of passive circuitry, and also made the chroma detection process considerably

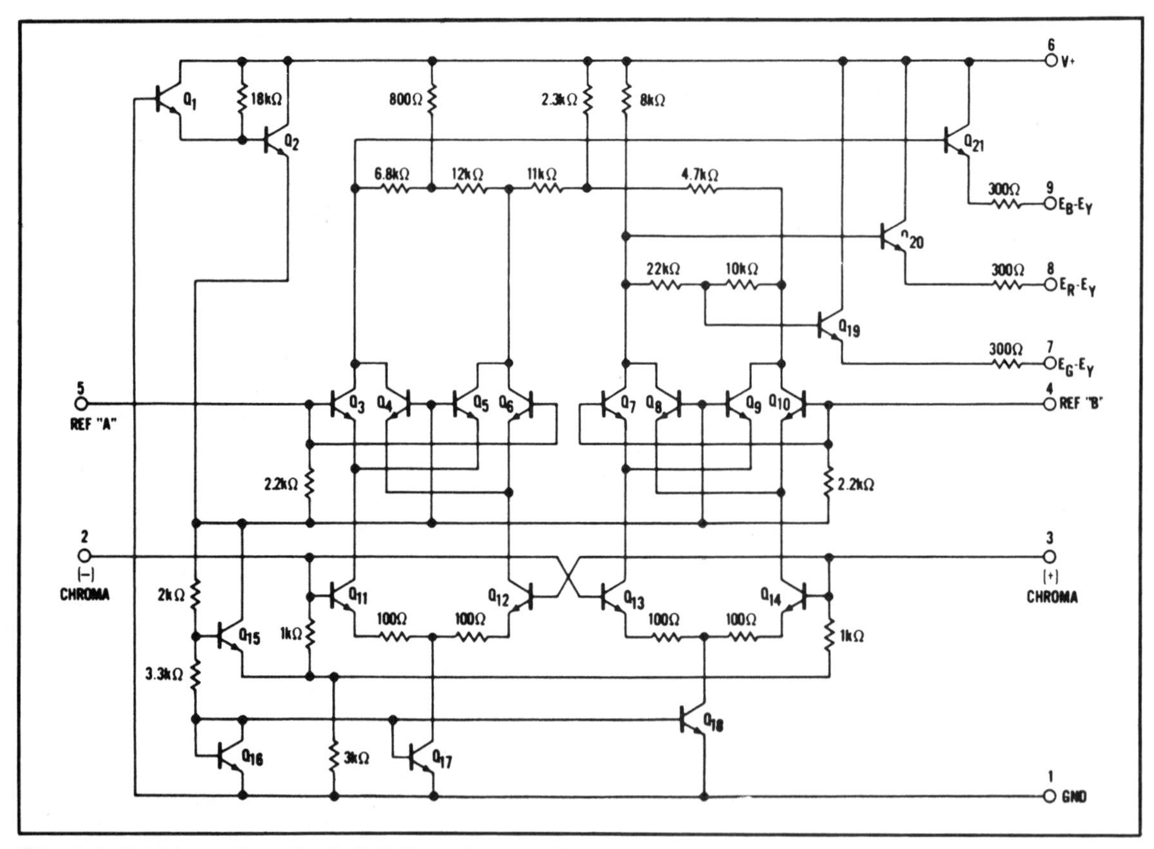

Fig. 11-3. Complete schematic of μA737E synchronous detector (courtesy Fairchild Semiconductor).

more accurate and predictable. Later the three vacuum tube CRT amplifiers were supplanted by power transistors directly supplying color signals to the three RGB cathodes. What has happened here is that luminance signals from the usual video delay line were routed to the emitters of the three CRT drivers and mixed with color in the emitter-base junctions of the three transistors. Now, with direct coupling to the CRT cathodes, gray scale tracking controls are now in the low-impedance emitter circuits of the three RGB amplifiers, resulting in a wider bandwidth; clamping and dc restoration are no longer required; common peaking is then possible with power dissipation shared; further CRT arc protection with grids at ac ground; and much of the old tube circuitry has been discarded, requiring considerably less power and much greater efficiency, along with fewer service requirements and ultimate cost savings.

As you will discover toward this and the receiver chapter's end, this *excellent* piece of engineering not only contributed to several varieties of chroma detection and automatic fleshtone correction, but also automated cathode-ray tube bias, removing CRT temperature drift problems for the life of the tube. It also made chroma processing much easier and now, with introduction of the comb filter, we are *beginning* to see the rapid evolution of truly broadband chroma-luminance combinations that will bring tremendous improvements in picture quality surpassing those now being broadcast. You may also observe that a great many major improvements in television reception and broadcasting have been made mainly by either RCA or Zenith.

MOTOROLA'S MC1323P

In the mid-1970s, Motorola Semiconductor announced a very flexible chroma demodulator arrangement designed for either conventional or possibly single-gun CRTs, which held some promise at the time; unfortunately, never fulfilled. This IC is a good example of chroma detector growing pains as well as what can be done with such circuits if the need arises. We might add, also, that although special research laboratories and the TV manufacturers, themselves, generated many of the new ideas, it was the *Semiconductor* houses that executed them and developed all the hybrid and monolithic integrated circuits in use today, adding many innovations and improvements while so doing.

Developed in two versions, with prime differences being three potentiometers and output connections, both will be described as one since the single gun (sometimes called "indexing") tube never fully materialized. The unique design, however, does allow independent adjustment of detector gains and demodulation axes so it can work with nonstandard phosphors and other possible problem areas.

A typical application diagram appears in Fig. 11-4, showing the addition of brightness, blanking, three-phase references from the chroma subcarrier, and potentiometer-controlled RGB-Y amplifier outputs. Note, also, the MC1323P has now become a 16-pin IC and is *vastly* more complex than its forerunners, and that a voltage regulator has been included.

This Motorola IC is known as a triple chroma demodulator since it has three separate doubly-balanced demodulators and three individual RGB-Y phase references, all controlled by either fixed or adjustable passive components.

A Vcc of 18 V supplies bias for both the voltage regulator and brightness amplifier, and the circuit itself has a 16-22 V dc operating window. Horizontal blanking is ac-coupled through pin 2 to the blanking amplifier, with chroma inputting through pin 3 and RGB-Y references appearing at different phase angles via the LC network at pins 4, 9, and 12. Each demodulator section is very much like the previous Fairchild version except there are active load transistors and not simply a group of resistors (Fig. 11-5).

Demodulators are balanced for signal and switching voltages to suppress both the carrier and undesirable demodulation products. When a switching potential appears at the base of Q20, for instance, Q17 and Q20, Q18 and Q19 will each conduct for half the subcarrier cycle. When chroma is applied to Q15 and Q16, a signal proportional to its phase and amplitude develops at A about dc

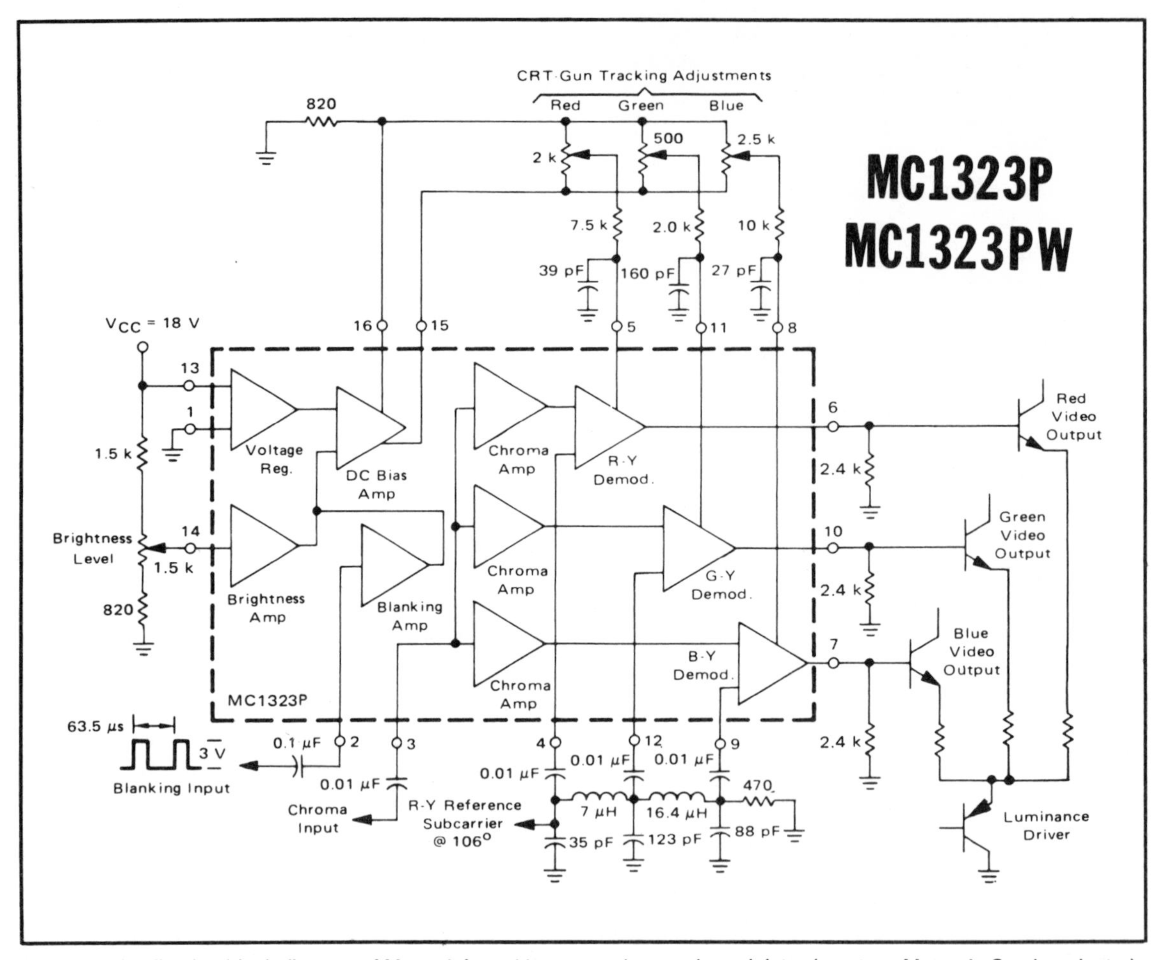

Fig. 11-4. Application block diagram of Motorola's multipurpose chroma demodulator (courtesy Motorola Semiconductor).

V_{REF}. Q25 then emitter-follows this signal for further video processing. Q24 draws equivalent base current with Q25 ensuring current balance among the Q17-Q20 doubly balanced-switches below. So that demodulation occurs on some well-defined axis, the switching subcarrier phase is chosen externally by the three sets of LC phase shift components already identified.

A 3-volt blanking input at pin 2 saturates a transistor for good blanking pulse amplitude, also preventing the demodulator switches from saturating during the 11 µsec blanking interval causing chroma shading at the beginning of line trace.

The MC1323 could also accept brightness information directly from the luma delay line, add its information just prior to the dc bias section and directly modulate the reference amplifier outputs, producing luma-chroma mix in the three demodulators and subsequent RGB outputs for picture tube cathodes.

RCA'S DYNAMIC FLESHTONE CHROMA DEMODULATOR

In early 1976 RCA announced a CA3137E chroma detector with "dynamic flesh-correction" and three color difference outputs, compatible with the CA3126Q chroma processor, which we'll be describing shortly. As the block diagram in Fig. 11-6 shows, this appears to be a "slightly" more complex circuit than those we've analyzed before. In

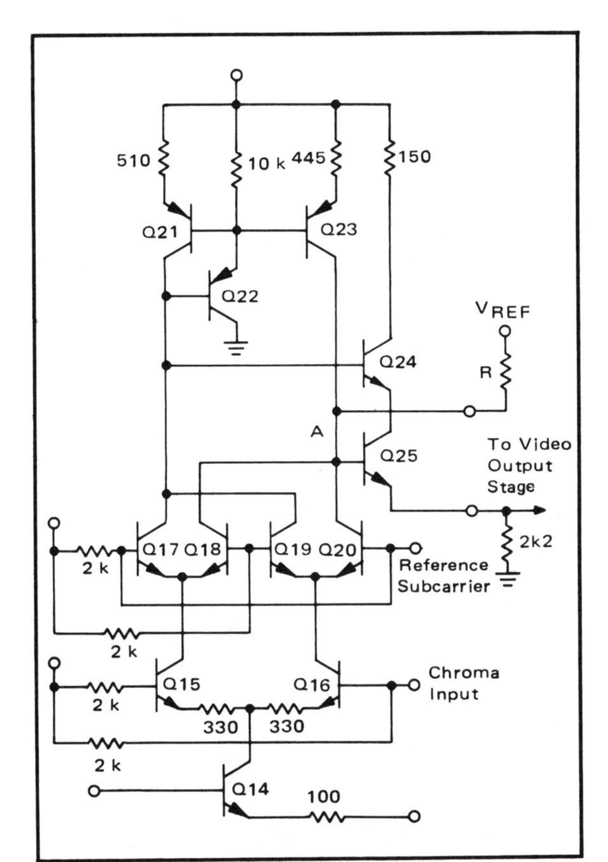

Fig. 11-5. An MC1323 demodulator with synchronous detection (courtesy Motorola Semiconductor).

reality, when you see the schematic, you'll wonder how RCA ever photo-masked this many functions on a chip. Then, it wasn't easy; today, with 28- and 40-pin ICs throughout the industry, you might begin to think this earlier attempt was relatively simple—or was it? Take care before rendering an opinion since most advances are evolutionary and often extract a price. One manufacturer had a frustrating experience with his big chroma chip for a long, long time—but it's finally coming around. With persistent orange-reds he did, indeed, pay the piper.

Fixed and dynamic flesh correction circuits, too, have had their share of problems, and it hasn't been until at least 1983-1984 that relatively simple monolithic methods have paid off. General Electric, on the other hand, has a sophisticated means of maintaining fleshtones without distortion, but considerable circuitry remains devoted to the task and competition claims there are still difficulties in low-level signal areas, even though we've noticed none.

To return to the CA3137E, this was state-of-the-art upon introduction and had adopted the concept for transmitter-type I and Q demodulators but with a matrix for the resulting three color difference signals. There was also a switch-disable flesh correction feedback circuit with some phase shifting and filtering, along with indicated dc feedback to keep faces at fleshtone hues without having to operator-adjust this function manually. This was a little different from resistor-preset circuits that "clamped" chroma and usually brightness, contrast, chroma levels, etc., so the operator could depend on static dc bias controls to keep a relatively usable picture on the screen. Unfortunately, oranges and reds were deliberately blended to show a broad range of fleshtones, probably spreading across 15 to 20 percent of the viewing screen—slightly inaccurate, to say the least.

In the CA3137E RCA tried valiantly to do all this *both* dynamically and accurately by changing the angles of chroma detection and effecting the process directly, along with feedback for proper fleshtones. At least it amounted to a good step forward and eventually resulted in RCA's famous line of ColorTrak® receivers that have remained on the market for a considerable period. Now, let's go to the actual "or equivalent" schematic and see how all this came about—for semiconductor manufacturers don't always "tell all" in their electronic diagrams (Fig. 11-7).

Chroma enters this IC through pin 3 and goes to the base of Q18 which, along with Q17, the other differential half, is biased by Q22. By differential action, one side or the other of the amplifier conducts and feeds the emitters of common base-tied amplifier Q19 and Q20. Signals are developed across R26-R27 for buffers F7 and F8 of the fleshtone corrector, and across R28 and R29 for F17 feeding the I-Q demodulators on one side and F19 on the other. At the same time, potentiometer bias for the chroma amplifier and saturation control is manually set, and diodes from the emitter of F13

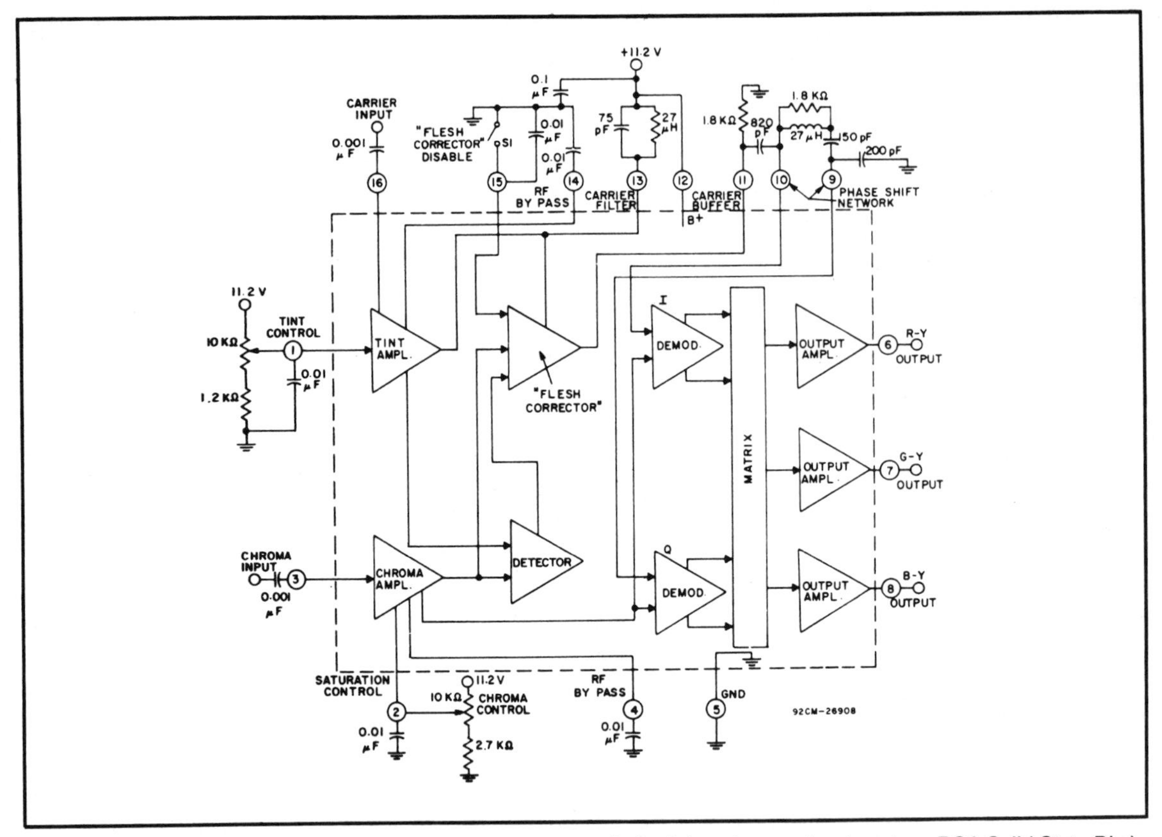

Fig. 11-6. RCA's CA3137E TV chroma demodulator with dynamic flesh (tone) correction (courtesy RCA Solid State Div.).

will conduct positively lessening Q9, Q20 conduction when this bias is exceeded.

The 3.58 MHz "carrier" enters the IC at pin 16 and is phase-aided by a manual potentiometer tint control at pin 1. Signals are coupled through follower F2 to the bases of differential amplifier Q2-F3, whose gain is controlled by its current source Q1 and the tint control. The output of Q2 then passes through Q4 to the bases of Q10-Q12 and also F6-Q6. Transistors Q10-Q11 and Q12-Q13 constitute a switch with cross-coupled inputs from amplifiers Q8 and Q9. By phase shifting of C1 and R3-R5, a control range of ± 45 degrees is produced for the two pairs of switches via the tint control.

With the tint control set for midrange, chroma phase in Q8-Q9 is compared with Q10 through Q13, producing an output usually the inverse of I chroma phase. With the flesh corrector switch open, removing ground from bases of current generators Q7, Q15, and Q57, the +I signal continues through chroma limiter F11-Q16 and on to the emitters of F9-Q14, with differential amplifier F6-Q6 becoming a limiter for the carrier. Pins 14 and 15 are the rf bypass, with flesh corrector output passing to F12 and the carrier buffer output at pin 11. Inputs to the two demodulators via pins 9 and 10 are through an 820 pF capacitor and an RLC I-Q phase-shift external network. If phase correction is not required, flesh corrector inputs are grounded and normal CW carrier flows to the demodulators through pin 11.

Chroma also reaches the I and Q demodulators from the chroma amplifiers via F17 for I and F19 for Q, while the Q28 through Q31 and Q35 through Q38 switches above receive the usual phase-shifted 3.58 MHz switching signals. Color is then developed across resistors R53-R54 and R59-R60 in

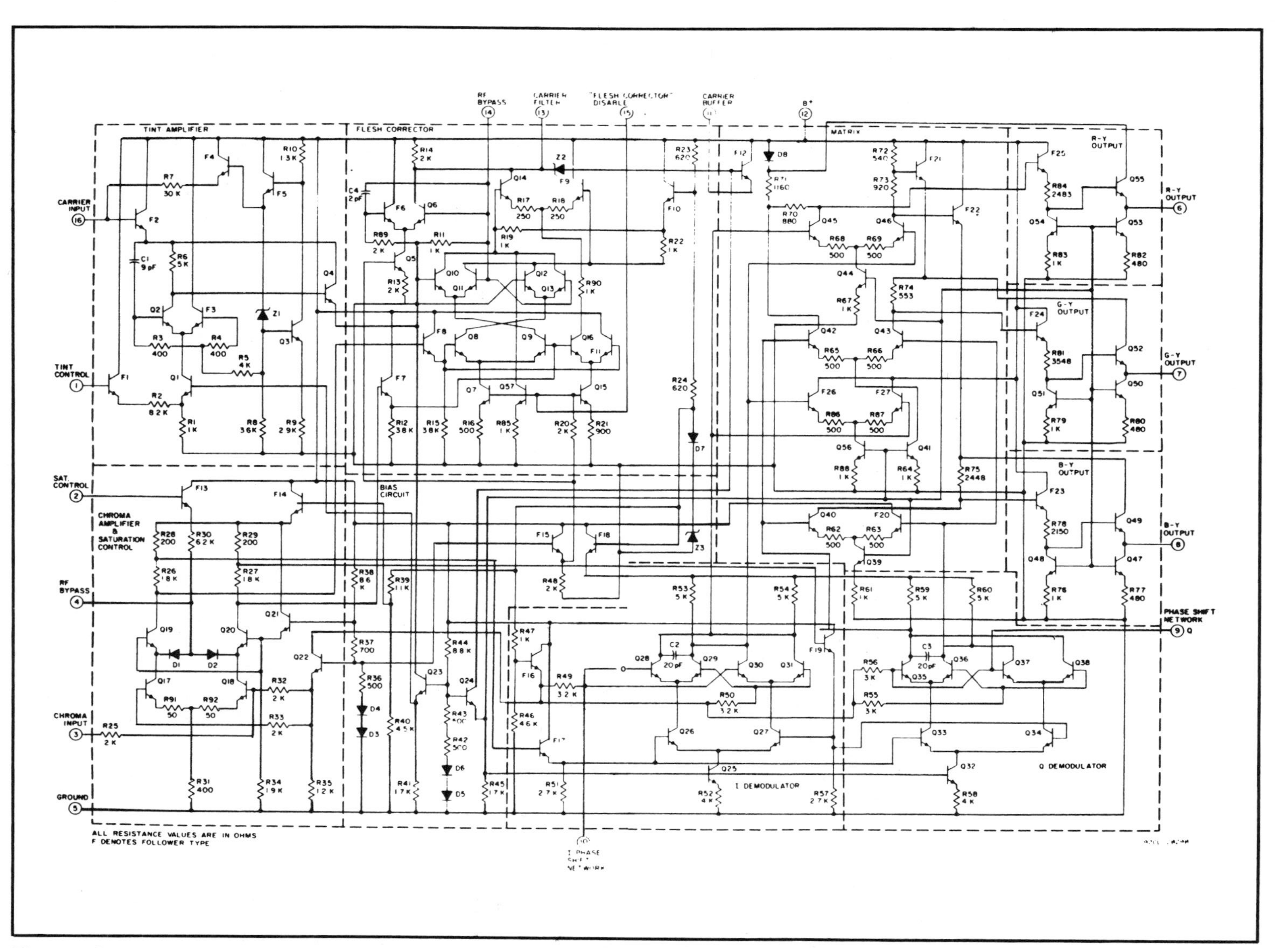

Fig. 11-7. Schematic of the complex CA3137E, then advanced I/Q demodulator with automatic color correction (courtesy RCA Solid State Div.).

the switch collectors for the three sets of differential amplifiers and matrix loads stacked above. These, in turn, furnish outputs for the three RGB-Y low impedance outputs at pins 6 through 8.

If your brow's a little sweaty after that one, think what a 40-pin job with triple the number of transistors and circuits can do. If one's available later on we'll tackle that too, and the state-of-the-art will have been rather thoroughly unfolded for all to see.

CHROMA PROCESSORS

The job of chroma processors begins immediately after composite video is detected and luminance becomes separated from chroma, which has *not* been detected. Originally, chroma was separated from luminance by a simple 3-4 MHz filter, while the luminance channel was blocked from receiving chroma by way of a 3.58 MHz series resonant trap. This worked reasonably well in the early days except that the luminance bandpass of such receivers was no more than 2.5 to 3 MHz, and chroma was restricted to 0.5 MHz for Q information and the same for I. This meant that *neither* adequate luminance nor chroma was being represented on the face of any of these early receiver picture tube faces and what you saw was considerably less than that commonly broadcast.

A glance at Fig. 11-8 easily describes the problems. As you can see, between vestigial sideband cutoff and the picture carrier (0 to −1.25 MHz) there is little or no transmitted information. But between 0 and 4.5 MHz are found both picture and the 3.579545 MHz chroma suppressed subcarrier, the latter being regenerated in the receiver with a crystal-controlled oscillator. Now, with 0.5 MHz additional information permitted on either side of the 3.6 MHz carrier, the full Q signal is available, but the I signal is not permitted to expand its single sideband. This means there's a loss of 1 MHz of I information that is broadcast but not received. And because luminance is rolled off at or before 3 MHz, there's at least 1 MHz of detail omitted.

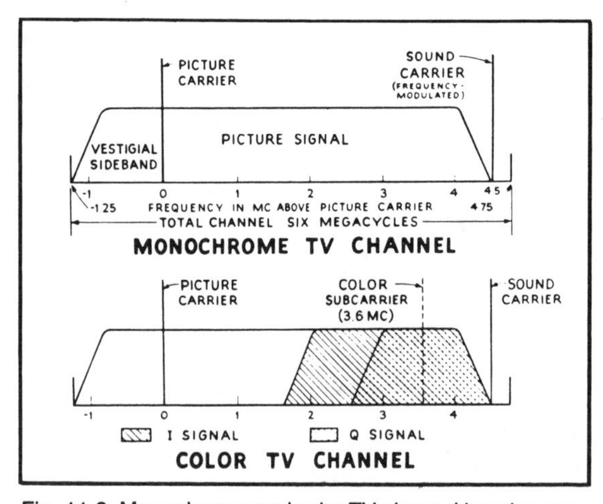

Fig. 11-8. Monochrome and color TV channel bandpasses.

Consequently, the receiver needed some sort of device to increase this luma-chroma resolution so that consumers might see relatively the same picture that was being broadcast. This new and highly effective device saw first use in 1978 and has, with improvements, been a considerable benefactor for at least luminance ever since. The 2 MHz chroma resolution, of course, did not come until 1984, and not many receivers are yet equipped in 1985. Nonetheless, we'll pretend there's full bandpass for both chroma and luminance in the circuits we're investigating, with all using either a glass delay line or charged-coupled device comb filter. All this will be carefully explained in the *video* processing chapter, with only chroma separation and combing treated here.

CHROMA COMB FILTER SEPARATION

Until the mid-1970s, chroma signal processing occupied a frequency window in color receivers between 3.1 and 4.1 MHz and operated as double sidebands at a max. modulating frequency of 500 kHz. With a trap in the luma channel tuned to the color subcarrier, all luminance above 3 MHz was lost, but there was little rejection of the I chroma information *below* 3.1 MHz, permitting ringing on the picture's vertical edges and being partially responsible for dot crawl. Worst ringing occurs (still does on 3 MHz non-comb filter sets) during highly saturated colors of varying hues and brightnesses. When this happens, the chroma-video trap is also often excited and Zebra stripes scintillate annoyingly.

Since chroma is always interleaved at half the 15,734 Hz line scanning frequency and the color subcarrier is an odd harmonic, color phase changes by 180 degrees on successive scanning lines. When summed, these chroma signals cancel leaving only luminance, and the combed (filtered) result becomes the *difference* in color information *between* adjacent (and following) lines. In separating the two, there has to be a time delay of one horizontal line, so that one will add while the other cancels. Some receivers, therefore, use a charge-coupled device delay line while others are satisfied with a glass block delay line, depending on source and philosophy.

Ideally, luminance and chroma separation occurs with a filter that matches luminance energy distribution as well as chroma sideband distribution, with their combined spectrum of frequencies represented by the teeth of a comb—hence the term "comb filter." Therefore, while luminance is cancelled, chroma is summed and so reaches the chroma circuits for subsequent passbanding, amplification and detection. A basic block diagram appears in Fig. 11-9 showing both simplified circuit as well as the combing action.

A GENTLE INTRODUCTION

So that the going won't be too difficult during transistions between initial and highly complex ICs, we'd best begin this discussion with an intermediate variety of chroma processor that does the basics

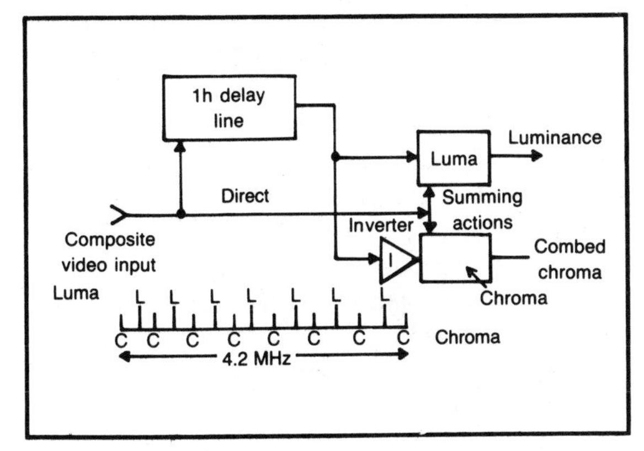

Fig. 11-9. Simplified block diagram of modern 4 MHz luma and 2 MHz chroma comb filter.

and little else to see the evolutionary stages of development. This approach should carry you gently into what has now become a virtual jungle of circuits and functions, now everyday operations for the better color television receivers. Fortunately or unfortunately, *nothing* in the way of color separation and detection is simple any more. Such inevitably occurs as receivers approach the plateau of excellence.

RCA's CA1398E is one of the forerunners of coming sophistication that can help blaze the path to better understanding. Introduced in 1973, it has (Fig. 11-10) a tuned transformer-coupled bandpass input to the chroma amplifier, an automatic color killer, color control, burst gating, a 3.58 MHz subcarrier oscillator, gain and hue controls, and on-chip regulator and bias circuits. Designed to mate with RCA's CA3125E amplifier and demodulator, the IC is coupled to a three-phase-shift reference network for a trio of RGB chroma detectors and transformer coupled to the three preceding amplifiers, with luminance entry into the CA3125E by way of a centertap on the coupling transformer's secondary. To the picture tube, therefore, goes standard red, blue, and green outputs with luminance already mixed.

The chroma amplifier uses a standard differential amplifier with constant current generator that also receives what remains of composite video between 3 and 4 MHz through the double-tuned coupling transformer. The gain of this amplifier is accurately controlled by an automatic color control whose threshold is manually set by the ACC filter and killer control, existing on the premise that burst amplitudes rise and fall according to chroma input. Burst, however, is introduced only during the 11 μsec horizontal blanking interval, keying the 3.58 MHz chroma receiver subcarrier, and generating an ACC dc varying potential. When there is no burst, a biased color killer switch conducts and shuts down any chroma output, thereby removing noise that might otherwise trigger succeeding chroma circuits.

Burst gating originates from a keyed horizontal flyback pulse that turns on the burst and fill-in gate to correct any frequency drift in the 3.579545 MHz

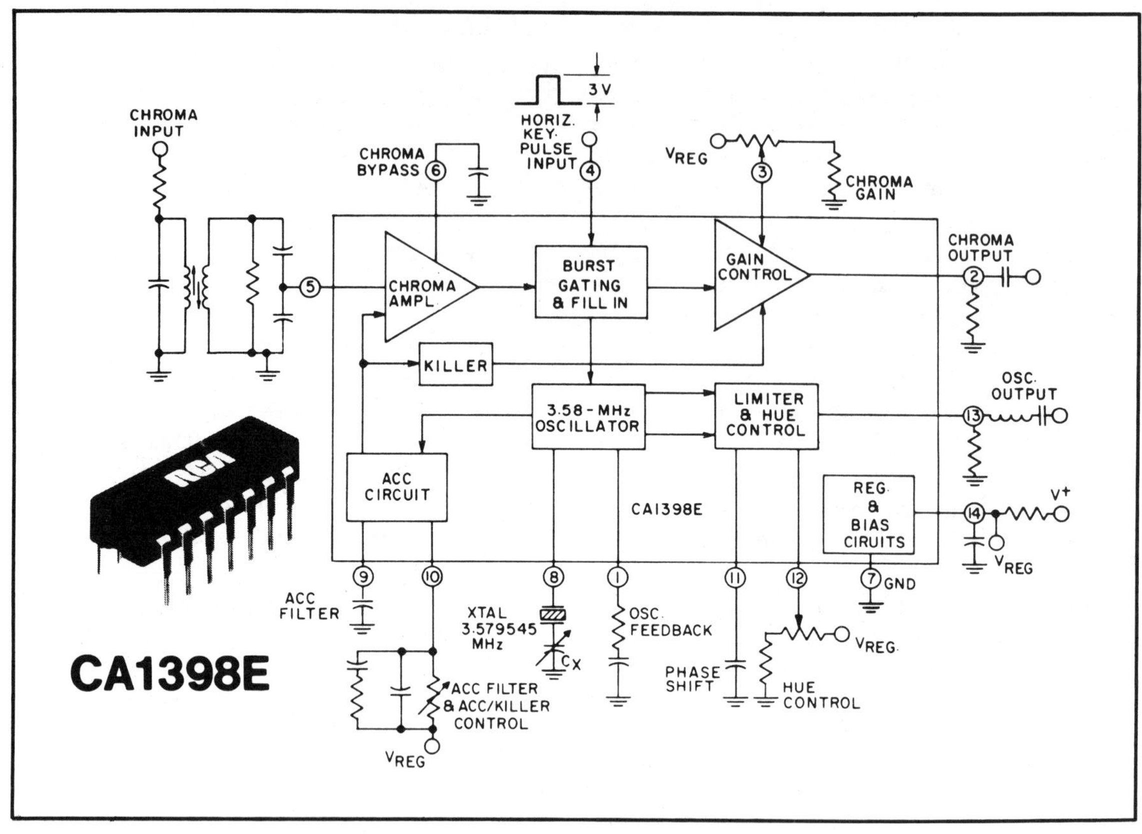

Fig. 11-10. Chroma processor with injection-locked oscillator and dc-controlled circuits (courtesy RCA Solid State Div.).

adjustable crystal-controlled oscillator (pin 8). Between gatings, a feedback network keeps the oscillator in conduction so the limiter and hue control may have 3.58 MHz CW energy and allow the oscillator output to remain continuous as it is coupled to reference points on the demodulator.

The second chroma amplifier is manually gain-controlled by a dc potentiometer and may be promptly turned off by the color killer when there is no incoming chroma. Voltage regulator and bias circuits consist of four transistors, three temperature control diodes and a base-emitter reference zener. Called shunt regulators, this array senses any changes in current drain and supplies more or less voltage/current potential to maintain reliable system operations throughout the IC.

At one time, you might like to know, most critical processor ICs had their individual voltage regulators in addition to that offered by the receiver's chassis. Today, however, with vastly improved pass-transistor load regulation, only a few external or internal transistors devoted to voltage control remain. With switch-mode power supplies and their usual pulse-width modulation control, a number of supplies may be well regulated from a single source. Naturally this all depends on the individual manufacturer, his suppliers, and the extent of IC deployment. With 28- and 40-pin ICs, possibly closer regulation may be maintained by internal chip regulators—a situation that could well vary with recent intermingling of digital and analog technology, on the way to all-IC receivers of the future. You should know a great deal more in the digital portion of the receiver chapter where most of these topics are discussed.

RCA'S CA3126Q CHROMA PROCESSOR

From here on in, the technical aspects of these

chroma processors are pretty intense and unless you have a vested interest in learning, their operational descriptions could become quite difficult and possibly confusing unless your electronics are reasonably advanced. Therefore, with that warning, we'll not spare the language and dive right into what makes a good chroma IC respond the way it does, including the stimuli. A block diagram of the chip is shown in Fig. 11-11, followed by its schematic in Fig. 11-12. As we go along both will be used from time to time as the explanation progresses.

RCA's CA3126Q is a monolithic IC with phase-locked subcarrier regeneration using sample and hold methods, ACC and color killer with overload protection, keyed chroma, internal phase shifting, and low-impedance outputs. Its only adjustment is crystal fine tuning for the 3.58 MHz subcarrier oscillator.

As the diagrams show, there is only a single chroma input through RC coupling, a potentiometer chroma gain control, an internal zener reference for an external shunt regulator, and 5 μsec keying pulse from the flyback (or thereabouts). All the remainder, except carrier and chroma outputs, are self contained and require no outside guidance or information. But as you can also see, there's nothing simple or basic about this IC other than it has just 16 pins connecting to the outside world.

Once chroma navigates the Q1-Q2 first chroma differential amplifier, it proceeds to the doubly balanced automatic frequency and phase control detector (AFPC), the automatic color detector (ACC), and an attenuator feeding the second chroma amplifier. In the AFPC, burst is compared with the reference (3.58 MHz) carrier so that any difference generates an error signal for proper chroma sync. The two sample and hold circuits receive this signal and, in

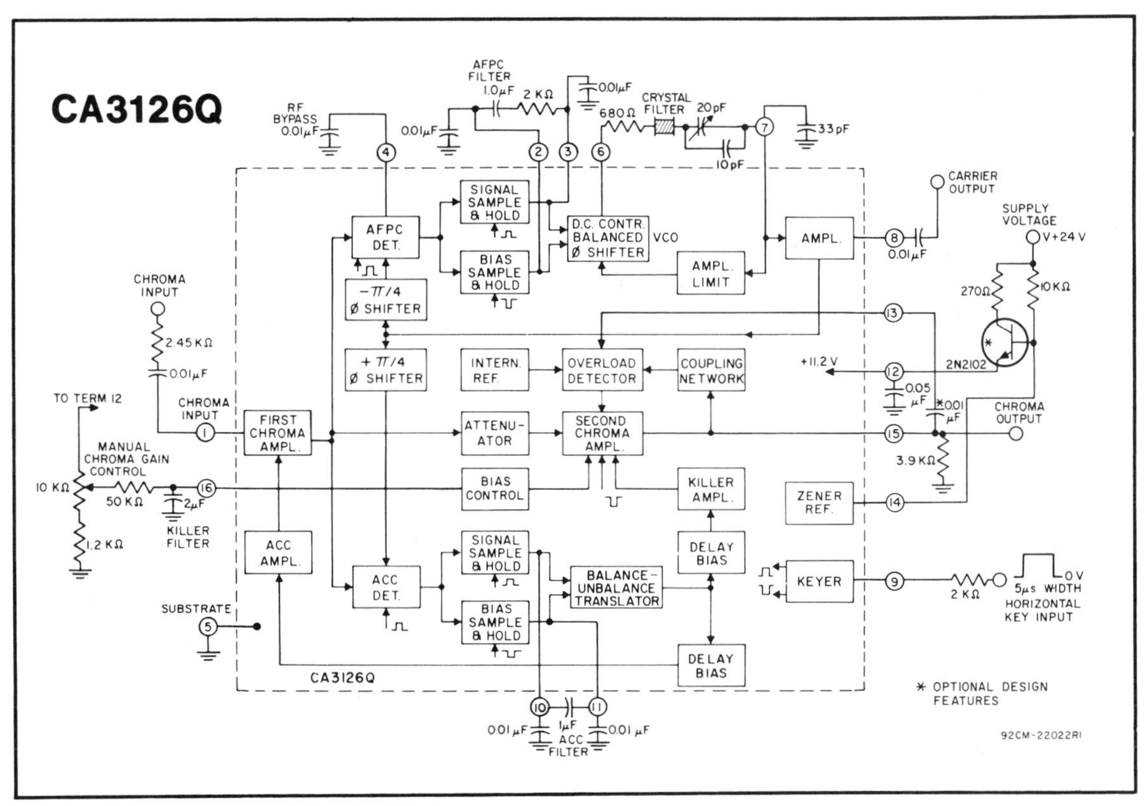

Fig. 11-11. Block diagram of RCA's more recent chroma processor featuring sample and hold techniques (courtesy RCA Solid State Div.).

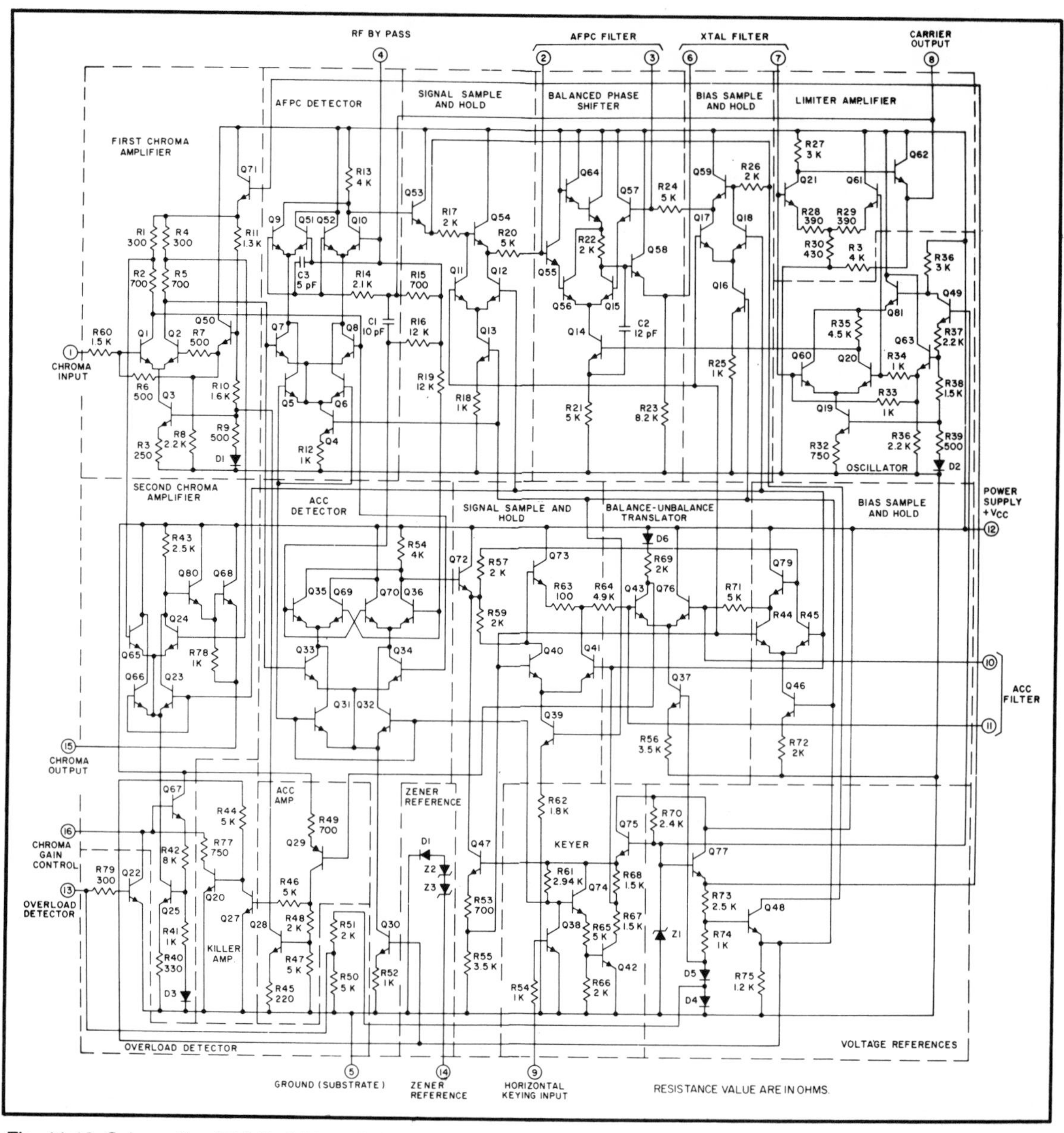

Fig. 11-12. Schematic of RCA's CA3126Q Chroma processor with all functions included—a study in medium complexity (courtesy RCA Solid State Div.).

turn, one samples the detected signal during horizontal keying and stores any peak error in the AFPC filter; the other detects during the horizontal scanning interval, supplying an accurate reference via additional storage. Error and bias information are then applied to opposing terminals of the differential phase control, synchronizing the reference frequency developed by voltage-controlled oscillator (VCO).

During phase and frequency detection (AFPC) and automatic chroma detection (ACC) a phase shifter ($\pm$ $\pi/4$) operates with feedback from the

3.58 MHz carrier amplifier. A low-pass filter (R14-C3) shifts carrier phase by $-\pi/4$ while a high-pass filter rotates carrier $+\pi/4$ for the ACC killer detector. The ACC uses similar signal/bias sample and hold circuits as does the AFPC, delivering the results into a balance-unbalance translator. The CW carrier, however, is in-phase with burst. When there is no burst, the balance-unbalance translator amplifier is off, Q29 is also off, and the ACC amplifier is nonconducting, allowing maximum current to the chroma input, with the second chroma amplifier also disabled.

With incoming chroma and amplification of burst, burst and chroma rise and fall proportionally, although burst is limited in amplitude. ACC begins to operate when the R47-R48 divider turns on Q28 as the potential increases removing bias at the chroma amplifier junction of R9 and R10.

The second chroma amplifier consists of Q65 and Q24 and is differentially driven from the collectors of the first chroma amplifier. It is controlled by three sources: the customer dc gain control; the killer detector; overload detector and keyer. During horizontal keying, the amplifier is interrupted by Q66-Q23, removing burst from the composite signal. Gain of the amplifier results from varying the current through Q25 via Q67 and the chroma gain control. Killer action results from Q29 action as described, and overload difficulties occur when signals with low burst-to-chroma ratios are received along with noise peaks. Q22, therefore remains off until its base potential rises to about 0.7 volt. It then begins conducting, lowering the base potential of Q67 and, therefore bias to Q25 and the second chroma amplifier. The keying circuit maintains AFPC and ACC detectors on during the keying interval and also disables the chroma output at the same time, preventing interaction between the two.

That's about the story of the CA3126Q. Following this rather detailed description, we'll probably have to revert once again mainly to block diagrams, since IC complexity will have risen to the point that schematics would become more confusing than helpful, sorry to say. But with only two or three more examples to go, this simplified version may become a welcome breather, even though much of the intense circuit design material may be missing. Regardless, we should regain a look at certain techniques in the sync countdown and luminance processor sections of the book which may help you with additional IC evaluation and familiarity.

Another industry problem has arisen, with the few U.S. manufacturers remaining, many ICs are now proprietary, custom-designed units that one television maker orders for his receivers alone. This means that the IC houses no longer have to publish extensive literature and all-encompassing application notes, making it more difficult to work with these monolithic circuits than ever before. Consequently, design information now comes directly from U.S. (not normally Japanese) manufacturers, and there are less than a half dozen of these with any market share. So good information does require considerable tenacity and hard-nosed research.

MOTOROLA'S VERY NEW TV COLOR PROCESSOR

Developed recently in Europe, Motorola Semiconductor is now marketing a 40-pin LSI device that operates with both PAL and NTSC and interfaces with the TDA3030B SECAM adaptor. Offered in two versions identified as TDA3301 and TDA3303, the TDA3301 will be described since it works with European and USA systems. The two ICs feature automatic black level setup, beam current limiting, a 4.43 or 3.58 MHz subcarrier crystal, have no oscillator adjustment, three on-screen displays (OSD), four dc user controls, typically 600 mW dissipation, and requires only a single 12 V supply.

The pinout (Fig. 11-13) and block diagram (Fig. 11-14) together offer a general but sufficient outline of what the TDA 3301 can and will do. With information generously supplied by Motorola, we'll try and work through the diagrams with you, section by section.

Timing logic for counters A and B is derived from the sandcastle input (27) and frame pulse input

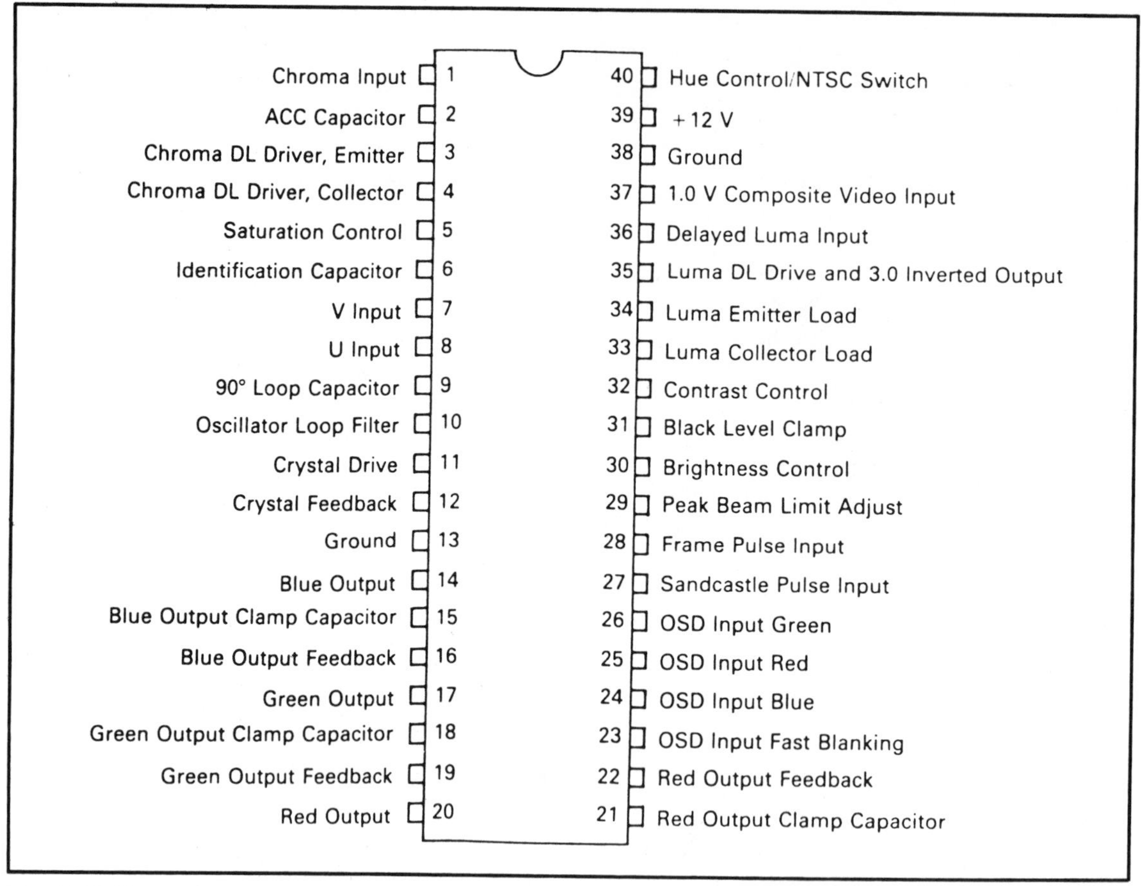

Fig. 11-13. Pinouts for Motorola's TDA3301 luma/chroma processor (courtesy Motorola Semiconductor).

(28) resulting in dual-edged triggered flip-flops. The trailing edge of the frame pulse initiates these actions and eventually provides control signals for luma/chroma blanking, beam current sampling, on-screen display blanking, and the brightness control. In turn, the blanking controller connects directly to the OSD black level clamp, CRT beam current clamp, and the color difference matrix and color killer. During burst-gate time, a feedback loop clamps the video signal at black level and ignores the contrast control making possible on-screen displays.

The brightness (brilliance) control operates by adding a pedestal to the outputs during cathode-ray tube beam current sampling, with minimum brilliance always at black level. Delayed luminance enters the IC through pin 36, is dc-level shifted, inverted and then continues on to the luma contrast portion where it is subject to actions of the automatic contrast controller, blanking and brightness. Then through a 3-way splitter, luminance information reaches the OSD black level clamp and contrast at OSD inputs (24, 25, and 26).

In the chroma section you will find the usual ACC, hue, color, and color killer controls, along with a color contrast and saturation section supplying a pair of chroma delay line drivers at pins 3 and 4. There's a burst gate, PAL bistable, color difference filters and matrix with fast blanking, a phase-locked 90-degree servo loop serving the reference 4.33 or 3.58 MHz VCO. During burst gating the hue control permits a $\pm$ 40° phase shift between burst and chroma, while VCO frequency is held constant by varying phase feedback. The

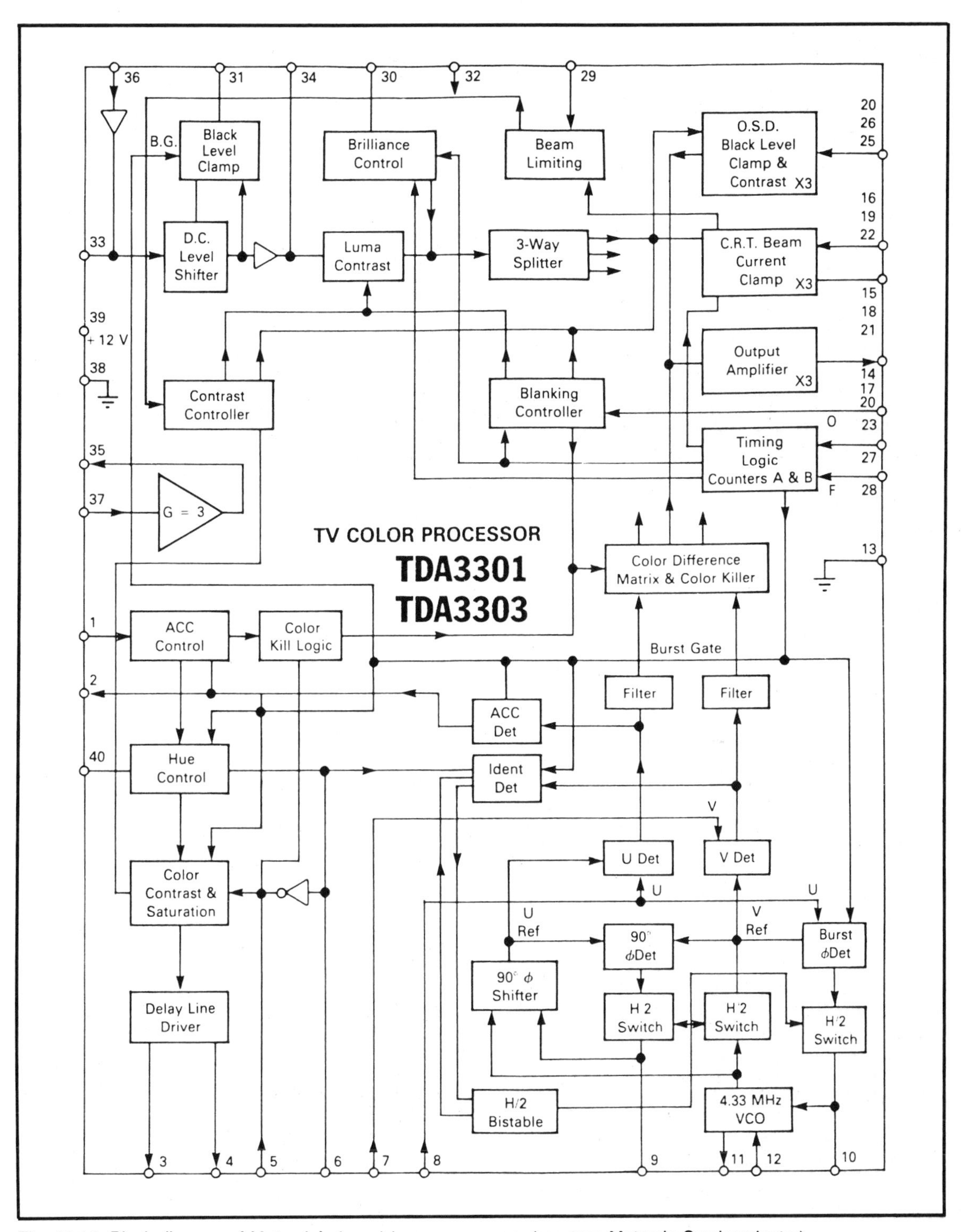

Fig. 11-14. Block diagram of Motorola's luma/chroma processor (courtesy Motorola Semiconductor).

U/V detectors shown on the diagram are European versions of I and Q here since their outputs then pass through filters directly to the color difference matrix and on to the RGB outputs at pins 14, 17, and 20.

In decoding PAL the chroma reference loop is locked to burst incoming from the PAL delay line matrix U channel, so there is no alternating component. H/2 switching is then done before the phase detector so errors occur at 7.8 kHz instead of dc. A commutator in the phase detector is driven from the PAL bistable and converts its ac signal to dc before the loop filter, developing no phase error. In decoding NTSC, this bistable is inhibited with only slightly less accurate phase control, but along with the manual hue control, the phase accuracy is satisfactory. For SECAM, the loop filter becomes grounded producing virtually no phase shift so the two synchronous U/V detectors work in-phase and not in quadrature. During picture transmissions, the two demodulators pass their outputs into basic RC filters, with color killing and blanking achieved by raising voltage levels above those of the color difference information. Two differential amplifiers do color difference matrixing which is added for G-Y.

LUMA/CHROMA PROCESSOR COMBINED

The most up-to-date integrated circuit presently available for description seems to be RCA's CA3234E luminance/chroma processor in a 28-pin plastic package—an IC that does it all for both video channels following the video detector and luminance delay line. This time, to illustrate the extent of such electronics operations, we'll show both block diagram and schematic, curtailing the overall description to essentials because we have covered a good deal of this before. But at the conclusion, you should have a pretty good idea of how analog television is developing at least for medium- and large-scale integration, and at the moment that's the principal intent. In the next section you'll see the point of all this, along with further development in the art of putting together one very large LSI package that really does everything.

Alongside the simplified block diagram we'll also include RCA's IC pinout terminal assignment to give you a better "feel" for the multitude of functions contained within (Fig. 11-15). Considering there are literally hundreds of transistors, diodes, and resistors in the package, you'll have to agree it's quite an achievement and was certainly not completed overnight. Something like this takes months and even years to design, troubleshoot, and place in a working receiver. From our standpoint, its magnitude is impressive.

As usual there are the voltage controlled oscillator, its automatic phase and chroma detector, tint control, and demodulator with RG and B matrix. The first chroma amplifier is ACC controlled and connects to a filterless color killer that's dc biased by chroma gain to turn off the second chroma amplifier when there is no incoming color. The second chroma amplifier is both automatically gain-controlled by luminance tracking and also manually by the operator.

The voltage-controlled 3.579545 MHz subcarrier oscillator regenerator is both RC and crystal controlled and locked to burst by the APC detector. Tone control outputs two quadrature references at 3.58 MHz which, when matrixed, supply a pair of reference voltages separated 100 degrees apart for the demodulators. Subcarrier harmonics are internally filtered, and demodulators have current mirrors to easily matrix with luma for RGB signals to NPN/PNP low impedance drivers.

Automatic tint control when not switch-defeated, changes the B-Y demodulator gain in response to sensing demodulator output currents. ACC gains are also changed to raise chroma near clipping levels in the second chroma amplifier for increased noise immunity, with an increase, also, in the killer threshold. Additional manual color controls are APC, tint, and saturation (Fig. 11-16).

Luminance signals input the IC following a conventional TV delay line so they will arrive at the matrix point simultaneously. But that's about all you'll find conventional in this sophisticated series of circuits. There are filters and traps at pins 2, 4 and 6 so that information is extracted on the first and second signal derivatives and from low-pass components. The first derivative generates a gating signal to signify luminance transitions which are

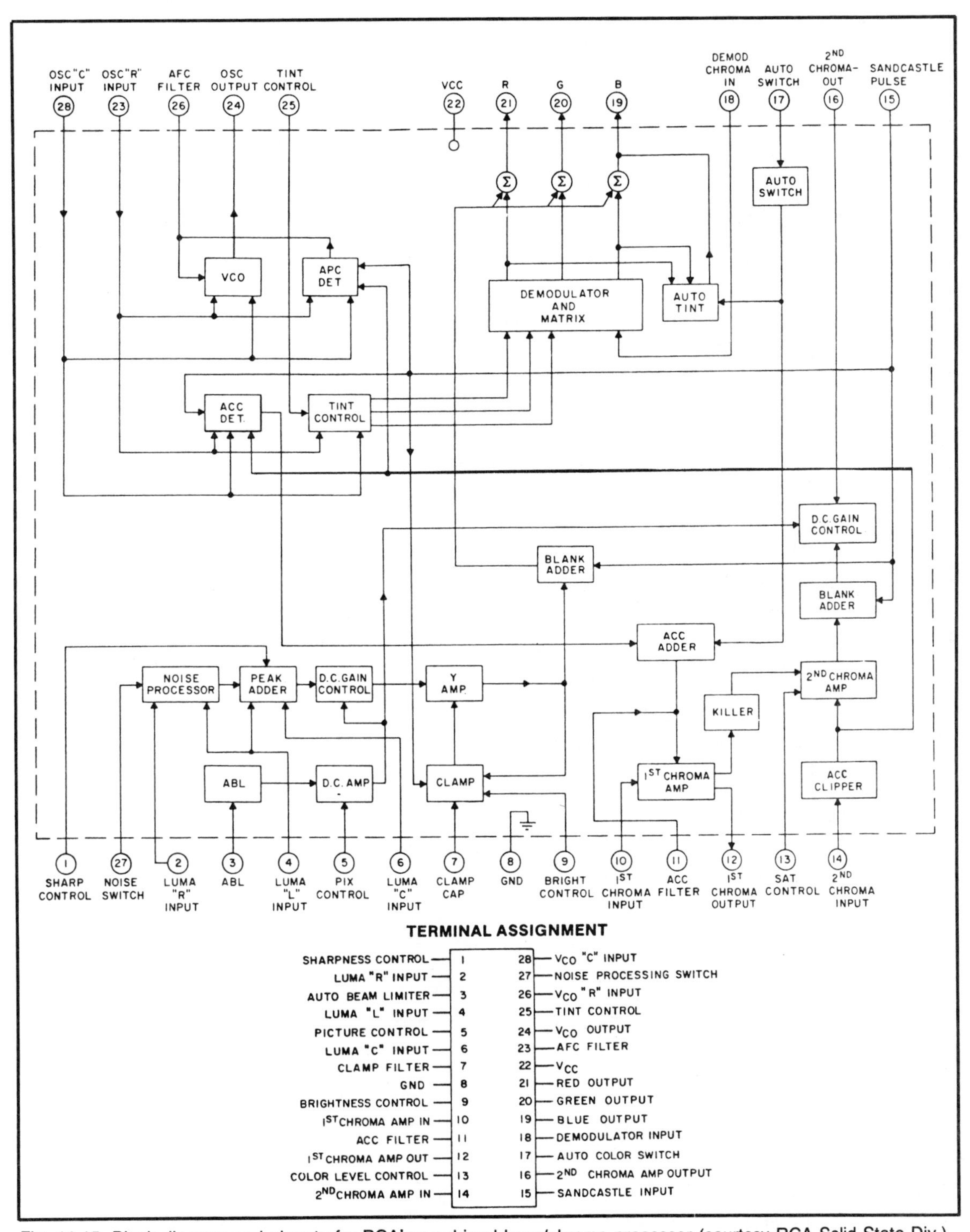

Fig. 11-15. Block diagram and pinouts for RCA's combined luma/chroma processor (courtesy RCA Solid State Div.).

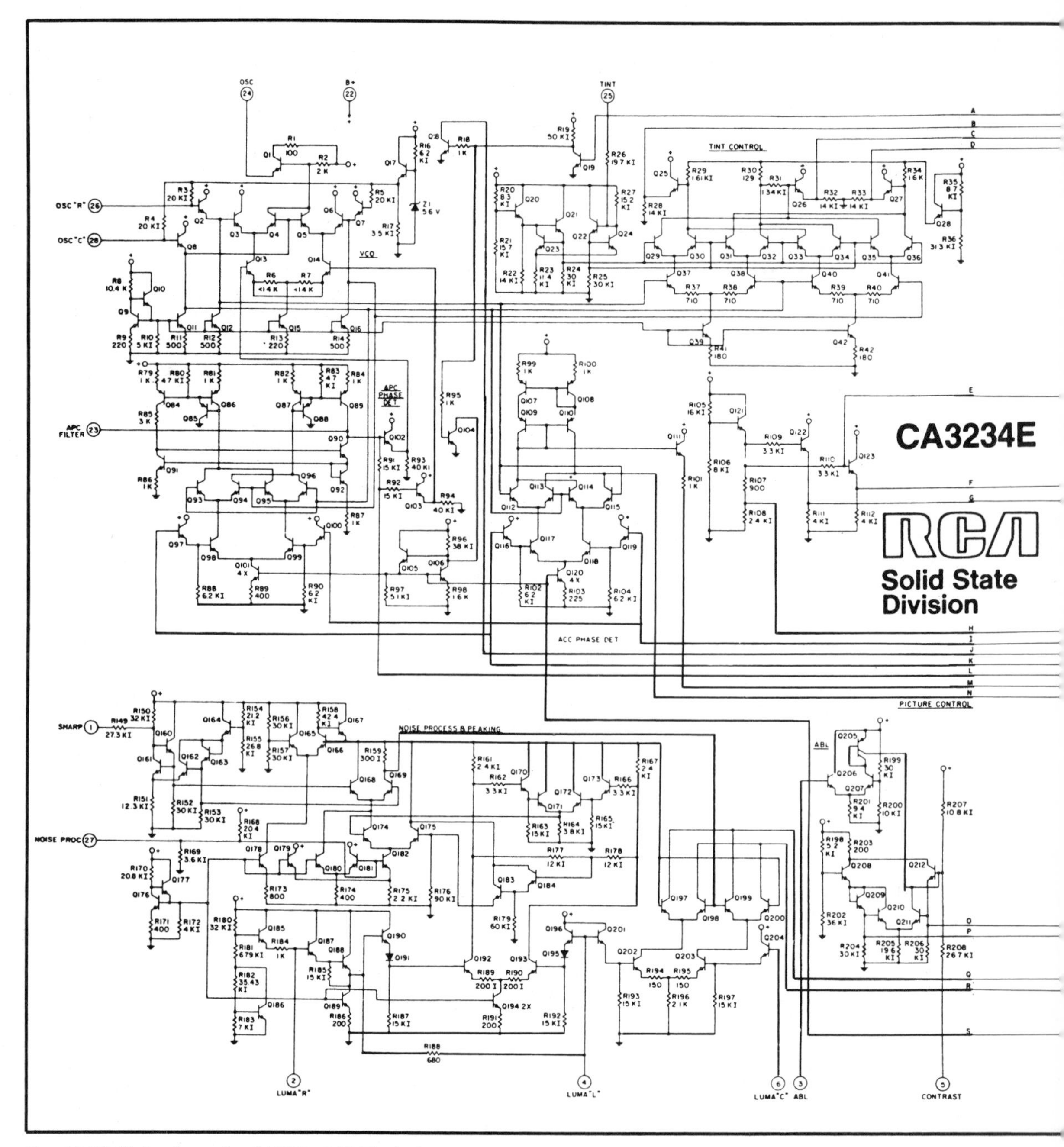

Fig. 11-16. Full schematic of RCA's LSI 28-pin Luma Processor (courtesy RCA Solid State Div.).

sensed, and peaking is adjusted 6 dB higher than ordinary video signals without transitions (black to white). This results in sharper video with less noise, which may be adjusted manually also by an operator potentiometer connected to the peak adder. Peaked luminance then passes to a gain control circuit that tracks with the second chroma amplifier and produces a 7:1 range from the manual

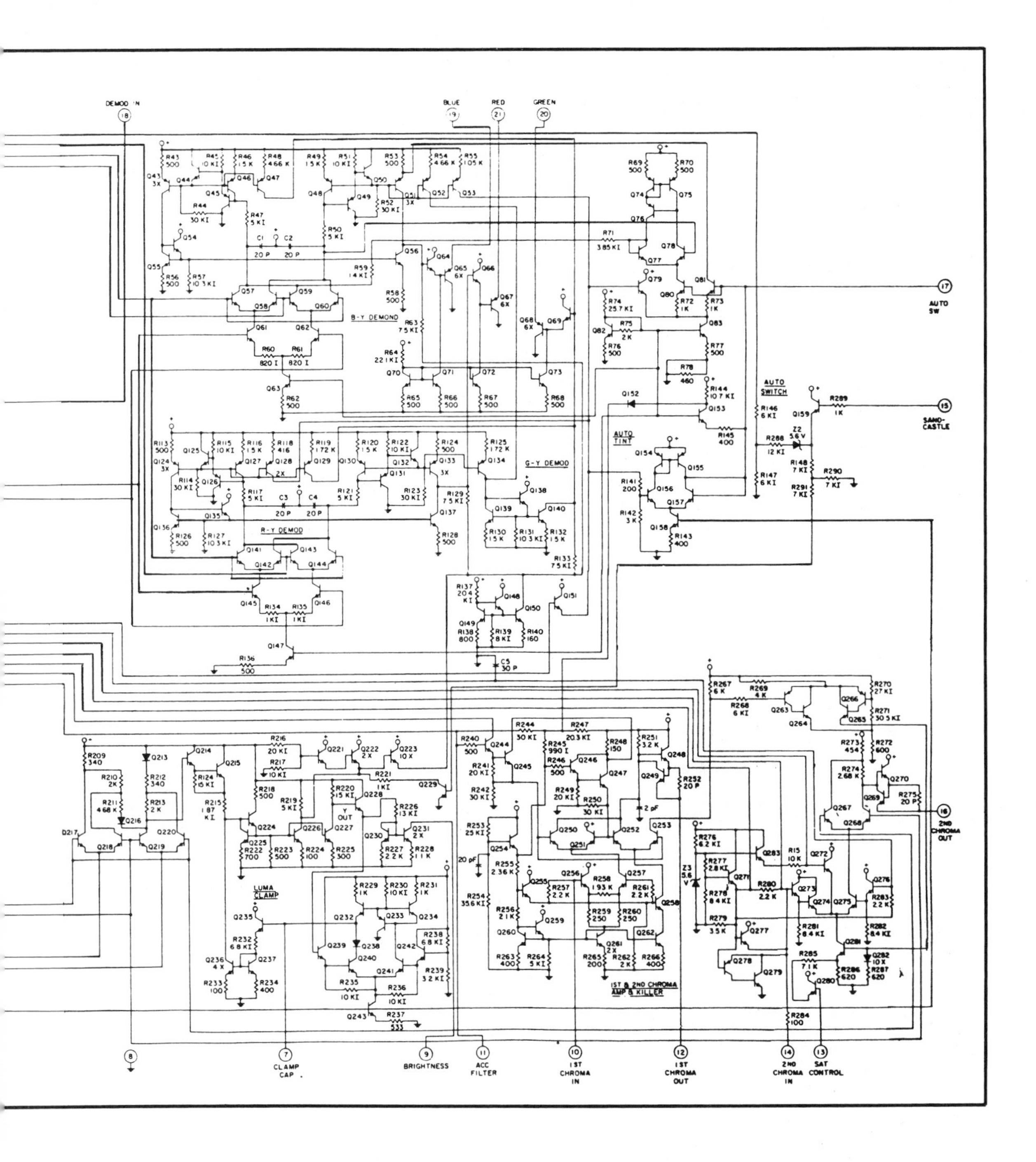

contrast control and its dc amplifier. An automatic brightness limiter monitors picture tube current and will limit luma and chroma gain when normal limits are exceeded. There's also a peak white limiter to ride gain on the luminance amplifier.

Luminance then passes to a variable dc level shifter and afterwards continues to the current mirror, governed by a variable dc shunt. One output of the current mirror goes to the RGB matrix and the other to control circuits gated by the burst gate where luminance information and brightness levels are compared to close the black level clamp loop. Blanking is applied to the current mirror as it enters the RGB matrix. External video controls are contrast, brightness, and noise. A sandcastle circuit in the form of a pedestal supplies sync information for the circuit through pin 15. Chroma input arrives via a 75 pF capacitor and centertapped 10 μH coil just before the luminance delay line limiting its bandpass to that of chroma information, rejecting luminance.

Once again, this is a pretty difficult LSI chip to understand unless you have carefully studied its predecessors. As you can see, much of these processes are evolutionary and, like the sandcastle circuit, build upon one another as advanced developments continue. Do they really contribute to better television receivers? You can bet whatever you'd like on the affirmative because of all the additional functions and feedback loops supplying so many worthwhile checks and balances, all within a 28-pin IC that in some receivers can be socket-mounted for relatively easy removal and substitution when and if breakdowns occur. Some engineers don't appreciate the possibility of eventual socket pin oxidization but the repairman is eternally grateful when time comes to change this chip for a new one. Obviously there are valid arguments both ways, depending on your bench or administrative positions.

EXTENDED I SIDEBAND DETECTION

In RCA's new system, both I and Q-phase information is demodulated equally, but with Q signals filtered to 500 kHz, as usual, and the I signal filtered to 1.5 MHz, its full bandwidth. This is especially important, according to Consumer Products Division engineer Ronald Keen when one picture is inserted or superimposed on another producing a colored border. Vertical portions of the insert will have paler saturation than top and bottom sections resulting from I-constricted bandwidths of the average television receiver.

With full I spectrum, chroma is now a relatively wideband subsystem and responds rapidly. The same description applies to a brightly colored piece of cloth, say a shirt. As distance between camera and material increases, these bright colors remain visible much longer than in the narrowband systems. However, to avoid quadrature distortion, the Q filter has to roll off very rapidly "without causing the group delay of the filter to increase dramatically."

Apparently RCA has done all this in its 2000 series receivers very successfully and, although most of us haven't had an extended opportunity to examine the effects on instruments, certainly the eye-effect is very pleasing, and one does see better colors in this significant achievement. So far, as we've emphasized before, RCA is the *only* full color, full bandwidth, and multichannel sound receiver on the market—at least up to mid-1985.

G.E.'S VIR

As explained before, VIR stands for vertical interval retrace, and is the color and black reference that has been reserved for line 19 during this approximately 1.4 millisecond blanking interval. Some years ago, General Electric recognized that this broadcast signal could be put to good use in television receivers, and devoted a good deal of engineering time to developing a suitable circuit. Using both sync and a line 19 recognition device, Portsmouth, Va. engineers produced a design, first in discrete transistors, and then had Matsushita package the electronics initially in two ICs, then compact it further into one. G.E. now simply calls it VIR, while Panasonic's Matsushita uses the name Color Pilot. In our humble opinion this has been the only reliable tint and color saturation, nondistorting solid-state IC available until RCA, after considerable effort, finally delivered full chroma bandwidth I-Q demodulation with color control feedback.

Being G.E.'s baby, however, we'll stick with the originator, using one dual diagram and an expanded single block to show how this highly effective color regulating device actually operates. Figure 11-17 illustrates the entire VIR broadcast signal in terms of IRE amplitude units and the usual

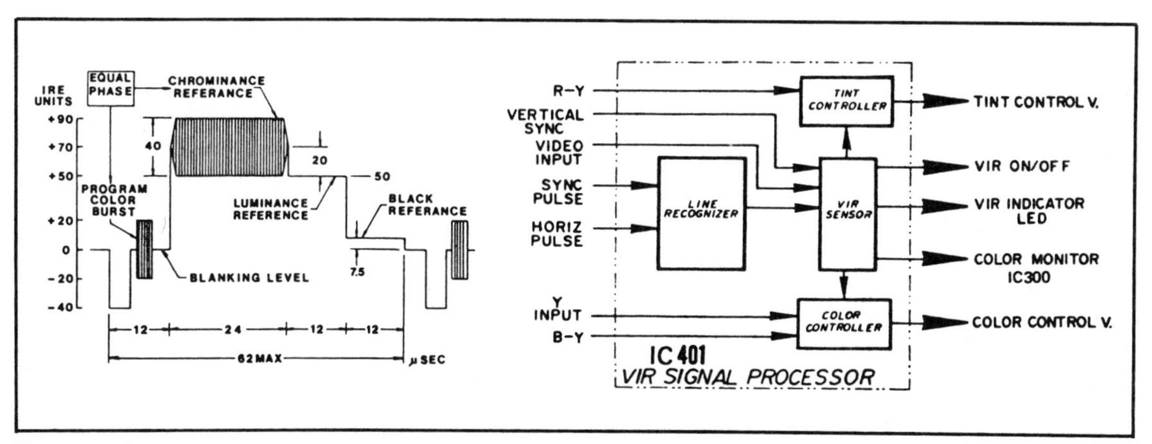

Fig. 11-17. Composite view of the VIR-transmitted signal and G.E.'s receiver block diagram (courtesy General Electric).

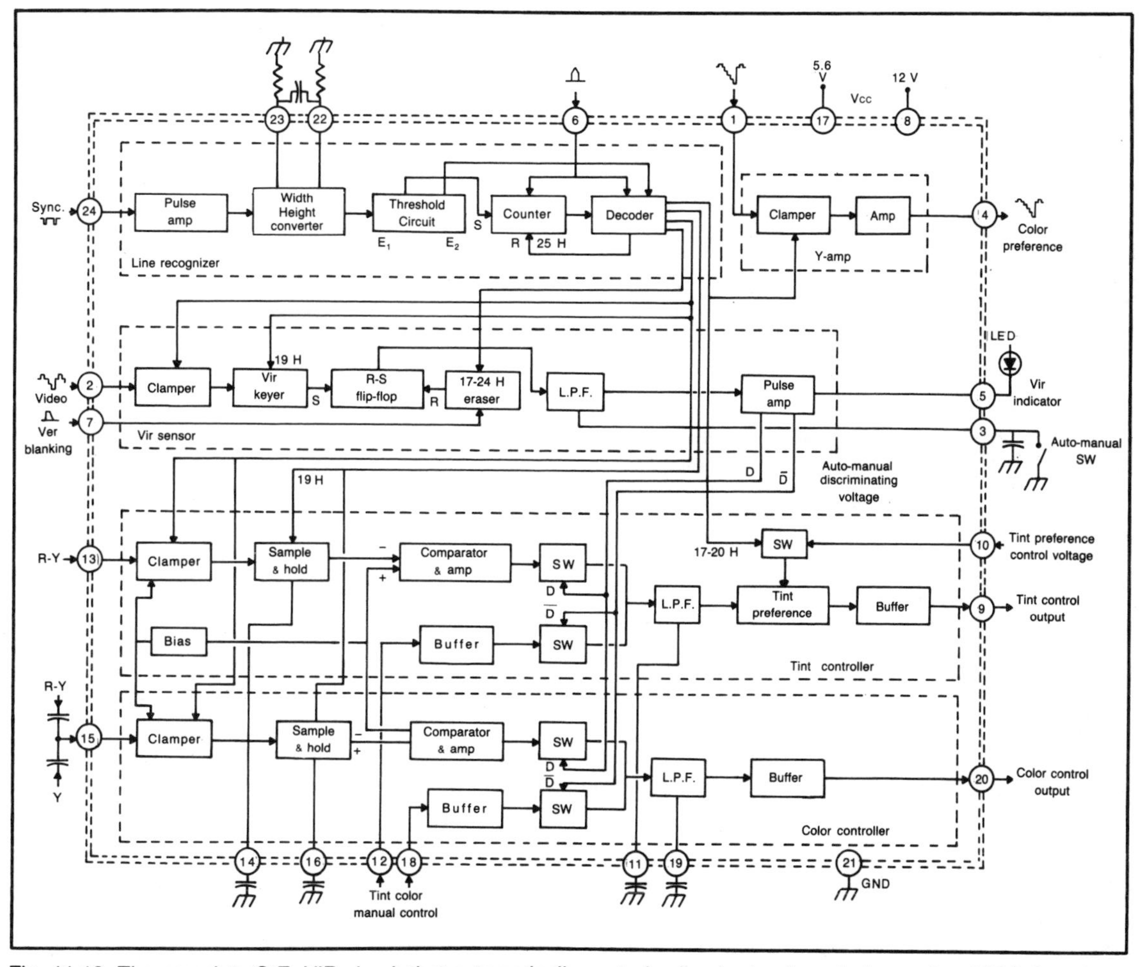

Fig. 11-18. The complete G.E. VIR circuit that automatically controls all color levels and phase when VIR is broadcast (courtesy General Electric).

63.5 μsec horizontal time base. Then below the signal is a gross block diagram of the AN5330 IC that does the job. Here a line recognizer receives both composite sync and a horizontal line pulse, so that when line 19 appears and video inputs are applied, the VIR sensor turns on and, in turn, generates signals to activate both tint and color control circuits. Vertical sync then triggers a reset for the VIR sensor. At this juncture, the color controller samples the B-Y and Y matrixed relative to VIR and then compares this sum to black level. Any differences produce a control voltage that results in B-Y equalling Y in relative levels.

Simultaneously, the tint controller samples the R-Y output during VIR and adjusts its level to black level with a control voltage if they are not equal. In the more complete block diagram of the AN5330 IC, itself, composite sync inputs are shaped by a pulse amplifier, delivered to a width-height converter, converting sync pulse width into proportional height (Fig. 11-18). This goes to a threshold circuit where outputs exceeding threshold develop into a pulse some 20 microseconds while another threshold separates vertical sync from composite sync. Threshold pulses develop a counter-enable pulse to set a horizontal flyback pulse counter. When line 19 is identified, a 25H reset pulse from the decoder stops the counter.

The VIR sensor develops control voltages denoting VIR signals or not, and switches the receiver to either automatic or manual. Fixed amplitude luminance is clamped to black level, then proceeds to the VIR keyer, which recognizes the luminance part of line 19. When the level exceeds threshold, the VIR sensor recognizes VIR and turns on the LED indicator, then switches tint and color controller circuits to automatic.

Luminance Y and blue minus luminance B-Y information are clamped so that their pedestal levels are equal, with line 19 keying sampling the signal during chroma reference. Sample and hold is next compared with the clamped voltage and the difference controls color. If color saturation is low, for instance, the comparator and amplifier generate a positive voltage, increasing the B-Y signal. With color absent, of course, VIR reverts to manual control.

The R-Y tint controller is quite similar to the color controller but has a tint preference circuit. The usual clamping occurs, chroma reference is sampled and held, compared with the clamped source, and a reference voltage goes to tint preference. This control voltage inverts and then shifts chroma phase so that chroma going to the R-Y demodulator remains at a 90 degree angle relative to chroma reference.

Briefly, but essentially, that's the G.E. VIR story, and it's a good one. These VIR circuits do, indeed, work well and are thoroughly tested and not found wanting. Furthermore, they continue to work well even in noise.

A ONE CHIP COLOR TV SET

At the moment our heading may be somewhat misleading since a single IC color television receiver hasn't really been announced yet, but you can bet it's on the way, and probably in the near future. A good many problems of impedance matching, chip sizes, power consumption and dissipation, bandwidth and IC sheet resistivity have largely been solved already, and the rest awaits just a few more steps before the compact color set follows on the heels of Zenith's "one chip" black and white receiver introduced in 1981 with the help of Motorola's Gerald Lunn and Michael McGinn. This particular IC included i-fs, AGC, video detector, noise processor, video processor, horizontal and vertical sync and 4.5 MHz intercarrier sound output, but not the sound i-f, detector, nor output. Otherwise, only the tuner(s), horizontal, vertical, and video outputs were discrete devices. Obviously, it's a great deal more difficult to design a considerably more complex color receiver to do at least the same things, but such a project is well underway as you will see in the forthcoming example. Credit here goes to M. Yoshitomi, H. Yamashita, K. Kojima and H. Takano of Mitsubishi Electric Corp., Kita-Itami Works in Japan and the IEEE Consumer Electronics Transactions for publishing their paper.

A TWO-CHIP COLOR RECEIVER

As the gentlemen of Mitsubishi relate, this particular effort has been conducted in three phases as outlined on the block diagram (Fig. 11-19). The first consisted of the usual separate sound, video i-f-detector-AGC, video and chroma processing, and sync separation and driving. The second try contained automatic fine tuning (AFT), video detector, chroma processing, and the sync treatments. The third step—this one about to be described—includes one IC for sound, video detection, AGC and i-fs, and a second that includes all video and chroma processing and the sync circuits. Note, of course, the tuner and RGB, sync outputs, and audio remain entities because of rf and power functions, respectively. These, however, under certain constraints might also become ICs as the effort progresses, or may not have to develop further if liquid crystal displays or their alternatives become feasible.

The two-chip advance appears practical as a result of transistor cells being reduced 60 percent, the number of IC chip elements increased now to probably over 1200 using ion implantation, a subsequent lessening of power consumption, sheet resistivity of 400 ohms, and transistor frequency response of one GHz. All these advances augur well for further developments with less and less peripheral components and more genuinely useful functions contained within. Here, we'll only print that portion of the receiver dealing with the video/chroma jungle circuit, while the other portion can be used in luminance processing to good advantage. So in Fig. 11-19 you see what Mitsubishi has done in this respect, most of which will seem rather familiar considering the history of the more recent ICs. Consequently, emphasis will be placed on IC fabrication and problem solutions rather than already well-known electronic functions.

As you can readily see, known operations such as ACC detection, killer, hue, black level, pedestal clamping, X-ray protection, and color difference outputs are all rather standard fare seen in most,

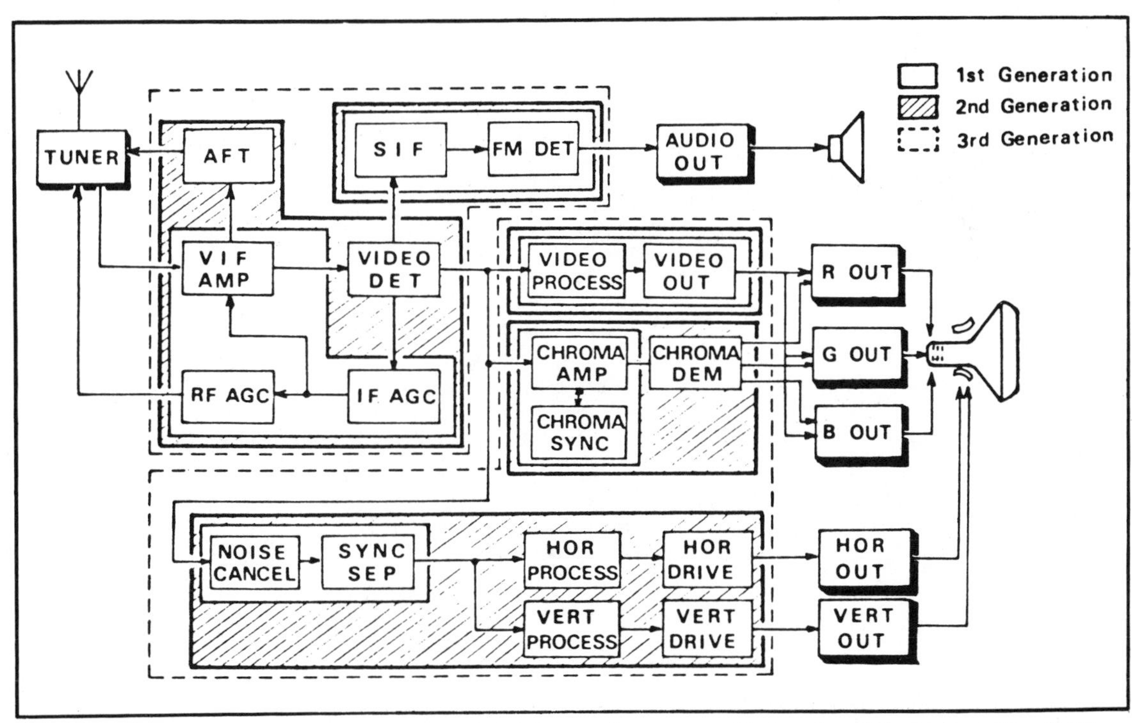

Fig. 11-19. Three generations of reducing color television receivers to two ICs (courtesy Mitsubishi and the IEEE).

if not all, the better color receivers, so we won't dwell on these to any extent unless there are some unique technical details worth revealing.

According to Mitsubishi there are approximately 1,000 active and passive devices on the color/luma chip. New additions are sync separator with automatic slice level control, AFC gain selector for weak signals and VCR inputs, X-ray protection to prevent smoking of the horizontal output transistor, and a wide video "tone" control. This latter permits a change of mixing ratio among normal video, another that enters via a high-pass filter, and double differential video, so that the picture will be at its sharpest, normal viewing, or low-frequency rolloff "soft," reminiscent of what U.S. receivers label as *sharpness* controls. The automatic slice level is said to furnish stable pictures even with ghosting since the sync pulse is automatically made deeper as secondary images tend to reduce its effect. The video/luma package is listed at 16mm^2 on a 42-pin dual in-line IC, with power dissipation of 600 mW.

Chapter 12

Sound—Conventional and Multichannel

FOR THE FIRST TIME IN MANY, MANY YEARS, it's a pleasure to originate a chapter on television sound. In the not too distant past, a television audio subsystem that boasted a response better than 6 or 8 kHz was either a phenomenon or a fake, usually the latter. But, with the coming of multichannel sound and much wider video bandwidths we are finally realizing the full potential of our receivers. Let's hope this carries over to VCRs as well, for a union of salutary picture and audio does make an ideal combination for us all. I only hope the general public will quickly learn to appreciate them.

One promise: we will not dwell on ancient ratio detectors and discriminators to take up space. These work fine in some FM radios, but have long since departed the trails of television and should never return, principally because of peak and quadrature detectors, which are able to conveniently and substantially do the job. In addition, they're easy to integrate with ICs, and this, today, is the way of the world. Further, as this chapter develops, we can probably prove with a spectrum analyzer that audio circuits have far outstripped their speaker counterparts, and external speakers for monitors and even monaural receivers is a beneficial way to go, especially if there are loudness (volume) and tone controls with worthwhile electronic backups. True, a set of good "boom boxes" are a trifle expensive, but if you're anything of an audiophile, then the outlay should be inconsequential. For concert or opera fans, this suggestion should translate into a *must.* Furthermore, when the TV isn't playing, there's lots of good music around, including an increasing quantity of 1940s big band sounds which are becoming very popular, especially on AM radio, soon to have its own agreeable stereo offerings. So, with FM, AM, *and* television parading all sorts of stereo for your immediate selection, including that from satellites, the lure should be irresistible. We hope that technical people *and* laymen all realize this great avenue of audible enjoyment and join the throng. The more listeners, the better programming everyone can expect.

MONOPHONIC DETECTORS AND AMPLIFIERS

Since *all* FM television sound has long been the province of integrated circuits, we'll go back to several of the earlier units, describe their inner workings as lucidly as possible, then bring you up to date with the latest available for this model year receivers. These should continue in vogue and engineering principles at least for the next several years, or until almost all the better sound circuits are full-fledged stereo. So considering a television receiver has a 10- to 15-year life by even conservative standards, such descriptions should bear the test of time for an extended period. Compared with AM and FM radio, there will still be mono sound for television long into the forseeable future, especially on local-originating programs such as talk shows, news, action remotes, and anyplace the second microphone adds little or nothing to either broadcast prestige or audience enjoyment—possibly in that order.

Initial Circuits

Based on what RCA identified originally as a gain-controlled operational amplifier, calling it an operational transconductance amplifier, our initial circuit is shown in Fig. 12-1. It has inverting and noninverting inputs, a bias amplifier current source, a complementary output through Q8 and Q10, active loads for operating transistors, and a regulated bias source for Q10- Q11. Take away a few of the temperature-compensating diodes and you have the fundamental design for many sound ICs even today. This, however, was little more than a simple, but effective amplifier, and a great deal more was needed to complete even initial requirements. These stages could be cascaded, but considerably more thought was needed to develop the necessary gain, detection, and final amplification for a workable audio integrated circuit. Furthermore, the principle of the differential amplifier—the basis for all linear ICs, even to this day—had to be expanded, plus solid-state capacitors, and impedance matcher-drivers for interstage coupling.

Such ICs, as you shall see, did not blossom overnight, but were the result of extremely diligent and intelligent engineering that continues with considerable intensity to this day. True analog is being superseded to some extent by digital, but cost, complexities, bandwidths, availability, and A/D converters will always leave a slot for analog almost everywhere.

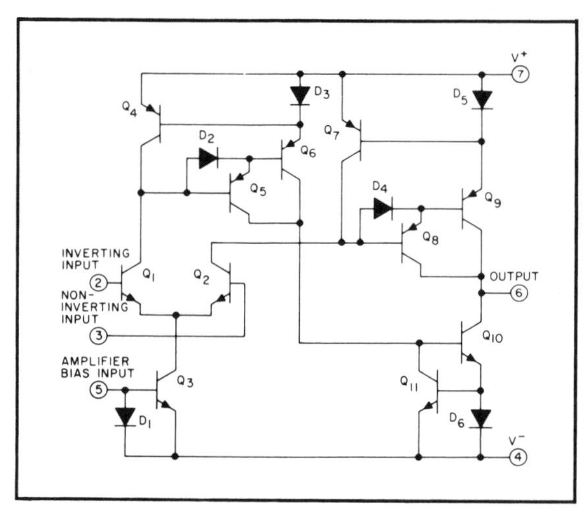

Fig. 12-1. An early transconductance amplifier, the basics for many linear ICs today (courtesy RCA Solid State).

The MC1351P

Motorola's MC1351P (for plastic) is a good example of early IC sound amplification and decoding development that illustrates some of the foregoing discussion. This 14-pin IC has the differential/limiting amplifiers and couplers, quadrature detector, audio preamplifier and driver, along with built-in voltage regulation. The coincidence discriminator needs only one RLC 90-degree phase-shift network and has short circuit protection (Fig. 12-2).

Inputs are through pin 4, with pins 5 and 6 at ac ground. The limiting amplifier is shown as three differential pairs having buffered (emitter follower) outputs not loading the preceding differential stage and driving the succeeding stage. Each pair has a gain of 10, or approximately 60 dB for all three. Pin 5 is decoupled to prevent gain-reducing feedback, and pin 6 is filtered since it terminates the temperature-stabilized voltage bias supply for the bases of the second and third differential amplifiers. Feed-

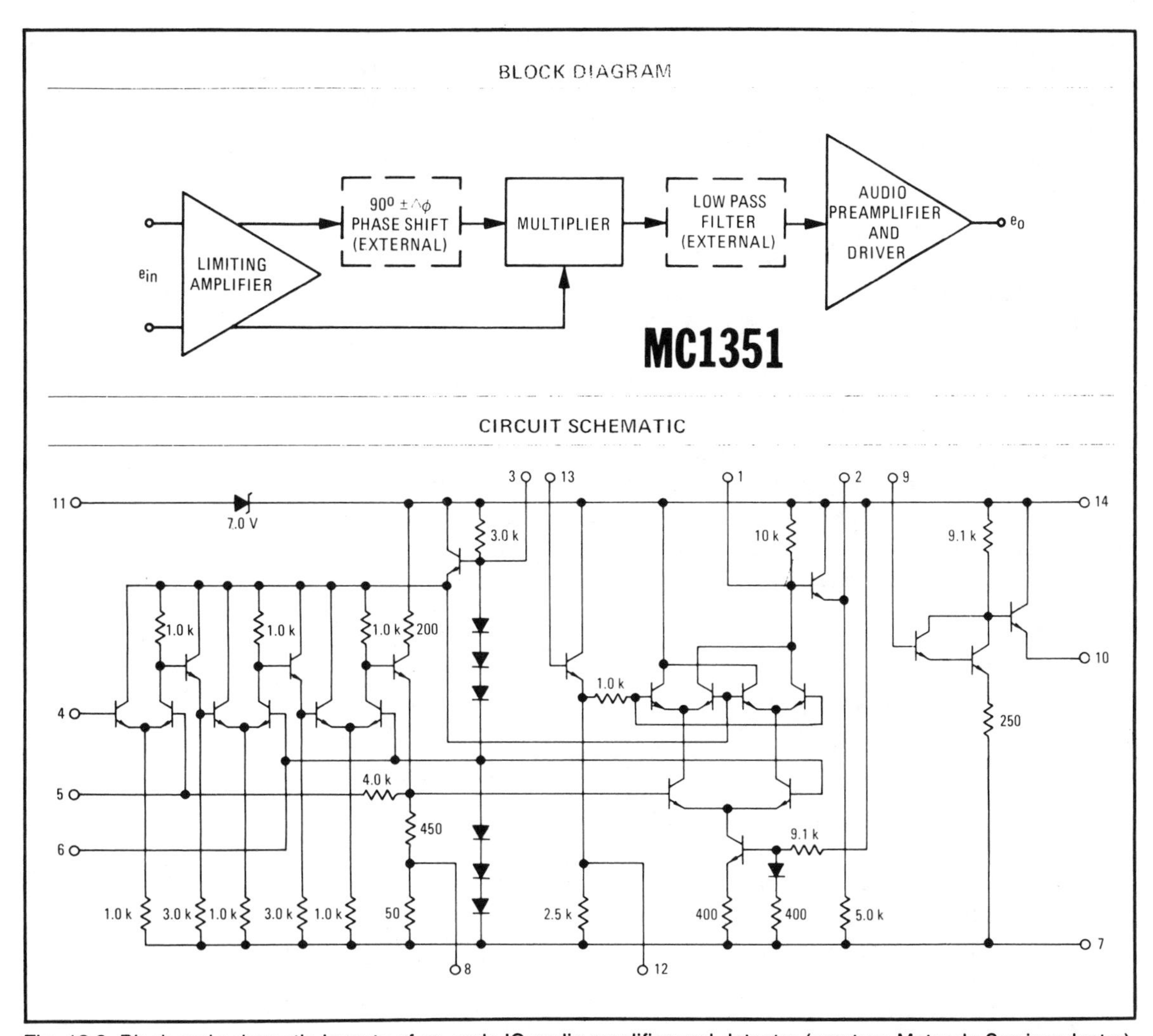

Fig. 12-2. Block and schematic layouts of an early IC audio amplifier and detector (courtesy Motorola Semiconductor).

back from the emitter of the third buffer is developed across the 4-kilohm resistor for the base of the first differential amplifier. The six temperature-controlling diodes also limit base drift of the shunt regulator transistor supplying collector operating potential for the three-stage limiting amplifier and the effective quadrature detector upcoming.

This quadrature detector has a differential input and two pairs of cross-coupled switching detectors above. One side of the input is biased from the six diodes' center point, while the other accepts amplified audio from the third buffer-emitter follower. Switching voltage is generated by an external LC tank circuit connected between pins 8, 13, and 3, that goes to the bases of the detectors through a 1 kilohm resistor. Tank circuit Q determines peak-to-peak separation of the detected output, noise performance, and detector output swing, and a Q of 24 will supply ±25 kHz deviation and a 200 kHz p-p separation following the detector.

When the switching signal is in 90-degree phase quadrature relative to the differential signal inputs, there is no output. But, with phase changes, full wave detected audio is developed across the 10 k resistor above and is emitter followed from pin 2 to pin 9, and thence through the Darlington output at pin 10. Detector product is 2 times the carrier frequency, with no fundamentals appearing at the detector's output. A volume control may be connected between pins 2 and 9, with feedback from an external power amplifier into pin 9. This external amplifier is usually a class-A device driving an 8-ohm speaker through a 10:1 turns transformer, and power dissipation for the entire chip averages 300 mW.

After we do several more of these audio ICs you'll fully understand how the technology develops and is sustained by vast improvements to electronics, packaging, and probably also cost. Details are not readily available, but plastic package sealing has come a long way since the early '70s when leakage put ICs out of commission long before their normal times.

RCA's CA3065

One of the very best known of these 1970 ICs was a pretty complex one by RCA that was used for many years by a considerable number of set makers and possessed some singular advantages. This one had an electronic attenuator and volume regulator, in addition to a dc volume control, tone control, and deemphasis filter. With all the additions, both schematic and block diagrams have expanded, so we'll use two figures (Figs. 12-3 and 12-4) to illustrate, and describe the operation from both perspectives. As you can see from the signal processing blocks, very few external components except a tank circuit, a couple of filters, and two controls are required. The remainder are now included in one way or another inside the 14-pin IC.

The three sets of limiting differential amplifiers are very similar to those previously described, except the bases of all three receive feedback from Q16. At the same time the bases of Q17 and Q20 are clamped by D7, with Q17 limiting any excess current out of Q16. Filtered and phase-compensated outputs from Q18-Q19 then pass through R24 to the base of Q22 and also to one end of the tank circuit at pin 9, which connects across to pin 10 and buffer Q27. Q23 and Q26 with C3 and C4 then become

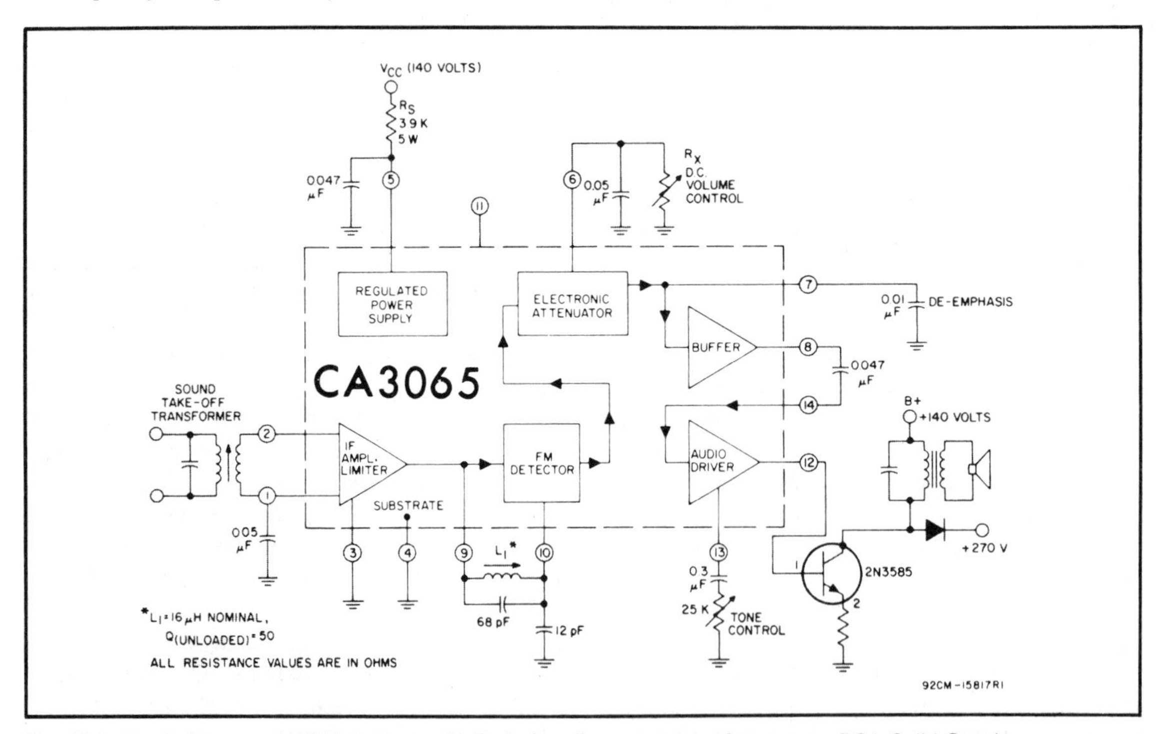

Fig. 12-3. Block diagram of RCA's more sophisticated audio processing IC (courtesy RCA Solid State).

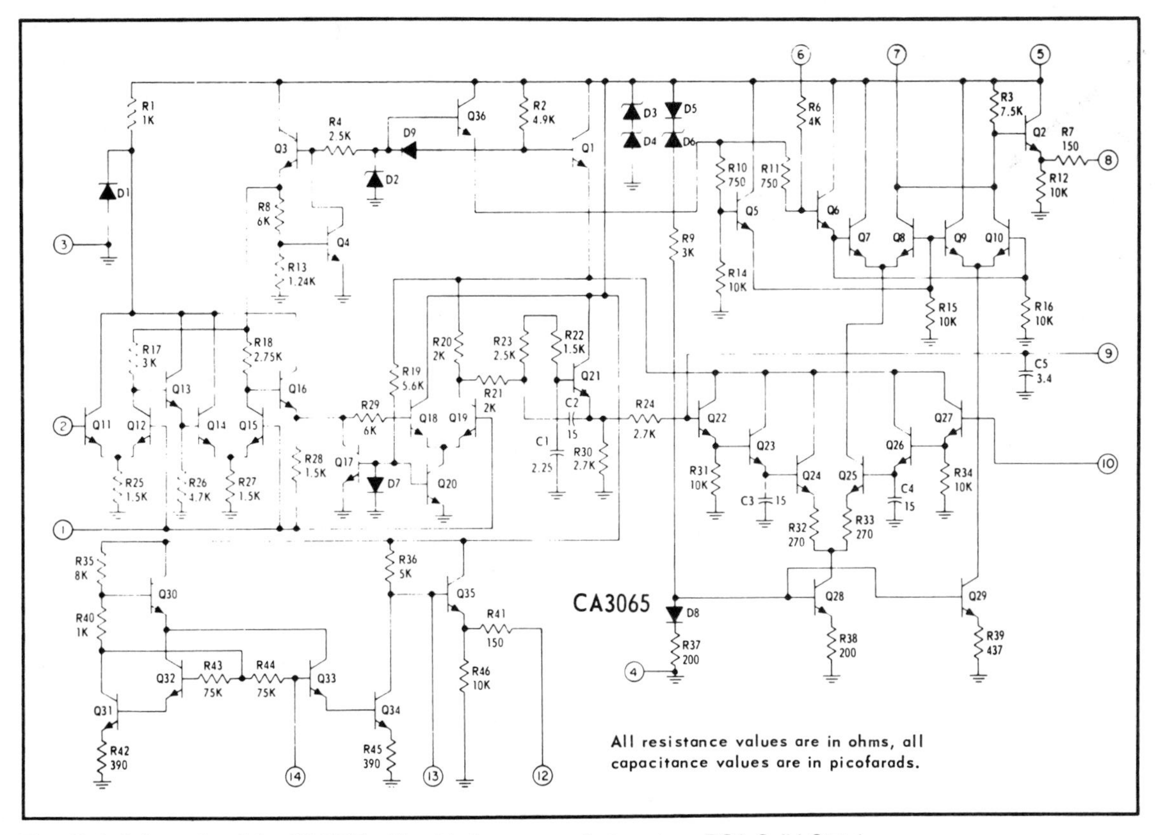

Fig. 12-4. Schematic of the CA3065 with added components (courtesy RCA Solid State).

peak detectors for differential amplifier Q24-Q25, with constant current emitter potential supplied by Q28.

Differential current from Q25 then flows to the emitters of Q7-Q8 as fully detected audio. At the same time, the emitters of Q9-Q10 are also current supplied by nonsignal source Q29. There is also fairly stiff voltage regulation from zener D2, D9, Q36, Q5 and Q6 for Q7 through Q10, which constitutes the electronic attenuator. With a dc volume control at pin 6 and the base of Q6, and a deemphasis filter at pin 7, regulated bias is set for the two pairs of differential amplifiers at whatever volume the operator requires.

These two differential amplifiers, except for their emitters, are actually in parallel, although bias for Q8-Q9 is fixed, while that for Q7-Q10 is variable. Therefore, signal amplitudes from Q25 below are automatically limited by the setting of the Q6 volume control and can't rise above that bias level because of limited differential amplifier conduction. Final output passes through follower Q2 and usually a class C inverter for a transformer and speaker after continuing through an audio driver below.

Pin 13 accommodates a dc tone control for the driver, while Q30 through Q32 constitutes a parallel-shunt regulator, maintaining the collector and base of Q33 within nonsaturation limits. Drops across R35 and R40, for instance will turn on Q30 more or less delivering greater or restricted current to the collector of Q32 and that of Q33. At the same time, drops across R40 affect both the bases of Q32 and Q33, allowing varying responses in these two transistors. Q31, however, responds to Q32 by controlling the drop across R40 with greater or lesser current demands, and so maintains bias between R43-R44 for the Q33 through Q35 driver circuit and a linear output.

Amplifier, Detector, and Power Output Combined

Before the final example, it might be worthwhile to look at one more intermediate 16-pin IC which had everything on a single chip, including a heatsink and a 3 W power output. Again, this is an RCA integrated circuit, identified as a CA3134, that is very similar to much of the CA3065 except for its power amplifier, unattended audio, and power output via a 100 μF capacitor directly into a 16-ohm speaker. In fact, except for some slight simplification in the audio amp, input and a couple of diodes added in the electronic attenuator, the two ICs are much the same—simply evolutionaries of one another.

Therefore, similar sections we'll treat lightly, and the new ones will receive due attention, considering the literature (or lack of it) identifying some fine design points, general design analysis might easily miss. (No application notes here.) Once again, because of size, the block and schematic diagrams will have to be displayed separately because the schematic, even being reduced, is too large for a composite drawing. The two are shown in Figs. 12-5 and 12-6. As you may surmise, the extent of these drawings will only increase, and this is why IC makers today supply little more than block diagrams (if that) for their wares.

As is apparent, 4.5 MHz intercarrier sound inputs between pins 15 (hot) and 14 (ac ground), continuing on into the familiar sets of differential amplifiers and their buffers, with peak detection taking place at the emitters of Q32 and Q40. Detected FM, however, is now current-channeled to pairs of differential amplifier emitters, Q33-Q34 and Q38-Q39, whose bases are connected to electronic attenuator volume controls, and are also power supply regulated. Active loads and biasing control the collectors of these two differential amps, which have two outputs: one of which connects between Q6 and Q8, receiving outputs from Q33; and the other is the usual electronic attenuator-con-

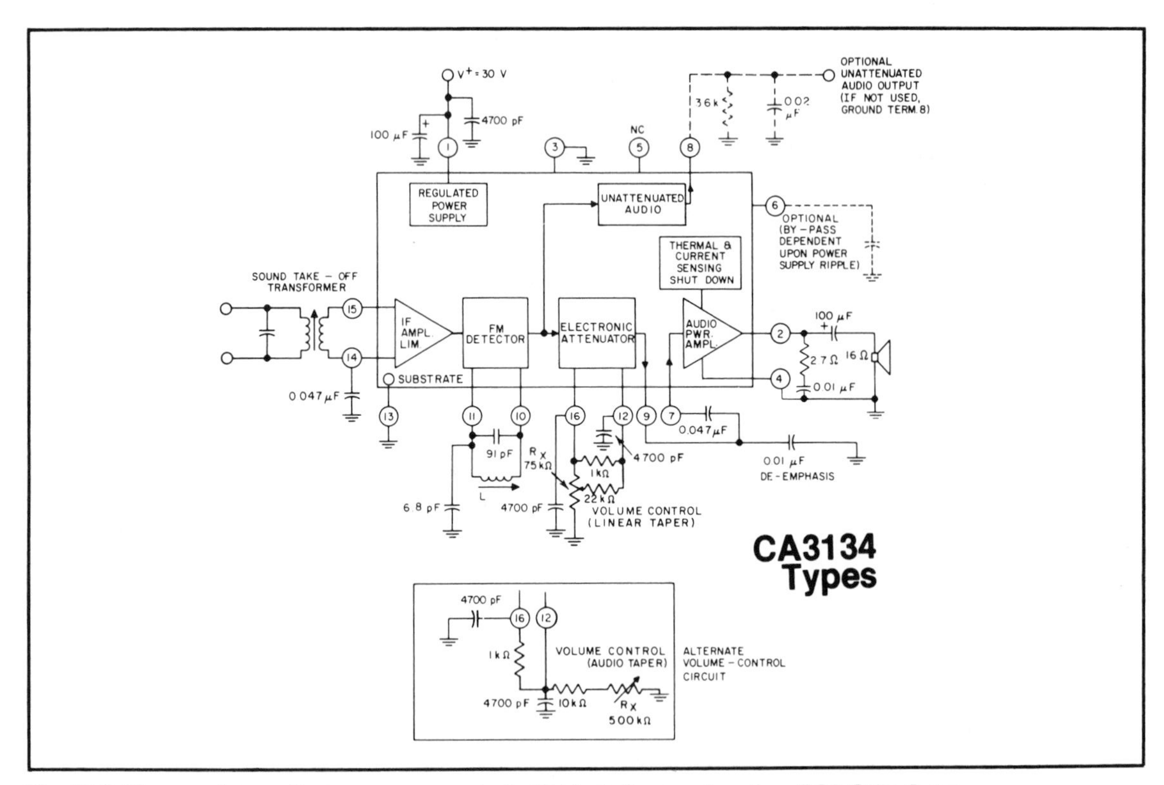

Fig. 12-5. The complete audio processor on a single IC block diagram (courtesy RCA Solid State).

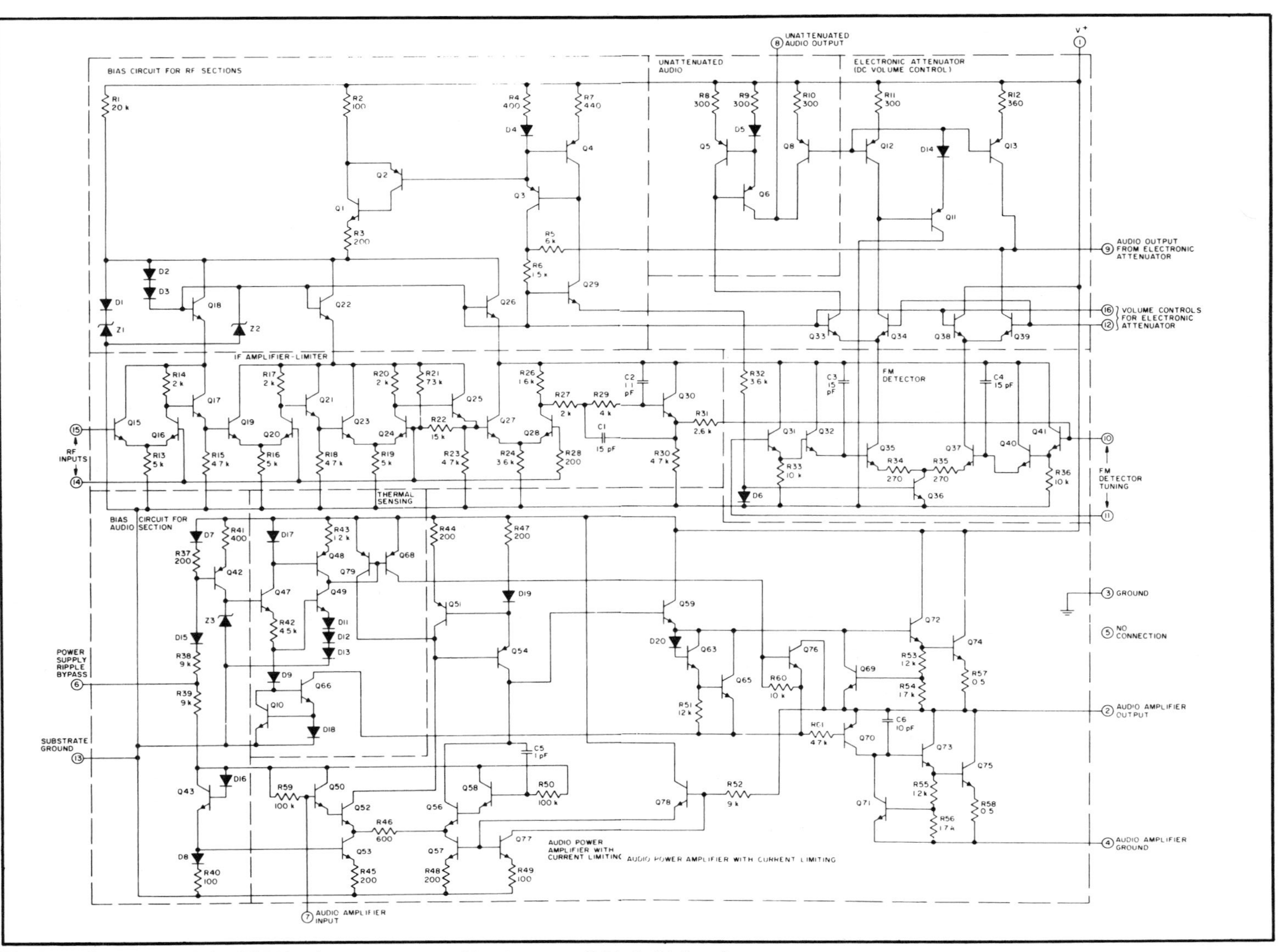

Fig. 12-6. Schematic of the CA3134 detector and amplifier (courtesy RCA Solid State).

trolled audio processed information through pin 9, returning to the IC through pin 7 and into the power amplifier. If the optional audio output is unused, terminal 8 should be grounded.

So far, so good. With regulated audio entering the audio input amplifier at pin 7 and R59, stacked transistor pair Q52 and Q53 supply a relatively medium impedance output to collector of Q51 and the base of Q54, with an active load supplied by Q79. Inverted audio from the collector of Q54 then goes to the base- emitter of Q59 in the power amplifier and transistors Q72-Q74 in neo-Darlington configuration, with feedback through Q78. Transistors Q70, Q71, Q73, and Q75 all provide power and some limiting, with Q69 across the Q72 output for shut down protection if required, similar to Q71 across Q73. Basic limiting occurs with Q76, Q70, and Q77, and Q78 all in the loop, with Q77 controlling any runaway output at pin 2 by shunting excess current toward ground. In other words, the harder Q78 conducts, Q77 does likewise drawing current through R52 and lowering all potential output to safe dissipation.

The power supply appearing in center left of the diagram is the familiar shunt regulator, with a single zener, and 8 temperature- compensating diodes. Tied base to collector, these stages conduct constantly, providing constant base bias for Q68, Q79, which in turn serve much of the remainder of the amplifier, keeping its potentials constant where required.

As we suggested earlier, this is probably the last schematic we'll be able to show in the linear audio monophonic section—and perhaps the multichannel sound portion also. The remainder are not only large but many are proprietary and the big manufacturers are not charitable about sharing their best designs with anyone.

A 7-Watt Audio Amplifier

Just to make the description series a tad more interesting, a Magnavox 19C1 chassis 7-watt audio amplifier is introduced, complete with bass, treble, gain, and individual power supply—all isolated from the ac line by a good, conventional power transformer. Remember those? Observe that the 30-33V power sources are unregulated, and only the + 12 V source has a zener with a 1 W limiting resistor (Fig. 12-7).

Voice and music inputs are through terminal 4, IC200, and then ac coupled into Q200 for inversion before entering IC202 through C210. At the same time, the noninverting (+) terminal of IC201 is biased by attenuator adjust R221, while the volume control through pin 5 adjusts the other half, with feedback through 68 k R223. As current output increases through IC201, IC202 is driven harder producing additional gain through IC202 for drivers Q301 and Q303. Variable resistor R313 then sets bias for Q305, which delivers paraphase outputs to the bases of the push-pull combinations of Q309-Q311 and Q307-Q313, producing power rated monophonic audio through 1000 μF capacitor C311 to a single speaker. Two such amplifiers, presumably, could handle stereo. Note all the careful filtering, biasing, and even emitter-to-base feedback through C309. It's a good example of a modern audio amplifier in current use, and decidedly puts Magnavox back in the audio-amplifier business.

Putting Them Together

As a parting shot in this monophonic audio section, let's put amplifier and detector together in an E51-56 Sylvania chassis and see if some parts aren't reasonably recognizable (Fig. 12-8). This, therefore, completes the picture . . . or should we say the *sound* picture for this portion. You're now pretty well up to date for mono, with all the newest developments to come in stereo/SAP that follows.

Conventional 41.25 MHz sound is taken off the collector of the third i-f and routed to a carefully biased Q100 amplifier and detector which selects the 4.5 MHz beat difference between it and the 45.75 MHz video carrier. After considerable filtering and a tuned L110 sound takeoff coil, the usual i-f limiting and amplification occurs, along with familiar quadrature detection, electronic attenuation, and buffered audio output in IC100.

Baseband audio is now ac coupled to IC101, the power amplifier processor and output, along with RC boost and filtering via R114 and C122. In addi-

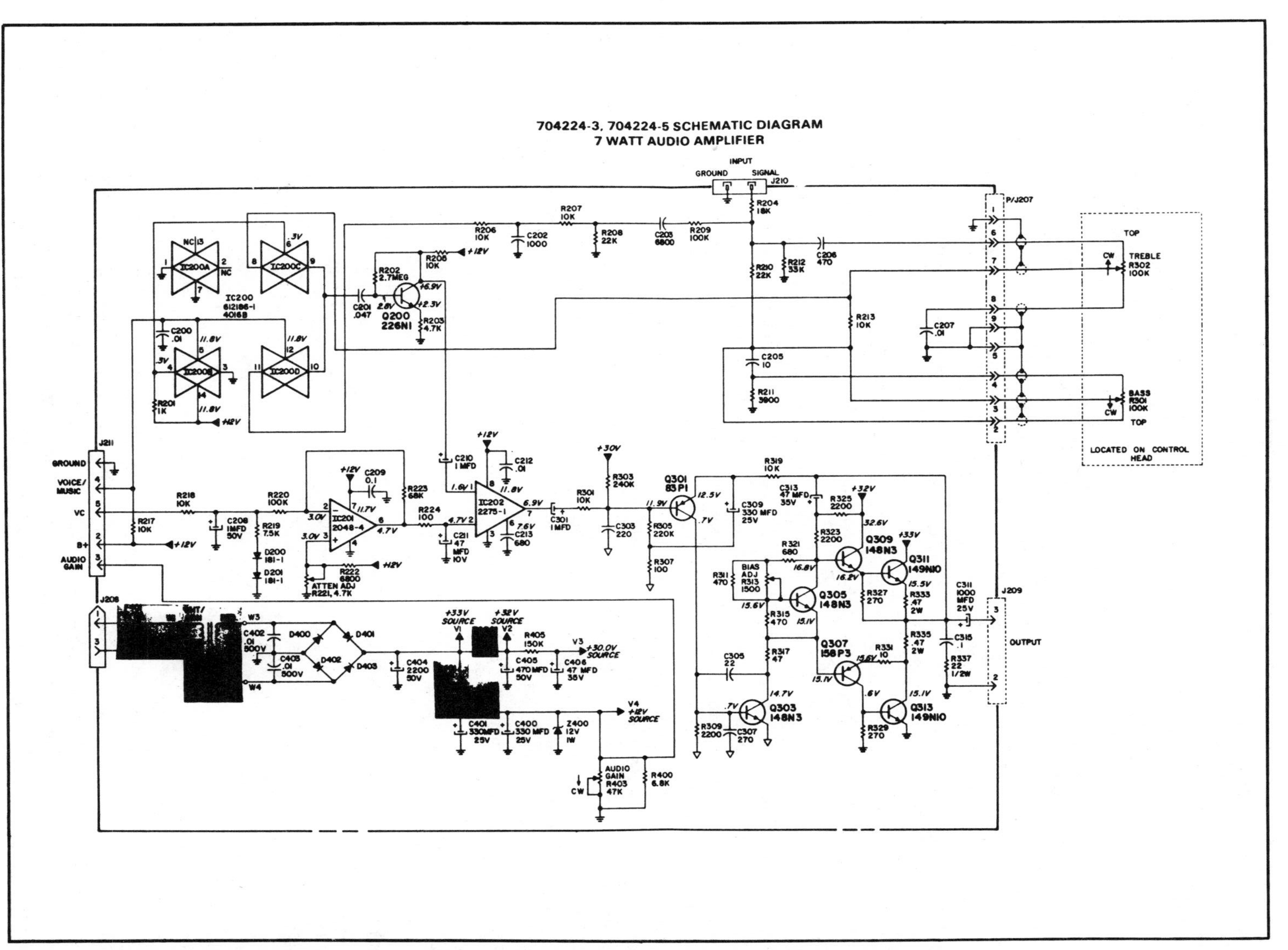

Fig. 12-7. A Magnavox 7-watt monophonic audio amplifier with its own power supply (courtesy NAP Electronics).

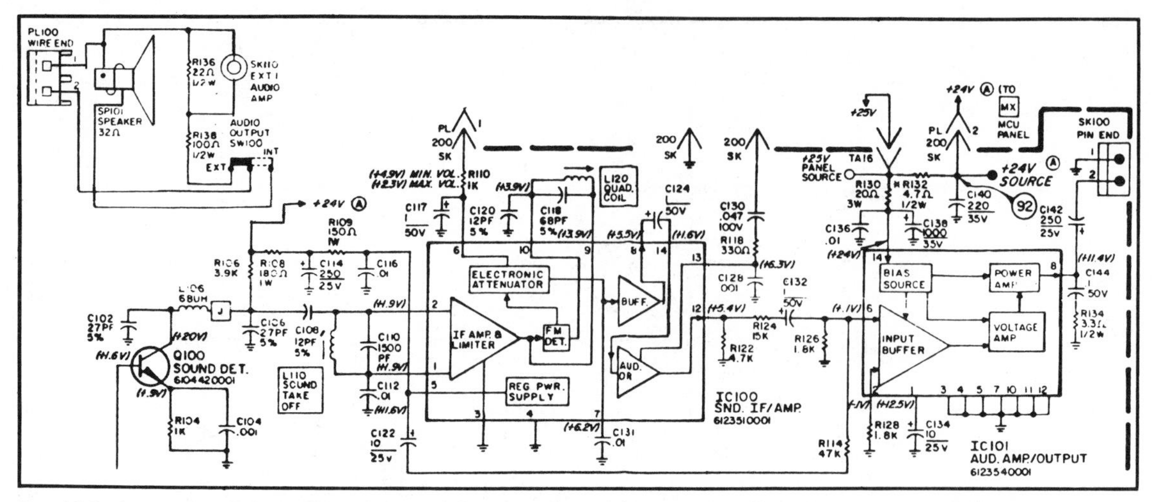

Fig. 12-8. A complete E51-56 Sylvania sound detector, demodulator, and audio output (courtesy NAP).

tion to an input buffer, there are also voltage and power amplifiers in this IC which, undoubtedly, delivers several watts either to a 32-ohm speaker coil or an external audio amplifier, as selected by SW100.

Considering that virtually all of the circuits in this 1- transistor, 2-IC sound section have been fully explained previously, there should be no compelling reason to repeat. Audio explanations hereafter will be considerably more complex and detailed than those previously described. Audio is, indeed, entering an advanced phase and a new era.

MULTICHANNEL SOUND

Approved by the Federal Communications Commission on the afternoon of March 29, 1984, Multichannel Television Sound, as submitted by the Broadcast Television Systems Committee of the Electronic Industries Association, is now going on the air in many cities through the United States and will be heard nationwide just as soon as additional broadcast equipment is available. As you will see, most major lines of TV receivers are already equipped with one of three outstanding systems, which will or won't prove themselves conclusively with the test of time. In most instances, however, it is expected all performance will justify a modest additional cash outlay, and by second generation manufacture, better ICs should be available to render appropriate service under almost any conceivable conditions that are bound to arise with the introduction of any new product or system. At the moment, a combination of Japanese and U.S. integrated circuits are carrying the load, with many others under active development here and in the future. For BTSC-dbx, as the FCC-approved system is now known, is probably the greatest advance in NTSC television since the advent of color and the comb filter. Especially is this true with the audio passband for stereo being widened to 50 kHz and linear audio carrier and baseband response available in excess of 100 kHz. You might think this is overkill, considering that most humans (or reasonable facsimiles) hardly hear much above 11- or 12-kilohertz; but it's been done to accommodate SAP and professional carriers that extend to 6.5 times the horizontal scan frequency of 15,734 Hz . . . more about this shortly.

The System

Based on Zenith's carefully considered and designed system for multichannel sound, plus the addition of companding (compressing and expanding) by dbx, Inc., Newton, Massachusetts to reduce the effects of buzz and noise levels, the pilot stereo frequency is the same as that of horizontal scan, with a main carrier deviation of 5 kHz.

The L + R main channel has a peak deviation

of 25 kHz with the usual 75 μsec preemphasis, while the L − R stereo subchannel is allowed a peak deviation of 50 kHz. When both are combined (interleaved) they still have a maximum deviation of 50 kHz.

Departing from the usual FM modulation in all TV sound and L + R, the L − R subchannel is level-encoded and produced as double sideband, suppressed carrier AM modulation at 2_h, or twice the horizontal scanning frequency (2 × 15,734) at 31.468 kHz. Regardless, each will have a maximum modulating frequency of 15 kHz. A Second Audio Program (SAP) channel has also been authorized, which is expected to be used mainly for voice traffic, primarily second language broadcasting. It is frequency limited to 10 kHz maximum, dbx encoded (companded), but has a peak carrier deviation of only 15 kHz and its carrier FM frequency is 5_h, or 78.670 kHz. In non-TV types of broadcasting you may have a voice or data channel at $6.5_h f$, the voice channel having 150 μsec preemphasis at maximum FM modulation of 3.4 kHz, or a data FSK channel with modulation of 1.5 kHz, each having an aural carrier peak deviation of 3 kHz. Were all carriers active, the aural carrier peak deviation cannot exceed 73 kHz for the entire transmitted group of signals. Such information is illustrated in Fig. 12-9 showing baseband frequencies in excess of 102 MHz, to which these BTSC-dbx stereo systems will respond without any problems. You will probably never contend with the 6.5_h carrier, but your stereo-SAP system better be flat to 80 kHz or more. All we've checked so far have exceeded that specification and probably will continue to do so unless cost cutting becomes an overriding factor. The usual FCC specifications normally apply strictly to transmitters and *not* receivers—so you never can tell without careful measurements, which entail spectrum analysis, to be sure.

Industry work began on systems study in 1979 seeking four objectives: 1) a compatible main channel; 2) a quality stereo subchannel; 3) a separate audio program (SAP) lesser quality subchannel for extra program services; 4) and a professional type subchannel for transmitter status telemetry, field crew cueing or other TV station services.

Transmission system groups interested in presenting their wares were invited to submit working samples, the CATV industry was asked to participate, and a single receiver was selected to evaluate the various decoders. Despite a good deal of brave talk, only three entries were received with equipment ready for testing: Sony Corp., representing the Electronic Industries Association of Japan; Telesonics Systems, Inc., and Zenith Radio Corporation. Tests were made in an EIA laboratory provided by Matsushita Industrial Co. in Chicago.

Sony proposed an all-FM system with maximum peak carrier deviation of 25 kHz; Telesonics offered an AM and FM system, with maximum peak carrier deviation of 50 kHz, in addition to three subchannels but with subcarriers at 5/2, 9/2, and 6 f_h; Zenith would deliver a full 15 kHz modulating frequency at maximum peak carrier

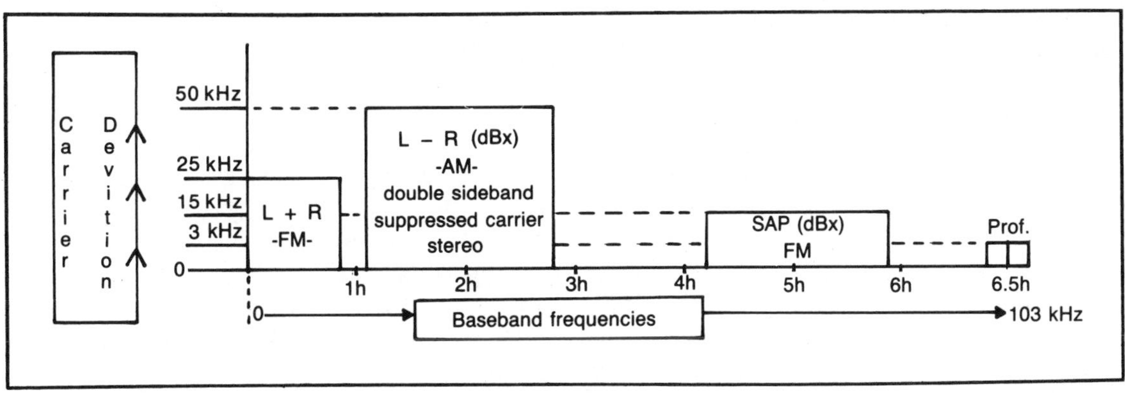

Fig. 12-9. BTSC-dBx stereo/SAP professional channel *system* showing carriers (with respect to the 15,734 Hz pilot) and modulation.

deviation of 50 kHz, and carriers at 2, 5, and 6.5 f_h. Noise reduction systems by CBS-CX, dbx, and Dolby were also investigated for quality and suitability. DBx was the winner here—so this is how the BTSC dbx combined system waschosen by the Electronic Industries Association of America.

BTSC findings were then presented to the Federal Communications Commission after its selection of the Zenith and dbx system in December 1983. Then in March 1984, a favorable decision by the FCC was rendered, substantially supporting the EIA's choice and permitting both broadcast and receiver manufacturers to proceed with all electronic requirements, and placing the system on the air. This hasn't been easy for either receiver or broadcast manufacturer, but engineering and installations are now settling down to refinements and improvements rather than initial instant chaos. Already there have been substantial gains over 1984-1985's struggling beginnings, especially for transmitters.

To keep the record straight, however, the FCC did *not* rule out any or all other competing systems, but specified that anyone using the 15,734 pilot subcarrier must broadcast and receive the BTSC-dbx system. So far no other multichannel sound contenders have appeared, nor are they expected to.

RECEIVERS

Out of all initial proposals generated *before* real production began, only three real systems seem able to survive:

- Quasi-split sound separated from video following the acoustical wave (SAW) filter and amplified *before* 4.5 MHz intercarrier detection.
- Standard demodulation to composite baseband and then transmission to the decoder.
- Sound detection *following* the video amplifier, but accompanied by extra noise reduction circuits.

Fortunately, we have examples of all three systems and, although, often incomplete, will offer a good idea of their methods and, in some instances, practical test results which, although not conclusive, do indicate there's sufficient bandwidth for successful multichannel stereo. And if you have that, in addition to some 30 dB separation, then you'll probably have listenable stereo, at least to some degree.

In all systems tested so far, 25 dB has been the *least* stereo separation encountered, with *channel* separation in excess of 50 dB. So far, those are pretty good numbers.

Various manufacturers are still using their favorite intermediate frequency amplifiers, with different twists to their systems, but more than one now has an AN6291 dbx-type decoder and an LM1884 stereo processor-separator tucked into a circuit board somewhere before the final audio amplifiers and, according to several broadcast manufacturers, these ICs are working very well and producing better than adequate stereo. Passing a broadcaster's critical appraisal shows, momentarily, that receivers are ahead of broadcasters, at least with satisfactorily operating equipment in the field.

Let's begin our receiver sound descriptions in reverse order, saving the more intricate of the systems for last. In this way you will probably derive more from a slow start than a fast one. Good or bad, the whole process is a little tricky for those who are not used to specialized audio—and this positively is, especially in the broadcast companding portion where dbx, Inc. isn't very helpful.

A National Semiconductor Version

With the aid of National Semiconductor's block diagram and a little propaganda for its own integrated circuits, let's start with this one since it is easy to understand and aptly illustrates the various steps in developing stereo and its complexities. The only portions they can't supply are tuner, SAW filter, sound trap, and the dbx expander which is currently furnished as two ICs identified as uPC1252H2 and uPC1253H2 by NEC or a single AN6291 by Matsushita. Both of these, we're told, work well, but the 2-chip approach has better voltage swings and greater outputs. (See Fig. 12-10.)

As you see, intercarrier sound is amplified and detected by the LM1965, goes into both the

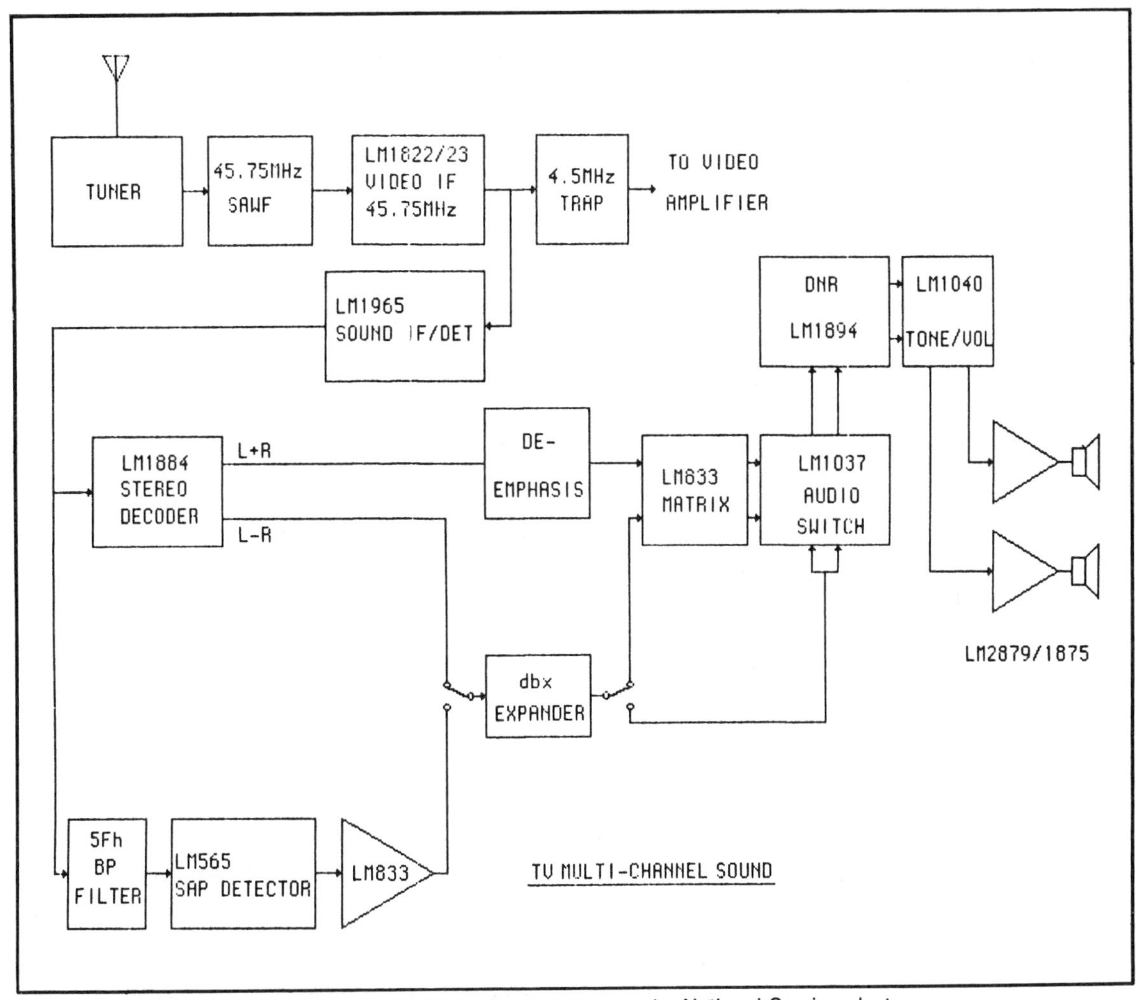

Fig. 12-10. A multichannel sound receiver block diagram as seen by National Semiconductor.

LM1884 L + R and L − R stereo decoder, is deemphasized (for L + R), expanded (L − R) in dbx and then joined in an LM833 matrix for the LM1037 audio switch and dynamic noise reduction LM1894, before reaching tone, volume, and output amplifier LM1040. Should the broadcast be bilingual, then the 5_h bandpass filter picks up this carrier which is rejected by the 2_h stereo decoder, and the LM565 SAP detector does its job for the LM833 and, once again, the dbx expander and the LM1037 audio switch and output.

National's LM1822 Video I-f

Containing a 5-stage gain-controlled i-f amplifier, a phase-locked loop video detector with noise inversion, AFC detector and gated AGC, this video i-f processor is now in use with good results. You will note there are comparatively few external components (Fig. 12-11), an AFC defeat, two potentiometer adjusts and four tank circuits whose frequencies may be inductively set. Differential i-f inputs aid rejection of undesirable rf coupling, with input gains depending inversely on source impedances which may vary between 500 and 2000 ohms, typical of SAW filters. An internal shunt regulator supplies operating voltage for the i-f block.

The voltage controlled oscillator (VCO) and its

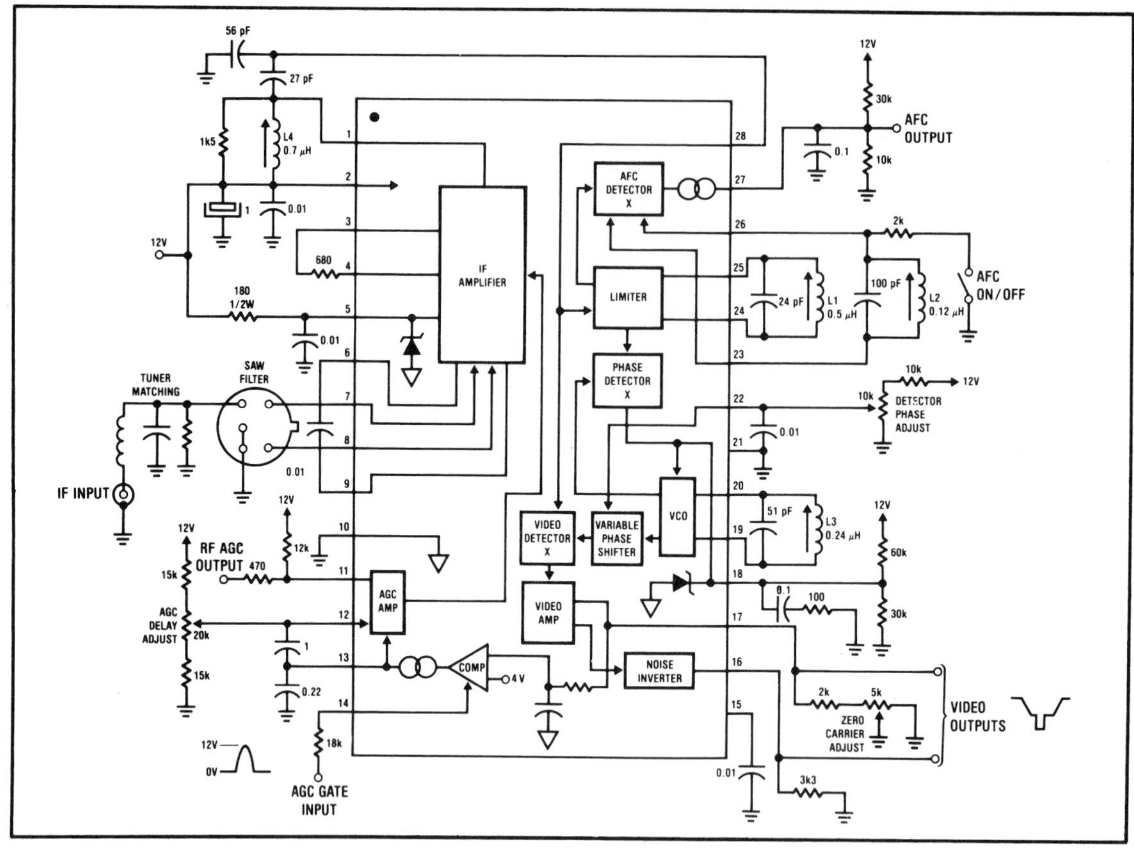

Fig. 12-11. National's LM1822 video i-f, AGC, noise inverter, AFC detector, and video demodulator (courtesy National Semiconductor).

variable phase shifter supply synchronous detection drive for the video detector, which also has dual video outputs (one noise inverted) along with zero carrier adjust. A parallel tank circuit becomes a tuned load for the limiting amplifier serving both AFC and phase detectors, with AFC output being a push-pull current source to stabilize the tuner's local oscillator for accurate channel tracking. Suitable for both home receiver and CATV applications, it will handle rf mixer inputs from 38.9, 45.75, 58.75, and 61.25 MHz at "selected points" on tuner response-SAW inputs.

The LM1884 Stereo Decoder

By far the most important of these National Semiconductor ICs, the 1884 stereo decoder, is specifically designed for TV/stereo, offers both L + R and L − R low-impedance outputs, with distortion typically at 0.1%. Suitable for TV, CATV, and stereo adapters, there is a PLL phase detector, buffer amplifier, voltage-to-current amplifier, an oscillator, dividers, pilot detector, Schmitt trigger, decoder, mono-stereo switch, and voltage regulator (Fig. 12-12).

It has low impedance L and R outputs, a mono stereo switch indicator, and gain ratio between L + R and L − R from −2 to +2 dB, and a dc mono to stereo shift of ± 20 mV max. Minimum pilot capture range at 25 mV amounts to ±5%, with less than 20 mV turn on and off.

As pilot tone arrives and mono switches to stereo, the decoder furnishes L − R to the output buffer which, operating with L + R, produces full stereo for final left and right matrixing elsewhere.

There's a stereo LED indicator, plus a manual stereo/mono switch and VCO outboard adjust.

The SAP detector is ordinary, and SAP may be either pulse counter detected or by PLL used as a discriminator. PLL in the LM565 has high loop gain and good hold in range and will follow sudden carrier phase changes and maintain lock.

Generally, the above takes care of the important ICs in the receive stereo circuits reasonably well and in enough depth to give you a "feel" for the general operation. The one part missing is the dbx contribution, and that we will describe next, all material being gathered from sources *other* than dbx, Inc.

A dBx Encoder/Decoder

About the best information we've found on dbx turned up in an IEEE Consumer Electronics paper given several years ago by engineers Shinohara, Fuse, and Yuji Komatsu of Nippon Electric, supplier of the two-chip compander used by much of the receiver industry.

According to them, the dbx compander has a range exceeding 100 dB and noise reduction of more than 30 dB in commercial applications. The two ICs described have a THD of 0.007%, a −94 dBV output noise level, and a wide dynamic range of more than 100 dB for consumer uses. The NEC block diagram is shown in Fig. 12-13.

The voltage-controlled amplifier (VCA) acts as a dynamic range compressor in the encoder and an expander in the decoder, with gains being controlled by the root-mean-square value level sensor (rms). It does this by furnishing a control voltage proportional to the rms value of the input signal into the voltage-controlled amplifier. The original dy-

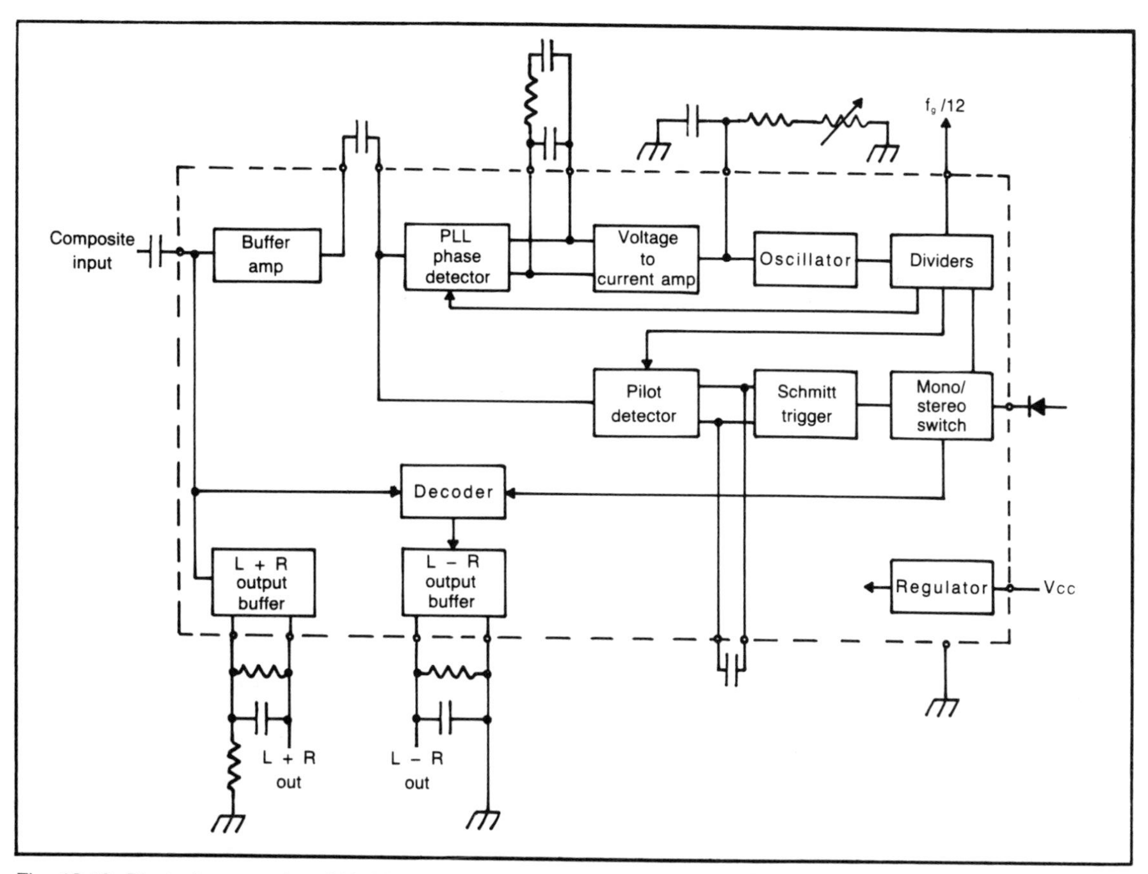

Fig. 12-12. Block diagram of an LM1884 stereo decoder (courtesy National Semiconductor).

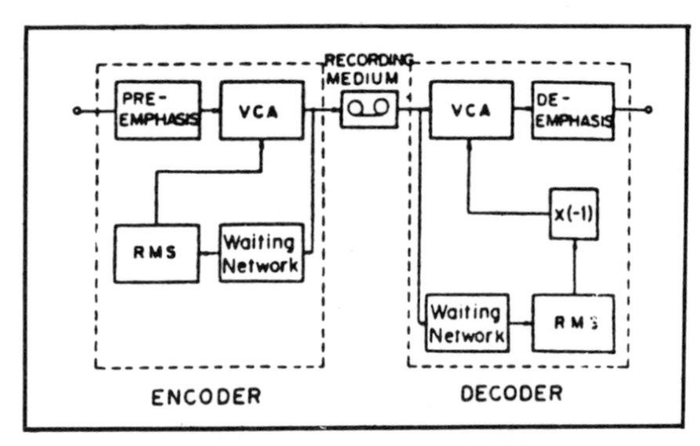

Fig. 12-13. Block diagram of dBx noise reduction system, both transmitting and receiving (courtesy NEC).

namic range reduces by 50% during encoding and doubles in decoding, achieving a very wide dynamic range.

The VCA IC is made up of a current mode amplifier, bias current generator, an idling current generator and a gain cell. In Fig. 12-14, transistors Q1 through Q8 constitute a gain cell, with Q3 furnishing a logarithmic output proportional to positive input current. By taking the anti-log of this signal, Q6, Q7 supply an output, reproducing the same output as input and delivering responses of low noise, low distortion and wide flat frequency response.

The RMS IC (No. 2) consists of a current mode rectifier, bias current generator, an amplifier, and output buffer (Fig. 12-15). D1, D2 and A1 are a log circuit with output proportional to the absolute value of the input. D3 and external capacitor C take the anti-log of this voltage and deliver an output proportional to the rms value of the input. Amplifier A2 then buffers this output to drive the VCA integrated circuit.

That's how compressors and expanders operate in dbx, permitting the BTSC (Zenith) stereo/SAP and dbx noise reduction circuits to operate as successfully as they do. Only stereo and SAP carrier information is companded, not professional channel(s).

With the entire system explained, you should be ready to look at two "live" schematic portions of different receivers now on the market showing how they distribute stereo/SAP sound. Both, by the way, have been air checked and perform admirably.

RCA's Multichannel Receiver

Already working in the marketplace with the 131/132 CTC series chassis, RCA's multichannel sound circuits, with the addition of National Semiconductor's dynamic noise reduction IC, are flourishing (Fig. 12-16). Extracting audio *after* the video detector in the same manner described earlier by National, the addition of a DNR™ has proved adequate to keep down spurs, buzz, and all the other audio problems that this method can generate. Audio engineering chief Tom Yost says there are few changes for the 1986 model year, and so we'll use the CTC 131/132 circuits for our description, which should be accurate enough to apply, generally, to the later versions.

All audio processing for these particular RCA chassis is done on three circuit boards entirely separate from the main chassis. These are involved immediately after a sound takeoff coil separates audio from video following the i-f/AGC processor, and eventually finds its way through buffer(s), etc. to the stereo broadcast board. Here, from the audio i-f buffer are two individual circuits: one of 45.75 MHz video carrier frequency, and the other of 41.25 MHz which, of course, is the audio carrier. In U1, these two frequencies are mixed, resulting in the

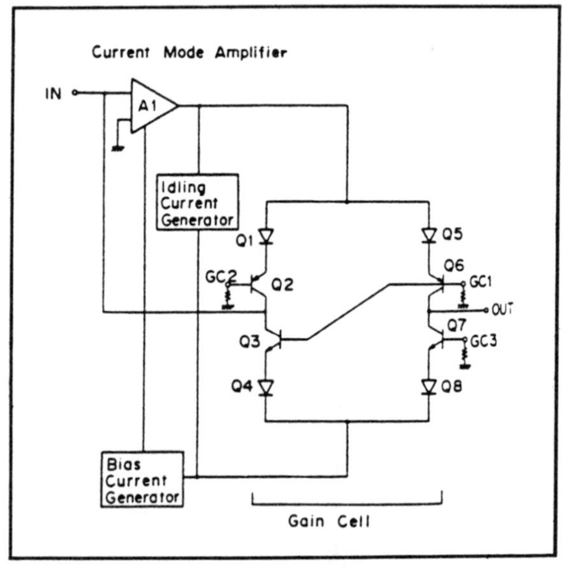

Fig. 12-14. Schematic of VCA voltage controlled amplifier (courtesy NEC).

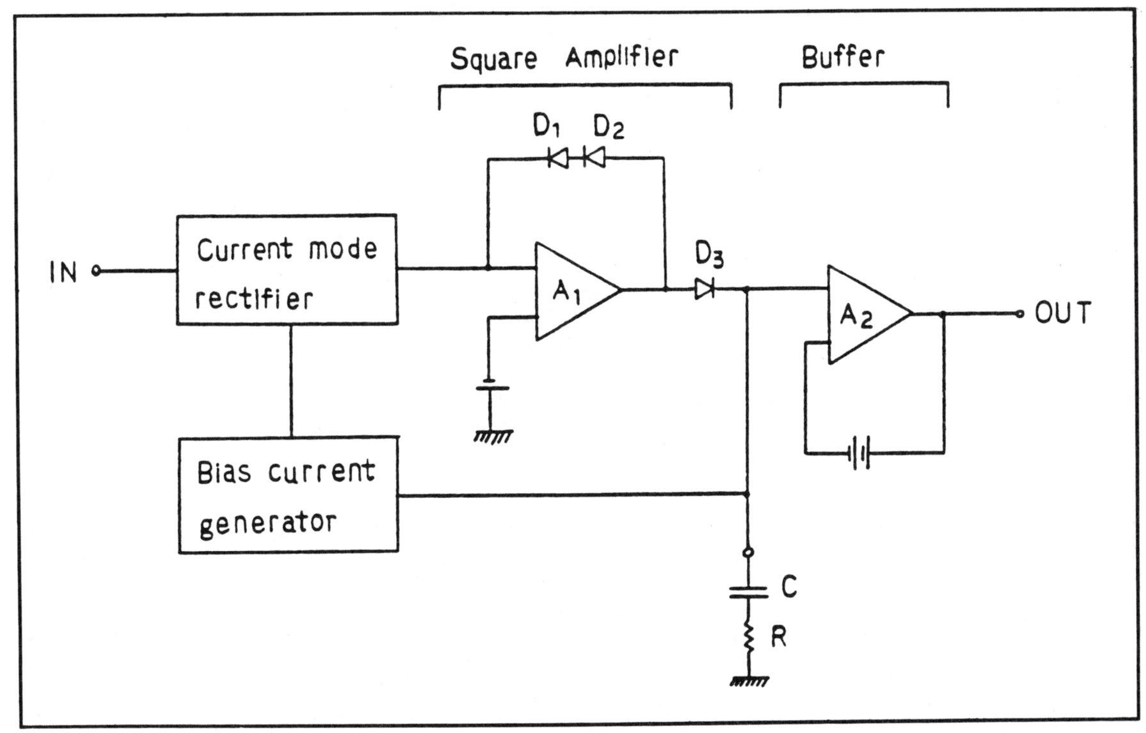

Fig. 12-15. Schematic of RMS integrated circuit (courtesy NEC).

usual intercarrier 4.5 MHz sound output. To clean up any harmonics or spurs, two 4.5 MHz bandpass traps operate and deliver limiter and bypass inputs to 4.5 MHz sound detector U2, which is usually an RCA peak detector as described earlier.

A biased bandpass reject amplifier with 1H 15,734 Hz stereo pilot carrier trapping next delivers paraphase outputs to the U3 stereo IC stereo detector. This trapping, however, affects U4 and not U3, where the pilot is both detected and used for a stereo lamp drive signal as well as doubling its frequency for the stereo suppressed subcarrier and its dual AM sidebands. A separate oscillator-adjust precisely sets the 31.468 kHz frequency.

In the meantime, additional 4H and 6H traps remove any possible undesirable professional channel interference, but permit SAP program FM non-stereo information to pass unobstructed into U4, where it is detected and amplifier for low-frequency audio and voice use whenever broadcast. Carrier frequency for second audio program SAP is 5 times 15,734 Hz, or 78.670 kHz.

With no incoming stereo, there is only L + R monophonic output which is processed by matrix amplifier Q13. But with incoming stereo, Q8 is also involved, and both signals are then routed to the expander board for further processing, along with SAP, if that signal is on the air.

National's Dynamic Noise Reduction IC

Here, on the audio switching assembly is a block diagram of the dynamic noise reduction IC we've been alluding to (Fig. 12- 17). Briefly, it has an R and L summing amplifier, gain amplifier, switch input, peak detector and two VCF circuits which, translated, stands for voltage-controlled amplifiers. A bandpass filter in this IC produces a low output from 20 Hz to 1 kHz, and a higher output at 7 kHz. Afterwards it is peak detected, delivering a dc voltage output proportion to signal input. Two internal filters then vary bandwidths with received control voltage, permitting essentially noise-free outputs as levels change.

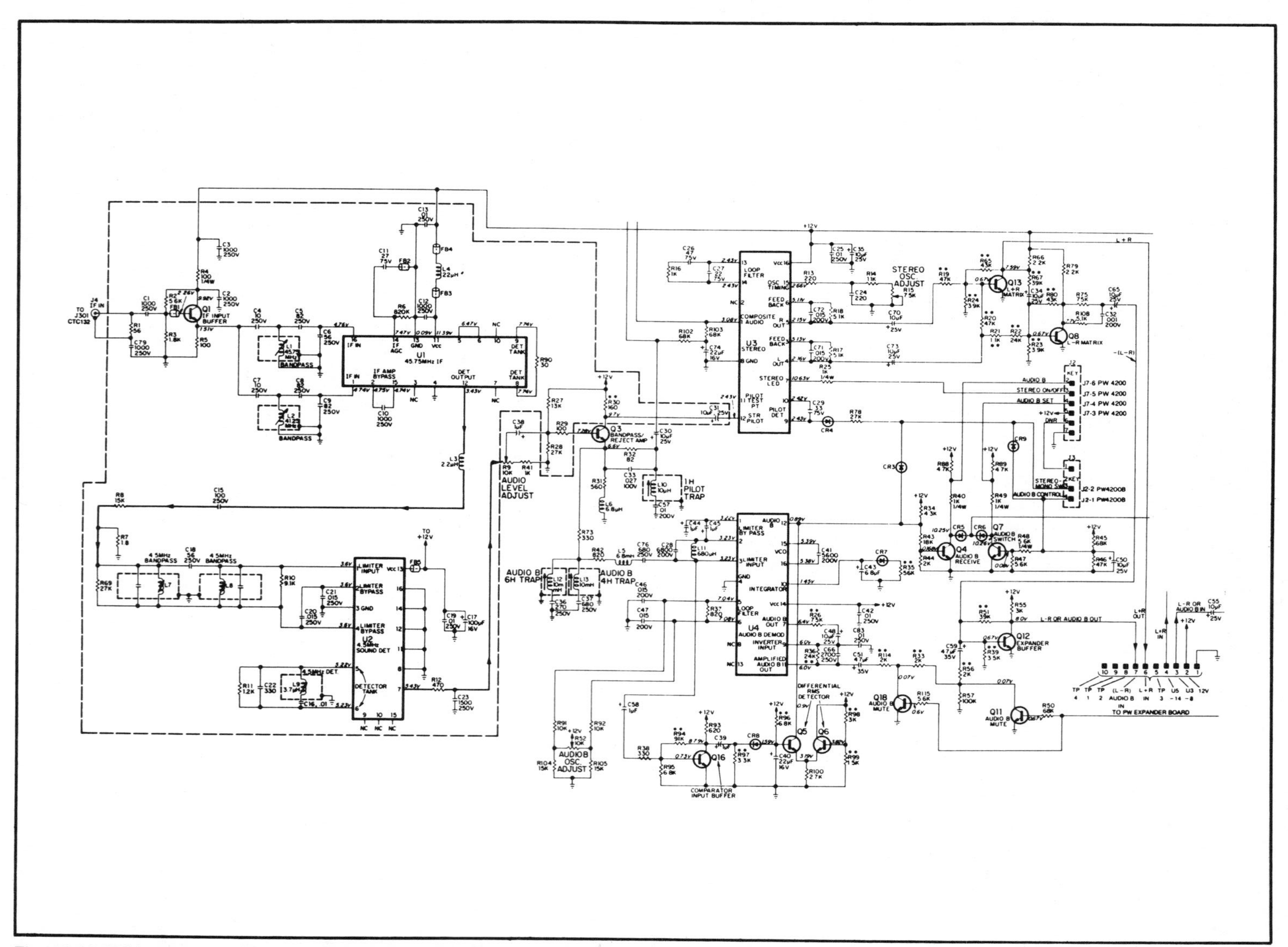

Fig. 12-16. RCA stereo system as shown schematically in current service literature (courtesy RCA Consumer Products).

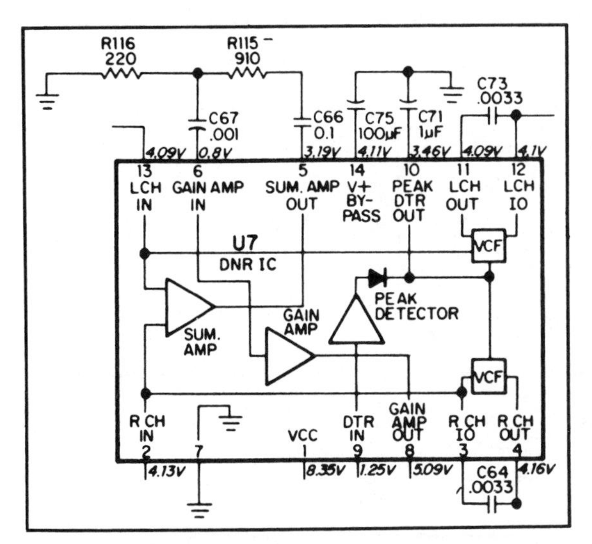

Fig. 12-17. National Semiconductor's LM1894 dynamic noise reduction IC as displayed on an RCA schematic (courtesy RCA Consumer Products).

Located before tone, volume, and balance controls, the LM1894 DNR™ circuit operates best at a 300 mV input and is easily biased using a blank tape and adjusting for threshold noise. When operating in a companding system such as this one, it must be positioned after the decoder since noise will reduce following companding.

There are two principles involved: noise levels in audio depend on system bandwidth, and also on the program material accompanying, low-level sound accentuates any noise, while high- volume programs tend to mask noise. These, then, are the principles of DNR, they compensate for such factors. Maximum noise rejection approximates 14 dB.

The LM1894 has two inputs and a bandwidth control. The two prime ones are through a low-pass filter and an integrating op amp. At the same time, dc feedback tempers the low-frequency gain, and above cutoff filter frequency, output decreases at −6 dB/octave due to the .0039 μF capacitor ch. 1 and ch. 3 outputs.

This control path develops a bandwidth control voltage paralleling the ear's sensitivity to noise with music. Right and left channels are added in the summing amplifier to maintain relationship, with capacitors between the summing output and peak detector determine frequency weighting, with gain set by both peak detector and gain amplifier, and the 0.1 μF capacitor at pin 10 controls attack and decay times. This voltage is then converted into proportional current for the two op amp blocks and two internal filters proceed to vary bandwidths with the received voltage. The lower noise outputs are between 20 Hz and 6+ kHz, and reducing bandwidth quiets noise.

The Magnavox Approach

Both Europe and Japan have instituted multichannel sound before our advent into this new medium and, consequently, have developed some electronics that are initial advances over what really wasn't available here. For Japan, the dbx IC decoder is a good example, and from Europe comes the quasi-parallel sound idea which some of our manufacturers are adopting to quiet buzz and other undesirable side effects.

The Magnavox system—since it is owned by Philips of the Netherlands—is an outgrowth of the European experience on the first design go-round and has, apparently, worked reasonably well. A block diagram of the tuner i-f sections appears in Fig. 12-18, and is used simply as example in leading up to the main discussion entailing a partial description of Magnavox's newest 25C5 chassis and its complete multichannel sound system. Of course, we've discussed this general idea before, so this is only a reminder of how sound and video can be separated successfully between tuner and i-f amplifiers and then processed in their respective subsystems. With NAP (North American Philips) and Magnavox working together, some innovative and very sound electrical designs are now appearing that will probably give the NAP Magnavox-Philco-Sylvania consortium a somewhat larger share of the market place.

In the block diagram, you see the double-humped i-f response, which includes both the 41.25 MHz audio and 45.75 MHz video carriers, with chroma subcarrier at 42.17 MHz indicated. With a controlled amount of video subcarrier and appropriate filtering, both audio and video carriers may still be processed to derive 4.5 MHz intercarrier sound without the undesirable sound effects

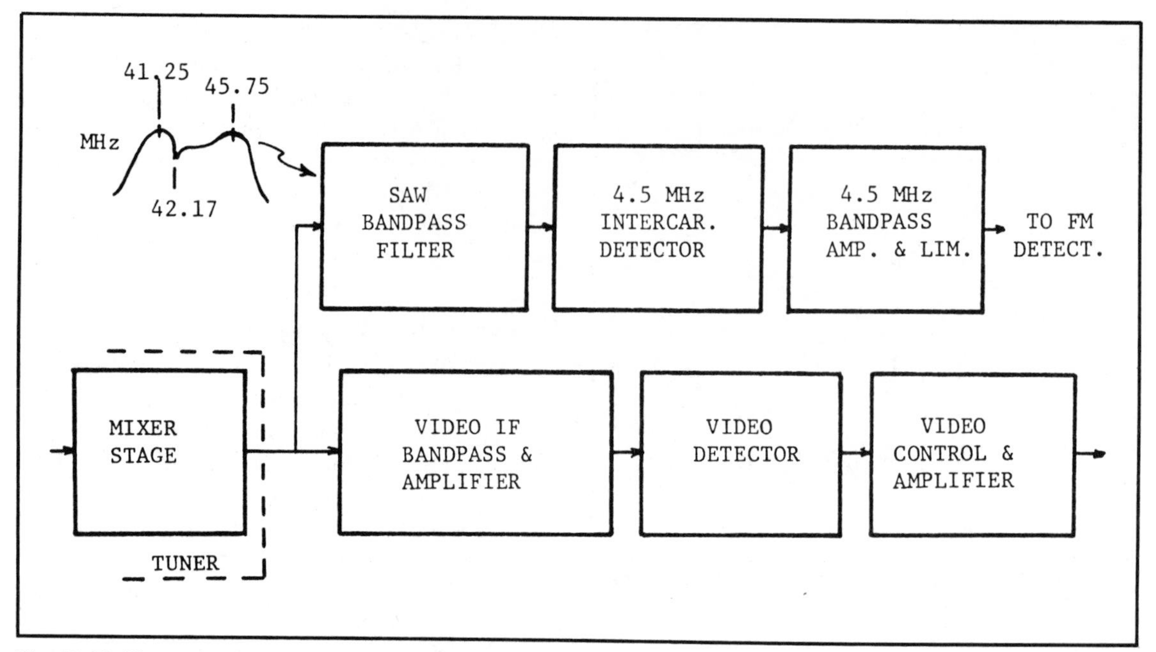

Fig. 12-18. Magnavox Quasi-parallel sound system—a simplified block diagram (courtesy NAP Consumer Electronics).

that will usually arise with standard aural detection somewhere around the 3rd i-f in most conventional color receivers. So far, this seems to be the most advantageous BTSC-dbx multichannel method of processing and, when properly implemented, does produce very good multichannel sound with little or no detectable interference. Whether most of the industry will adopt this system remains questionable at this time, and it will probably be a matter of several years before trial and error, personal preferences, and cost cutters find the most expedient way. Even then, some may continue to try and detect/demodulate sound the conventional way, but just add wideband circuits thereafter to handle the 50 kHz expanded modulation and the 5H SAP carrier. As we've already indicated, multichannel sound still has some settling to do before several final systems are fully selected— it may take a while.

In the 25C5 chassis Magnavox multichannel sound system (Fig. 12- 19) that follows, we won't repeat all the intricate details but simply give an overview of what's going on throughout the decoder.

With the 4.5 MHz intercarrier already set up and available for processing, sound is first filtered and then applied to dual limiting amplifiers where L + R is frequency demodulated and L − R is amplified, along with the 15,734 Hz pilot carrier for the multiplex stereo decoder. Should the signal have a 78 kHz carrier instead, then it is routed to the IC2 FM SAP carrier decoder, which is switched to accept this information, and then FM is quadrature detected and ready for the quad-input amplifier and eventual dbx expansion in IC201.

On the other hand, if doubling the horizontal scan frequency pilot produces a 31 kHz carrier, then IC1 knows this is an L − R stereo signal and begins to operate accordingly. At the same time, SAP is trapped by series resonant circuit C7 and L1, so that none of this carrier will contaminate the 50 kHz deviated stereo.

Considering the various dividers, we would expect the VCO frequency to be set at 62.963 kHz, divided by 2 to 31.5 kHz, and one more time to 15.7 kHz. These dividers would then establish both pilot and stereo frequencies, permit pilot sensing, energize the stereo indicator, flip the stereo/pilot switch, and begin the multiplex decoding for L − R.

As you can see, L + R and L − R outputs appear at the IC1 outputs, entering quad input amplifier IC6 for general amplification and passage to the IC3 analog multiplexer/demultiplexer from L − R and IC7, the quad input amplifier from L + R. L + R and SAP are then also routed to the multiplexer/demux IC, but only L − R is further processed in the IC201 noise limiter and expander, which is the dbx portion of this large decoder. Mono and stereo signals are then multiplexed and delivered to amplifier and speaker outputs as left and right stereo information.

Of course, when stereo is not received, only L + R is processed without all the attendant detection and multiplexing actions and is just issued as monophonic audio through *both* the right and left channels.

In IC201, note the rms and current-controlled amplifiers (CCA) as explained earlier in the chapter by the Japanese IC developers. All this should have given you a fairly worthwhile working knowledge of multichannel sound. There will be a great deal more in the way of engineering development as time goes by—but at least the foregoing is a good start, and it all works fairly satisfactorily.

THE BROADCAST END

With four or five manufacturers making promises, but most delivering little if any equipment as of mid-1985, we were able to turn to Modulation Sciences and its Vice President of Engineering, Eric Small for highly usable and well-thought-out information. As you will see in the photographs, Modulation Sciences does have both stereo and SAP units ready for shipment and accepts orders on a 60-day ARO basis. Furthermore, the dbx companding boards are of its own design accompanied by built-in monitoring facilities, eliminating an additional piece of equipment. They *obviously* deserve a plug. A front/back photo of the TSG™ television stereo generator is shown in Fig. 12-20. Note metered modulation and audio sections, along with a LED readout for attenuation and the usual function lamps for all the various conditions. On the back are secure-type terminal strips and BNC connectors for all audio controls and video in/outs plus composite outputs. A 5 A fuse protects the entire assembly.

A functional outline of the generator is shown in Fig. 12-21. Sum and difference signals are first amplified and then level-set for their matrix, which has either L/R or sum/difference switch positions. If L and R outputs, these signals are low-pass filtered and then broadbanded by voltage controlled amplifiers. Broadband control with feedback and high frequency control accept inputs from the stereo generator, which also has balanced composite and unbalanced composite outputs, in addition to a half-dozen metered signals ready for measurement. Both channels are further amplified, especially for high frequencies, can be gated for control, and continue through additional loudness voltage controlled amplifiers to the L + R matrix and its outputs. There's also a loudness control voltage, which affects both of the loudness amplifiers.

On the left, for L + R there is the required 75 μsec preemphasis, phase compensation and a low-pass filter. For L − R, there's also a preemphasis bypass block in shunt with the dbx encoder and another low-pass filter. With composite video insuring positive sync lock for the alarm, loudness control, meter, and master clock, all these signals go into the stereo generator for further processing. Of course, there's a stereo mode input as well as SAP and the PRO professional channel at 6.5H, if that's desired. However, two other pieces of equipment are needed to handle SAP or the low-level data or voice channel. The power supply furnishes ± 15 V, ± 20 V, and a separate 10 V output.

A control card on the right of the diagram shows all the indicator and metered signals available as well as the two controllers and the front panel loudness control step and reset. Measured are peak deviation of the L + R pilot, L − R, SAP and PRO—all available from the front panel and also remotely. And initial setup and performance verification need only a wideband aural demodulator, a spectrum analyzer and oscilloscope. A phase meter and phase alarm offer continuous time relation indications between left and right channels, and the alarm will sound if L/R reversal occurs.

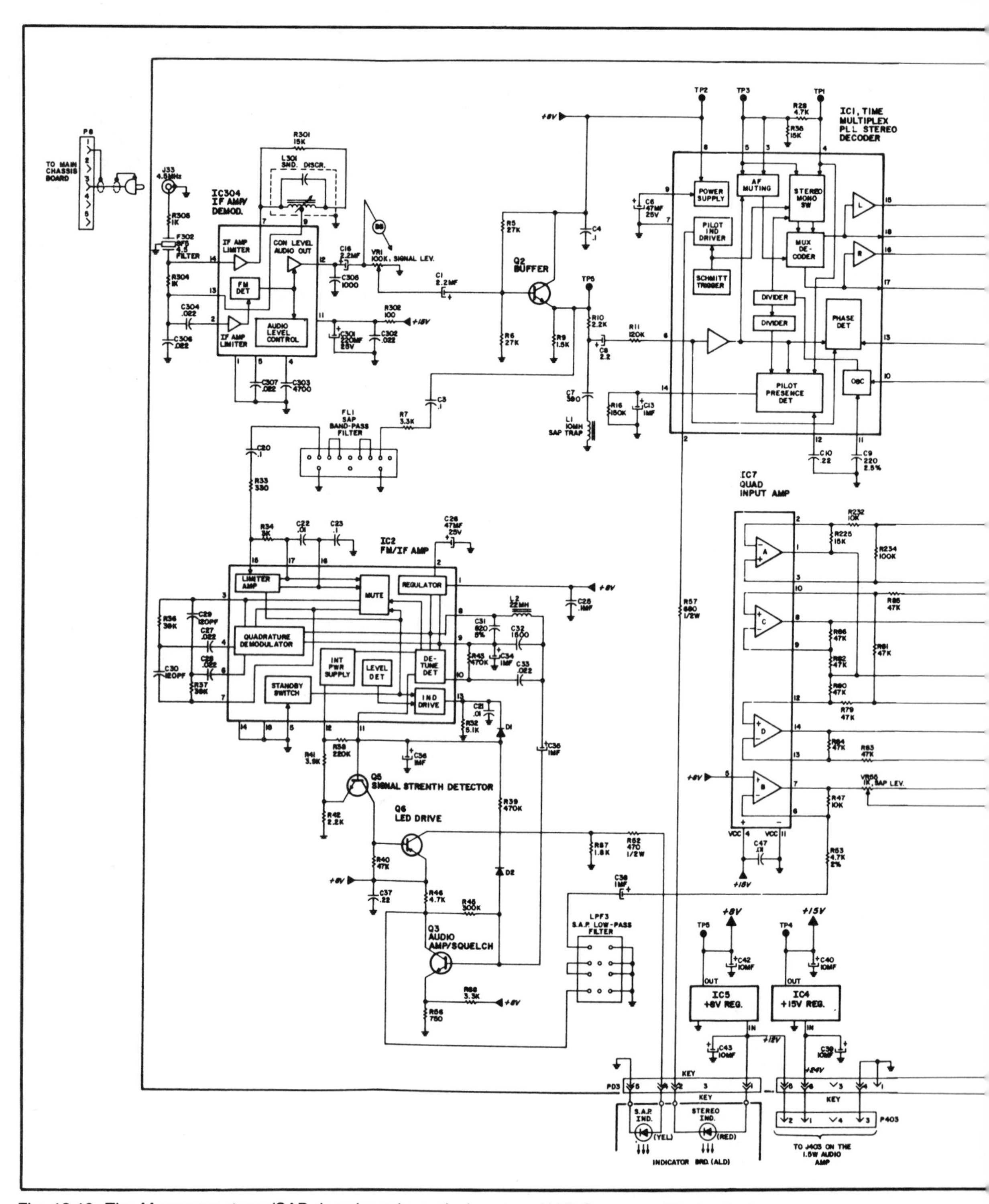

Fig. 12-19. The Magnavox stereo/SAP decoder schematic (courtesy NAP Consumer Electronics).

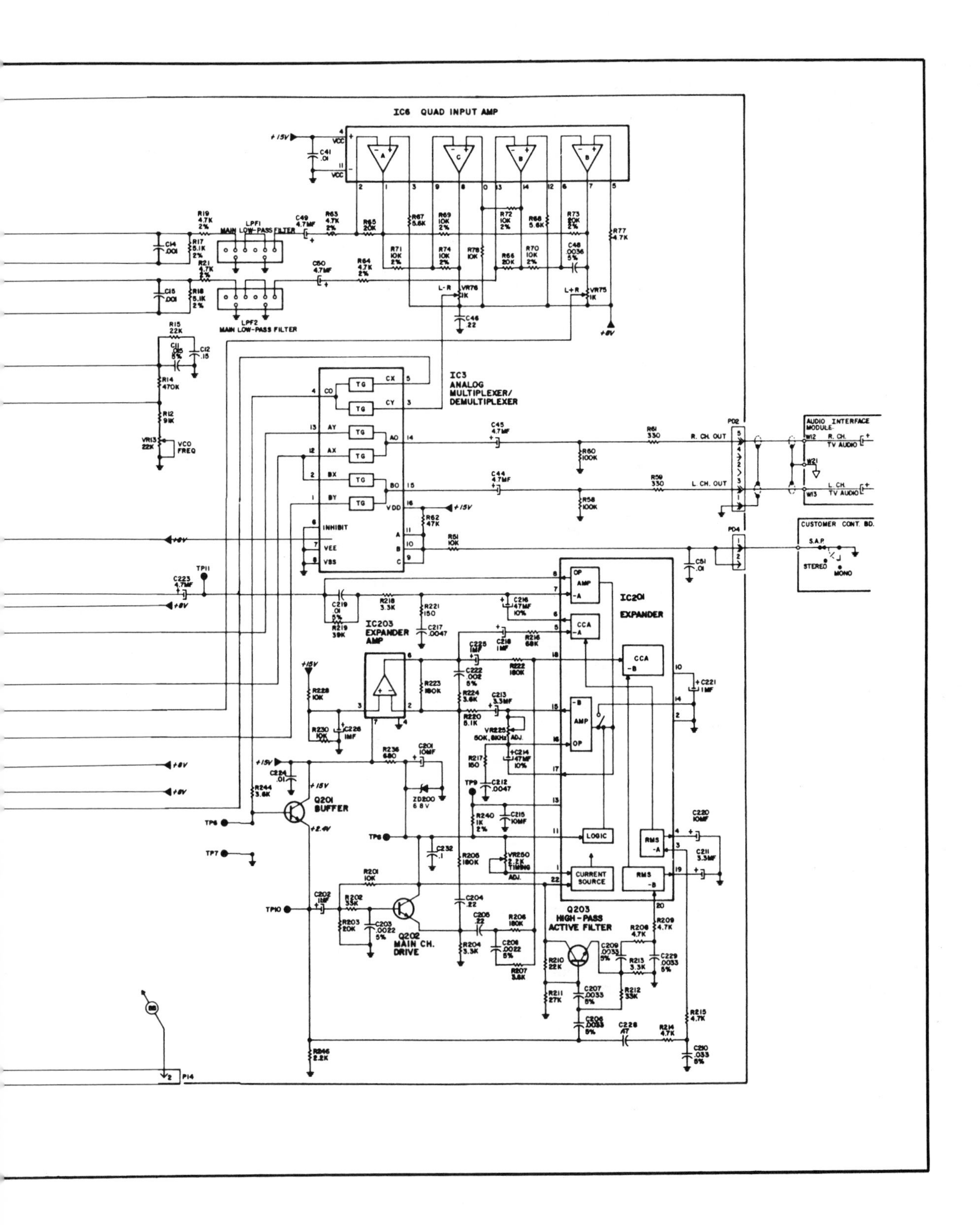
IC6 QUAD INPUT AMP
LPF1 MAIN LOW-PASS FILTER
LPF2 MAIN LOW-PASS FILTER
IC3 ANALOG MULTIPLEXER/ DEMULTIPLEXER
IC203 EXPANDER AMP
IC201 EXPANDER
Q201 BUFFER
Q202 MAIN CH. DRIVE
Q203 HIGH-PASS ACTIVE FILTER
AUDIO INTERFACE MODULE
CUSTOMER CONT. BD.
R. CH. OUT
L. CH. OUT
S.A.P.
STEREO
MONO

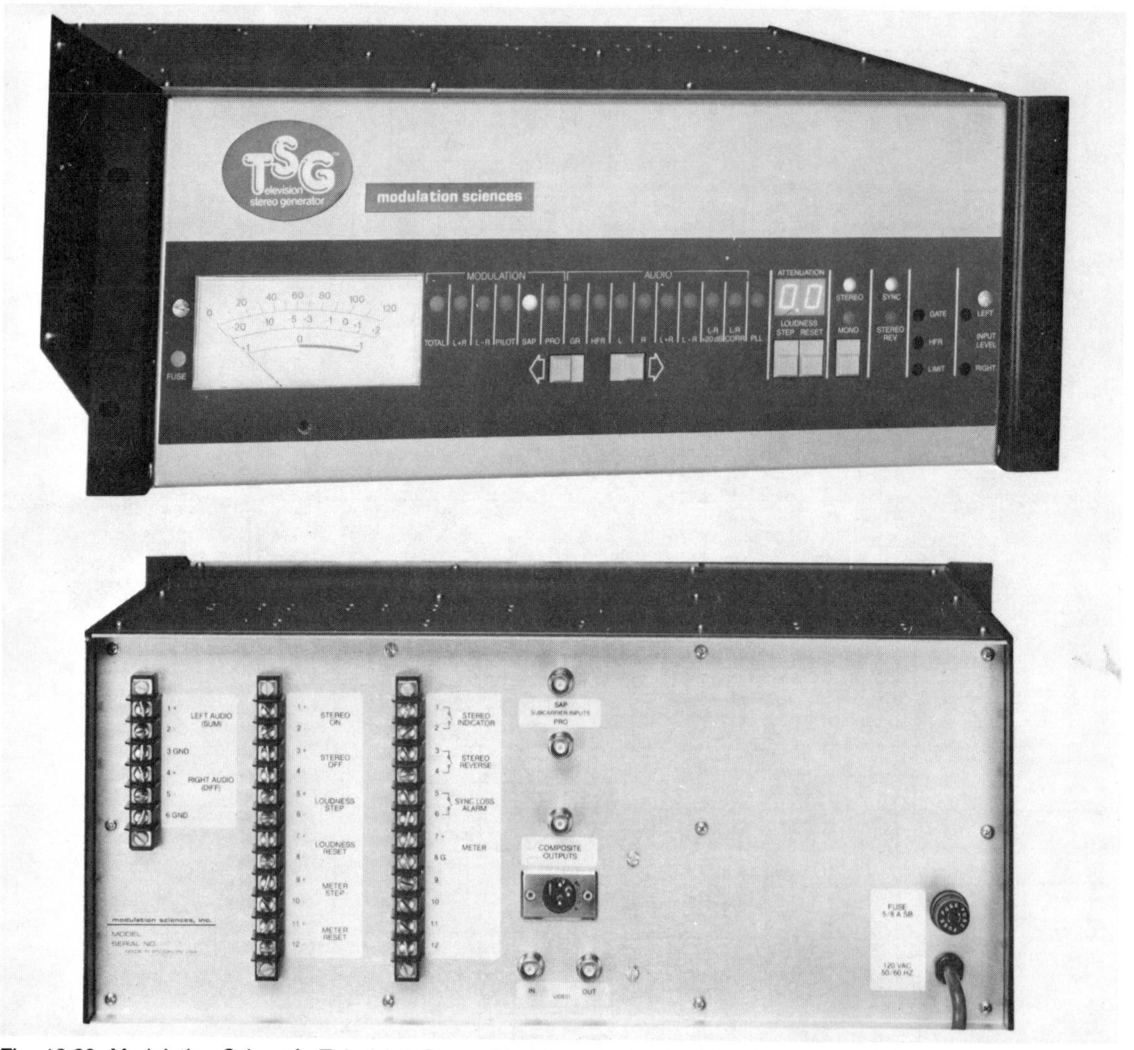

Fig. 12-20. Modulation Science's Television Stereo Sound Generator (courtesy Modulation Sciences).

While some TV broadcasters would like to transmit sum and difference stereo signals, Modulation Sciences recommends that stereo should be generated from right and left inputs, and this is why there are dual inputs to the TSG generator. With composite video input, separate horizontal drive becomes unnecessary, and the exciter sync lock then derives from the overall video signal which includes standard sync pulses. Should composite video drive be lost, the unit automatically reverts to monaural and on-board jumpers select either right or left inputs for mono. Modulation Sciences is also licensed by dbx to produce its own encoder boards, ensuring maximum broadcast performance for companding and noise reduction. In addition, the equipment has built-in full audio processing, eliminating the need for stereo compressors and limiters prior to the generator. It includes noise gating, loudness control and platform gain.

TV Sidekicks I and II are available, too, as separate equipment (See Fig. 12-22) to accommodate the SAP and PRO services, if required. A block diagram of the system contained in the TSCA-189 is illustrated in Fig. 12-23. This unit also has a sync

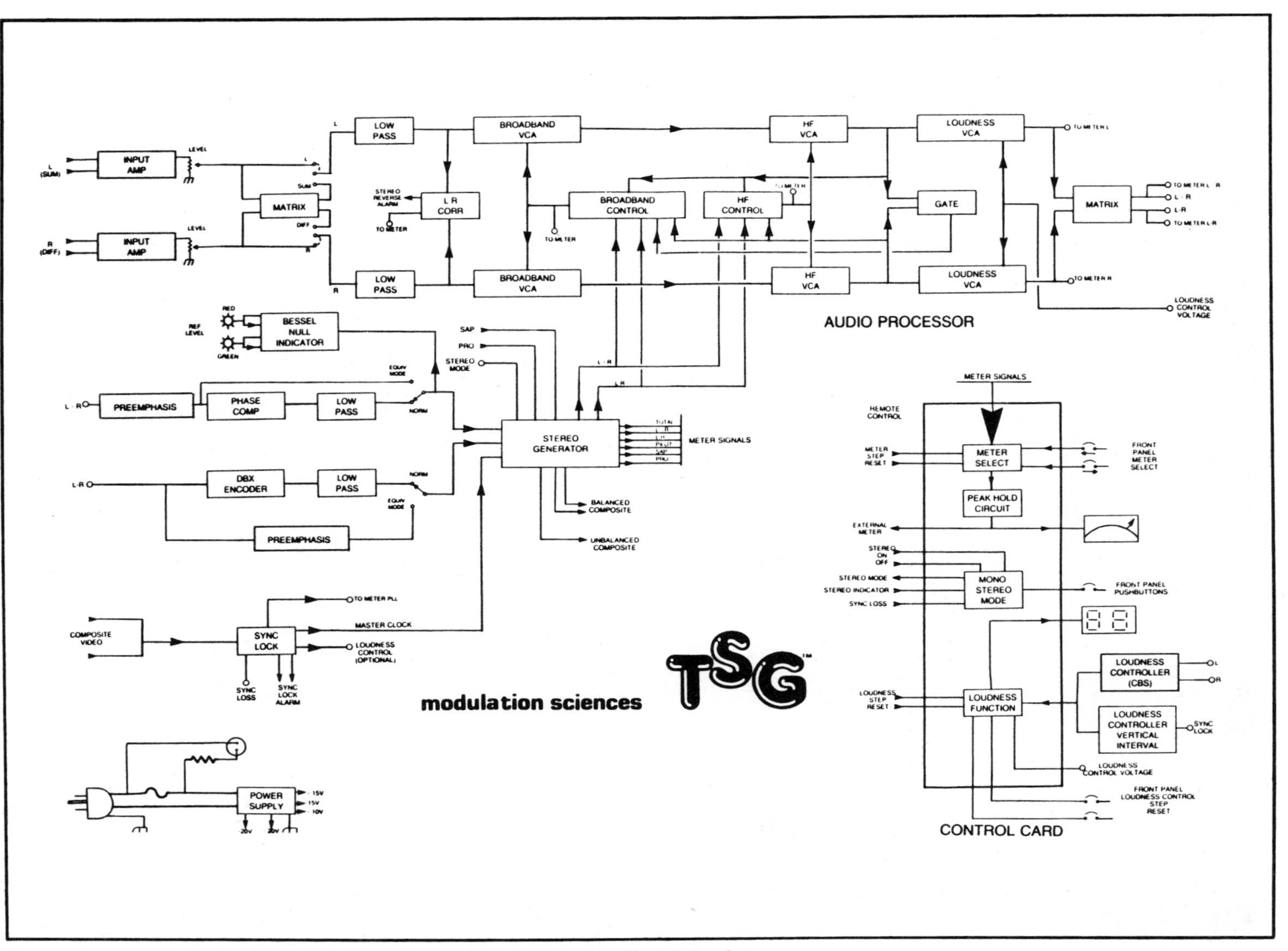

Fig. 12-21. Stereo exciter/generator for multichannel TV sound (courtesy Modulation Sciences).

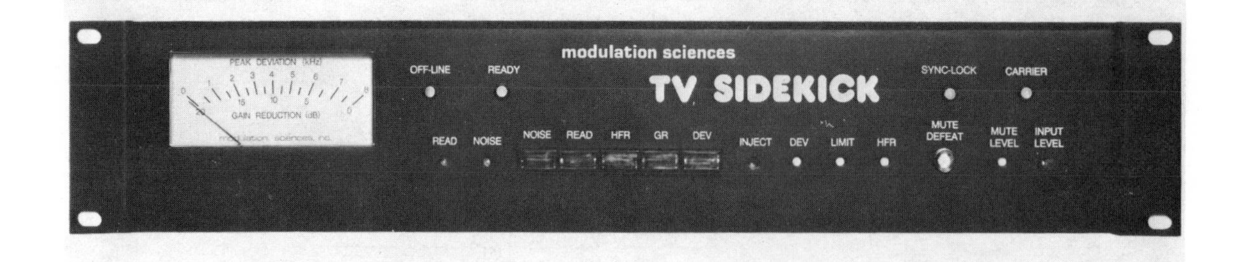

Fig. 12-22. Photo of Sidekicks I or II accommodate either SAP or PRO channels, respectively (courtesy Modulation Sciences).

stripper, dbx encoder, frequency and threshold control, mute control, compressor, noise generator and broadband limiter. At the top of the diagram there's a meter readout for the compressor, high- frequency limiter, encoder, and synchronous AM. Full specificatons may be obtained from Modulation Sciences, Inc., 115 Myrtle Ave., Brooklyn, NY 11201.

Will BTSC-dbx multichannel sound stereo/SAP work with any transmitter? Unfortunately not! V.P. Eric Small advises that only a direct FM type aural exciter will work with stereo TV; serrasoid or indirect exciters will not. You can tell the difference if there are multipliers in the oscillator chain. But if the oscillator frequency is that of the carrier, then you're safe. As Mr. Small also observes, there are probably no vacuum tubes in any direct FM exciter. If you have this problem, however, all is not lost. Several companies are able to supply retrofit direct FM exciters for most television transmitters.

Mr. Small says broadcasters must also beware of intercarrier phase modulation (ICPM) that takes place when "the visual is amplitude modulated." Any stray video is detected by the aural

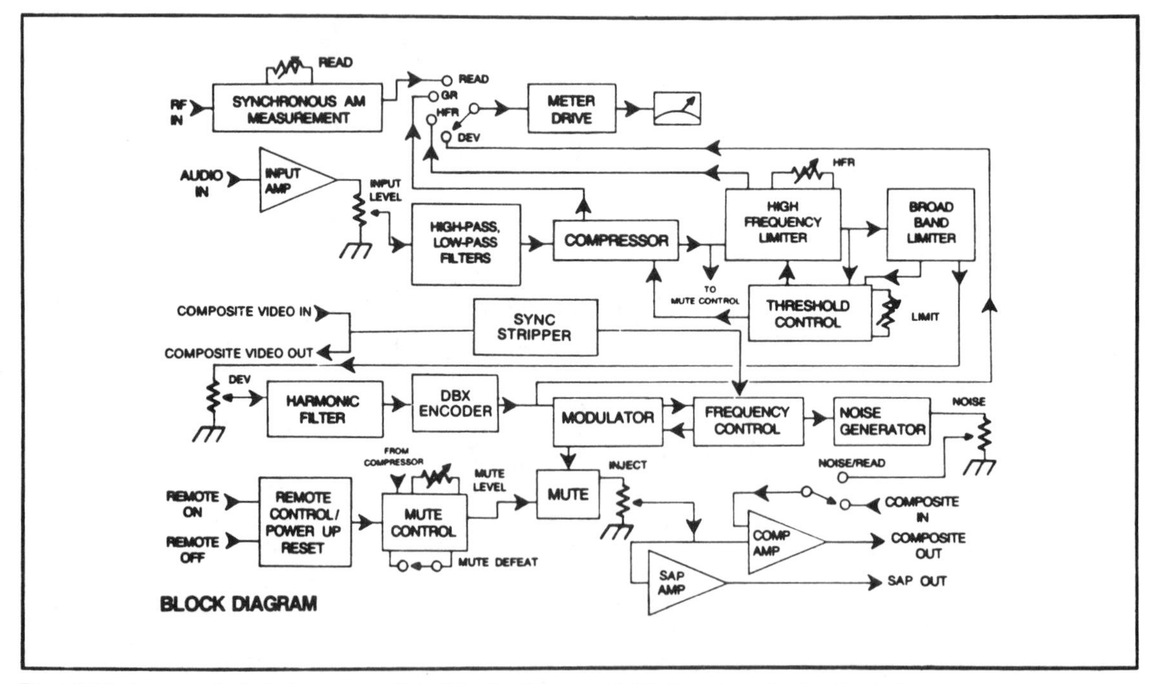

Fig. 12-23. A general block diagram of Sidekicks for SAP and PRO (courtesy Modulation Sciences).

demodulator and causes interference. Notch diplexers, apparently, don't significantly degrade stereo signal separation, and wideband FM can be obtained by a few simple modifications in many instances.

That should give you a reasonable idea of what constitutes a new multichannel sound stereo generator or exciter. It won't be long before the entire country will be enjoying a picture-sound musical programs as never before. This is truly an exciting medium and there are many, many new and worthwhile products to go with it.

Chapter 13

High Voltage

YOU MIGHT WONDER WHY A COMPLETELY separate chapter has been devoted to this topic? Obviously it's important, but besides that, high-voltage supplies for better-built cathode-ray tubes and their all-significant regulation has become of prime concern as baseband and RGB inputs with bandwidths of up to 15 MHz are rapidly becoming part of the better television receivers. At these high frequencies, unless regulation is virtually precise, horizontal raster side pull will certainly show and jeopardize a receiver's otherwise good reputation. Therefore, also centering controls and even hold controls on some of the very new TV/monitor receivers with outstanding specifications have been included, just in case.

How high is high voltage? In the 19- through 25-inch garden variety of contemporary receivers, it can vary between 25 and 30 kV at approximately 1 to 1.5 mA of accompanying current. Used for both CRT accelerating and divided-down focus potential(s), high voltage simply "draws" electrons from the CRT's cathodes, through its various grids and columinating lenses so that they strike the RGB-coated phosphor screen and produce various light color shadings, depending on varying cathode potentials. Federal regulations also require a high-voltage safety device that will abort voltage/current runaway conditions that might result in a catastrophic fire. Therefore, all modern receivers should have such a provision in their design, and most, if not all, certainly do. So from the standpoints of HV generation, regulation, and safety, we will try and give you some worthwhile working information, using primarily *real* receivers and their various design and executing schemes. We'll begin with the basics, as usual, and work up for better understandings of prior and existing problems and their solutions.

HV BASICS

In the beginning and in the ending, high voltage consists of only five elements: a tube or transistor output (driver), a damper diode, one high-voltage (flyback) transformer with multiple windings, one or more rectifiers, and filters for dc smoothing (Fig. 13-1). Actually, you may look at

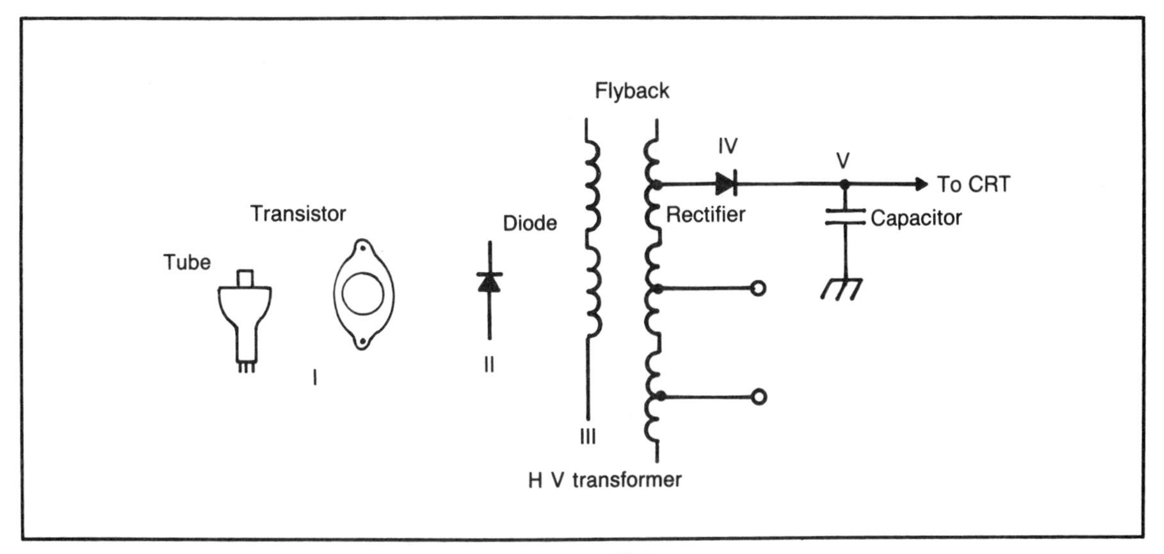

Fig. 13-1. The five elements of any high-voltage system are still the same.

high-voltage simply as another dc power supply, even though it has considerably greater potential but much less current than its lower-valued brothers and sisters. Regardless of the TV system in which it's used, high voltage does precisely the same thing with identical results: electron acceleration and raster illumination. During the days of vacuum tubes, "no high voltage" was one of the *most* common complaints heard. Today, with current pulses and sturdy transistors carrying the load, the old hot-bulb, finger-burning horizontal output tube is hardly a memory—although frequent replacements netted the repair people many an easy buck and an ultra-fast service call. Pull out the old tube, put in a new one and the picture sprang to life as if by magic. Then, a ready bill for $25 or $30 and you were on your way to the next one, picking up over $20 in the process. With transistors, unfortunately, you can't see electrode plates glowing red or nice white holes in a blown tube, so it's usually back to the shop for examination and repairs. There's more to be said in the final chapter on troubleshooting.

What many in the television business still don't understand is how this rather simple system works for just about any and all receivers and why. We'll try and explain in very basic terms using a good example to illustrate.

Initial Tube Types

Just for old time's sake and a little nostalgic history, we thought that a couple of waveforms and a short tube-type description might help to illustrate tube operations and problems as they existed yesterday.

In Fig. 13-2 you see an old-fashioned horizontal driver (really a discharge tube) horizontal output, flyback transformer, etc. that's the basis for this type of early system; that we might add, is *totally* unregulated. The driver receives peaked (actually differentiated) pulses that are shaped by the RC network between its plate and the heavily biased grid of the horizontal output. C_s normally charges to B+ via R_L, but when the driver conducts it offers a low impedance path to its cathode potential and C_s partially discharges before the tube once more cuts off. Because R_s and C_s form a long time constant, C_s then charges slowly forming the sawtooth from A to B. As the tube once again conducts, plate voltage drops from B to C, but rises immediately again to the potential remaining between C and D. C_s then begins its charging cycle anew. Because of heavy grid bias, the horizontal output tube will not conduct until it is above the cut off dotted line, which is always positive. When this occurs, horizontal output plate current flows in the primary coils of the flyback (high voltage) transformer creating a magnetic field about them, and inducing voltages in the secondary. When the output tube cuts off, the field collapses and a counter current is formed, resonating these coils at approximately 71 kHz, a half cycle of which amounts to about 7 microseconds and executes horizontal retrace so the process can begin all over again. Flyback (now you understand the term) oscillations would continue for several very undesirable and highly visible cycles (seem as narrow vertical lines on the CRT) if this period was not damped. Therefore diode "D" was added between the horizontal output and primary of the transformer to remove such oscillations whenever they occurred.

With flyback current reversal, a very considerable pulse of voltage is generated that, when rectified and aquadag-filtered by an equivalent 500 μF capacitor on the coating of the picture tube, becomes high voltage, focus voltage, and usually B-boost. Current through the secondary (in this instance) also drives the horizontal deflection coils.

In essence, removing a few trapezoidal drive voltages and substituting them with square waves serves the transistor configuration just as well. Components are slightly different, but final actions are all the same. Flyback transformer time (current buildup and collapse) becomes a horizontal blanking period permitting both trace cutoff and startup as well as allowing the introduction of a horizontal sync pulse and color burst on its back porch. In color transmitters, of course, the horizontal blanking period has been fixed at 11.1 μsec to easily accommodate both burst and pulse. Receivers often allow as much as 12 μsec, cutting off just a little of the horizontal sweep.

That's just about as basic as one can become in explaining what goes on in generating high voltage. So you probably don't need further specifics—certainly not in vacuum tubes. However, there is some theory of operation here and there which might be of benefit to those in design and an aid also to serious readers or students who would like to master the intricacies of high-voltage development and receiver sweeps. In earlier days it helped your author and should do the same for you since the principle transcends time and remains applicable yesterday or today, and undoubtedly tomorrow. The horizontal output continues as a current device and must be driven hard to produce the few or many (additional dc sources) outputs required. It is, indeed, an interesting circuit as you will see.

HV Semiconductors

With the blessed movement away from vacuum

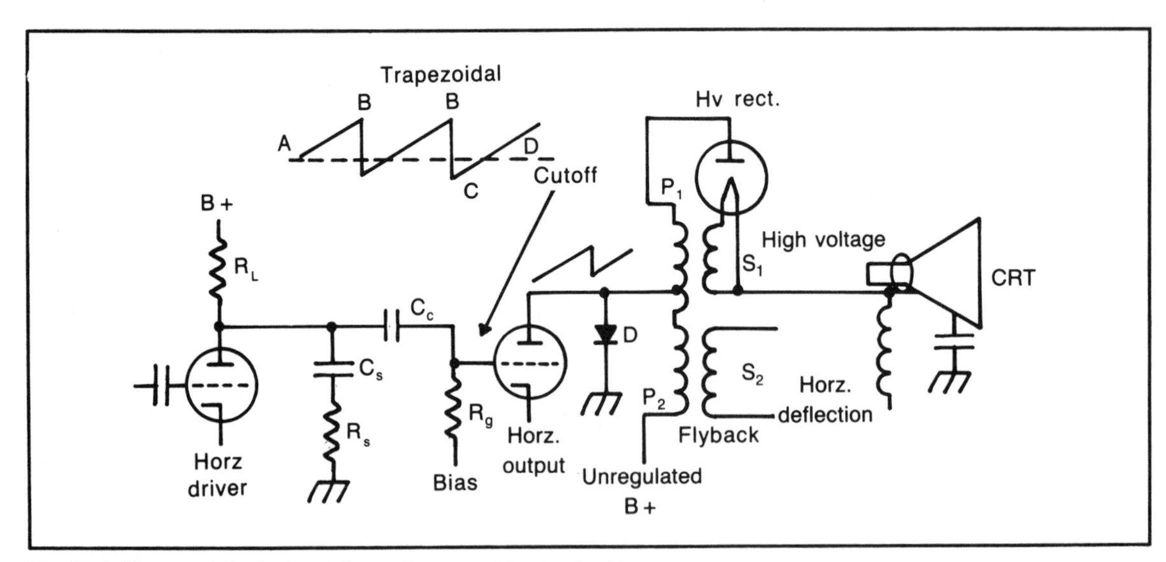

Fig. 13-2. Vacuum tube horizontal waveforms and basic circuit.

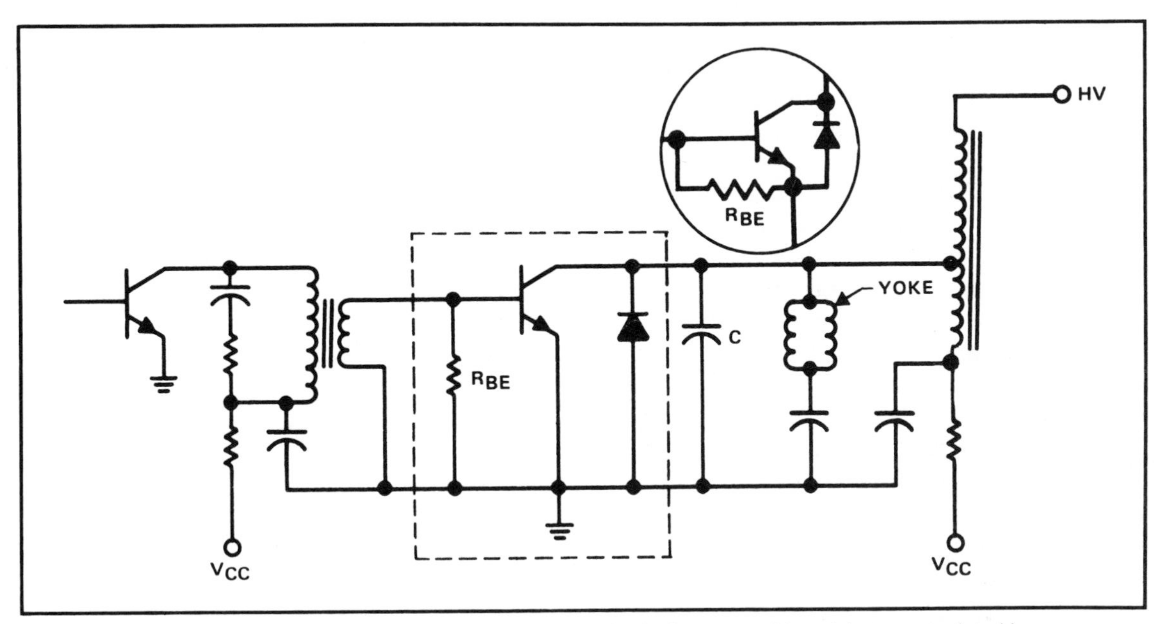

Fig. 13-3. Typical single semiconductor horizontal output with both damper and transistor encapsulated in same case.

tubes (anywhere), there are also constraints on horizontal output and rectifying devices that merit serious consideration even though most of us are not in engineering design. However, if we are aware of the more serious shortcomings, it will be much easier to examine their successes and failures and compensate for eventual problems that are sure to occur with age, heat, and the unwelcome presence of warm air moisture known to most of us as *humidity*. Like automobiles, it isn't the cold weather that ages the faithful four-wheeled steed, it's water suspended in air that does the job to transistors, integrated circuits, and the various power packs. Humidity, with its attendant oxidization, rust, and corrosion is a certified killer of most metals and their derivatives. Unfortunately, we can't paint metal transistors with zinc chromate nor seal plastic ICs completely so they will never leak.

Coping with transient problems in the damper-output-rectifier portion of the receiver has now become a manageable problem due to advances in manufacture, better thermal junction characteristics, computer design, and solid specifications allowing devices of this category considerable leeway in handling loads and other stresses encountered. Nonetheless, all high-voltage components in the final design need to be potted and molded, then placed in some sort of box to be voltage and thermally cycled to observe normal operating and any marginal runaway conditions that in any "first cut" are bound to occur. Should there be additional heat factors from other receiver circuits such as video and vertical outputs, these will have to be accounted for as well. Advanced junction temperature means power loss in addition to ambient temperatures surrounding the transformer winding, its core, potting, and attached components.

Since about 1976, the damper diode has often been included in single transistor-driven horizontal deflection circuits within the same case as the actual power transistor. A typical circuit appears in Fig. 13-3 that is still in use today. It shows the usual transistor driver (on the left) followed by a current coupling and amplification transformer, a base resistor, the horizontal output transistor and accompanying diode, a filter capacitor, the deflection yoke and the high-voltage flyback transformer. One of the first to do this was Texas Instruments in its series TIP69 through TIP310 with various volt/ampere ratings ranging from 1200 V at 2.5 A to 1400 V at 7.5 A.

When the horizontal output ceases to conduct, the induced flyback energy charges capacitor C, resulting in a pulse of high voltage for use by the HV rectifiers and CRT focus. As the circuit attempts to ring because of the transformer primary inductance, and falling voltage passes through zero on its way negative, the damper diode is forward biased. Conduction follows with current rising instantaneously to maximum value and will continue until most of the flyback energy is dissipated, all the while pulling down the collector of the output transistor. On the next drive cycle, the output transistor is once again forward biased, the damper is back biased and the entire procedure begins its cycle of cutoff and conduction, at the same time driving the deflection yoke, with its S-correcting capacitor, so that complete horizontal deflection and flyback will take place. The damper diode has supplied up to 40 percent of the required deflection energy, with the transistor and other components delivering the rest.

Texas Instruments maintains that circuit performance of single transistor-diode combinations is virtually identical to that using discrete components. Drive requirements are the same and on impedance amounts to only one or two ohms, with base-emitter resistance having little effect on output transistor base drive. Thermal impedance and safe operating ratings of the same as those for discretes, and the increased power dissipation is said to be "well within" device and package limitations, even though inclusion of the damper increases power dissipation some 45 percent. Apparently, most of these combined damper-transistor packs will easily retrofit most receivers still having external dampers with, possibly, less spurious radiation. For those making the change, just be sure that power dissipation and drive characteristics either match or exceed the original. An oscilloscope check near the flyback—which we'll describe later— might be good insurance to see that all's well. Remember, that horizontal deflection requires the greatest amount of power in the television receiver and must be fully capable of handling requirements under all sorts of ambient conditions. This power dissipation results from saturation losses, base losses, and fall-time losses. Saturation and base losses take place during the 52.4 microseconds of horizontal scan, while the fall time and flyback of 11.1 microseconds are affected by high voltages and currents at turnoff. The output transistor also has a positive temperature coefficient and current multiplies as fall time increases. In the end, if the semiconductor isn't within design specifications, you may well experience thermal runaway as the transistor turns a cherry red and shorts or opens and conveniently dies. Naturally, the basic concept is to design a circuit that is essentially self-stabilizing, with excellent collector regulation, which will also be extended to any boost, HV, and focus voltages, as well. In newer receivers—especially the TV/monitor series—you'll see this objective carried out. Meanwhile, let's look at another variety of semiconductor as the horizontal output in a somewhat different application.

The silicon-controlled rectifier (SCR) is coming into its own as never before among video consumer products, principally in low-voltage power supplies, but has had its moments in high voltage also. So let's take just a short review of a circuit that RCA Soverville designed for 90° deflection in-line color picture tube receivers in the not-too-dim past.

As shown in Fig. 13-4, this is a reactor-commutating variety that uses both SCRs and diodes, along with inductors to supply energy for both high voltage as well as a pincushion transformer, B-boost and +30 V for lesser signal stages. Following in the footsteps of its larger (and older) forerunners, this circuit is very, very different from the simpler single- and dual-stage transistorized equivalents, which were pretty difficult to service until their peculiarities were thoroughly understood.

The main idea behind this design is to eliminate any power transformer and regulate high voltage and any extra small signal voltages by using a reactive regulator to store energy in a commutating capacitor. In this instance, the circuit also operates from a +120 V transistor regulator circuit for more stable results.

The trace resonant circuit consists of SCR1/diode 1, the LY yoke and the two 1.5 and 1 μF capacitors in parallel, along with T2 and an LC

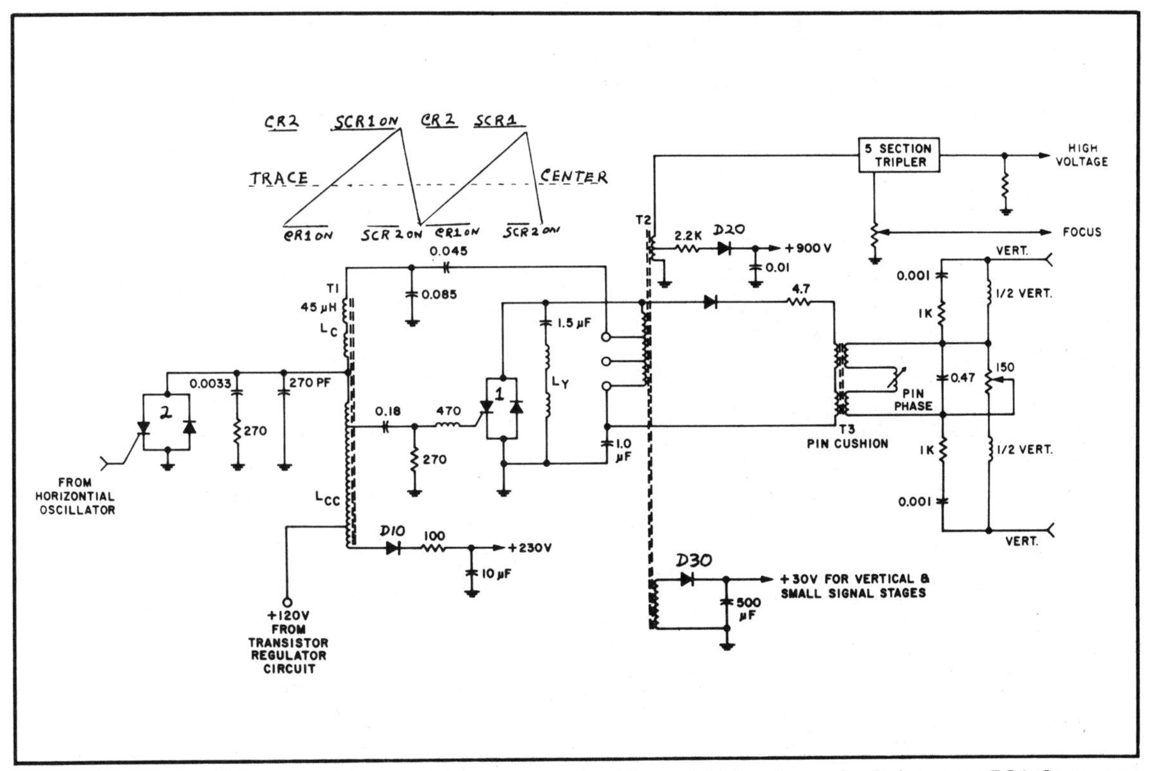

Fig. 13-4. A special SCR-commutating saturable reactor yoke drive and high-voltage circuit (courtesy RCA Consumer Products).

input to gate of SCR1. Retrace includes the same two capacitors, the yoke, the 0.00033 μF and T1. Input inductors L_c and L_{cc} are wound on the same cores so that stray fields from the two halves cancel. During charge and discharge times energy stored in the various capacitors, including the 0.045 and 0.085 μF units, discharges and then recharges as SCR/diode units 1 and 2 turn on and off with timing from the horizontal oscillator into unit 2. Their times of conduction are shown in the sawtooth current waveform representing the deflection yoke drive. Note the Trace Center, followed by the various designators for circuit timing. Saturable reaction, resulting in reverse response to lesser or larger current drains, provides additional voltage regulation. A pincushion transformer with guide diode in parallel with the deflection yoke and having both vertical and horizontal inputs offers top/side/bottom raster correction, and a five-section tripler at the top of T2 delivers the required high voltage. Diodes D10, D20, and D30 provide high/low B+ and 900 V B-boost for the receiver's other needs. Note they are all half-wave since pulsating dc is easily filtered at the horizontal frequency.

High-Voltage Multipliers

Following a single capacitor, often the aquadag coating on the picture tube, and a single rectifier came the various types of high-voltage triplers and quadruplers. Initially made of selenium, these units eventually melded into the silicon family and became quite reliable and often easy to substitute. They accepted a relatively low voltage from the flyback and by a number of interlocking diodes and capacitors tripled or quadrupled this voltage until it was sufficient for CRT acceleration. Naturally there are some losses along the way, so the output isn't precisely some exact multiple of the input, but something less.

Sometime ago, engineer M.E. Buechel of Varo Semiconductor made a thorough study of such quad-triplers and devised a pair of diagrams and basic equations which are of considerable interest. You see them in Fig. 13-5.

Voltage V1 arrives from the flyback transformer as positive-going and V2 (below the line) as negative-going. Diode D1 conducts as it is biased on by V1 charging C1 to some voltage equal to V1 plus dc. C_{n-1} then becomes a coupling capacitor between points A and B, as D_{n-1} conducts on the negative portion. On the next flyback pulse, D_n conducts charging C_n to a voltage value of V1 plus V2. The total high voltage then becomes

$$V_{out} = dc + \frac{(N+1)}{2} V1 + \frac{(N-1)}{2} V2$$

for odd numbers of rectifying diodes. If you were to use the terminal between D1 and D_{n-1} for focus, that would amount to

$$V_F = V1 + dc$$

In the event there are an even number of diodes in the multiplier, the equation becomes

$$V_{out} = N/2(V1 + V2)$$

and the focus voltage amounts to $V_F = V1 + V2$.

Apply such diagrams and simple equations to any and all high-voltage schemes such as these and you'll have the answers. The only real difference in the two configurations is that when D1 conducts in the even number multiplier, it charges C1 equal to V2 minus dc, with the total positive peak at B equal to V1 + V2. However, in the even diode circuit, the focus voltage amounts to an integral part of the output and will often track the output with load changes. When enclosing such high-voltage sources, the encapsulating material is all-important since it insulates diodes from one another and also any external components nearby which could result in high-voltage arcs called corona. Occasional arcing probably can be withstood, but any constant repetition will destroy the multiplier in short order.

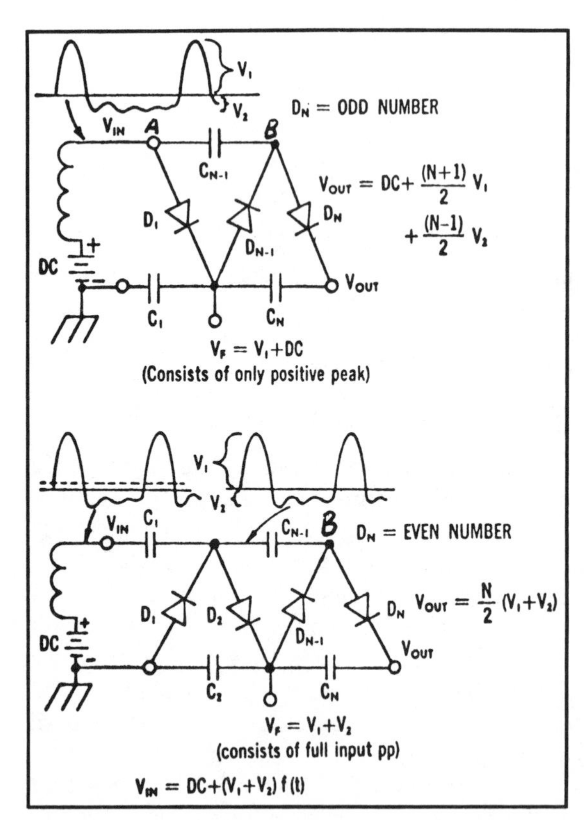

Fig. 13-5. Odd and even-numbered diodes in high-voltage multiplier circuits (courtesy Varo Semiconductor).

Radiation Checks

Any radiation must be held to a minimum of less than 100 μV between 0.45 and 25 MHz for rf; X-radiation specs. are 0.5 milliroentgens per hour; and 500 μA is maximum external chassis/cabinet leakage allowed by OSHA, UL, and CSA. Having all three figures should keep most receiver owners-repairers out of trouble if specifications are not exceeded.

While on the subject of radiation, we might add that this could become a problem with the initial batch of digital TV receivers having relatively high-voltage swings. For instance, you'll probably not be able to use rabbit ear antennas too successfully with some models. This has necessitated heavy metal shielding over much or all of the digitizers. The following, therefore, may be worthwhile to those making qualitative measurements on this problem.

Rf spray signals are usually more annoying than regular impulse noise which most agc circuits can handle reasonably well, especially those with noise limiters. But rf spray energy is synchronous with agc-gated horizontal sync pulses, actually defeating the normal advantage of keyed agc which works on sync tips only.

Usually, this rf is generated by fast turn ons and offs of the high-voltage rectifier at exactly the 15,734 Hz horizontal timing rate. So if this occurs, that particular energy will develop on tips of the horizontal sync pulses and be quite visible with an oscilloscope. Normally such interference appears on low VHF channels only since you're mainly picking up harmonics or modulation through the tuner and not through the i-f shielding.

Therefore, if you want to be sure of your problem, tune your receiver to channels 6 or 3 or some convenient nonbroadcast spot and put in a very weak synthesized carrier and monitor horizontal sync with an oscilloscope. Your tormentor just might show up *after* the video detector. By trying different rf filters, etc., this problem may be defeated. Otherwise, disconnect the rabbit ears and buy yourself a good outside antenna—those are about the only reasonable solutions other than another TV receiver. Interference problems every now and then are, indeed, tough! If you're looking for real trouble try an avid CBer with a linear within 100 yards.

Stacked Diodes

With flyback transformers becoming smaller and smaller because of advanced rectifier applications and better ways of building up high voltage, the stacked diode assembly within the transformer housing has become a considerable step forward in both reliability and space saving. No longer are there large separate flybacks with external triplers and quadruplers because industry has found a convenient way to package the entire assembly in a single piece, saving both money and precious chassis real estate (Fig 13-6).

The flyback-diode combination we've selected for your information came from by Varo Semiconductor of Garland, Texas. It's compact, well insulated, and has a focus tap supplying 25 percent of the high voltage at 0 mA. As each diode conducts

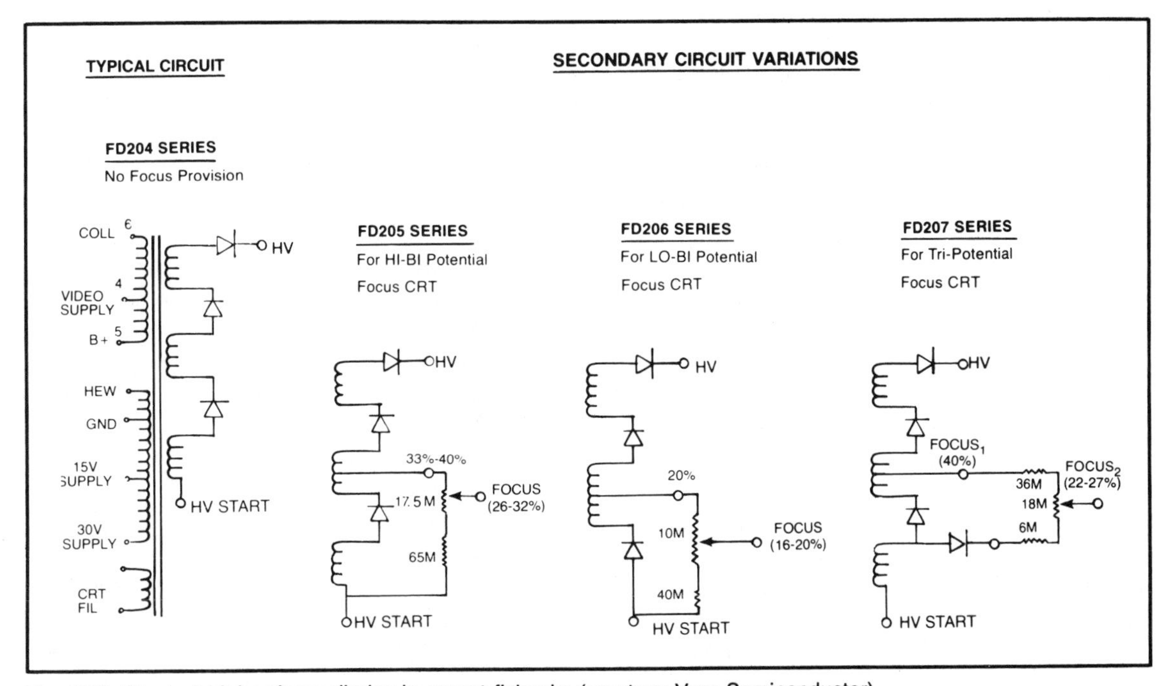

Fig. 13-6. Stacked high-voltage diodes in recent flybacks (courtesy Varo Semiconductor).

across connected series coils, final value output is reached and maintained by flyback current-reversal excitation and the CRT load that's always present with the receiver in operation. Prime problem with many of these all-in-one flybacks has been shorting diodes and an occasional condition of shorted turns within the windings. But this happens less and less with design and field experience, so that reliability here has also markedly increased and is an occasional rather than an aggravated one. Servicing, also, has improved considerably because repairs consist of a total replacement unit, and only a few solder connections or screws need be disturbed—the rest is usually encapsulated and can't be seen or manipulated.

Flyback Tuning

Some of this may well be passe compared with the latest developments, but historically it's been of considerable importance and certainly millions of television receivers still percolating in urban and suburban homes continue to use specially tuned flybacks.

Third harmonic tuning has been a mainstay of the industry because of its peak-voltage reduction in the primary (and output transistor collectors) of flybacks, but doesn't aid large screen color receiver HV regulation. Shunt regulators with feedback were popular with the old tube receivers, but potential X-radiation was always imminent and the semiconductor horizontal output acts much more like a switch than a current ramp driver.

The solution, at the time was to tune these transformers to the fifth harmonic since tests showed that the least primary voltage amplitude occurred at the fifth, *provided* overall harmonics were kept at minimum anyway. This latter results from using large values of leakage inductance and small distributed capacitances—or larger coil layers with fewer turns per layer. Such design also reduces power dissipation during collector-current fall times. The system then exhibits an harmonic content of about 3 percent of the untuned amplitude, a broader waveform top causing longer HV rectifier conduction along with lesser peak currents, and an impedance of about 1/2 that of a third harmonic tuned flyback at high beam current.

Today's high voltage circuits, as you will see shortly, use highly regulated B+ to maintain most of the HV regulation properties as well as some special transistorized circuits to help out. Good or bad, it isn't just a single transistor driver that bothers TV set makers today, it's overall performance of the entire system that's important, not just an isolated part. Nonetheless, modern flybacks are often tuned to the 7th, 9th, or 11th harmonics, depending on sweeps, retrace times, transients, bandwidths, and even cathode-electron drives. With monitors and television receivers of high quality beginning to predominate, practically anything is possible, depending on the specific application. Generalizations are quite difficult considering the huge number of varying requirements scattered throughout the video field today. So we'll have to take them item by item wherever and whenever they appear, and depending on how much information is available. Sometimes, believe it or not, the manufacturers really don't care to talk—especially if they've designed a winner.

MODERN HIGH VOLTAGE SYSTEMS

Now that the preliminaries are over, we can move on to up-to-date assemblies involving considerably more complex electronics designed for greater efficiency, higher voltages, better reliability, tighter X-radiation protection, and considerably better B+ regulation for both the receiver and HV. All this has come about because of high-quality receivers and monitors required for full luma-chroma bandwidth broadcast reception as well as the considerably more stringent requirements posed by computers, display devices, and even video games. For bandwidths up to 8 MHz video and over 100 kHz audio are no longer uncommon and these specifications literally demand one very good monitor/receiver to handle them. Further, if and when high definition television turns the corner, we are very likely to see even greater bandwidths and other receiver changes needed to receive and process it.

So as you consider the following explanations and view even more complex diagrams, television won't stop here by a long shot. There'll be considerably more upcoming even in the next several years. In colloquial language, the "back burner's burnin" to accommodate present and future needs now appearing on the market or are potentially ready. At the moment, we'll begin with Zenith's A-line sweep system and work from there.

Zenith's New Horizontal Sweep

Mounted in their 9-351 module (one of the rare TV manufacturers who still have such things), the new system will be installed in 13- and 19-inch receivers using an RCA-developed Coty tube and deflection yoke system (Fig. 13-7). The sweep receives operating B+ from a well-regulated series-pass power supply with no external adjustments. Regulation derives from a zener emitter reference diode in a second transistor which senses any change in output voltage and returns compensating current drive to the base of the Darlington-connected pass transistor (not illustrated). Regulation is said to be ± 1 V over the receiver's normal ac operating range.

Horizontal drive pulses enter the system through pin 2 and 3F3, reaching the base of pre-driver QX3202, and are amplified for the QX3206 horizontal driver and its secondary centertapped current drive transformer. Horizontal output transistor Q3028 and its built-in damper combination turn on and off at the usual 15,734 Hz horizontal rate, delivering current to terminal N of the flyback. As this process continues, stacked diodes in the primary generate 22 kV of high voltage as well as a focus supply in between. CRT filament voltage emanates from HOT terminals S and R, with dc bias supplied from the +26 V source that's filtered by the two shunt capacitors. There's also an automatic brightness limiter circuit that begins at HOT terminal 0, is filtered and dc assisted by the 12 Vdc supply and limits excess CRT current drive when high-voltage levels become excessive.

Like most other color receivers today, this one also has a shutdown circuit that kills high voltage if it becomes excessive. A 60-volt pulse is nominally received through diode CR3206, is rectified and filtered by the assorted capacitors and resistors, and arrives as dc at the cathode of zener CRX3204. Should this pulse at any time exceed breakdown for CRX3204, shutdown transistor QX3207 will conduct lowering the base back voltage on QX3204, resulting in conduction of this PNP transistor. When that occurs, diodes CR3209 and CR3211 also conduct, shunting the horizontal drive signal through RX3209, the base of QX3207 and its emitter to virtual ground. This promptly turns off high voltage, preventing runaway or undesirable excess.

This is probably as simple a circuit as you'll run across among the shutdown community since some of them are quite extensive and somewhat deliberately contrived, or so it seems. Some, also, are current sensitive as well and will activate on both over-voltage and excess-current conditions. We may find one of these as the explanations continue and more complex circuits are considered. If so, its operation will be dutifully investigated and explained. In any event, excess current through a limiting resistor would result in a voltage drop that would grow as more current flowed, turning on some transistor that would either disable the horizontal drive or shut down the output transistor as just described. Such a circuit isn't a big thing, but certainly will prevent undesirable high-voltage conditions that could result in catastrophic failures, including receiver fires. This is why the better receivers are usually bargains, along with their other more advanced features which always adds to better sound and good viewing. A poor television set is a sorry thing!

The Sylvania E30 Shutdown

Since high voltage generation is relatively straightforward in most instances, we won't cover it in this Sylvania chassis explanation, but devote complete attention to the shutdown circuit which generically, has, in the past, given both engineers and technicians trouble in design as well as troubleshooting. Here's the circuit (Fig. 13-8) and here's how it works.

It is designed to monitor the +220-volt source from the integrated flyback transformer (IFT) produced as a result of horizontal scan rectification and varies accordingly. Should this voltage rise beyond

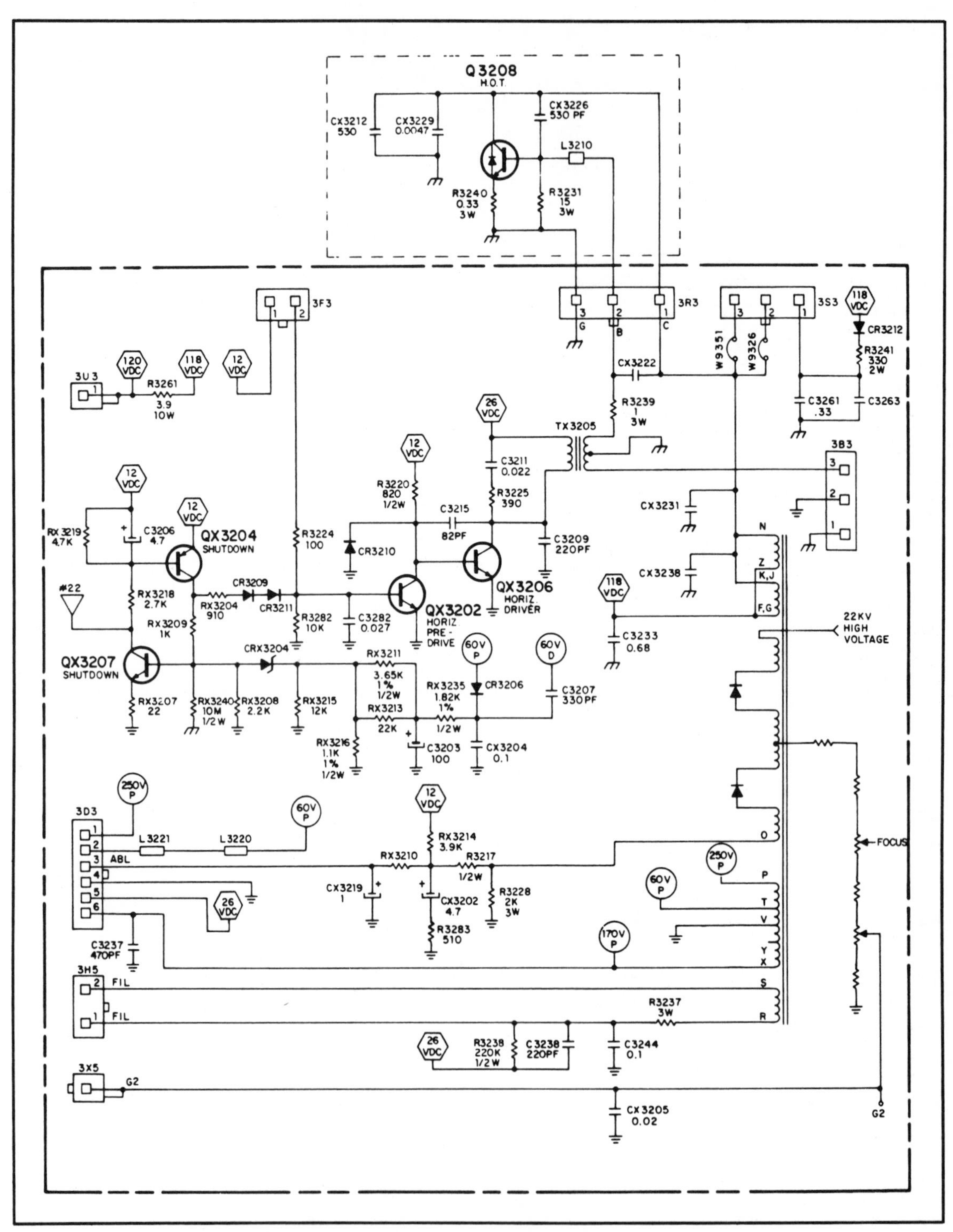

Fig. 13-7. Zenith special shutdown and high-voltage circuit (courtesy Zenith Electronics Corp.).

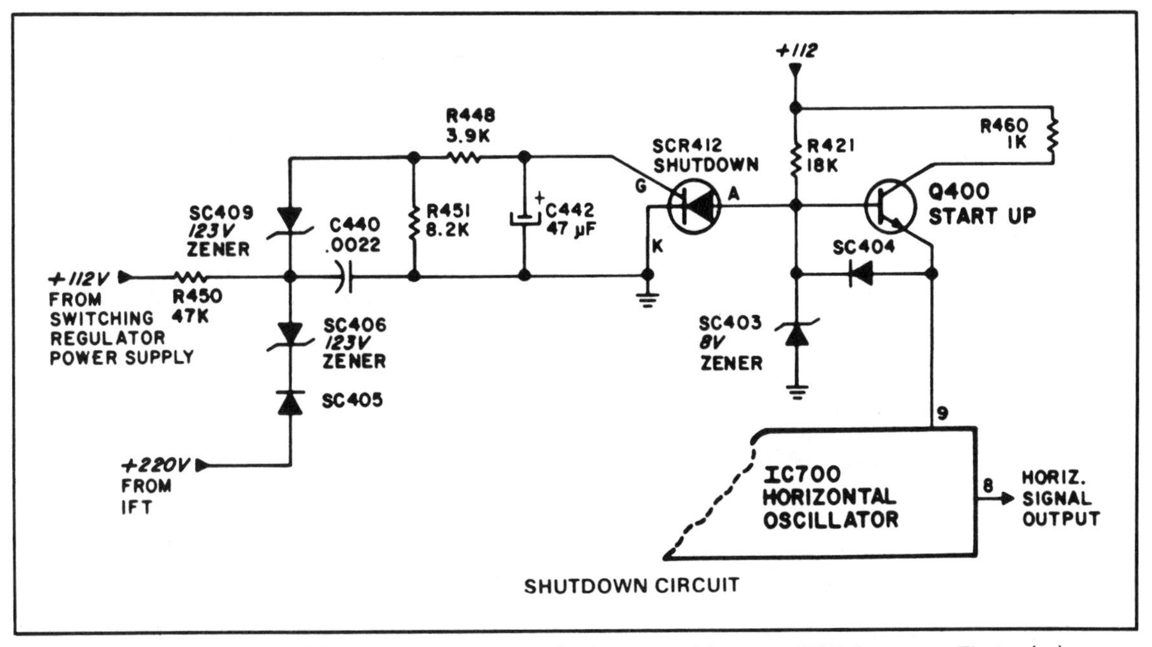

Fig. 13-8. Sylvania's effective SCR shutdown that is totally dc operated (courtesy NAP Consumer Electronics).

+246 volts, the horizontal oscillator is turned off. When either the +112 V or the +220 V sources cause their respective SC409 or SC406 zeners to avalanche, gate current reaches SCR412 which turns on, removing current from the base of start up transistor Q400, turning off the 8-volt zener and clamping the B+ input to the IC700 horizontal oscillator at a very low potential. With little or no operating voltage, of course, the horizontal oscillator collapses and high voltage is dutifully turned off until the faulty condition has been removed or repaired. Observe that this is a dc-operated circuit all the way, even to the gate turn-on firing. When SCR412 turns on, current flows freely between anode and cathode and will not stop until the receiver is turned off, removing the existing potential from its anode. After turn-on, any effective gate action ceases to exist in an SCR and this is why the receiver must be shut down before any attempt can be made to find the problem. You may also discover that a defective or tender SCR is the entire cause of your problem and not the voltages it is supposedly protecting. The switching regulator power supply may give you one direction since the +112 V drops promptly to 90 V.

Another SCR Overvolt Protection

This is a relatively simple extension of the previous Sylvania description of using zeners, diodes and an SCR to effectively disable high voltage when it rises beyond normal limits. This particular circuit belongs to an RCA CTC 118 that's been around for several years and will continue to function for a few more. Therefore, an explanation of what and where it is protecting becomes germane to the timely topic. In this particular instance, however, the action involves both B+ *and* the horizontal oscillator tank circuit with its 100 µH adjustable coil and shunt capacitor (Fig. 13-9). For action, however, you'll find it both simple and effective so we won't dwell heavily on the circuit description.

There are dual inputs into the shutdown circuit, one from flyback sampling pulse XRP, and the other from emitter resistor R418 in the emitter of horizontal output transistor Q402 and its combined damper diode. The sampling pulse enters the circuit through CR409, is rectified, filtered, and then divided down by R416 and R430 to about 24.5 volts. Should it rise above 27 V zener CR406, this diode avalanches and gates on SCR401. Current flows through R411 and CR405 and the horizontal oscil-

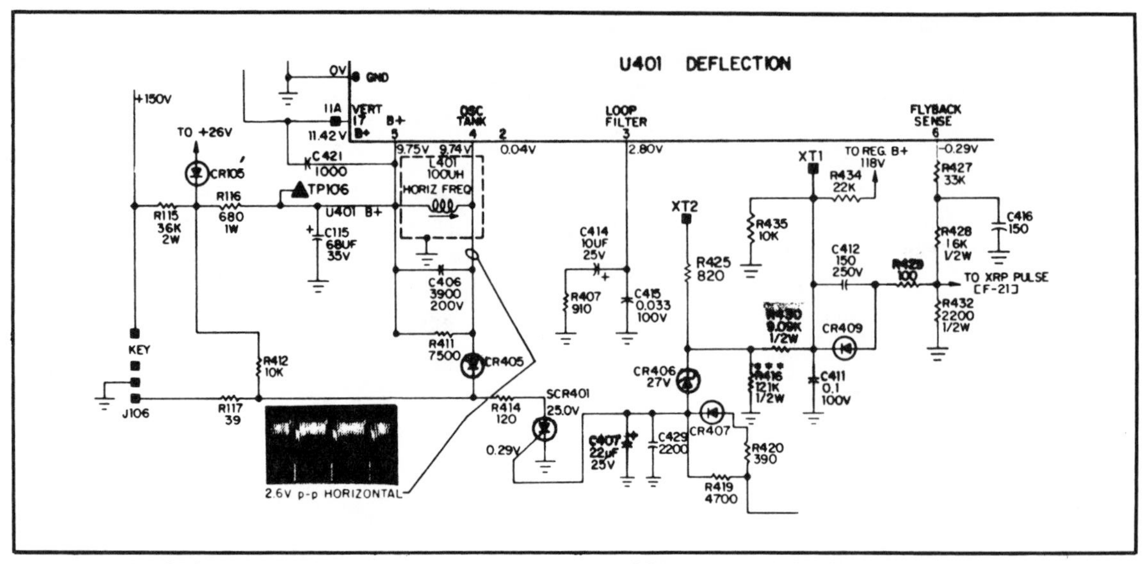

Fig. 13-9. RCA CTC 118 overvoltage protection circuit—another SCR (courtesy RCA Consumer Electronics).

lator in IC401 is immediately disabled, causing high-voltage shutdown. The SCR, of course, remains in conduction until its anode current is shut off by one means or another, including set turnoff.

In the second instance, current through the horizontal output transistor is monitored as it flows through R418 at the junction of R420 and R419. Should it rise above preset limits, CR407 will conduct and turn on SCR401 again, and once more kill the horizontal oscillator through its tank circuit. In this way, both current and voltage are checked continuously so that neither will get out of hand.

For a relatively noncomplex function with only a few components and very direct electronics, this overvolt protection circuit does an excellent job at minimum expense. Someone did some good thinking here and came up with very satisfactory answers. We'll have to presume the circuit works well for lack of disparaging field complaints. And so will you.

General Electric's PC Chassis

These receivers also have effective high-voltage shut-down circuits, which are sensitive to both out-of-bounds high voltage and CRT beam current. Figure 13-10 illustrates dual inputs from the high voltage area, with the anode of SCR990 connected indirectly to the prime regulated +116 V supply that's divided down to 12.7 volts for this purpose. One input is from the HVT across R996, with rectification and filtering by C993 and shunt diode Y991. This is the input to the cathode of SCR990. The other flyback input is sensed by R990, rectified by Y992, filtered and applied to the divider connected to the cathode of zener Y993.

Should high voltage exceed +115 volts, zener Y993 will conduct. Simultaneously, resistors R935

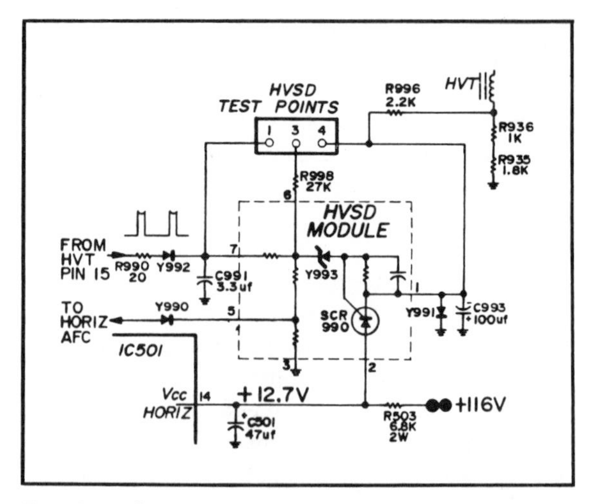

Fig. 13-10. General Electric's PC chassis shutdown is innovative (courtesy General Electric).

and R936 are monitoring CRT beam current, and as it increases, Y991 conducts harder placing a more negative potential on the cathode of SCR990. When the gate, cathode and anode of SCR990 reach their allotted potentials, SCR990 will conduct and short all VCC supply voltage for the horizontal oscillator to ground. If the shutdown circuit loses its high-voltage input sample, diode Y990 conducts and causes the receiver to lose horizontal sync.

What you have here, then, actually is three-way protection and should very adequately serve the purpose. There are several rather large electrolytics in this circuit that may dry out and give you problems over the years, but the general design coupled with considerable ingenuity merits a "well done" for its effectiveness all-'round protective features. You may want to study this one again to let the interacting features sink in. In balancing one portion of this circuit against the other, you do have some interesting electronics.

Another RCA Shutdown

This one's a little different in execution but achieves pretty much the same results, disabling the television receiver. This particular X-ray protector is found in the deluxe CTC 121 first introduced in 1983. Therefore, a great many of these sets remain in the field, and a good description of what goes on may be helpful to those contending with it. In the several years we've had ours, however, there's not been a whimper (Fig. 13-11).

This circuit looks pretty complex, especially with all the diodes, but it's not that difficult and can be explained rather simply. A 45-volt pulse from the set's flyback is rectified by CR402 so that the emitter of Q401 sees a bias of around 27 volts. With its base clamped at about 28 volts by zener CR404, which sets the potential for the collector of X-Ray latch Q402, and also one of the bias supplies for U401, the deflection and regulator control, Q401's base is more positive than its emitter so there's no conduction. Should excess high voltage appear from the flyback, the Q401 emitter voltage rises and Q401 eventually will conduct. This forward biases Q402, and CR410 at the base of horizontal driver Q403 (not shown) fires, latching the horizontal

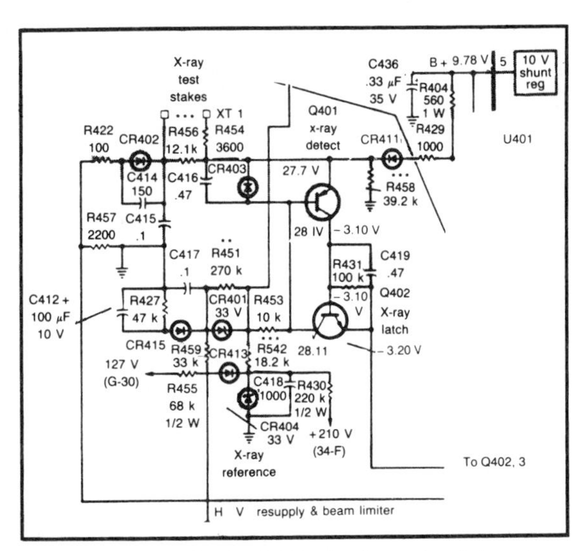

Fig. 13-11. The RCA CTC 121 shutdown circuit (courtesy RCA Consumer Products).

driver on in hard conduction. Under these circumstances there is no on/off horizontal and flyback drive so flyback-generated voltages turn off disabling the receiver.

According to RCA, HV resupply current is also sampled by dividers R451 and R459 that also goes to one terminal of an error amplifier in U401. Should excess current across R459 result in conduction of CR401, then Q401 and Q402 once more latch the horizontal driver into hard conduction, once again disabling the receiver. In this particular receiver, the X-ray protection circuit is more important than ever since half-wave diodes and filters supply operating potentials for the 210 V source, the +19 V, and +24 V sources, the horizontal driver, screen voltages, and high voltage. If horizontal drive is ever eliminated, you can see that the entire receiver will quit cold.

Conclusion

All the foregoing should have given you enough material in this chapter to understand just about any high voltage circuit in the books. We might have dwelt lightly more on some of the various pulse-width regulator systems, but when you've seen the basic ones, the remainder are just slight variations in one form or another.

Chapter 14

Cable Television

AN INTEGRAL PART OF THE TELEVISION PICture throughout the United States and in many countries abroad has become universally known as CATV. Originally identified as a community antenna service way back in 1948, it was then made up of a master antenna system serving those communities that could not receive established broadcast programs with home antennas. The first system, according to the National Cable TV Association, was thought to have been established by John Walson of Service Elect. Co., Mahanoy, Pennsylvania.

After a number of years, CATV subscribers wanted more than just regular TV station broadcast signals and operators began taking first microwave transmissions and, later, satellite C band service. There were some periods of harsh regulations during the 1960s, but regulations were relaxed to a certain extent by 1972, and in 1975 satellite communications began, along with pay CATV.

NCTA relates that the first U.S. domestic satellite, Westar 1, was launched April 1974 by Western Union, followed by RCA's Satcom 1 in December 1975. Since then, Satcoms III (lost), Satcom IIIR (Cable Net 1), and Satcom IV have all found their places in geosynchronous orbit 22.3 thousand miles directly above the equator at West Longitude positions 79°, 136°, 131°, and 83°, in that order, with all sorts of new and replacement "birds" being sent aloft in 1984 and 1985. For complete information, please refer to *Satellite Communications* (TAB Book No. 1632).

Now cable television features Cable News Network, National Teletext Magazine Keyfax, financial news, world-wide sports, special feature full-length movies, and a great deal more to keep subscribers happy all over the country. Today, in fact, the various polling and rating companies say there are between 35.8 and 37.2 million basic cable subscribers, penetrating some 42% of all TV households, and 19.6 million pay cable subscribers. CATV, as you can see, has become very big business and is increasing by an estimated 265,000 to 300,000 households per month. Even the very big satellite earth station (TVRO) business that's booming especially among out-of-the-way areas in the west and south, has had no appreciable effect on this substantial growth.

This latter is especially true because of the addition of educational facilities, uncut off-Broadway and Las Vegas extravaganzas, concerts and pageants, while two way cable and shop-at-home services are being rapidly expanded in many localities. It's also possible for subscribers to sign up for home protection services such as fire, police, intrusion, and medical services, where a single button push or computer signal will summon prompt help. Special children's programming has aided considerably in so many homes needing juvenile entertainment or baby sitting. Later, as more universities make degree and home study programs available, this is expected to increase the value of CATV substantially even over its worth today.

As opposed to the satellite's general entertainment or religious fare, CATV offers highly specialized programs that directly supplement many business and family needs. That's why both the television receive only (TVRO) and cable television (CATV) will survive very well in their own mediums and prosper. Eventually, we may even vote, bank, work, and play on CATV, or some reasonable facsimile, thereof. City traffic jams foretell many evident events that should be occurring in increasing magnitudes well before the end of this century. This is why the TV video explosion all fits into said scheme with an elegance never imagined before the 1980s. Indeed, almost all types of communications will most certainly endure because they're needed to supply services and entertainment to every diversified segment of the population.

CATV MUSTCARRY DELAYED

Cable TV will *not* be required to carry multichannel sound at this time, and the FCC will only "*monitor* development of MTS technology, its use, demand and delivery to the cable consumer." Taking a big load off CATV's back by this Memorandum Opinion and Order of February 8, 1985, the Commission's action will permit CATV to develop the necessary engineering and hardware required for stereo TV to be successfully introduced to many or most homes now serviced by the medium. Not only are video/audio carriers affected in bringing CATV up to snuff in handling multichannel, but CATV converters, especially the settop variety, cause problems resulting from unstable and bandlimited components. Further, in FM, random noise power density increases 6 dB/octave and is 8 dB greater at the 5H SAP subcarrier than at the 2H stereo subcarrier. Interestingly enough, buzz—the bugaboo of multichannel sound—is weak in the 2H region and often will *not* appear through small speakers. Split sound demodulators, in *any* TV setup are unsuitable for cable, too, because of settop converters whose noisy local oscillators can put synchronous FM noise on both sound and picture carriers. Unfortunately, weak signals are often problems when combined with TV receiver noise, and buzz plus crosstalk factors will have to be removed from the cable systems before satisfactory transmissions are realized.

Meanwhile, the FCC has instructed its Mass Media Bureau to annually report on the following: broadcast transmissions of MTS; consumer purchases of MTS receivers; and CATV MTS capabilities. When the "Commission decides to revisit this issue, a formal record can be developed to receive information relating to MTS use and development at that time." Essentially, the Commission is saying that what cable must carry should be decided by the marketplace and not an arbitrary Federal edict—a reasonable decision considering all of the many factors involved and perhaps the relatively slow growth of broadcast multichannel sound, at least during 1986 and 1987. Thereafter, technology and public demand can take this new technology anywhere. By then, too, CATV should be able to cope, but probably only in the newer and better-equipped installations. In the meantime, although broadcast networks are committed to stereo service, responsible industry associations predict that MTS penetration will not rise above 18 to 44 percent by 1989, with only limited stereo programming available in 1985-1986.

For CATV, this means ample time should be available to equip cable systems that are progressive and possibly prod others who are not. In light of the Mass Media Bureau annual report instructions, it's evident the FCC doesn't expect CATV to become a SAP-stereo sound carrier for some time to come. One estimate placed the cost of upgrading cable for MTS to exceed $700 million should this service become a Federal requirement.

Nor do the proceedings before the FCC indicate major TV network complete adoption of TV stereo "in the near future." Electronic Industries Association and the National Association of Broadcasters predict a 10 percent MTS receiver penetration by the middle of 1985, and only between 18 and 44 percent by 1989. By mid-1985, only about a dozen TV stations are on the air with multichannel sound (MTS) and only limited stereo programming is available, most of it not specifically produced for broadcast television. Nonetheless, CATV realizes that MTS *will* become an enormous factor in TV entertainment and the industry is slowly preparing to face the challenge.

In view of the inevitable, we thought that up-to-date-information from a major CATV equipment supplier might aid in understanding both problem and cure. Yes, there are head ends and amplifiers even today that will handle multichannel TV sound, and scramble many signals to boot. So we went to General Instrument in Hatboro, Pennsylvania, for some of the answers.

CABLE TECHNOLOGY

CATV is actually the distribution of rf television signals and information (some of it two-way) to subscribers with television receivers by way of coaxial cables or optical fibers. Usually there's a flat fee for so many channels, then there are other channels which are scrambled and an additional fee is charged for this extra service. Signal pickup for CATV may come from broadcast stations, microwave, or satellite earth terminals, or all three, depending on the individual CATV setup. However, more and more dependence is now placed on satellite service since additional network programming, special services, and other events not usually received by ordinary means is being uplinked to geosynchronous satellites some 22.3 thousand miles above the earth and positioned directly above the equator. As the earth turns, so does the satellite, and therefore a receive dish and electronics may access one particular satellite from a single location without any further movement other than its original position emplacement. In addition, studio broadcast quality pictures are possible from the satellite's frequency-translated downlink, which will often help compensate for much of the usual signal deterioration in many of the operational cable systems today. Of course, people with 3 MHz luma amplifier systems probably won't notice, but the full definition group with excellent receivers will.

At any rate, as of the start of 1985, Spacenet 1, Satcom 3R, Westar V, Galaxy 1, Comstar D4, Anik D1, Satcom 4, and Westar III, were all serving U.S. CATV with excellent programming and all sorts of desirable services. If a quick count serves correctly, these total 93 channels of programming, and that ought to be enough for anyone, and an extra village or two in addition.

As of now, the cable television industry is the largest private user of domestic satellites in the country and will undoubtedly be accessing more later this year or next as services and satellite numbers expand. So far, however, cable is only using the 4/6 GHz C band spectrum since virtually all commercial traffic is still carried on this medium. With fewer and fewer parking spaces available, it may be that cable will have to use some Ku band later, but at the moment that's only speculation since we're not privy to programmers' long-range planning.

Video, however, isn't the only successful medium for CATV, data transmissions of every description must also be included, especially videotex which can deliver graphics and text to homes, and TV receivers with appropriate decoders are able to use this service for advertising, education, impaired hearing captioning, financial transactions, electronic mail, home computing, and other services under development. Looking further into the future, we would certainly expect education and homework assignments to be very much a part of such traffic also.

Before and during CATV startup, a very considerable investment is required to get the system on line. Until this is done, any worthwhile revenue is not forthcoming. Often as much or more than seven years is required before startup owners can begin to recover any portion of their costly investment, and then operating costs, subscriber numbers, and overall income must be carefully managed to make the system work.

Initial outlays involve head end equipment, distribution services, subscriber settop converters, and the usual office and studio-warehouse facilities. Subscriber interactive terminals, for instance, can require an outlay of $300-$400 at minimum, and will probably increase allowing for both complexity and inflation. In short, depending on the extent and quality of the cable system, startup money may well amount to a great deal of capital. Now with better video and audio incoming from satellite transmissions and multichannel sound, many existing systems will require a good deal more money to keep their plants competitive and subscribers satisfied. It's obviously *not* an inexpensive undertaking, nor will mounting labor costs and new technology reduce capital outlays and system maintenance. Consequently, subscriber costs are sure to mount.

SYSTEM DESCRIPTION

This is a general system description touching only the highlights and leaving specifics to whatever information is supplied on the latest available equipment. Therefore don't expect anything more here than generalities that could apply to any reasonable installation in existence. Our reasoning is that updates can and do occur almost monthly, and a good, basic system description is worth more than trying to describe specifics which everyone may not have or want to know about. Consequently, the imprecise approach.

Consisting primarily of a head-end receiver-remodulator installation and a coaxial distribution network, the head end intercepts and receives incoming signals from a number of sources, including the usual microwave and broadcast sources, as well as one or more fixed satellite antennas, depending on whatever SatSystem's being accessed. Dish sizes can range from 5 to 9 meters, depending on local reception, noise conditions, and EIRP which, translated, means effective isotropic radiated power, and usually amounts to some 30-36 dBW in the better signal areas. Your received signal, however, depends on carrier-to-noise power for some specified bandwidth. Of course, other factors enter into this also and downlink C/N involves the net system gain over noise temperature (G/T), receiver bandwidth, and Boltzmann's constant. Clear view space loss to the 22.3 kilomile-positioned satellite alone totals approximately 196 dB. Some multiple system owners (MSOs) are transmitting these satellite signals via microwave link to subsidiary head ends which are known as fan-out links and are usually only 10-60 miles long compared with backbone systems having lengths of 200-400 miles. FM rather than AM transmissions, because of noise, are often the preferred medium here.

The Head End

This complex provides all the downconverting, signal conditioning, scrambling, channel assignation, and amplification/modulation required for the entire system to function. There are bandpass and bandstop filters, converters, mixers and modulators involved; and in addition to outside program sources, there are also taped and local studio programs which the head end will originate itself, depending on management decisions and available equipment. Pilot carriers for automatic gain control (agc) will be generated, too, and any head end may also have microwave transmit facilities for one purpose or another.

Trunk Cabling

Once downconverted and demodulated video/audio information is ready for distribution, it is remodulated and usually distributed over trunk cables that may run up to 10 miles or so. Branching off from the trunk are distribution cables, and then subscriber "drops" connect to these. As signals travel down the trunks they are, of course, attenuated and require amplifiers at certain intervals to restore signal levels to their original values. Such occurs, naturally, because of inherent wire resistance and temperature increases during hot weather from their normally set adjustments at 68° Fahrenheit. Temperature decreases also affect cable responses and there has to be some means of keeping amplifiers from either being overdriven or subject to undesirable signal loss. Automatically controlled gain amplifiers are the answer, with a high-frequency pilot riding shotgun on the gain control detector and a low-frequency pilot working on a slope detector. Detector outputs are then compared to dc references, and any deviation from these references returns as an error signal. These error signals are amplified and control variable attenuators for a constant amplifier output during the high pilot frequency. Slope control detectors also have dc reference emissions and comparisons and operate on a feedback to the third rf amplifier whose gain then remains at a constant level. Therefore, signal amplitudes, levels and tilt are maintained over the temperature extremes during winter and summer operations.

Trunks must also feed distribution cables, and another set of amplifiers known as bridging amplifiers connect the two. These have the same general electrical characteristics as deluxe trunk amplifiers, but noise figures are often not as stringent and they can operate at higher levels. Outputs normally pass through power splitters, serving the designed number of distribution cables throughout designated parts of the system.

Distribution cables are highly important since it is from these that subscriber drop cables—one for each customer's home— originate. Taps on the distribution cables mate with the various drop cables, each using a directional coupler for smallest level signal attenuation, impedance matching, and to prevent any video/audio feedback. Should signal levels diminish to the point of being questionable, distribution amplifiers may be installed, several in series, to increase the number of drops and provide usable signals. However, since high output levels increase both noise and distortion far in excess of the trunk amplifiers, there is a decided limit to the number that can be used, and signal quality must be carefully monitored to prevent obvious degradation. SWR, noise, amplitudes, cable quality, couplings, head end originations, and maintenance all enter into the factor of quality video/audio/text reception. Unfortunately, some cable signals are rough!

Subscriber Drops

The system is complete as it interconnects with the customer. Pickup begins with a tap which may serve several subscribers or possibly only one—depending on quality and design. It consists of a directional coupler, an isolation network, probably a dc blocking capacitor and some sort of output port. Separate isolation circuits or power splitters may divide the CATV distribution signals for multiple subscribers. Rarely exceeding a few hundred feet, this is the least expensive and most lossy of all system cables, usually amounting to braid-shielded RG-59, unless there is excessive noise or broadcast signals in the area. If so, then double aluminum foil and braid are often used to provide as much as 60 percent shielding and 100 dB isolation. Belden, for instance, makes both Duobond™ and Duofoil™ CATV cables with shielding effectiveness of at least 100 dB and relatively little loss compared to much that's on the market. Transfer impedance (shield performance), constant wire impedance, characteristic line terminations, and cable covering all enter into final CATV performance as seen by the subscriber. Lossy cable with poor propagation characteristics has delivered many an inadequate signal because of unsatisfactory conductors, leaky couplers, and external interference. With time, these drop cables and their connectors do deteriorate, flex, corrode, and oxidize in one way or another, although usually so gradually that the subscriber hardly notices until his TV picture becomes unbearable—then the problem is often blamed on the poor television until it's proven otherwise. It's best to check another receiver in known good working order. For severe problems use a spectrum analyzer, system sweep, or cable fault locator. Any and all of these can ordinarily clear up line and amplifier problems anywhere in the system.

Settop Converters

Convenient for subscribers without CATV-equipped TV tuners, but headaches in some respects, these will convert the various cable signals to a single VHF channel input to your old or less expensive TV receiver. Important, since a cable system may have 60 or so channels, such converters are also a source of CATV revenue in the monthly service charge as well as containing descrambling circuits for special channel programs which the television receiver may not have (legitimately). Should locally-broadcast programs be repeated over cable, good shielding in high signal areas is mandatory because of secondary images caused by time delay differences between broadcast and cable-regenerated telecasts, each on the same channel. Since converters are made especially for CATV systems, such interference should be less if it does, indeed, exist. Bandpass abilities may, however, deliver something less than the usual 4.2 MHz luminance signal ordinarily telecast. Also, multichannel sound is often affected by these converters' local oscillators, and better units may have to be built when *Mustcarry* becomes law sometime before the year 2,000. Meanwhile, phase-locked loop, all electronic tuning converters are even now on the market, and additional upgrading with time is expected as advanced electronics require. A good cable system, we might add, produces excellent pictures if properly maintained with instrument-checked service.

General Instrument's Jerrold Division has a number of these converters which they say will meet all system configurations from 9 to 66 channels, with various levels of descrambling, plus remote control units that permit antenna/cable switching, channel entry programming, and any possible program selection available.

Talk-Back Systems

These are divided into dual cable, split frequency, and a separate phone line. When such facilities are made available, the regular head-end transmit cable is termed "downstream" while subscriber talk-back is called "upstream"—almost sounds like spawning salmon talk. At any rate, if the return transmits on the same cable it has different carrier frequencies from those downstream. Of course, if two cables are used, all signals are separate anyway, and any phone line is just that and nothing more, but may only handle voice and/or data.

If dual transmissions are in order, 5 to 30 MHz portions are assigned often four channels of upstream return information, complete with trunk, bridging, and distribution amplifiers, where trunks may operate at high gain for video or low gain for data traffic. Bridging amplifiers also have a higher gain mode for video, while distribution amplifiers are used when long cable runs are required before reaching the big trunks and their amplifiers. Several types of attenuators and equalizers are installed to make both trunk and distribution lines relatively equal in signal handling ability. Unless new amplifiers have recently been engineered, *separate* uni-directional amplifiers must be installed to bring the signals back, along with the necessary filters, etc. With all the good engineering and computer-aided design floating around, it would seem that two-way amplifiers even in the same box could be satisfactorily developed at less cost than singles—of course some economic factor may also be present of which we are unaware.

For dual cable systems, costs may not be more than 50% additional over a single cable operation, at least for laying the wire. However, it does have the most capability, capacity, and flexibility, and probably will be used considerably more in business and specifically populated areas with high incomes rather than general service throughout any type of random consumer base.

In some new communities, however, police, fire, burgular, and fire security services are offered on a pay basis to everyone who has these requirements. It is probably in these categories that CATV has very considerable room to expand, at least until other two-way services become popular in the later 1980s.

For system analysis it's about as far as we should be involved in CATV for this particular book. Were it devoted exclusively to that topic, then a system layout procedure, along with levels, amplification, impedance-matching and all the rest would be appropriate. This specific type of information, however, is available from at least a half-dozen manufacturers who, if you're a serious prospect, will even do the design for you if, of course, you agree to buy their equipment; which, really, is only fair. What we have tried to do is give you an overview of all that's involved without too much detailed depth, which might not mean too much to most casual readers.

There are, however, one or two particular items more that can make interesting reading, so we'll continue with these and close the discussion. There'll be something more on satellite downlinks in the troubleshooting chapter in the event you have more than passing interest here—admittedly it is one of our favorite topics with both commercial and consumer earth terminals becoming so much a part of our suburban and rural landscape.

Addressability

Means the cable operators ability to control individual subscriber service from the head end. It offers prompt service changes, reduces tampering problems with converter computer control by identifying and "tagging" each channel so that specific programming may go to the authorized subscriber. Customer service and the controller system is illustrated in Fig. 14-1, showing keyboards, computer, memories, and converter- modulators. There are STARCOM® controllers for both one and two-way systems, both using Digital Equipment computers and Jerrold-written software, or the new TOCOM/ACS-1000, configured to work with the H-P 1000 A600 Series computers which will accommodate additional system growth when needed.

The addressable controller may also interface with the billing computer, reducing operator cost and saving dual data entry with special software supplied, if necessary. Of course, scrambling devices are easily controlled with several type-selected methods.

Scrambling

Done in various ways, many of which dear to the originators, but usually with broad-category similarities, scramblers may affect audio, video, or both, and include sync inversion or compression in the process. There's even tiering at various levels, plus programmable energizing pulses to which the decoder will recognize and respond. We won't go into the various methods at length since this is only one chapter in a television book, but one scrambling system does stand out.

Jerrold calls it Trimode and says it's an effective means of protecting pay service against CATV

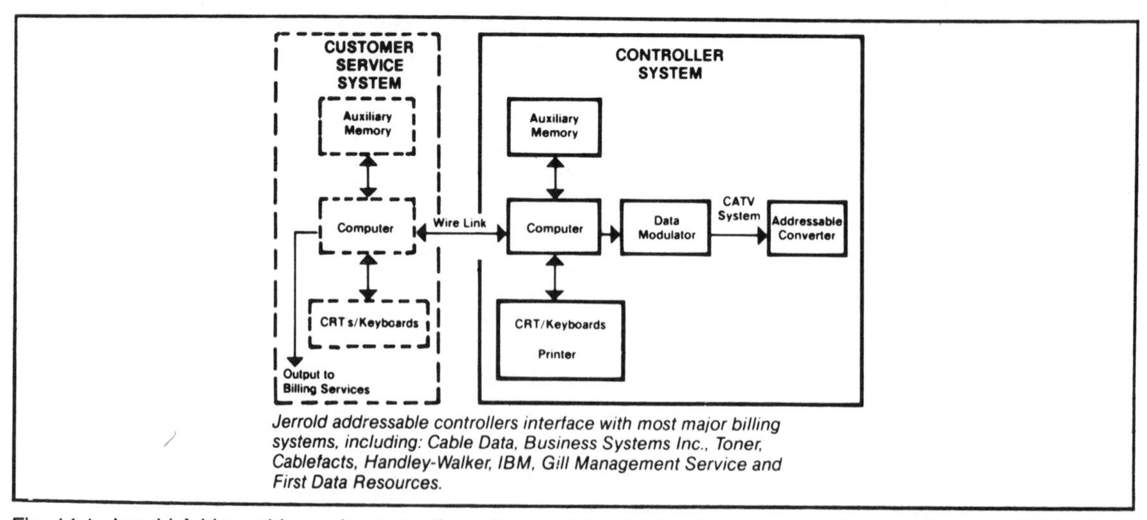

Fig. 14-1. Jerrold Addressable service controller selects individual subscriber programming (courtesy Jerrold Div. Gen. Instruments).

piracy. Scrambling at 6 dB, 10 dB, and 6/10 dB levels, it defeats signal thieves having 1- level electronics and can randomly switch between these levels in clear mode at one of eight operator-selected time intervals. Said to be a sync suppression scrambling system, it operates through Jerrold's digital scrambler/encoder, offering state-of-the-art signal security.

TYPICAL HEAD END

A typical head-end configuration has been supplied the Jerrold Division of General Instruments consisting of a pair of head-end combiners, with antennas outside and video/audio processing units inside. On the left (as shown in Fig. 14-2) are the VHF, UHF and microwave inputs, with their preamplifiers and receivers, while on the right are several VHF channel inputs in addition to pay TV and the program switcher. In the center are the three directional couplers and combiners, a pilot carrier generator, return filter, and the trunk cable output. Signal level meters and television monitors are also evident. Such audio/video may either be processed directly on the same carrier as received or converted to other carrier frequencies, depending on the possibility of co-channel or crosstalk interference (a consideration all CATV head ends must be conscious of when the facility is planned). There's probably nothing more annoying than co-channel sweep-through as one carrier interferes with another on the same channel. Without careful shielding and planning, this can happen to any cable service anywhere, doing considerable harm to both video reproduction and community reputation. There are even conditions where cable must use offset carriers to prevent same-channel interaction, often presenting problems for the receiving television receivers unless they have rather wideband AFCs to pull-in the non-standard frequencies. Fortunately, today, most of the better receivers do have this ability, and so the CATV station that used to be a perpetual problem isn't anymore because of advanced receivers introduced since about 1982-1983.

Multichannel Sound

Jerrold/G.I. also has developed i-f and rf-baseband processors for BTSC-dbx multichannel sound that can be received either from satellites or off-the-air broadcasts. There's even provision for a videotape recorder, plus modulator that will interface with the other two inputs to the combining network and trunk feeder lines. In all three instances, stereo/SAP is either processed directly on on received carrier or remodulated for transmission down the line, with decoding up to individual television receivers. All this is shown in Fig. 14-3, but not in great detail, just the basic outline of what the triple output/input system should look like.

At the subscriber drop, Jerrold has provided us with a block of a Starsound MTS Adapter configuration that may be of further aid in picturing the multichannel sound system as it is received in any subscriber's home (Fig. 14-4). Shown are both plain and descrambling configurations, with plain and descrambling converters connected directly to the cable drops. Observe that the descrambling converter will feed rf video directly to the television receiver, while the plain converter supplies all signals first to the Starsound Adapter, which then channels its outputs as rf video and baseband audio to the TV and external audio amplifiers and speakers, respectively. With the video rf feed in the descrambler configuration, it's implied that audio is rf modulated to the TV as well, and output from the Starsound is baseband. This would imply a TV/monitor setup in the descrambler configuration if one wanted to take advantage of inherently better audio into the TV amplifiers, with the TV possibly having external speakers of its own.

It's been our experience that internal TV speakers—especially in table model receivers, don't cut the mustard. Therefore, it's always best to invest in a good pair of wood-encased speakers to provide best effects in multichannel stereo sound—monophonic, too, for that matter since some of the better television receivers have a "stereo effect" arrangement that can be quite suggestive of dual-channel sound.

The Trunk Amplifier

Once all this processing has taken place at the head end, signals are placed on a high-power trunk

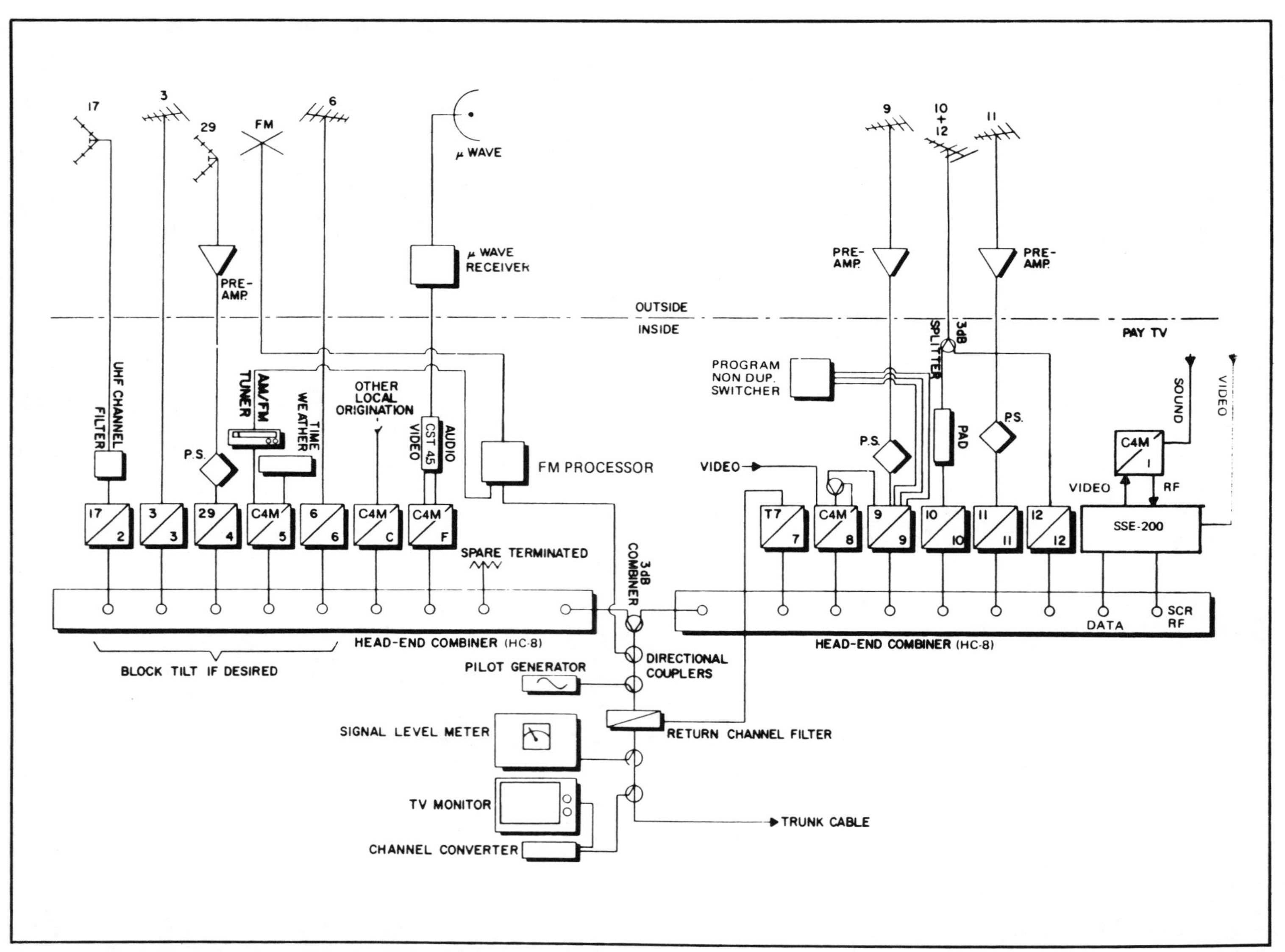

Fig. 14-2. Typical head-end configuration (courtesy Jerrold Div. Gen. Inst.).

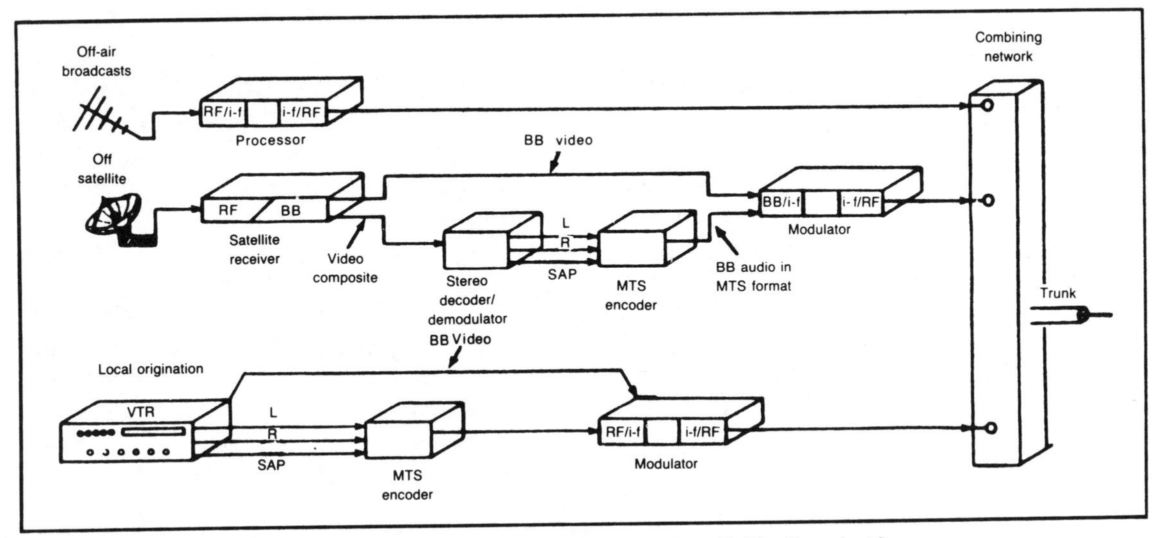

Fig. 14-3. Multichannel television sound headend system (courtesy Jerrold Div. Gen. Inst.).

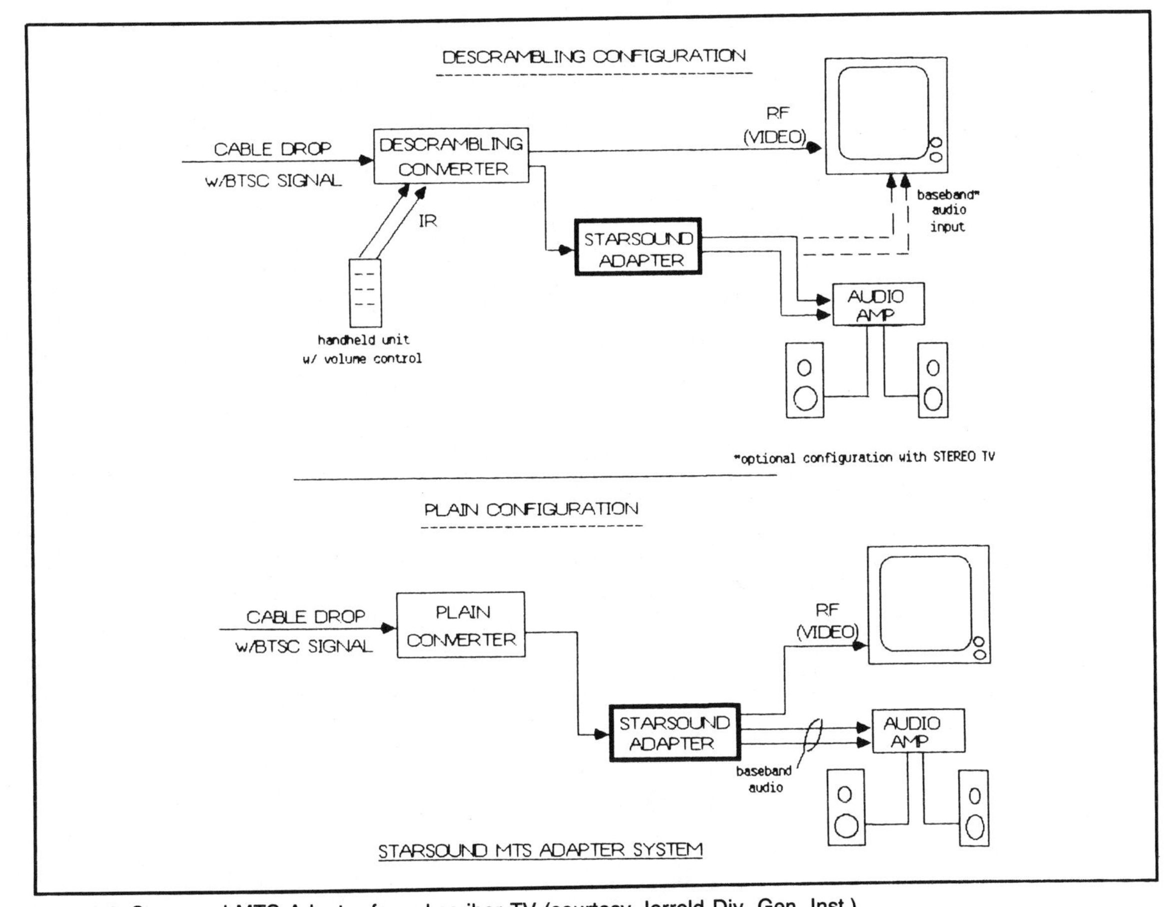

Fig. 14-4. Starsound MTS Adapter for subscriber TV (courtesy Jerrold Div. Gen. Inst.).

amplifier and delivered to bridgers, filters, and equalizers for the remainder of the system. As you see in Fig. 14-5, padders and equalizers initially prepare the trunk amplifier input, with feedback from a bridger from the lower end. In this way, a relatively linear signal emanates from the output filter into the truck amplifier output with considerable driving power for subscribers downstream.

Newer signal processors have SAW input filters for adjacent channel attenuation, loss of signal sensing and i-f switching, standby 30 V backup power, much improved intermodulation and cross modulation, very low spurs, with non-phaselock, incremental phaselock, or harmonic phaselock modes. Both picture and sound are independently amplified and agc-corrected so that sound and picture levels may be independently adjusted and processed. Modulators now are scrambling compatible, white-level limiters to prevent over modulation, there are plug-in output modules for any channel using plug-in output converters, complete with circuit isolators that will shut off designated circuits for testing. There are also amplifiers today having 80-channel capacities at 550 MHz, complete with redundancy options, forward and return trunk amplifiers, bridgers, and heavy surge protection. There's also a dual pilot return amplifier, and 30 dB directional coupler test points and advanced

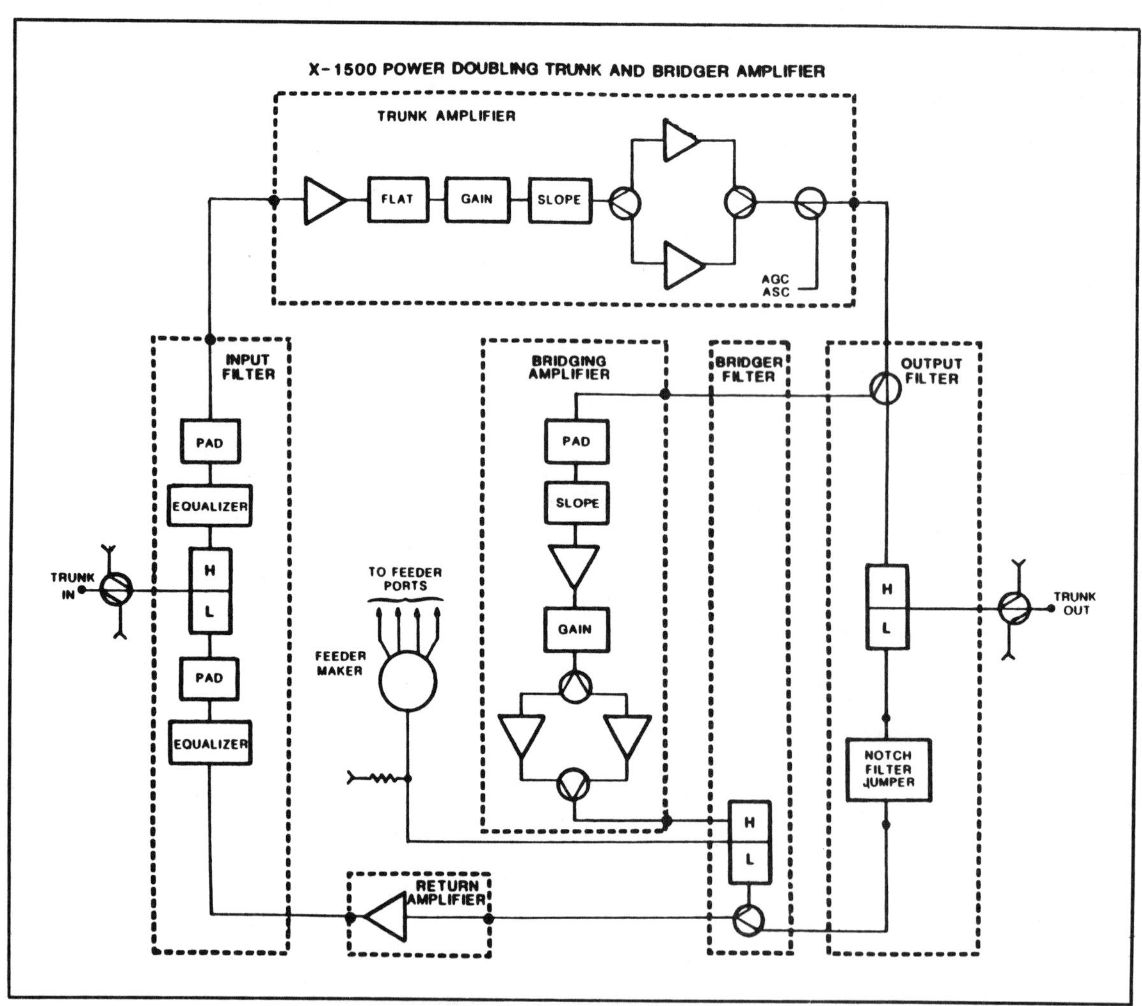

Fig. 14-5. The trunk amplifier configuration (courtesy Jerrold Div. Gen. Inst.).

status monitoring.

All this shows that cable TV isn't standing still as new developments are introduced that affect its services. True, all the foregoing equipment description applies to Jerrold, but undoubtedly there are other manufacturers who are at least attempting to modernize their systems also. We would hope and expect that it won't be too long before some of the better cable networks offer multichannel sound to all those who have a need. All the above indicates subscribers haven't long to wait.

Measurements

Nominal measurements for CATV—as in many other rf transmission means—are made in both decibels (dB) with reference to 1 milliwatt or 1 millivolt. As you will see there is a *considerable* difference.

The standard term for level measurements in CATV is dBmV; and dBmV means some level with reference to 1 millivolt (mV) rms (root-mean-square) across 75 ohms. The 75 ohms being the normal, practical video impedance of lines and amplifiers accepted by U.S. industry in all visual aural *carrier* transmissions. Note that it is *not* demodulated video baseband, but strictly rf. Rms, of course, is 2.828 of any peak-to-peak sinewave signal. In terms of logarithms,

$$\text{dBmV} = 20 \log_{10} \text{rms mV}$$

with the 1 mV reference used because it is usually the signal level a television set must receive to display a noise-free picture. Field-strength meters, for instance, are all calibrated in dBmV, so it's a handy measurement to remember. And 1 mV across 75 ohms becomes 0 dBmV.

While the above is a good reference, it does not take into account current, since at rf this is usually very small. But in order to establish any sort of power levels with direct measurements, current must usually have a piece of the action. Therefore, once more we return to one-thousandth of something to establish a specific level. This time, it's a milliwatt. Now we can express power ratios, just like voltage and current ratios, but this time:

$$\text{dBm} = 10 \log_{10} \text{rms mW}$$

Is there a relationship between the voltage and power measurements?

Certainly:

$$0 \text{ dBmV} = -48.75 \text{ dBm}$$

and you can work that around any way you want to, remembering that these measurements are universal, including satellite signals, microwave, and most other rf carrier signals. However, at higher frequencies, impedances are usually standardized at 50 ohms, and so there has to be a translation between the two. For instance (Fig. 14-6):

$$\text{dBm (75 ohms)} = \text{dBm (50 ohms)} + 5.72 \text{ dB}$$

and

$$\text{dBmV (75 ohms)} = \text{dBm (50 ohms)} + 54.47 \text{ dB}$$

Also, if you'd like to translate dBmV into plain watts,

$$\text{dBmW (75 ohms)} = \text{dBmV (75 ohms)} + 48.75 \text{ dB}$$

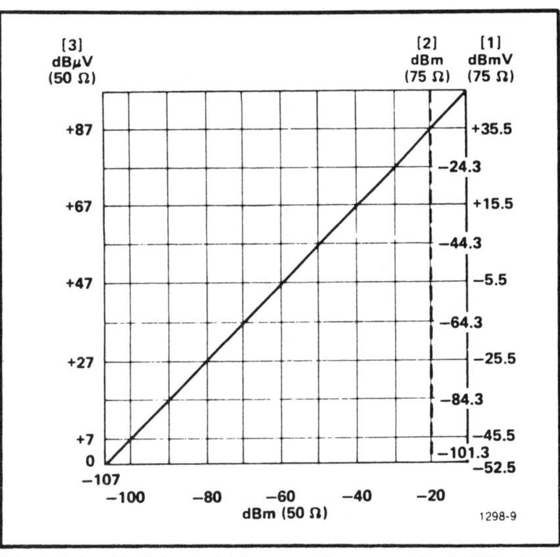

Fig. 14-6. dBm/dBV dBμV conversion graph for 50-75 ohms (courtesy Tektronix).

You can also put these in terms of dBµV (microvolts) if you find it helpful. For that and the other conversions, we'll borrow another convenient chart from Tektronix (see Fig. 14-7).

One other useful relationship is dBm to dBW, and that looks like this:

$$1 \text{ dBW} = 30 \text{ dBm}$$

But remember that when you're working with logarithms you don't divide or multiply, you add or subtract. Therefore if you have a -100 dBm reading and want to convert this to dB watts, then,

$$\text{dBW} = -100 - 30 = -130$$

Therefore, you can factually say that dBW is *always* 30 dB below dBm. In the next chapter we'll also have a short discussion of spectrum analyzer and oscilloscope waveform measurements. For some of you this may prove helpful—for others, it may be old hat until we combine oscilloscope *and* spectrum analysis displays on a single photograph. That often piques anyone's interest, just like Mastercard commercials. It may also be possible to do some signature analysis checks on the digital receivers *if* we can get past all the soldered-in shielding that surrounds the logic circuit board(s).

SCRAMBLING

This information probably doesn't belong exclusively in the CATV chapter, but both satellites and subscribers (including cable) are affected here, and it's as good a place as any to expound on some very new information that will become important to everyone in the video-audio viewing industry as a few months and years progress. What we're talking about here is the Linkabit™ system that was only announced during the summer of 1985 and is still being applied to the various satellites as they decide that paid viewing is better than open skies. Whether the Congress will continue to hold still for open sky satellite broadcast exclusions probably won't be known for some time, but they do give the private homeowner the benefit of the a doubt whenever there isn't scrambling. There is now a bill before that august body to place a moratorium on all transponder scrambling for another two years.

At any rate, as far as a reference book will permit, here's the latest information on such moves, taken directly from unedited copy for *Modern Electronics* magazine, for which I am the satellite and video contributing editor.

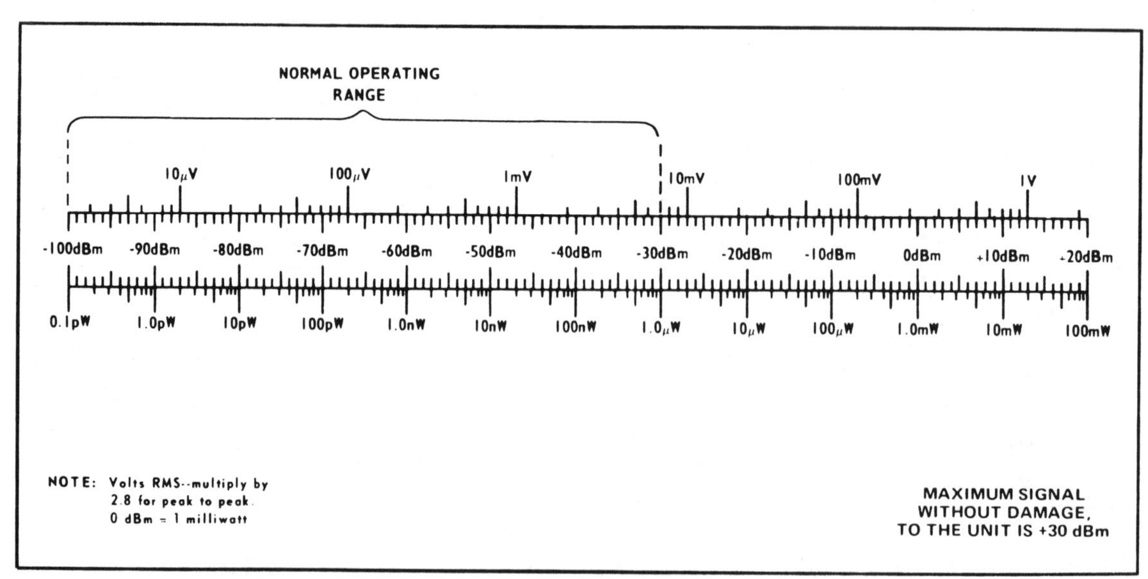

Fig. 14-7. Power and voltage conversion bar graph used with spectrum analyzers (courtesy Tektronix).

LINKABIT™ SUPPLIES THE SYSTEM FOR CATV AND DBS

Signal Scrambling Is Here—Courtesy M/A-COM Linkabit™

By Stan Prentiss

Like it or not, sound and sight scrambling has already begun to reach the airwaves, skywaves, and CATV as specialized movieland programs encipher outputs for public pay TV. Furthermore, since the system is addressable with a unique 56-bit key for each descrambler, the degree of difficulty in pirating access is said to be 72,057,590,000,000,000 possible key combinations, or 72,057,590 × 10^9 permutations—trial and error searches, to you.

Designed and built by satellite and microwave giant M/A-COM and its Linkabit™, Inc. division VideoCipher™ uses a Data Encryption Standard (DES) algorithm of the National Bureau of Standards, reportedly producing a descrambled signal equal to or better than the plain-language original even in relatively noisy environments. System applications are CATV, SMATV, satellite DBS, and private networks, with up to 56 tiers of independent programming which may be altered upon command, depending on system operation and/or customer billing intervals. Descrambler costs to CATV are $495 and $395 to independent subscribers, plus the usual monthly fee—naturally!

Video has sync removed and a portion inverted, plus the centering of 3.58 MHz color burst at some non-standard line position, while audio is transmitted as a pair of digital channels during horizontal blanking. The audio is encrypted and combined with error coding bits for extremely high security and quality.

How and Why It Works

There are actually three VideoCipher versions available, with VideoCipher IV—a modification of VideoCipher II—scheduled for reliable cable and terrestrial consumer applications. Most of our information, however, concerns VideoCipher II, and this is the system described.

It transmits secure address and control information to the various descramblers as part of the video signal, providing complete control at points of origination. Organized in related program categories, descramblers will respond up to 56 independent tiers and can pass only programs specifically authorized. Individually addressed control messages may be received at the rate of 250, 000 per hour, along with a 56-bit authorization word and a monthly access key. A 56-bit program mask work is also sent, and if the mask and descrambler's authorizations have common 1s digits, program viewing is permitted.

The service is further broken down into DBS (direct broadcast service) commands and CATV commands, so that all DBS channels may be activated by separate program masks, while another 56-bit program mask controls descramblers for separate service elements on each CATV channel, entirely independent of DBS.

Addressing and control data transmits as 63 kilobits during the nominal 11-microsecond horizontal blanking interval in the form of multilevel key structures. Each subscriber has a special unit address with certain keys, while secure retention of critical descrambler information stored in nonvolatile memory, permits rapid restoration of service following interruptions or power outages.

Each descrambler is preloaded with its X-Y coordinates based on your local post office zip code. If descramblers are not affected by up to 32 imposed and independent blackout regions, they then permit the showing of scheduled programs, all else being in order.

You may also receive personal messages such as electronic birthday cards, personal stock exchange quotations, etc., and view up to 256 pages of text information per channel. These may show program guides, headline news, sports, and so forth, and even notices of unauthorized ongoing or coming events.

While all this percolates, the descrambler keeps tracks of your available credit for impulse purchasing of programs, which may either be increased or decreased, depending on how the account is handled. Then there's infrared remote control that will permit impulse pay-per-view selections as well as on-screen program displays for ratings, cost, and your present line of credit, in addition to second language programming.

Descrambler Interface

The Videocipher™ 2000E accommodates both baseband video/audio outputs as well as 70 MHz LNA-downconverter inputs to TVRO receivers, including baseband demodulation. At baseband it receives both unclamped, deemphasized, composite video outputs, as well as clamped video and stereo or monural outputs from the receiver to permit viewing of non-scrambled programming. It can also remodulate the VHF output for any TV receiver. In addition to the stand alone version for the subscriber, there is also a 2000R integrated receiver/descrambler available from M/A-COM for all those who would want to be there "fustest with the mostest."

Interface and descrambler specifications as furnished by Linkabit™ are shown in Table 14-1. Also see Figs. 14-8 and 14-9.

B-NTSC

The newest wrinkle in the more highly advertised group of "scramblers" is nothing more than standard National Television Systems Committee common old TV we've had since 1954 with M/A-COM's Linkabit™ (or (Videocipher II) added. Good enough for terrestrial all these years, RCA looked at the encoding market and decided that for compatibility, cost, transmission, receiving, and an already in-place standard, B-NTSC was also good enough for satellite transmissions and receptions, too.

So with Ku band satellites K-1, K-2, and K3 ascending into geosynchronous orbit between late 1985 and 1988, whoever buys or rents transponder space on any of the three can have either regular luma-color or Linkabit scrambled, according to intent or cost. Marrying the two results in line and field sync stripping, video inversion, and digitizing the two 6.2 and 6.8 MHz audio signals, the latter to be transmitted during the horizontal blanking period along with the 3.58 MHz chroma subcarrier. Sync is then sent during the vertical blanking period and recovered for proper receiver reinsertion.

Linkabit™ is said to service some 250 channels of information along with a 56-bit word authorization for customer access. At this writing, the customer price still remains $395, plus the usual monthly user fees for whatever entertainment

Table 14-1. VideoCipher II DBS Descrambler Specifications.

Video	
FREQUENCY RESPONSE	±1.0 dB, DC TO 4.2 mHz
DIFFERENTIAL GAIN	3% OR BETTER, 10% - 90% APL
DIFFERENTIAL PHASE	3° OR BETTER, 10% - 90% APL
SHORT TIME DISTORTION	3% OR BETTER
LINE TIME DISTORTION	1% OR BETTER
FIELD TIME DISTORTION	3% OR BETTER
SIGNAL-TO-NOISE RATIO (WEIGHTED)	57 dB (P-P SIGNAL TO RMS NOISE)
SYNC JITTER	± NSEC OR LESS
H AND V BLANKING	FCC (CLOSE TO RS-170A)
Audio	
FREQUENCY RESPONSE	±1.0 dB OR BETTER, 20 Hz to 15 kHz
DYNAMIC RANGE (UNWEIGHTED S/N RATIO)	75 dB MINIMUM, 80 dB TYPICAL
TOTAL HARMONIC DISTORTION	0.5% OR BETTER AT 1 kHz
IM DISTORTION	0.5% OR BETTER
NUMBER OF CHANNELS	MONAURAL, LEFT, RIGHT

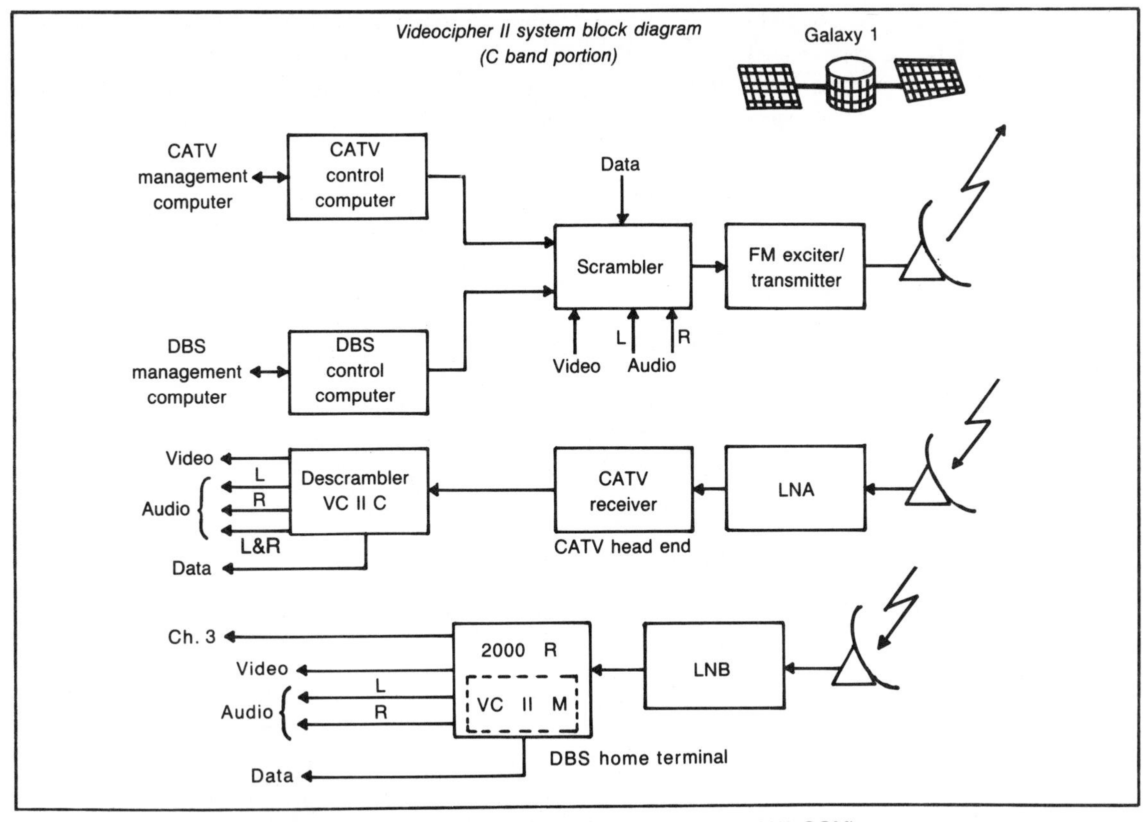

Fig. 14-8. Linkabit's Videocipher II system block diagram—C Band (courtesy M/A-COM).

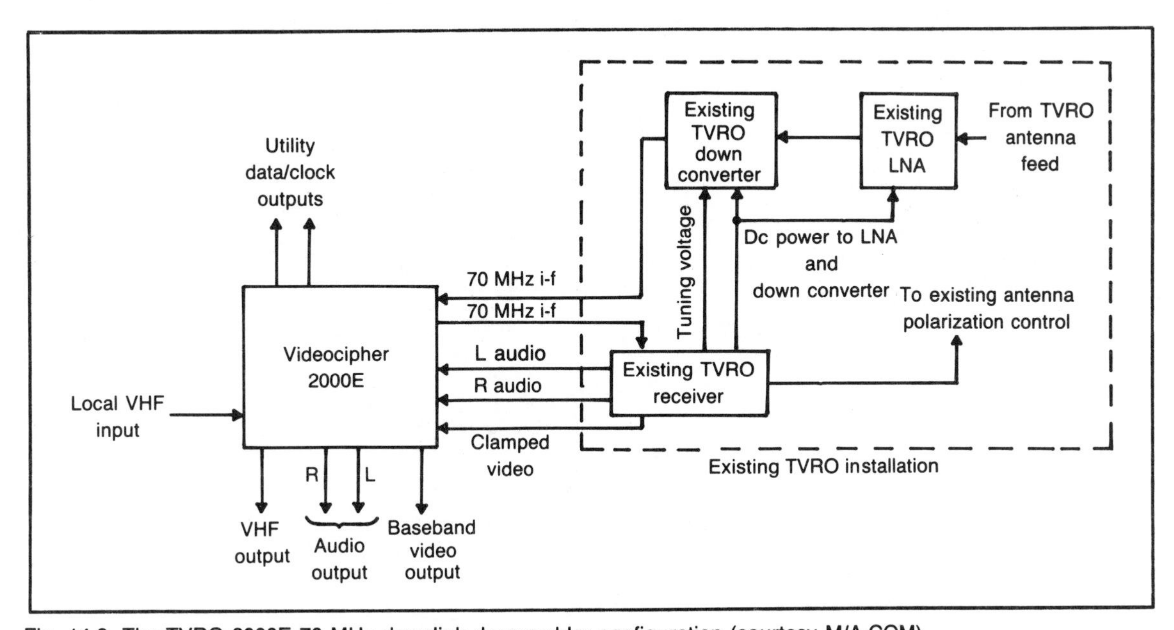

Fig. 14-9. The TVRO 2000E 70 MHz downlink descrambler configuration (courtesy M/A-COM).

medium that's using it. We would expect a brief flurry of scrambling, but followed by "open" skies when several million private earth stations are finally on the ground. Advertisers have yet to ignore a potentially lucrative audience, especially when they can paint a good looking picture, vending hundreds and thousands of products while so doing.

Chapter 15

Troubleshooting Analog and Digital TV, and Satellite Earth Stations

BECAUSE THERE IS NO SUCH THING AS A COMplete troubleshooting guide or procedure for everything electrical, we thought bits and pieces here and there might be more useful than a treatise on some one particular topic that might not be useful to everyone. So we'll range everywhere from stereo to satellites, with whatever digital procedures can be generated at this early stage. As you are well aware, a product usually has to be on the market for a while before serious troubles appear, and digital has just begun to surface anywhere, let alone over some measurable, statistical period. So in this respect, we'll do our best with whatever seems to suit the occasion. As for analog and satellite receivers, there's much more to work on; and in analog TV we'll have a wealth of information for you which should stand the test of time—provided you're willing to use signal generators and oscilloscopes to do the job. Simple multimeters and ohmmeters are completely off campus and will *not* be considered in the procedures.

With both analog and digital circuits now in almost every television receiver, only fairly complex-signal and basic dc-level analysis is at all worthwhile attempting. A multimeter measurement of some remote test point almost tells you nothing since so much direct coupling is involved that one receiver section will affect the other almost indiscriminately and offer insurmountable difficulties if you try and rely on vacuum tube tactics. Except for revitalizing a stereo generator, we won't talk about vacuum tubes at all; they're just a little too far in the past and not at all likely in the future.

THE STATE OF SOLID STATE

It's probably wise to start with a short discussion of what to expect in today's world of solid state before continuing on into complex circuits that have all these ingredients mixed within. Solid-state basics are relatively simple, it's the applications and relationships that are difficult to master. But, once you have the appropriate fundamentals, you have a good chance of surmounting the rest. Without basics, however, there's no way that either engineers or technicians can cope with some of the most complex electronics the consumer industry has ever known. Eventually with digital we may, and probably will, reach the state of automated readouts that can say "change this" or "substitute that;" but in the meantime, capacitors, resistors, ICs and just plain old transistors will be with us for sometime to come. Know their proclivities (tendencies) and you're on the 50-yard line already. Resist, and every kickoff goes to the end zone with no possible runback.

Always remember that semiconductors, as opposed to vacuum tubes, are current-operated devices since it is current flow across their junctions that produces switching, amplification, mixing, clipping, detection, or what have you. Their loads may be either passive or active (other transistors), with passive or active filtering also. Before we get into that, it's well to look at several categories of solid-state devices that you'll be meeting and hopefully conquering in every sphere of solid-state products on today's markets.

Bipolar Transistors

EBCs are just as easy as ABCs; actually easier, because there are only three letters in the bipolar alphabet instead of the 26 in English. Furthermore, the meaning is totally specific and what these transistors do is unmistakable: they either mainly switch or amplify, with the first often called *logic* and the second *linear.*

In Fig. 15-1, you see the physical diagram and symbol for a diode. When two semiconductor types are joined together, some of the N-type electrons cross the junction and combine with P-type holes while holes also cross and recombine with free electrons in the crystaline structure of the semiconductor. Charge carriers diffuse from one side to the other primarily because of thermal energy and this

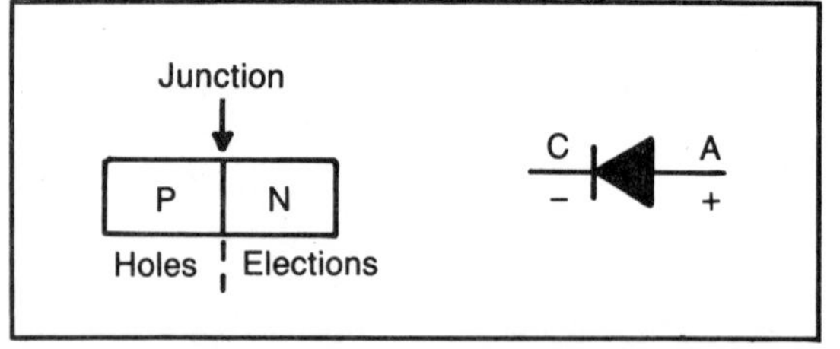

Fig. 15-1. Diode semiconductor formation and symbol.

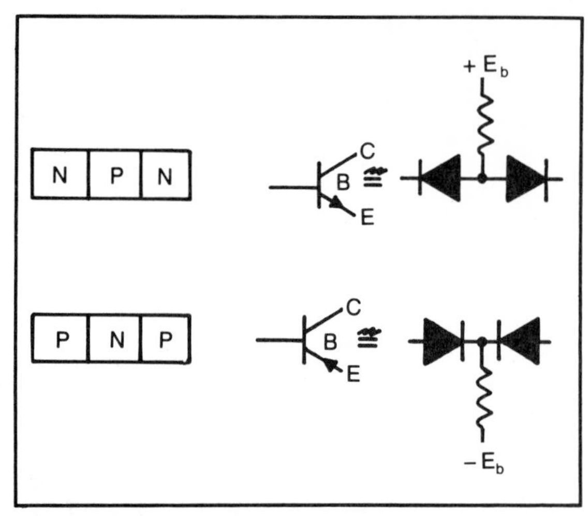

Fig. 15-2. Bipolar NPN/PNP transistors and symbols are similar to back-to-back diodes but dissimilar in performance.

movement is called diffusion current, resulting in a potential barrier which, at quiescence, forms an energy barrier preventing further carrier diffusion across the junction. It is not until a potential such as *forward bias* is applied that current begins to flow in this or any other *bipolar* semiconductor. As you will see shortly, some field-effect transistors react differently.

Should you apply reverse bias to a diode, the energy barrier (or space charge region as it is sometimes called) widens and very little *reverse* current flows. With forward bias, just as soon as a small 0.5 to 0.7 voltage drop across a silicon diode or 0.2 V drop across a germanium diode is overcome, then current flows freely, depending on the biasing potential. The cathode of any diode must be negative with respect to the anode by at least the aforesaid drops to make it conduct. Otherwise it just sits there and looks pretty or ugly, depending on your elation or frustration.

Butt a pair of diodes head-to-head and you have a full-fledged apparent transistor. Now there are *three* junctions with which to contend (Fig. 15-2). Called the *emitter, base,* and *collector,* they are your EBCs of all bipolar semiconductors whether linear analog or digital logic. Like all Gaul being divided into three parts, *all* bipolar semiconductors are likewise partitioned. Doping during the formation process, determines the characteristics of these transistors, which then have controlled impurity levels and spacings on either side of the junctions. Obviously there are many ways to form a transistor, but your interest lies in how it works not the physics of manufacture.

This results in a new wrinkle. Since in NPN transistors the majority carriers are electrons, collectors of these transistors must be positively biased to attract them. Similarly, PNP transistors have holes as majority carriers and collectors must be negatively biased to attract them also. So now, as shown in Fig. 15-3, you have the base and emitter *forward* biased, and the collector *reverse* biased. This rule holds true for any of the three most useful configurations: common emitter; common base; and emitter follower (or common collector). Biasing for all three appears in the diagram and, as you see, is uniform regardless of their functions, with current flow always in the direction of the arrow.

The three types of configurations have specific uses: the common *emitter* has *both* current and voltage gains with input and output impedances between 20-5 k and 50-50 k ohms, and power gains on the order of 40 dB; the *common collector* (or emitter follower) has high input and low output impedances (depending on the emitter resistor), with less than unity voltage gain and is used primarily as a current driver for other circuits; the *common base* has an input impedance usually less than 100 ohms, and an output impedance of from 1 k to 1 megohm, but little current and a power gain of possibly as much as 30 dB. This is possible because of the low-input and high- output impedances, but must always be considered in relative terms rather than in some power driver configuration, which it is not. In all instances, the inputs forward bias these three transistors, but only the common emitter will invert, while the other two will not. As you can see, therefore, each has specific uses and is placed in a circuit for some very particular reason. Further, every different type of transistor has special characteristics such as gain, power handling ability, bandwidth, and switching speeds and *must* be replaced with an identical or reasonably equivalent type if you want original performance.

By the way, if you wanted the back-to-back diodes in Fig. 15-2 to operate, you'd have to place a bias voltage at their junctions as well as operational incentives at their inputs and outputs. So they don't operate like transistors at all, and in addition, there'd be no gain and at least a drop of 1 volt if you even managed to get a signal through. Diodes, then, are for clipping, clamping, temperature compensation, signal guidance, and rectification. Special diodes with selected breakdown characteristics are called zeners and are used as reference of voltage regulators, and sometimes couplers in integrated circuits. Once again, they are to be replaced only with the same type semiconductors if you want their circuits or special functions to work. Zener knees and power handling abilities are especially important.

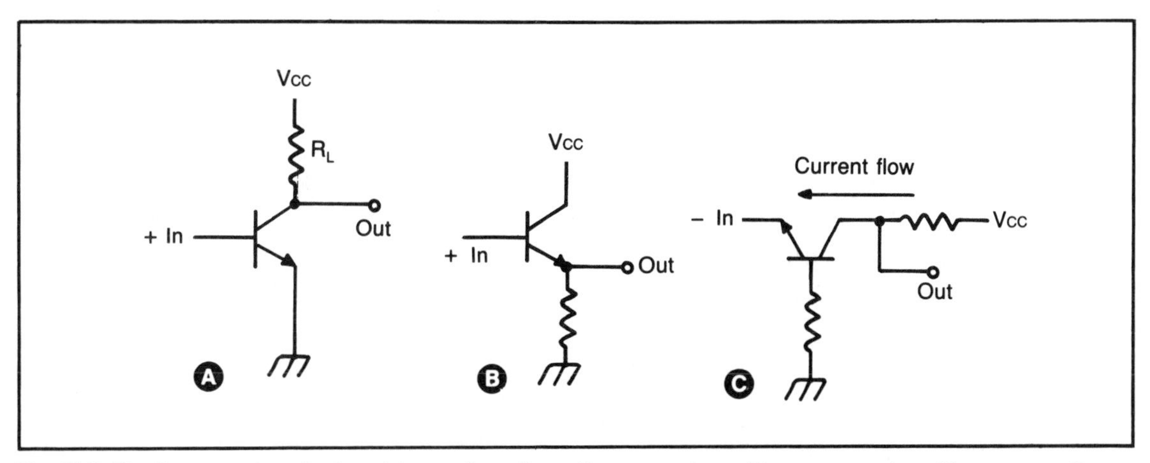

Fig. 15-3. The three usual *analog* transistor configurations: A) common base; B) common emitter; C) common collector.

MOS Field-Effect Transistors

These are especially important in the analog portion of the television business because they're the rf amplifiers and mixers most often used in tuners. High-input impedances and square-law characteristics make them particularly suitable as voltage amplifiers. Constructed as single-channel, depletion (always on) or enhancement (required forward bias) types, their equivalent electrodes to transistors are source, gate, and drain: EBC = SGD, with the base normally tied to source in the usually-used configuration (see Fig. 15-4). Those found in television tuners usually have two gates, one for agc control, and are always depletion mode types requiring pinchoff to be shut down. They offer low power consumption, ruggedness, and low cross modulation with only one charge carrier and a single channel. Gates in these devices have metal control electrodes and act as a charge-storage or control element between source and drain. When *reverse* biased it will deplete charge carriers in the conducting channel.

The normally off enhancement MOSFET mode you're not concerned with because it's not ordinarily used in consumer products. Just like transistors, there are common source, source follower, and common gate circuits; but you'll only normally see the common source variety used in your applications. In the common gate configuration, both base and gate are usually referenced to ground for N-channel devices.

Thyristors

These consist of alternate layers of P- and N-type semiconductor silicon that may be placed in series to form a number of different thyristors. Your basic types are *diacs, triacs,* and *SCRs.*

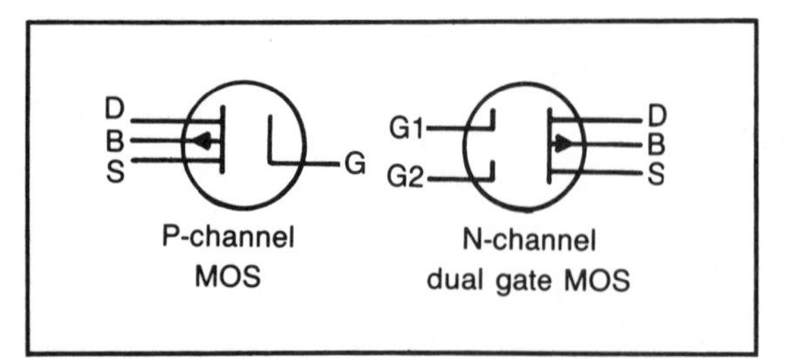

Fig. 15-4. Single and dual-gate metal-oxide semiconductor transistor MOSFETS.

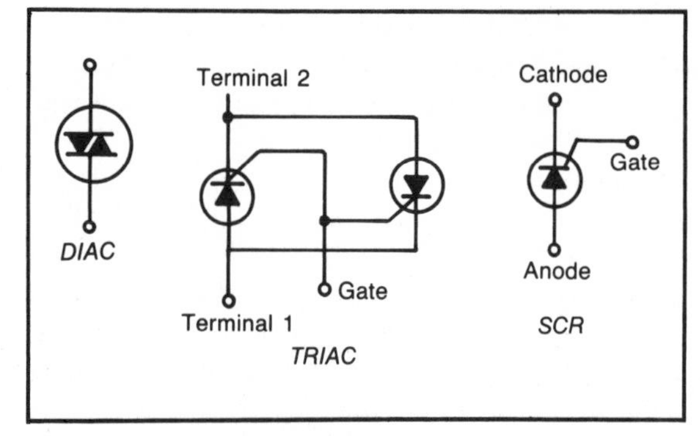

Fig. 15-5. The diac, triac, and SCR thyristor family with which consumer TV is concerned.

The Diac (Fig. 15-5) has three layers, two electrodes, what appear to be two back-to-back diodes in parallel, and a "floating" base. Normally used as triggering devices, impressed voltages must reach the breakover point before diac avalanche. Current *increases* substantially with decreasing voltage and may be switched *on* with either polarity of applied voltage. It normally operates at 60 Hz.

The Triac, like the SCR has three electrodes, two terminals and a gate, and will operate on either voltage polarity applied to the appropriate main terminal. It both blocks voltage-current in the off state and conducts in the on state. Breakover voltage can be controlled by positive or negative current pulses to the common gate. Usually there aren't many of these around.

The SCR is the most important of all the thyristors used in television. Used as a voltage regulator, supply, and safety cutoff device, the silicon-controlled rectifier is virtually everywhere, especially in the cheaper receivers. It is a three-electrode, four-layer device; blocks current in the reverse direction and transmits large amounts of current in the forward direction once the gate electrode is sufficiently forward biased.

Once turn-on occurs, the SCR becomes a closed switch, with latching current less than twice the holding current, and plain or sophisticated (phase shift) triggers applied. At the end of conduction, SCR forward current reduces below holding current or to zero, some reverse current flows, the gate recovers, and the SCR then turns off, waiting for another gate trigger to turn it on. Note that the gate *cannot* turn the SCR off. You do have to be careful of radio frequency interference with these triacs, as well as that which may be carried over power lines. Effective filters are necessary, and may be of the ordinary LC type to attenuate harmonics and reduce noise interference.

Linears

Now used in all video and audio signal processing, linear integrated circuits have been a real boon to designers and the semiconductor industry alike for a new market, cost reductions, more and better electronics, and high repeatibility. Since these have been thoroughly covered in the video and chroma processing chapters, we won't laboriously repeat, other than to say that the differential amplifier, peak and/or synchronous detectors, and emitter followers are the backbones of all analog ICs. A differential amp is shown in Fig. 15-6 with two inputs and two outputs. It can have only one input and one output if need be, and the outputs may be either inverting, as shown, or noninverting if the signal is coupled emitter-to-emitter and then amplified. Dual circles in any drawing represent a current generator, usually another transistor which is shown here instead.

LOGIC

For many of you, this is new state-of-the-art that's actually old as the hills. But, it's only recently been applied to the television—first in tuner control, then the usual volume, picture, color potentiometer controls, and now to the entire receiver following the video detector. Once you're familiar with its operation "you're going to like it."

We like to think of logic as including shift registers, memories, microprocessors, LED drivers, stepping controls, countdown circuits, A/D and D/A converters, gates, flip-flops, and all the rest. In short, anything that operates between the "rails", as our English cousins call them, meaning on/off voltage swings from 0 to VCC. Actually this isn't possible because there are load resistors and internal transistor resistances. So we're really talking about some 80+ percent swing, depending on driving and passive components already in the circuit. The one outstanding exception would be complementary MOS (CMOS) logic where N and P elements form alternate loads as these transistors switch states from high to low, according to inputs.

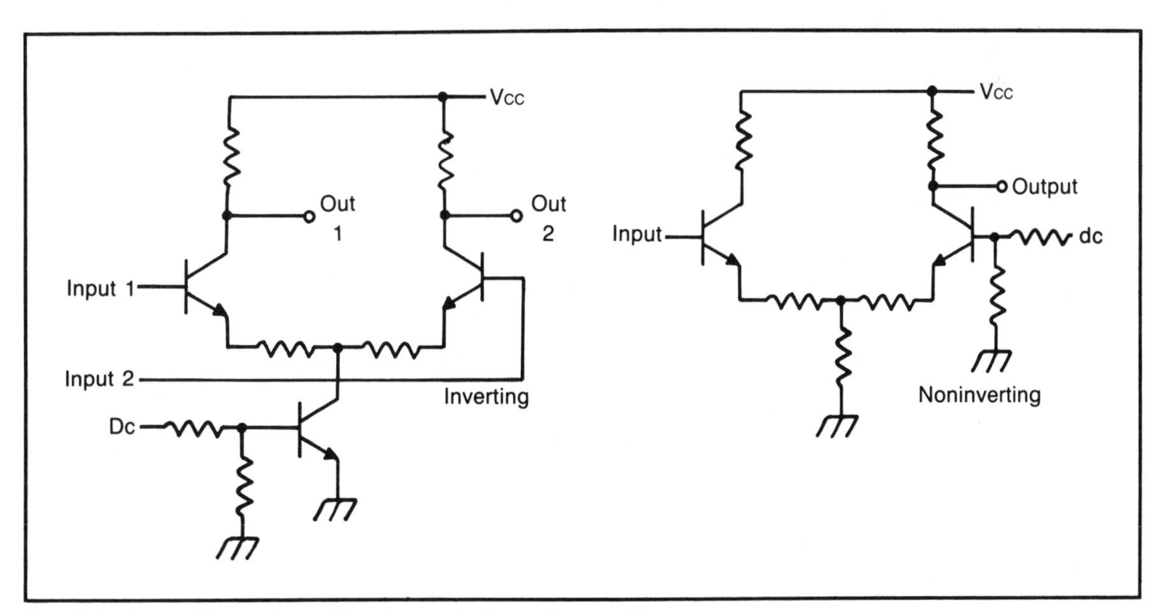

Fig. 15-6. Basic differential amplifier that made linear ICs possible.

All of this, however, is pretty much covered symbolically, so we really won't have to be too concerned with the individual semiconductors and precisely what they do.

With the introduction of medium-scale and large-scale integration to consumer/and, you'll seldom see a logic schematic any more, and often you're fortunate to have even a representative block diagram. One television manufacturer tells us he can have production logic ICs in house 12 weeks after inception. With even more automated equipment coming in the future, that time could, eventually, be cut in half. So, if you don't think 1s and 0s are here to stay, ask the TV people who are going in that direction as fast as their computers and the West German Intermetall ITT Division will take them.

Gates

The basis for *all* logic begins with gates. Even flip-flops can be made of gates. In fact AND, NAND, OR and NOR gates practically do everything we want in the way of basics. Just add enough inputs for one good output, and your introduction to logic is off and switching with impunity. Obviously, there's more to this particular science than just these four, but in the 130+ years since George Boole of Dublin University first wrote about it, his *Investigation of The Laws of Thought* has traveled a long way. With computers now governing our very lives and daily welfare, logic has decidedly come of age for everyone.

Having said all that, let's look at a few hardware examples, then a few symbols, etc., before swinging into the combined science and art of finding fault(s) where it or they really exist.

AND/OR gates are the basics and are represented by diodes (Fig. 15-7). In the case of the AND gate, two positive A and B inputs have to overcome the back bias from B+ to produce an output. With the OR gate, any positive (in this case) input produces an output, and only *no* input produces an equivalent zero output. The AND gate, sometimes called a coincidence gate, may be found in some logic occasionally, and so may the OR gate. Neither, however, are especially flexible and offer no voltage or current gain, so they're rarely used.

Supplanted by the NOR and NAND transistorized gates, however, these and flip flops have become the backbone of logic wherever it's used. Here, current gain and accompanying inversions are apparent and because of the transistor, utility has been increased a thousand fold. The two gates are illustrated in Fig. 15-8, along with their truth tables. Observe the conditions for an output; otherwise, voltages remain at Vcc and constitute a 1 output. But whenever these transistors conduct, they force voltage drops across the load resistors resulting in a logic 0 (zero). Also observe that there is an equivalency between the NOR and NAND gates, which is known as De Morgan's theorem.

$$\overline{A} + \overline{B} = \overline{A \bullet B}$$

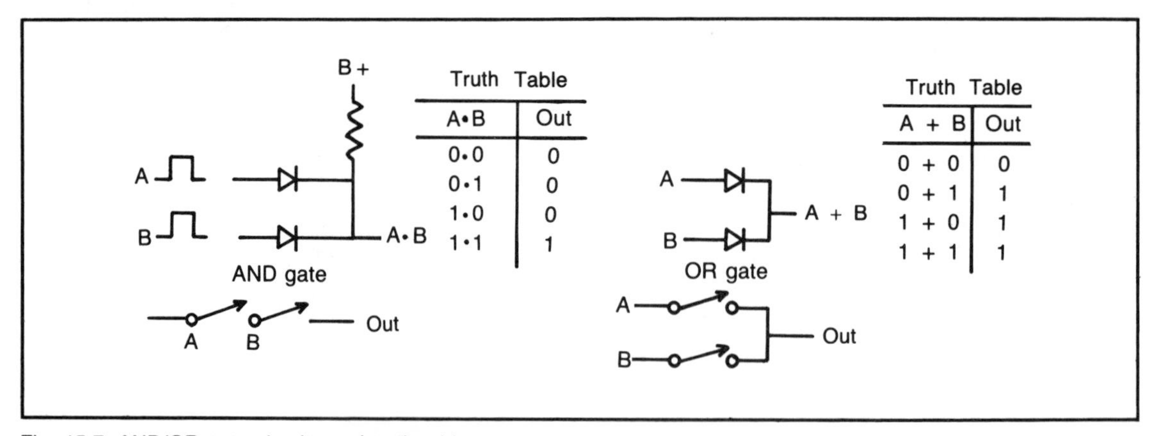

Fig. 15-7. AND/OR gate circuits and truth tables.

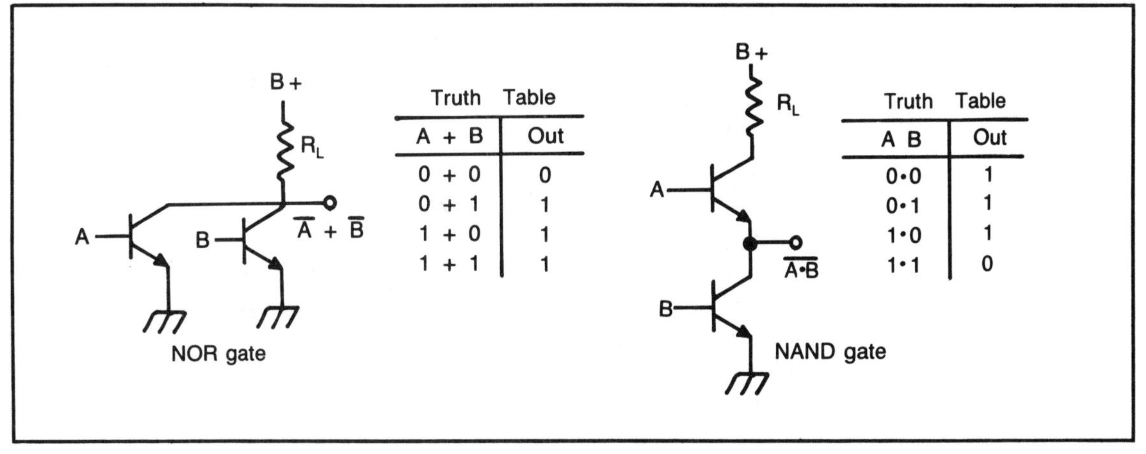

Fig. 15-8. With NOR and NAND gates comes flexibility and switching transistors.

Which, in logic, says that in the language of double negation:

$$\overline{\overline{A}} = A \qquad \overline{A + B} = \overline{A}\,\overline{B}$$
$$\overline{\overline{A}\,\overline{B}} = A + B \qquad \overline{\overline{A} + \overline{B}} = AB$$
$$\overline{AB} = \overline{A} + \overline{B}$$

OK, you're not logic designers and therefore we'll not proceed further, but this is an indication of what can be done with a very flexible and exciting medium—one where you're on the threshold of entering and in which you'll either sink or swim with its mastery. But do look at the combined truth table in Fig. 15-9: they're all there!

Flip-Flops

There are all sorts of these around including clocked NAND gates, Type D and master J-K slave flip-flops. The NAND clocked gate is an interesting one, and so is the J-K. These are used in shift registers, ripple counters, dividers, etc. and are a totally necessary part of the logic process (Fig. 15-10).

In the clocked NAND gate version you have both set and reset inputs going into NAND gates which must be coincident with the clock. Their outputs are connected to another pair of gates which are cross coupled to produce 1s and 0s outputs. Any change of state is *inhibited* (rejected) without a positive clock input. This gives somewhat more control than firing pulses into a plain set-reset flip flop which the two cross-coupled gates represent.

For additional control and to prevent any ambiguities in outputs, the J-K flip flop has come into use, with two AND gate inputs and internal master-slave conditions that prevent false triggering. When neither J or K inputs are high, the clock pulse will

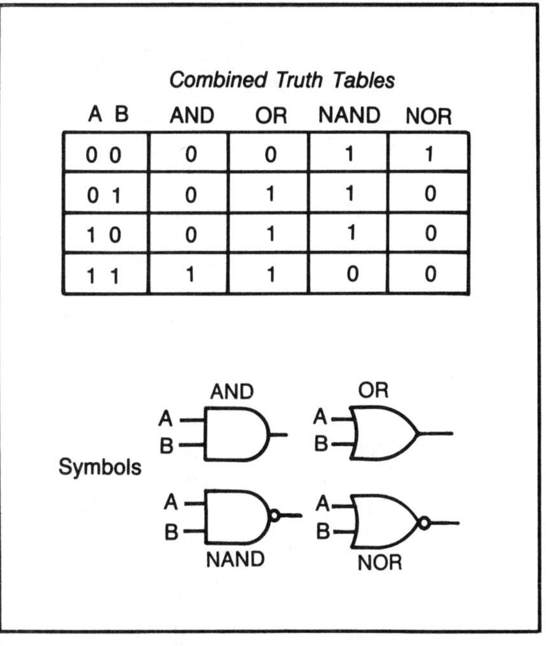

Combined Truth Tables

A B	AND	OR	NAND	NOR
0 0	0	0	1	1
0 1	0	1	1	0
1 0	0	1	1	0
1 1	1	1	0	0

Fig. 15-9. The combined truth table for all basic logic gates—note obvious inversions.

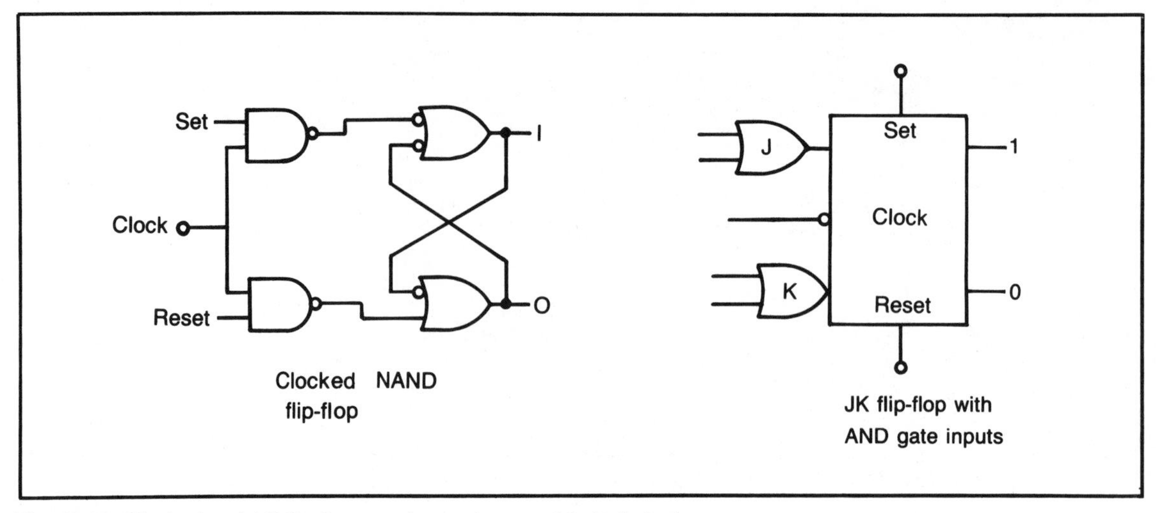

Fig. 15-10. Clocked and J-K flip-flops are integral parts of today's logic.

not result in a change of states, and when both J and K inputs are high, the clock pulse will cause a change in states from low to high or vice versa. There is *no* intermediate condition.

Binary Notation & Symbols

The remainder of the logic introduction consists of just a few elementary symbols for the various logic elements and a binary conversion from decimal to familiarize you with the procedure.

Decimal	Binary
0	0
1	01
2	10
3	11
4	100
5	101
6	110
7	111
8	1000
9	1001
10	1010
11	1011
12	1100
13	1101
14	1110
15	1111

Binary Coded Decimal

(in 8,4,2,1 code)

Similar, except when grouped, and then each group stands for a number part. For instance: 525

In binary coded decimal it would be

5	2	5
0101	0010	0101

Easy to use but requires more bits to represents a number.

After decimal 15, the binary 1s and 0s are grouped in blocks of four so that large numbers may be conveniently handled. For instance if 525 scanning lines were to be represented in binary, they would become 0010 0000 1101—all based on powers of 2 right up to the left-most 0010, which amounts to 512, with 1101 filling in the remaining 13 for a total of 525. Of course, there are many other coding schemes such as gray, octal, other digital basings, and so forth, which you don't really have to know.

For this portion of the chapter, the various illustrative symbols are really all that's left and you have most of those already. The remainder are really combinations of gates and flip flops to accomplish the enormous number of logic functions that are readily available. Although the various types of logic hardware such as T²L, ECL, RTL, DTL, Schottky, etc. are all being used, only MOS and T²L really concern you, and these have already been illustrated. Additional logic circuits are listed in Fig. 15-11, in the event they might be of interest, but they are not necessarily used in consumer products—at least those of today.

TROUBLESHOOTING PROPER

After that introduction to analog and digital logic and circuits you should have something of an idea of what goes on in many of these circuits so that knowledge may be applied and a reasonable answer obtained. Unfortunately, even in integrated circuits, you won't always find a definite "yes" or "no" answer, and the hardest of the lot are the *maybes.* You don't need this information for the obvious: a short or open kills your circuit and destroys any signal processing along with it. And, with so much direct coupling, what would otherwise be minor difficulties in ac-joined circuits can prove a nasty headache in dc-coupled ones. As we move from one sector to another in the television receiver we'll try and point out critical portions of certain circuits and a few good check points of what to look for. Let's begin with power supplies.

Power Supplies

Whether half wave, full wave, bridge rectifier, regulated or unregulated, the voltage output is the first thing checked. Bad filter capacitors and rectifiers both reduce voltage and, what's more important, current output. A lower voltage at other times

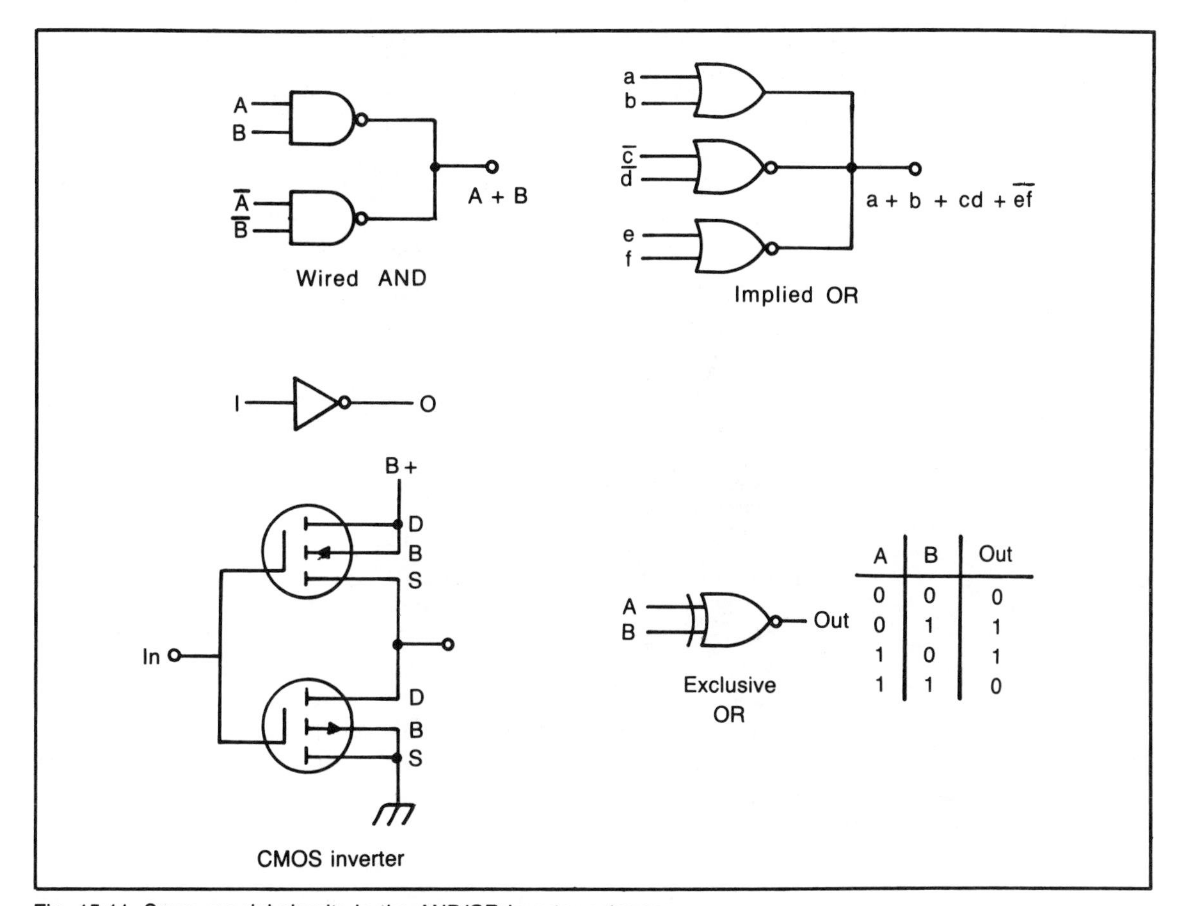

A	B	Out
0	0	0
0	1	1
1	0	1
1	1	0

Fig. 15-11. Some special circuits in the AND/OR inverter category.

means there's too much current flowing into somewhere, upsetting the entire power supply. If you have no idea of what that current should be, the drop across a 5 or 10-ohm resistor at the output, and using Ohm's I = E/R law, should give you a good indication. If you don't have any idea of current drain, then module removal or possible branch disconnects can help. If output should return to normal when some part is removed, then you have the problem source ready for a "fix."

Should the power supply itself be at fault, here's how to go about that: Let's say you have a full-wave supply with the usual π-shaped capacitor-inductor (choke)-capacitor filtering. Set your oscilloscope's time base for 2 msec/div., put a probe on the input and see what appears. At 2 msec/div. you're seeing a good sawtooth of voltage at some 2 V (top trace) that recurs approximately every 8 milliseconds. Therefore 8×10^{-3} divided into 1 amounts to 125 Hz (round figures) and this means you have full-wave rectification, and the bridge or full-wave circuit is operating (frequency-wise) as it should. Also, at 1 V/div., you have no more than 4-volts sawtooth amplitude resulting from the charge and discharge of the input capacitor (Fig. 15-12).

In the lower trace, there appears just a tiny ripple. Using a special 8 × 10 on this particular oscilloscope, it turns out that the lower (or output) trace has just 1.5 millivolts of ripple, and that's just fine for a transistorized regulated supply. If the input filter or output filter, for that matter, had produced a ragged, lower-valued waveform, either the filters could be faulty or some diode in the rectifier assembly would be faulty. Were this a half-wave supply, your time base reading would have doubled and the output measured 60 Hz, but with considerably more ripple.

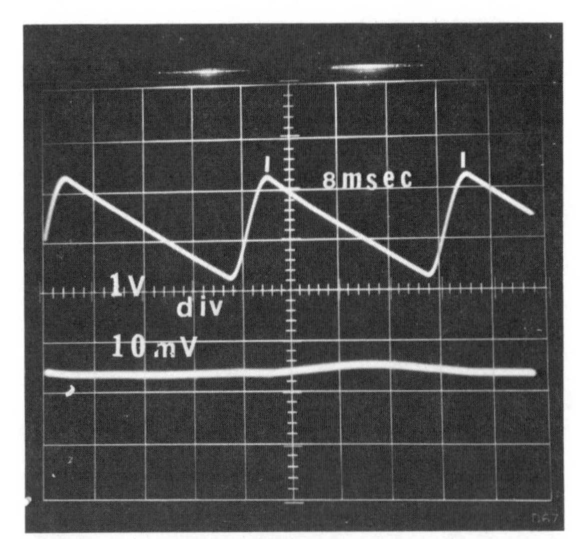

Fig. 15-12. Input and output filtered waveform ripple in a full-wave power supply. Output ripple should only be in low millivolts.

To restate, in power supplies you normally have just two problems: either the filters have dried out (seldom shorted) and are not working, or one of the rectifier diodes is bad and requires replacement. If, however, the output was low, but all waveforms relatively adequate, then there's current drain from some other part of the receiver and you'll either have to pull modules or find another means to disconnect.

This is especially true when there are multiple power supply connections, often originating from different sources that are not necessarily related. What you see in Fig. 15-13 is an excellent example, including a start up (initial) voltage, pulse-width modulator, chopper, and high-voltage shutdown. We won't go into excruciating details, but this is a triple plus Zenith chassis, having a considerable number of varied functions on a single M10 module.

Ordinarily, you'd pull this module and return it to the distributor for exchange at a relatively reasonable price. There may be some diodes, big capacitors, or a single transistor that's the problem and the module is repairable on quick turn around. There are also pulse-width modulator waveforms available for oscilloscope viewing that may help with faulty locations. Unfortunately, these big, omnibus supplies are no piece of cake and, with high voltage included, you're usually better off exchanging it for a rebuilt or new unit. In a sense, that's a copout, but in the service business time is money, and there's no use wasting time if a replacement is handy.

Shut Down Circuits

Many of these are alike, and most either kill

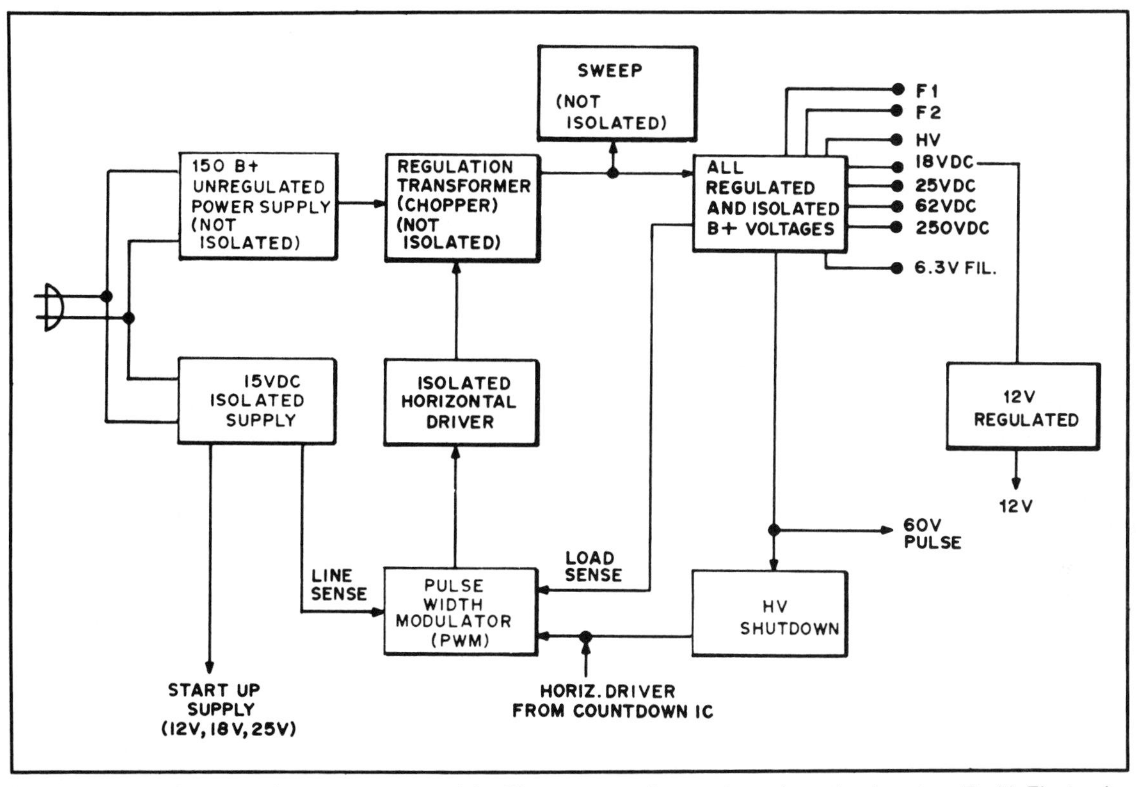

Fig. 15-13. Zenith's electronic power sentry module. Change or repair—you have the option (courtesy Zenith Electronics Corp.).

high voltage completely or throw the horizontal oscillator out of sync, virtually amounting to the same thing. Figure 15-14 shows a good view of these transistorized sensing circuits. Used for sweep failures and X-ray protection, it only works when high voltage increases about 6 kV or when there's a large potential difference between hot-cold portions of the chassis, which is a simple resistance-voltage trigger.

A sample of HV is rectified by CR3353, which charges C3351, subsequently discharging through six multi-resistor networks leading to the cathode of zener CRX3355. Should high voltage increase beyond normal limits, rectified and filtered dc forces send zener into breakdown, turning on transistor QX3351, which lowers the base potential of QX3338, turning that PNP transistor on also. When this occurs, current flows from emitter to collector and forward biases CR3354 *and* CR3337, shutting down both the pulse-width modulator and the horizontal countdown IC. The brightness limiter is also affected by dc flowing through the shutdown transistor's emitter when QX3351 is conducting.

When there are shutdown conditions, the receiver will have to be turned off and then power applied through a variac, with both waveforms and voltages monitored into that circuit. As you increase ac you'll usually begin to see the trouble source and can work from there. Rarely do components in the shut-down circuit fail all by themselves. You'll just have to monitor all input voltages using digital multimeters and dual-trace oscilloscopes to locate the troubles—and sometimes there are several. The bipotential resistor is not shown to the base of QX3351 because it is used only in the late chassis.

Linear Integrated Circuits

These are the analog i-fs, chroma, and

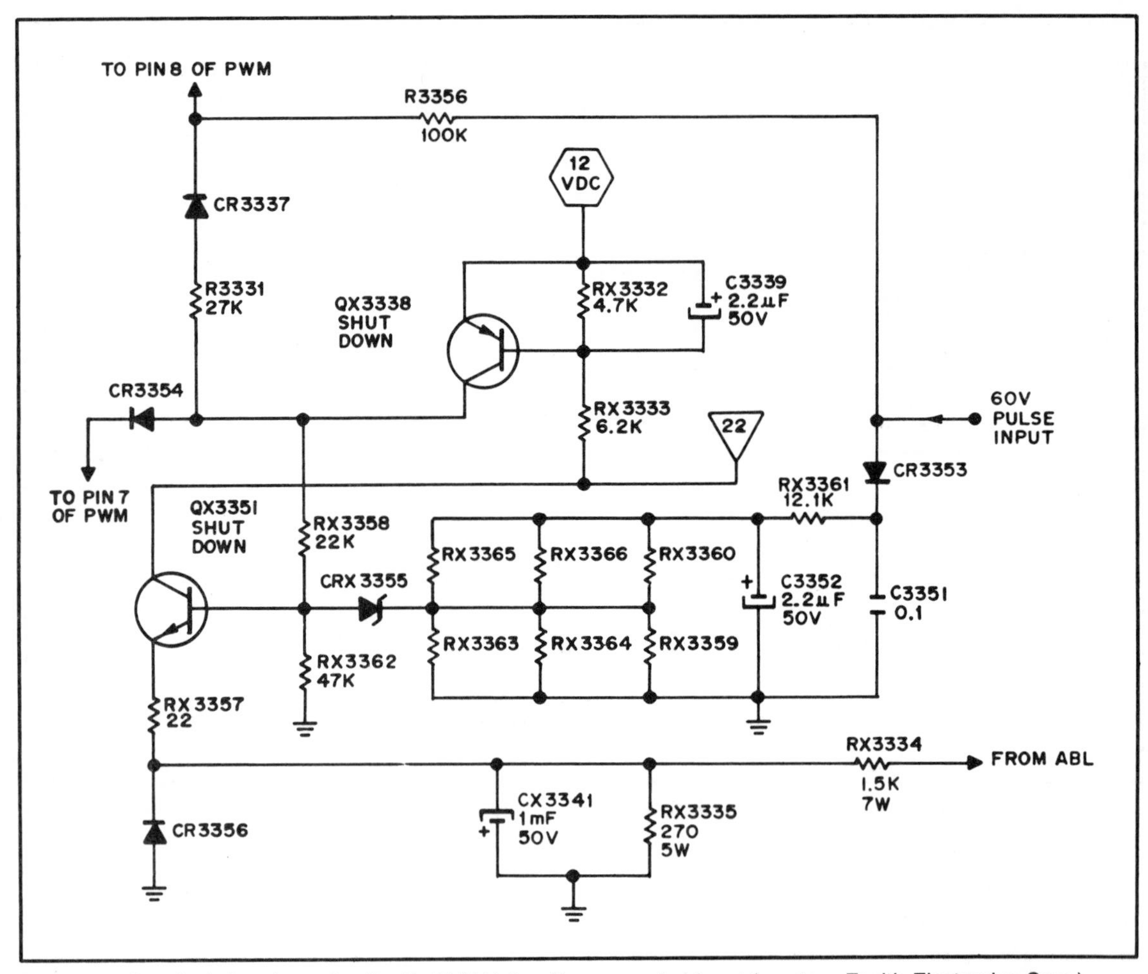

Fig. 15-14. A typical shut-down circuit with 60 V high voltage sampled input (courtesy Zenith Electronics Corp.).

luminance processors that will still be with us in the less expensive receivers for many years to come. Usually operating from 6-12 V sources, the dc and ac inputs are the *only* means of analyzing their troubles, unless one is shorting and warm or hot to the touch. Here you may *not* depend on simple meter readings to find faults. Signal in, signal out will work 1,000 times better and quicker than any analog or digital multimeter. Unfortunately, ICs often have varying resistances and can respond just a little differently to various stimuli, meaning that those tenths or hundreds of some voltage means little or nothing. Here, if your power supply is adequate, there's small ripple, you have an adequate signal voltage incoming and little outgoing, then that IC needs to be changed. On the other hand, a lower than normal dc-coupled input may result from the circuit before rather than the suspected IC itself. However, should there be ac-coupling and there's little input signal, then either the capacitor, itself, is bad (you'll see excessive dc) or the integrated circuit has given up the ghost.

Once in a while, however, there are several inputs and outputs affected in these linear ICs and you'll just have to sweat the various signal and dc paths until you find the problem. Sorry, but there's no other way to do analog ICs other than careful analysis. Part of the solution, of course, is understanding the circuit and applying this knowledge to troubleshooting. That's why we've tried to give you as much theory as possible throughout the book. Linear ICs are just another collection of diodes, resistors, and transistors, and something has to make them work. (See Fig. 15-15).

Power Outputs

All these are the usual EBC transistors, normally requiring considerable drive voltages, high voltage swings, and often have emitter couplings or feedbacks associated. Where HOTs (horizontal output transistors) are concerned, the horizontal oscillator *must* be on its 15,734 Hz frequency, the driver transistor and its normal current driver transformer have to be operating as a rectangular waveform driver, and power supplies—especially the collector supplies—should be correct and usually very well regulated because this often constitutes the *only* means of high-voltage regulation.

RGB luma-chroma video outputs, too, will often do the color and fine detail picture mixing before signals enter the picture tubes. All must arrive at the emitter-base mixing point in relative amplitudes and at certain dc levels. If not, you'll have saturation problems that may erroneously point to a perfectly good picture tube. Luminance amplifiers must do their jobs just as chroma demodulators must do theirs. It's impossible to produce well-balanced color pictures otherwise. Use a gated rainbow generator for your signal source.

Audio outputs often have feedback from their

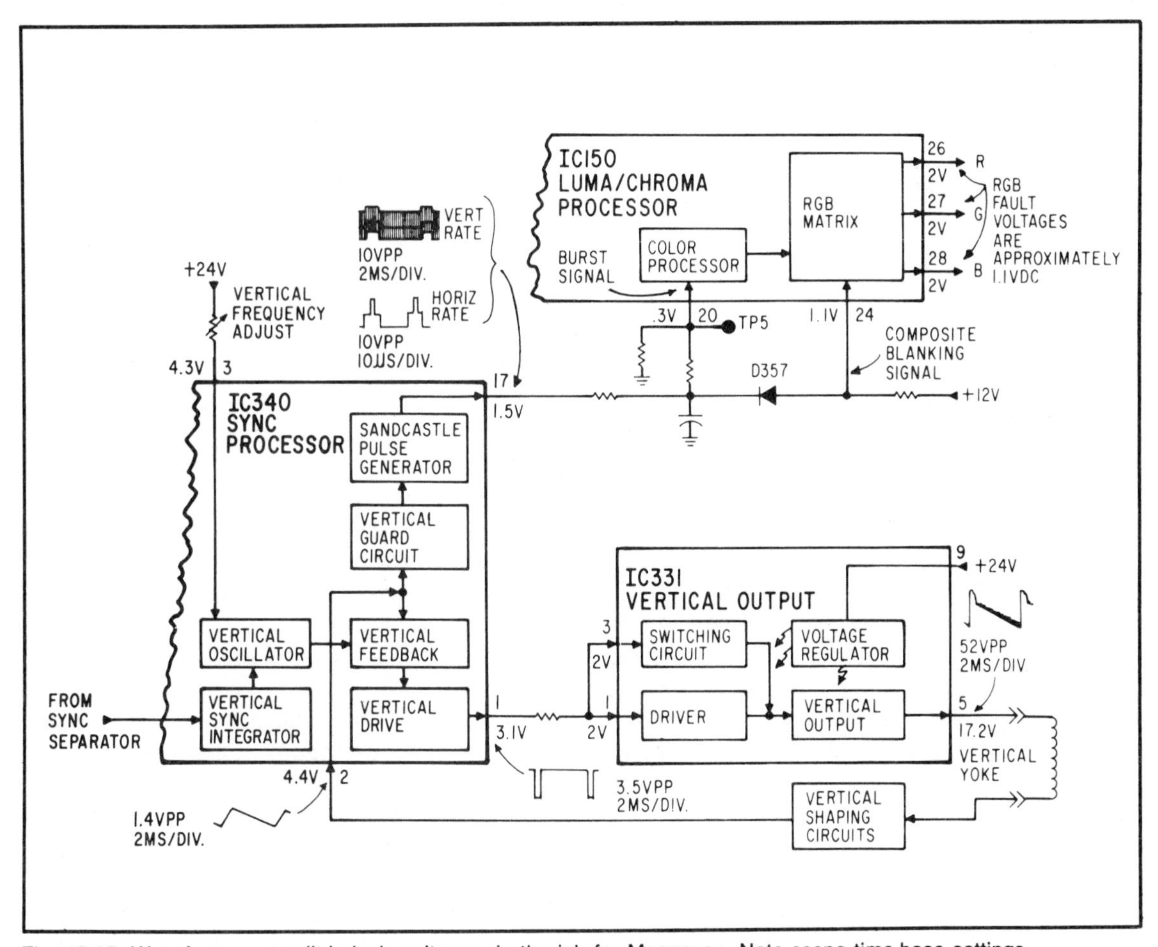

Fig. 15-15. Waveforms not split-hair dc voltages do the job for Magnavox. Note scope time-base settings.

emitters, and signal distortion or motorboating may result if this feedback path is obstructed. Here you'll want to use a sine-wave generator either through rf or baseband and trace the signal back from any audio output to discover the trouble.

Sorry to say, audio problems will become doubly complex with the addition of BTSC-dbx, or multichannel sound. There'll be two sound detectors then, plus all sorts of additional carriers and highly complex stereo detection and buzz quieting. Only a special multichannel sound signal source will help here—another absolutely necessary piece of equipment in addition to oscilloscopes, NTSC and gated rainbow color bar and multi-function generators. The tender touch won't help a bit here. Buy good equipment and learn to use it now! Incidentally, oscilloscopes with 5 V/div tops won't really help since 5 × 8 × 10 is only 400 volts full range and this seldom covers vertical and horizontal outputs, B-boost, and *none* of the older tube or hybrid receivers. You'll want 10 to 20 V/div. at least for all around service work. The 8 and 10 mentioned are 8 vertical graticule divisions on the scope and the 10 stands for a 10× probe. You'll also want 3% accuracy, too.

Digital Count-Down Circuits

Here you can use a 5 V/div. oscilloscope if you were to limit your servicing to this IC. However, you won't, so do what you can with the rest. We had hoped to find a fairly simple example of a digital countdown IC but among the new receivers there just aren't any. So we must refer back to previous general explanations of what such circuits do and let that suffice. However, when working with such rather large and now highly complex ICs, do look at all inputs and coincidence circuits or pinouts first before tackling anything else. Often, waveforms have to be almost absolute for these ICs to do the job, and you positively cannot rely on basic dc voltages. Fortunately, such ICs don't go bad often, but when they do, you usually have total sync failure for either vertical or horizontal circuits. If the problem's horizontal, you'll very likely have neither high voltage nor raster. Vertical sync or output collapse ordinarily results in deflection falling to only a horizontal line across the screen. If so, then vertical drive voltages *and* final outputs will (or should) quickly tell the story.

Many of these composite sync ICs have a local oscillator (LO) that may be set by a potentiometer. It's always a good idea to use an electronic counter to put them right on frequency since aging and component drift will change their frequencies anyhow and you want as little of this as possible. Be sure, however, that the impedance of your counter doesn't upset the time constants of the oscillator. An oscilloscope check of such oscillators helps, too, with their amplitudes and fairly clean waveforms, general operating conditions should be reasonably apparent.

Alignments

We are speaking of i-f alignments and *not* tuners. Leave these to the factories which are going to repair most tuners anyway. They have test racks and experience you will never have, and it really isn't worth your time trying. Intermediate-frequency amplifiers are a somewhat different story, especially if you have a quality sweep generator with accurately calibrated markers and a good X-Y scope to go with it. You *can* align the new receivers just as well as the old ones, especially if you go into the tuner output i-f connection directly with the new sets and through the UHF input on the tuner with older models. The difference is that the new receivers have considerably more gain, and the tuner approach will overload.

At any rate, there are three principal reasons for any alignment that isn't absolutely necessary: 1) you have plenty of time and need the exercise; 2) there's a fat buck for your trouble; 3) you have something to prove. Otherwise, don't even bother your curly head.

However, for your kicks and mine, we did align—or look at the alignment curve—on a brand new G.E. BC chassis which looked simple enough, even with asynchronous detector. Sometimes these detectors give you problems (Zenith's used to) and at other times they don't; apparently the difference in design. At any rate, this G.E.'s alignment instructions said to shunt the detector oscillator tank circuit with a 100-ohm resistor. Sorry, but that didn't work. So we hooked up the bias potentiometer according to Hoyle, worked around with the sweep width and amplitude until there was a reasonable representation of a swept i-f response curve. Naturally, all shields had to come off first and the chassis underside exposed to get to the various test points.

But to make a long story short, injecting the sweep directly into the chassis tuner input did the job, including the SAW filter, and the response as shown in Fig. 15-16 looks pretty nice. G.E. specified the chroma and video carriers at 37 percent on the curve, and we could have widened the response slightly to put the 47.25 MHz lower adjacent channel sound trap a little closer to the waveform's skirt, but overall, G.E.'s factory job wasn't bad at all. A quick calculation will show the chroma and video markers are almost exactly at 40 percent—and that isn't bad for a quickie. Center top marker, as always, lies at 44 MHz which is the center of our U.S. i-f response curves as specified by the NTSC. Figure 15-17 shows precisely what *not* to do with an alignment. It shows saturation, sweep generation mistuning, and other waveform distortions.

When undertaking *any* i-f alignment try and follow the manufacturer's instrument to the letter. Then, if that doesn't work, innovate a little but

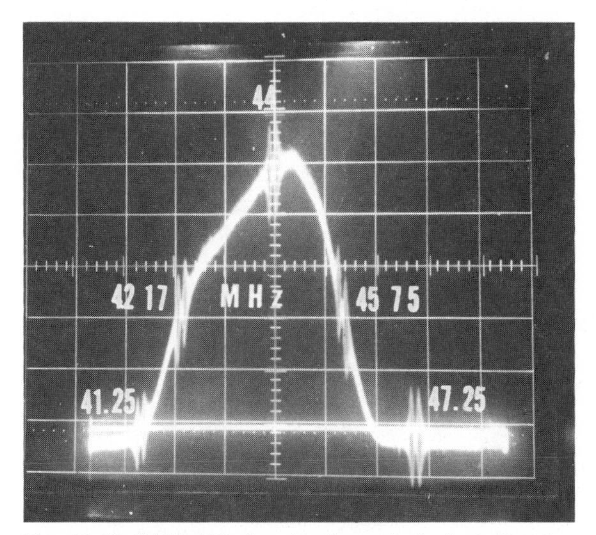

Fig. 15-16. G.E.'s BC chassis aligns nicely, including the usual transistor dog-leg characteristic.

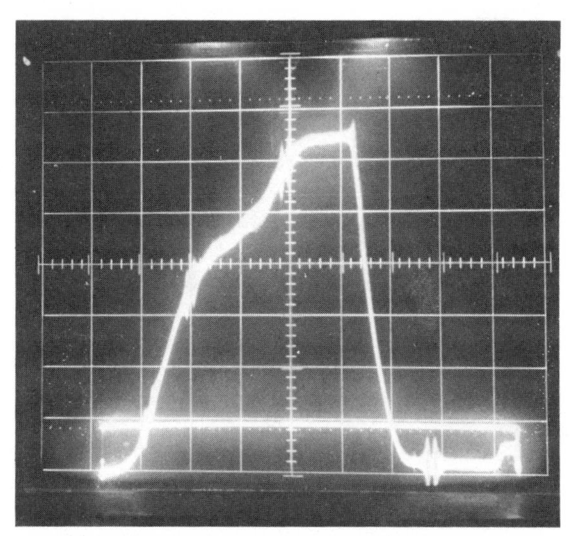
Fig. 15-17. Don't allow alignment enthusiasm to saturate and distort like this one.

remember that today's ICs are tender and the agc substitute steady-state voltage must be set exactly right, and your scope should be switched on dc amplifiers to hold the waveform steady. By all means use a potentiometer for the agc dc substitution.

Good alignments do matter a great deal—but you better know your business and have exceptional equipment to back them up. For once the i-f is out of kilter, color disappears, sound is garbled, and there's little recognizeable picture. True, we all make mistakes, but you might like to avoid this one because customers can become somewhat irate upon occasion. Inversely, an old receiver that's badly in need of such attention can come alive with tender loving care. Just apply agc gently (filtered if necessary), watch the amplitude of your sweep generator output and its markers, be positive that all "grounds" are at the same common, then use your oscilloscope X-Y amplifiers to do the rest. The only really miserable receivers to align, were the two-tube sets of some years back on which a certain company tried to save dimes by deliberately neglecting to add the third i-f. Some of the old timers will know exactly the one we mean. Its lineage had four hairy legs and howled at the moon—whenever energized.

SATELLITE EARTH STATIONS

Just a word or two about television receive only (TVRO) satellite earth stations—an electronics discipline some of you have already entered, and more are sure to be engaged, with the million plus numbers for consumers alone already on the ground. Basically, this has become a substitution business, with a drifting or otherwise defective receiver returned for repairs, and low-noise amplifiers are block-down converters treated much the same way.

Frankly, we wouldn't expect you to do much with the GasFet gigahertz technology of all of these signal preamplifiers, and normally there's no trouble in the signal feeds. As certain test equipment becomes available for the receivers, you may want to try a hand here since you're working largely in the megahertz spectrum, with which most of you are by now familiar. However, to do anything worthwhile, you must have *all* the necessary service literature, and you'd be wise to deal with *no* company that doesn't supply exactly what you want.

Yes, many of these satellite receivers have UHF-type tuners and other associated components with which you should have a working knowledge. The bandwidths are a little wider, initially, but they are reduced quite rapidly to about 4 MHz after final detection. Naturally, you're going to find some "tricks of the trade" here and there, and a few specialized components, but in several years, satellite receiver repair could very well become a profitable business, and much easier to work with than, say, a hybrid analog and digital television receiver—especially those with heavy shields around all the digital digs and possibly not enough around analog areas.

Replacement of cheap, warped, sagging and bent satellite dishes, too, should begin to make the dollars flow by 1990 or 1995, and some new designs by then may make sales and service even more attractive. We would strongly recommend a look-see at this new and fascinating electronics phenomenom, and the acquisition of sufficient electronics equipment to handle whatever comes along. There are several pieces of test gear we'd like to recommend immediately for service and precision evaluation of this part of our booming electronics industry: one is Wiltron's 6600 series of sweep-signal generators, and Tektronix's 494 spectrum analyzers. We're talking close to $50 thousand for both pieces, but they're pure gold, and any satellite electronics station can't very well do without either (Figs. 15-18 and 15-19).

You should know that our recommendation is a direct result of use rather than reading some catalog, and both have served very well on the field and in the laboratory. With just a little more FM bandwidth, the Wiltron sweepers can really do it all—at least the signal generating portion. (See Fig.

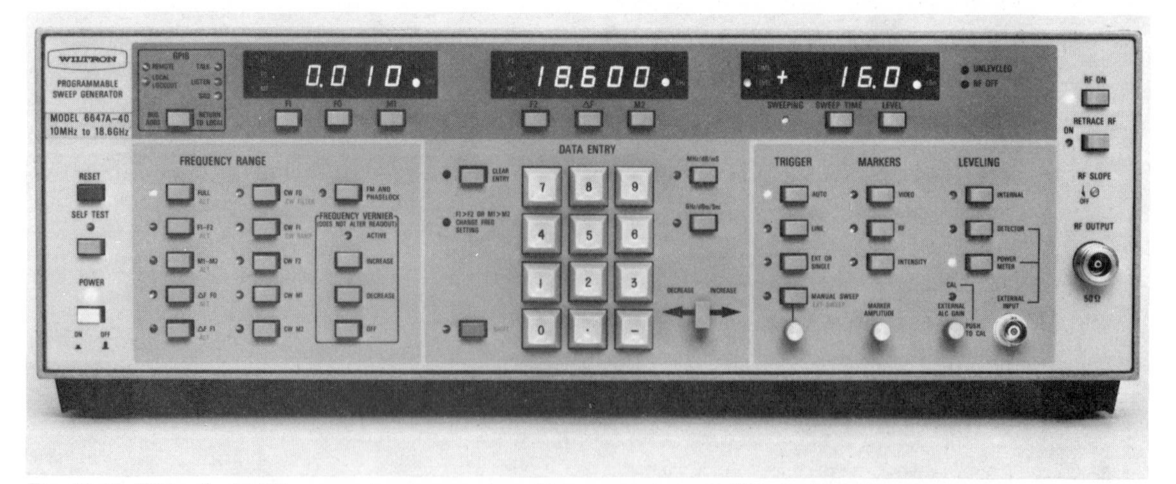

Fig. 15-18. Wiltron's 6647A sweep-signal generator with a range from 10 MHz to 18.6 GHz (courtesy Wiltron).

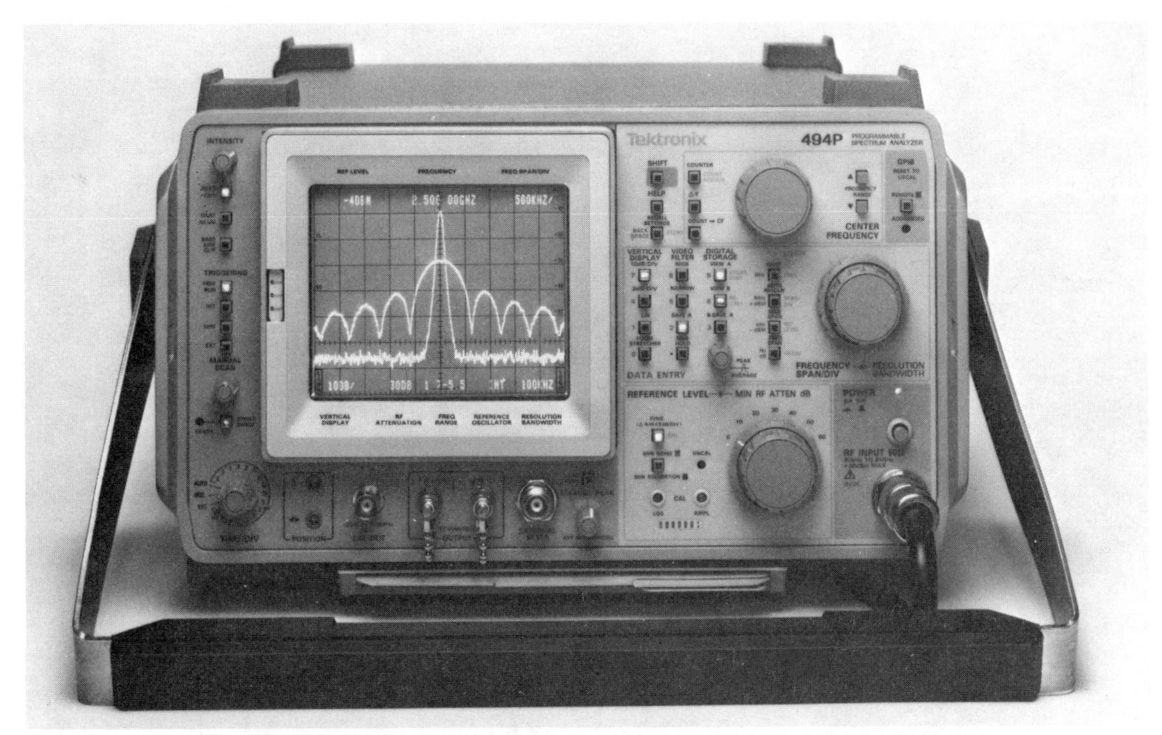

Fig. 15-19. Tektronix splendid 494P programmable, portable spectrum analyzer (courtesy Tektronix).

15-18). On the other hand, the spectrum of signals between 50 kHz and 21 GHz are always visible with Tektronix' 494 (Fig. 15-19). Transponders from Galaxy are visible via a Regency LNB from a test point near Annapolis, Maryland. Note some differences in amplitude (Fig. 15-20). The weaker ones just may give you sparklies, which is nothing more than low-signal noise.

Should sparklies bother you, then a low-noise block downconverter *and* a preamplifying LNA could furnish *another* 14-18 dB to work with. Just start with an LNA under 100 degrees K and then graduate to the LNA-B combination. They could work wonders, especially if both are supplied by Aventek, whom we find highly reliable and doing one of the very best jobs in the privately-owned earth-terminal business with such components (Fig. 15-21). But be very careful of how you mix and *don't* match. Gigahertz components and very low-signal amplitudes are sensitive to anything, and just a little extra SWR (standing wave ratio) or impedance mismatch can play hob with incoming audio and video from 22.3 kilomiles in the heavens. In extreme situations, go to a 12-foot dish with about 42 dB of gain, along with good quality lead-in and short

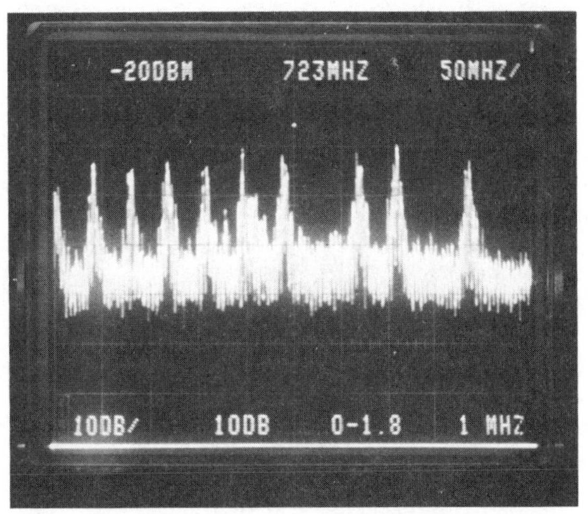

Fig. 15-20. Don't think that Satellite transponders aren't visible—they are, but only through the eyes of a *good* spectrum analyzer.

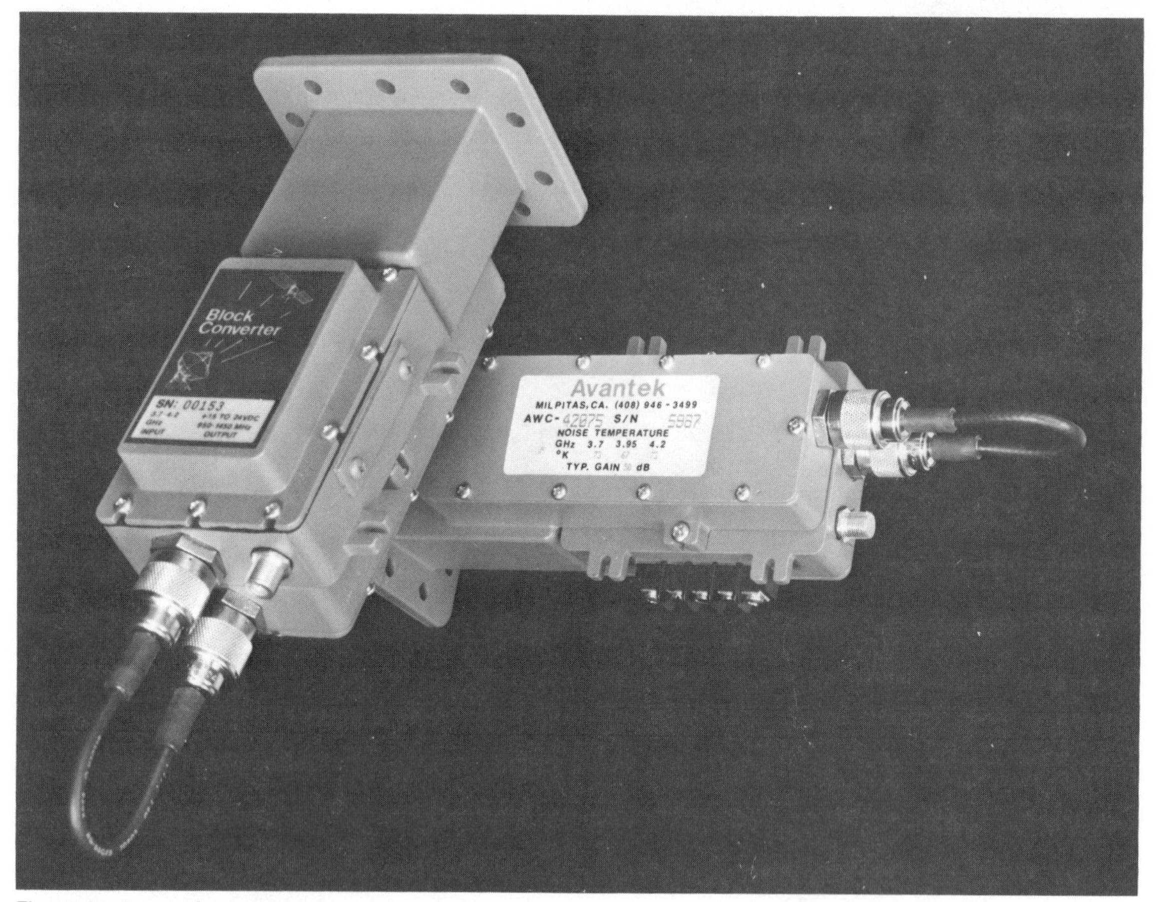

Fig. 15-21. Avantek's new block converter and companion 3.7 to 4.2 GHz low-noise amplifier for C-band satellite downlink reception. Combination gives some 68 dB of gain for any 950-1450 MHz i-f satellite receiver.

lengths to prevent additional loss. If all this doesn't solve the sparklie problem, just thank the would-be customer graciously and move on—the cost will be considerably less than 40 probable callbacks and the accompanying aggravation.

So far, preferred downconverter frequencies from the LNB are 950 to 1450 MHz, because these will also access most of the Ku band DBS-type block downconverters, too, making your receiver able to handle both C and Ku bands as it suits your convenience. Under certain circumstances the 450 to 950 MHz group may run into UHF TV interference in the big cities, although, conceivably, good notch filtering could avoid that problem also. There aren't enough of the latter category on the market yet to evaluate that possibility precisely.

As for satellite transponders, the readout in Fig. 15-20 pretty well tells its own story. Coming out of a buffered tap on the LNB, following a 50 dB gain LNA, we're looking at this group at a mean reading of about 45 dB down (−20 dBm is the reference), each horizontal division amounts to 50 MHz, center frequency 723 MHz, and from −20 dBm you count 10 dB/division, with a resolution of 1 MHz. Give you a fair idea of what a spectrum analyzer can do? You know immediately that if your receiver can handle signals at −50 dBm then you have plenty of drive to work with under *all* conditions, rain, snow, or shine. Transponders toward the center of any 500 MHz band have the best amplitudes, while those on the edges are often the weaker—whether C or Ku bands. It just works out that way, apparently because of internal satellite transponder amplification.

We have yet to try an analysis on a single transponder, but with a super-super spectrum analyzer, it could well be possible if sensitivities and resolution were both markedly superior.

What you really want to know is the receiver's response, its bandwidth (in excess of 20+ MHz), and if full value video and audio signals are available at both baseband and rf outputs. If not, send the critter back for a worthwhile replacement. You do not cheat on good satellite reception—and get away with it! This really goes for television receivers also, a spectrum video, chroma, audio, rf and baseband TV/monitor can't be beat. Ask RCA about its 2000 series that's been in production for a couple of years. They can't make enough of them, even at close to the $1,000 or more retail price. In

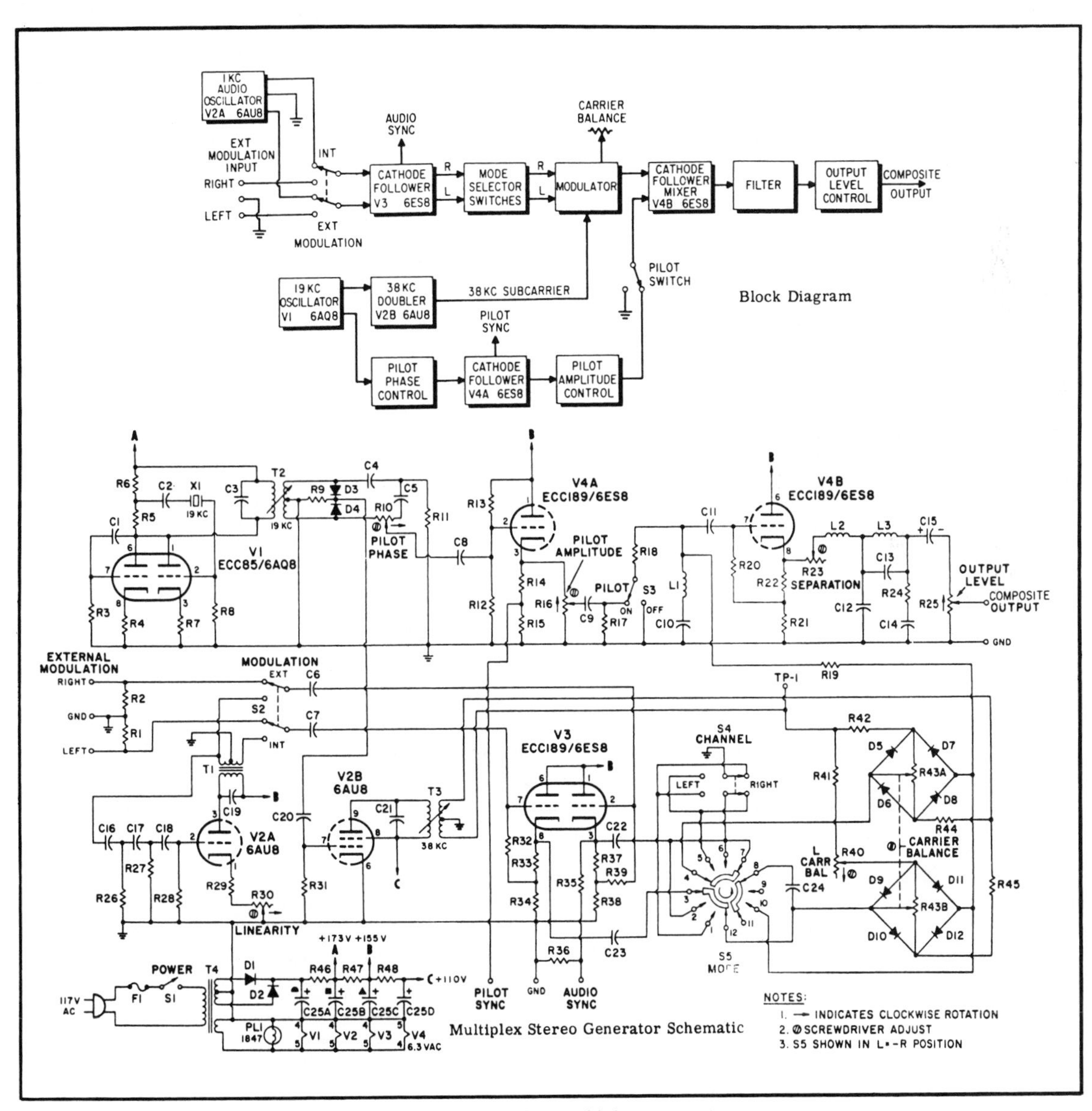

Fig. 15-22. Block and schematic diagrams of old Precision multiplex generator.

video, quality is always recognized and appreciated!

TEST EQUIPMENT

This can also have problems, but if you grit your teeth and dig in, a good many seemingly insurmountable problems can be overcome in much less time than you may think. Let's look at this example.

A Multiplex Stereo Generator

Sometimes (not often) it pays to exhume a dusty recluse from the storage shelf, re-tube, check voltages and waveforms (if given), and execute a very careful alignment before returning the instrument to active duty. Normally we would *never* recommend this procedure for any piece of equipment approaching the venerable age of 10, but in this instance, a stereo generator is still just that, even though there have been improvements since this one was born in the glory days of Precision Apparatus Co., November 1962.

If you care to remember or recall, vacuum tubes have just about tripled or quadrupled in price since those early days and transformers for almost any test equipment are often one-of-a-kind. So do not proceed with one of these oldies unless you're certain your problems are but tubes, possibly a handfull of capacitors and a resistor or two. The way to do this is gamble a few bucks on a few valves—as our British cousins call them—then begin to hunt and peck critical coupling, blocking, and filter capacitors for go/no-go.

As an example, in vacuum tube equipment, dc voltages are on the order of 150 to 350 volts and a fair amount of current to go with it. So coupling capacitors will normally have different voltages on either side, along with signal passage to accompany. Filters, if not functioning, often drop voltage levels, permit excess ripple, or allow transient interference to appear in critical waveforms. If shorted, of course, resistors will probably burn and, in coupling functions, a high dc grid potential will often heat up tube plates to a cherry red. Usually, the greatest problem with resistors are not usually value changes but catastropic burnouts from too much current or nasty little flecks of dirt and more current paths through the carbon wipers in potentiometers that do all sorts of peculiar things—as you shall see.

Theory of Operation

Because you really can't repair anything unless you know how it works, let's take a few lines and segue through the combined block diagram and schematic so you can see what's what (Fig. 15- 22).

V2A constitutes a triode blocking, phase-shift oscillator that will deliver *both* right and left 1 kHz oscillations simultaneously. With modulation switch S2 in the internal position, these oscillations are coupled to the grids of dual triode follower V3, whose cathode outputs establish both the audio sync voltage as well as the right and left channel signals selected by Channel switch S4. When S2 is switched over to external modulation, outside voltages are ac-coupled in place of the V2 output.

V1 is the crystal-controlled 19 kHz Pilot oscillator with tuned primary-secondary coupling transformer T2 that supplies both the pilot signal and, through V2B, generates the 38 kHz carrier for the two waiting diode bridge modulators, D5 through D12. Transformer T3 with tapped secondary doubles the 19 kHz signal which is suppressed in the two modulators and only the multiplex signal

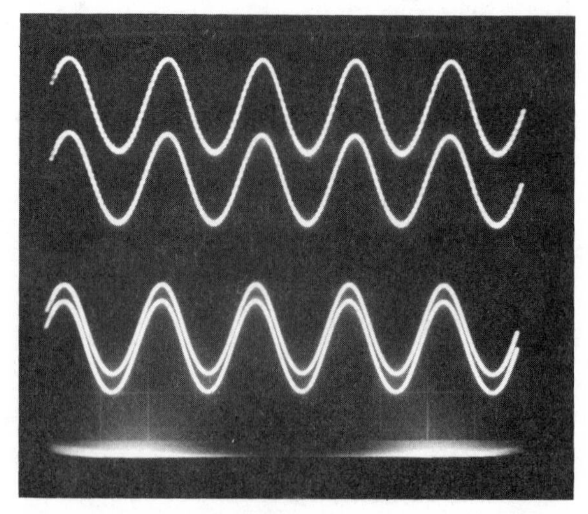

Fig. 15-23. Misadjusted and proper adjustment of bridge modulator.

in square wave form appears at the two bridge outputs. Through R19 these signals are coupled to the grid of V4B where they are filtered by series resonant L1 and C10, and then combined or not via S3 with the pilot signal. Tube V4B acts as both a summer and cathode follower for the RLC-compensated and filtered composite signal output. The level-control potentiometer then regulates amplitudes of the final signal by operator selection.

With actually three signals generated, transformers and phase- shifting networks require careful tuning and exceptional stability to produce the correctly multiplexed information. Whenever tackling one of these "critters," be sure all adjustments are correct before looking for problems that may not really exist.

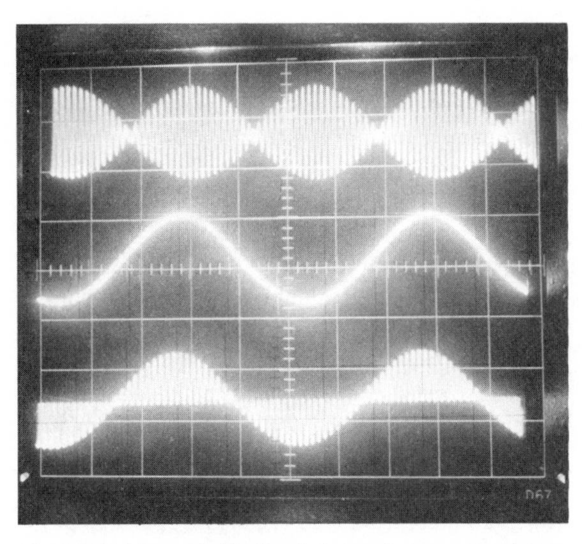

Fig. 15-25. It's a good feeling to have all stereo waveforms appearing precisely, where L = −R, L = R and L or R.

The Hunt Begins

After re-tubing this 4-valve (two are dual purpose) generator, you should poke around with an oscilloscope and see what its ac and dc amplifiers tell you as the hunt begins. For in such equipment, tolerances are normally 15 to 20-percent.

What we noted was a varying level output with a peculiar setting for the R23 separation control. Further, modulated waveforms were nonsymmetrical, and amplitudes of the two sets of bridge rectifiers for carrier balances were unequal. (See Fig. 15-23).

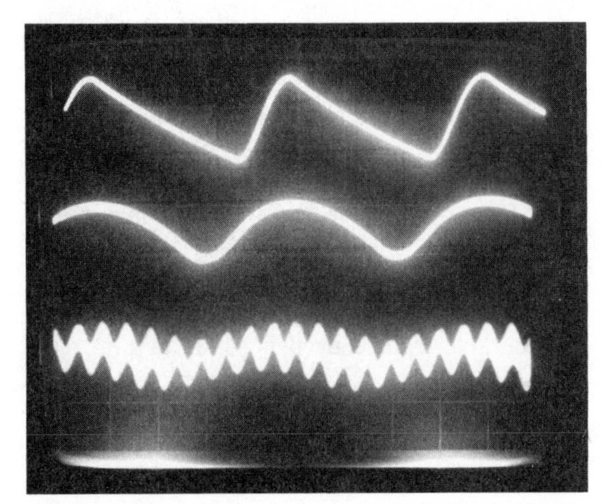

Fig. 15-24. Always investigate power supply filtering *before* repairs.

After establishing that the fullwave power supply was, indeed, adequately operational with no more than 8 volts of sawtooth at first filter C25A, only 500 mV of ripple at C25C and less than 100 mV at C25C, we began examining the V4A, B pilot and composite output triodes to make sure we didn't have real problems (Fig. 15-24).

An oscilloscope check on either side of C22 with ac amplifiers confirmed that some dc leakage was occurring between V3's cathode and the S5 mode switch, driving this tube harder than it should have been. The second "problem" resulted from an amplitude imbalance in adjustments of the two bridges via R43A, B. Once these two minor imperfections were resolved, spot checks of the remainder revealed little else in want of repair. Figure 15-24 illustrates improper balance and cross modulation (top trace), while the bottom trace has only to be merged (dual trace) just a little more to assume complete waveform balance and integrity. We might add that one of the two "concentric" potentiometers was dirty and would permit intermittent amplitudes at any time, giving the composite output a miserable appearance.

Following re-tubing and the short repair interval, we allow a 6- hour tube "burn-in" before final calibration. This consisted of seven steps, including the external addition of a phase compensating RC network to compensate for pilot phase error. Dc voltages, by the way, are acceptable in this equipment if they are within ± 20 percent—and that's considerable leeway. As you may have suspected, there are no waveforms and only dc voltages at the various tube electrodes for V1 through V4.

We thought you might like to see the resulting signals outputs after the search and rescue was completed (Fig. 15-25). It does restore one's confidence somewhat to see what can be done with a little care, chastening, and electronic chemistry. True, we're expecting a newer instrument than the old E-490 to enter the fray shortly, but in the meantime and more, this old Precision Apparatus piece responds and works well, with a 40 dB stereo separation that's pretty good on anyone's backup shelf.

If you have a bit of spare time, try something like this yourself, the results may surprise you! Conversely, a sad piece of equipment having both bad components and poor design, could convert useful optimism and results into total, catastropic defeat. Therefore, pick your targets carefully and allow for considerable success if your deflated ego needs a genuine lift.

Measurement Tips

Reverse voltages for silicon-controlled rectifiers in high-voltage applications such as HV transformers may be rather easily measured as follows:

Hold an ordinary oscilloscope probe approximately one inch away from the tertiary winding of the flyback transformer and proceed through misadjustment of the horizontal oscillator, line voltage changes, or power supply regulation. The *total* amplitude of the waveform plus any ringing between pulses is your reverse voltage requirement. In addition, you may also observe damper ringing action as well as horizontal on/off timing and anything else connected with horizontal flyback action. Just don't let the 10:1 probe touch a live portion of the former or you'll buy at least another probe or an oscilloscope front end.

Scopes should be at least five times (5 ×) faster than any waveform being measured to prevent significant error. And 10:1 probes must be calibrated to present true waveshapes. You also will *not* want to try and measure fast rising currents across any old series resistor—only a coaxial current sampling resistor is recommended, and that's a commercial part. However, if you have a current probe that specs within measurement limits, then by all means use it instead. Such a current probes are also useful in automobile electronics—particularly Ford products—to evaluate many electronic ignition systems. We have found such probes thoroughly useful in many otherwise unmeasureable situations, and they are accurate. Tektronix is a good source.

The use of direct probes with small (often 10 k) series resistors is always discouraged. The far superior 10:1 probe should always be favored since its frequency response is a great deal better and a 9-megohm in series with a 1 megohm shunt presents 10 megohms impedance to any circuit, and therefore less loading. Probes, also, should be considerably faster than the oscilloscope they're isolating. For example, a 50 MHz scope can easily use a 100 MHz 10:1 probe for better measurements and higher accuracies. But woe be unto he or she who mismatches probe and oscilloscope capacitances. If you do, you're in serious measurement trouble. The idea of any 10:1 probe is to lessen 'scope impedance rather than increase it and generate long trailing edges on rectangular waveforms.

RADIOS

Believe it or not, radios will become even more a part of your servicing trade as multichannel sound and AM stereo begin to appear in substantial numbers. Even basic audio investigations may help.

While by no means an "expert" nor practicing servicer on these audio products, there are several characteristic faults occurring that may save you considerable trouble by being able to spot them *before* unsoldering a bunch of connections and shotgunning more than a few semiconductors. Primarily, these tips are directed toward the discrete transistor radio with one or two ICs rather than a one or two chip "box" that deserves to be shotgunned.

Once you wrestle the radio out of the dash, even the stereos can be hooked up to some type of speaker and signal-traced to the general trouble area. Naturally you'll have to know whether the problem is FM or AM, front end, or amplifiers. Occasionally a blown fuse or bad antenna plug confuses the issue, but you should certainly be able to find those and refurbish or replace without hesitation. One means a dead radio and the other excessive noise and little or no reception.

Out of the auto, a piece of reasonably long wire will serve as an antenna in any decent signal area, and a signal tracer's speaker will let you listen at the other end. In between, switch from FM to AM and see which is at fault. Then take the rf or signal tracer probe and go through the unit, beginning with the output amplifiers and working backwards—it shouldn't be that difficult to find. Where there are partial problems, a signal generator set somewhere between 0.55-1.6 MHz for AM and 88 to 108 MHz for FM, loaded with 1 kHz modulation, will usually find its way to the trouble spot which you can see with a decently sensitive oscilloscope. Just watch where the waveform signal distorts or waveform amplitude decreases significantly and without obvious cause. These varieties of troubles are usually simpler than you think and are often rather easily repairable. Just remember that the standard semiconductor usually works up to 90 percent of its maximum potential until the day it dies. Sure, there are leakage and partial problems, but these are the exception rather than the rule.

Now we come to the *nasty*. Named the "intermittent" for good reason, these are the troubles that come and go randomly without rhyme or reason. Radio voltages check good, yet the thing aborts its signal half the time. If you have a reliable incoming signal with good antenna and secure hookup, power is getting to the radio as it should, and the thing plays perfectly when operating, then suspect a faulty ground or bad solder joint. Look for excessive flux and also probe the printed circuit board with finger and fiber probe. Soon you'll find a spot that cuts output off and on. But the true trouble may be considerably removed from that area and you'll have to keep digging.

Yes, you resolder any suspicious-looking connections, use a magnifier to find cracked boards, and wiggle anything that can be moved. But, above all, don't forget a securing nut or screw, especially if it's part of an electrical ground. Ten to one when this simple solution appears, you've found your problem. After securely tightening, then apply a drop of your wife's fingernail polish to said screw and the surface it's securing. There should be no further problems, along with tidings of great joy to the motor vehicle operator. Music always hath charm to calm the heaving beast and breast.

These same principles apply directly to your super audio amps and decoders in TV. However you'll need certain new equipment for multichannel sound, and one excellent, high-sensitivity oscilloscope. If you don't use these you won't find the troubles. Here, radiomen *only* need *not* apply.

Appendix

Channel Utilization

CHANNEL UTILIZATION

§ 73.606 Table of assignments.

(a) *General.* The following table of assignments contains the channels assigned to the listed communities in the United States, its Territories, and possessions. Channels designated with an asterisk are assigned for use by noncommercial educational broadcast stations only. A station on a channel identified by a plus or minus mark is required to operate with its carrier frequencies offset 10 kc/s above or below, respectively, the normal carrier frequencies.

Channel No.	Frequency band (MHz)
2	54-60
3	60-66
4	66-72
5	76-82
6	82-88
7	174-180
8	180-186
9	186-192
10	192-198
11	198-204
12	204-210
13	210-216
14	470-476
15	476-482
16	482-488
17	488-494
18	494-500
19	500-506
20	506-512
21	512-518
22	518-524
23	524-530
24	530-536
25	536-542
26	542-548
27	548-554
28	554-560
29	560-566
30	566-572
31	572-578
32	578-584
33	584-590
34	590-596
35	596-602
36	602-608
37	608-614
38	614-620
39	620-626
40	626-632
41	632-638
42	638-644
43	644-650
44	650-656
45	656-662
46	662-668
47	668-674
48	674-680
49	680-686
50	686-692
51	692-698
52	698-704
53	704-710
54	710-716
55	716-722
56	722-728
57	728-734

58	734-740
59	740-746
60	746-752
61	752-758
62	758-764
63	764-770
64	770-776
65	776-782
66	782-788
67	788-794
68	794-800
69	800-806

ALABAMA

	Channel No.
Andalusia	*2–
Anniston	40–
Birmingham	6–, *10–, 13–, 21, 42+, *62+, 68–
Decatur	
Demopolis	*41
Dothan	4, 18, *39+, 60–
Florence	15, 26, *36–
Gadsden	44+, 60
Huntsville	19, *25+, 31+, 48–
Huntsville-Decatur	54
Louisville	*43+
Mobile	5+, 10+, 15+, 21+, *31, *42
Montgomery	12, 20, *26+, 32, 45–
Munford	*7–, *16–
Opelika	66
Selma	8, 29–
Tuscaloosa	17, 33, *39–
Tuscumbia	47–

ALASKA

	Channel No.
Anchorage	2–, 4–, *7–, 11, 13–
Bethel	*4
Dillingham	10
Fairbanks	2+, 4+, 7+, *9+, 11+, 13+
Juneau	*3, 8, 10
Ketchikan	2, 4, *9
Seward	3–, 9–
Sitka	13

ARIZONA

	Channel No.
Ajo	*23–
Coolidge	*43
Douglas	3, *28
Flagstaff	2, 13, *16
Globe	*14+
Green Valley	46
Holbrook	*18+
Kingman	6–, *14–
McNary	*22+
Mesa	12–
Nogales	*16+
Page	*17
Parker	*17–
Phoenix	3+, 5–, *8+, 10–, 15–, 21, 33, *39
Prescott	7, *19
Safford	*23+
Sierra Vista	58
Tucson	4–, *6+, 9–, 13–, 18–, *27–, 40
Tucson-Nogales	[2]11
Yuma	11–, 13+, *16–

ARKANSAS

	Channel No.
Arkadelphia	*9+
Batesville	*17
El Dorado	10–, 18–, *30+
Fayetteville	*13–, 36
Fort Smith	5–, 24+, 40–
Harrison	31+
Hot Springs	*20, 26
Jonesboro	8–, *19+
Little Rock	*2–, 4, 7–, 11, 16–, *36
Mountain Home	43+
Mountain View	*6–
Pine Bluff	25–, 38–
Russellville	*28+

CALIFORNIA

	Channel No.
Alturas	13+
Anaheim	56–
Arcata	23
Bakersfield	17, 23–, 29, *39–
Barstow	*35+
Bishop	*14–
Blythe	*22–
Brawley	*26
Chico	12–, [1]*18, 24+, *30–
Coalinga	*27–
Concord	42
Corona	52
Cotati	*22–
El Centro	7+, 9+
Eureka	3–, 6–, *13–
Fort Bragg	[1]*17+
Fresno	*18+, 24, 30+, 43, 47, 53, 59
Hanford	21
Indio	[1]*19+
Los Angeles	2, 4, 5, 7, 9, 11, 13, 22, *28, 34, *58–, *68–
Modesto	19–, *23+
Oroville	28
Oxnard	63+
Palm Springs	36–, 42
Redding	7, *9, [1]16
Ridgecrest	*25
Riverside	46, 62
Sacramento	3, *6, 10, 29–, 31–, 40–
Salinas Monterey	8+, 35–, 46–, *56, 67–
San Bernadino	18–, *24–, 30
San Diego	8, 10, *15, 39, 51, 69
San Francisco	2+, 4–, 5+, 7–, *9+, 14+, 20–, 26–, *32+, 38, 44–
San Jose	11+, 36, 48–, *54, 65
San Luis Obispo	6+, *15+
San Mateo	*60
Santa Anna	40, *50–
Santa Barbara	220
Santa Cruz	[1]*16–
Santa Maria	12+
Santa Rosa	50–, *62
Stockton	13+, 58, 64
Susanville	*14
Tulare	26+
Vallejo-Fairfield	66
Ventura	236, 264, 296A
Visalia	*49
Watsonville	*25+
Yreka City	*20+

COLORADO

	Channel No.
Alamosa	*16
Boulder	*12, 14
Colorado Springs	11, 13, 21
Craig	*16+
Denver	2, 4–, *6–, 7, 9–, 20, 31, *41
Durango	6+, *20–
Fort Collins	22–
Glenwood Springs	3–, *19+
Grand Junction	5–, 8–, *18+
Gunnison	*17–
La Junta	*22+
Lamar	12–, *14–
Leadville	*15–
Montrose	10+, *22
Pueblo	5, *8, 26+, 32–
Salida	*23+
Sterling	3, *18+
Trinidad	*24

CONNECTICUT

	Channel No.
Bridgeport	43–, *49–
Hartford	3+, 18–, *24, 61+
New Britain	30+
New Haven	8, 59+, 55
New London	26+
Norwich	*53
Waterbury	20

DELAWARE

	Channel No.
Dover	*34
Seaford	38, *64
Wilmington	*12, 61

DISTRICT OF COLUMBIA

	Channel No.
Washington	4–, 5–, 7+, 9, 14–, 20+, *26–, *32+, 50

FLORIDA

	Channel No.
Boca Raton	*63
Bradenton	*19
Cape Coral	36
Clearwater	22
Cocoa	*18–, 52
Daytona Beach	2–, 26
Fort Lauderdale	51
Fort Myers	11+, 20+, *30
Fort Pierce	*21–, 34
Fort Walton Beach	35, 53
Gainesville	*5–, 20
Hollywood	69
Jacksonville	4+, *7, 12+, 17, 30+, 47–, *59
Key West	16+, 22+
Lake City	*41
Lakeland	32
Leesburg	*45–, 55
Madison	*36–
Marianna	*16+
Melbourne	43+, 56
Miami	*2, 4, 6, 7–, 10+, *17–, 23–, 33, 39, 45+
Naples	26–
New Smyrna Beach	*15+
Ocala	*29, 51–
Orange Park	25–
Orlando	6–, 9, *24–, 35+, and 65
Palatka	*42
Panama City	7+, 13, *22+, 28–
Pensacola	3–, *23, 33+, 44
St. Petersburg	10–, 38, 44+
Sarasota	40
Sebring	*48
Tallahassee	*11–, 27+, 40+
Tampa	*3, 8–, 13–, *16, 28
West Palm Beach	5, 12, 29+, *42+, 61

GEORGIA

	Channel No.
Albany	10, 19–, 31–
Ashburn	*23+
Athens	*8–, 34
Atlanta	2, 5–, 11+, 17–, *30, 36, 46–, *57+, 69
Augusta	6+, 12–, 26, 54–
Brunswick	21+
Carrollton	*49–
Carnesville	*52
Cedartown	*65–
Chatsworth	*18–
Cochran	*15
Columbus	3, 9+, *28, 38+, *48, 54+
Dawson	*25
Draketown	*27–
Elberton	*60+
Flintstone	*51–
Lafayette	*35
Macon	13+, 24+, 41+, *47+
Pelham	*14–
Rome	14+
Royston	*22+
Savannah	3, *9–, 11, 22, 28–
Thomasville	6
Toccoa	32–, *68–
Valdosta	*33, 44–
Vidalia	*18+
Warm Springs	
Waycross	*8+
Wrens	*20–
Young Harris	*50+

HAWAII

	Channel No.
Hilo (Hawaii)	2, *4, 9, 11, 13, 14+, 20+, 26+, *32+, *38+
Honolulu (Oahu)	2+, 4–, 9–, *11+, 13–, 14, 20, 26, 32, *38, *44
Lihue (Kauai)	3+, *8–, 10+, 12–, 15–, *21–, *27–, *67
Wailuku (Maui)	3, 7, *10, 12, 15, 21, *27, *33

IDAHO

	Channel No.
Boise	2, *4+, 7, 14
Burley	*17+

Caldwell	9–
Coeur d'Alene	*26+
Grangeville	*15–
Idaho Falls	3, 8+, 20, *33+
Filer	*19–
Lewiston	3–
Moscow	*12–
Nampa	6, 12+
Preston	*28
Pocatello	6–, *10, 15, 25+, 31–
Sandpoint	*16+
Twin Falls	11, *13–
Weiser	*17

ILLINOIS

	Channel No.
Aurora	60
Bloomington	43
Carbondale	*8
Champaign	3+, 15–
Chicago	2–, 5, 7, 9+, *11, *20, 26, 32, 38–, 44
Danville	68
Decatur	17, 23–
DeKalb	*33, *48–
East St. Louis	46
Edwardsville	*18–
Elgin	*66+
Freeport	23, *65–
Galesburg	63
Harrisburg	3
Jacksonville	*14
Joliet	*14–, 66+
Kankakee	*54–
LaSalle	35
Macomb	*22+
Marion	27
Moline	8, *24–
Mount Vernon	13+
Olney	*16–
Peoria	19, 25+, 31+, 47–, *59+
Quincy	10–, 16+, *27+
Rockford	13, 17–, 39
Rock Island	4+
Springfield	20+, 49–, 55+, *65+
Streator	*64+
Urbana	*12–, 27–
Vandalia	*21

INDIANA

	Channel No.
Anderson	67+
Angola	63
Bloomington	4, *30–, 63+
Elkhart	28+
Evansville	7, *9+, 14–, 25–, 44–
Fort Wayne	15+, 21+, 33–, *39–, 55
Gary	*50, 56+
Hammond	62+
Indianapolis	6, 8–, 13–, *20–, 40, 59–, *69
Kokomo	29–
Lafayette	18, *24
Madison	*60+
Marion	23
Muncie	*17+, 49
Richmond	43+
South Bend	16, 22, *34–, 46
Terre Haute	2+, 10, *26–, 38
Vincennes	*22–

IOWA

	Channel No.
Ames	5, 23–, *34+
Burlington	26–, *57–
Carroll	*18–, 30+
Cedar Rapids	2, 9–, 28+
Centerville	*31–
Council Bluffs	*32
Davenport	6+, 18+, 30–, *36+
Decorah	*14+
Des Moines	8–, *11+, 13–, 17+, *43–, 63–, 69
Dubuque	16–, *29–, 40–
Estherville	*49+
Fort Dodge	*21, 50+
Fort Madison	*38+
High Point	*14–
Iowa City	*12+, 20–
Keokuk	*44+
Keosauqua	*54+
Lansing	*41+
Mason City	3+, *24+
Mount Ayr	*25–
Ottumwa	15+, *33–
Red Oak	*36
Rock Rapids	*25+
Sibley	*33
Sioux City	4–, 9, 14, *27–
Spirit Lake	*38
Waterloo	7+, 22–, *32–

KANSAS

	Channel No.
Chanute	*30+
Colby	4
Columbus	*34+
Dodge City	6+, *21–
Emporia	*25+
Garden City	11+, 13–, *18
Goodland	10
Great Bend	2
Hays	7–, *9
Hutchinson	*8, 12, 36+
Lakin	*3
Lawrence	38
Manhattan	*21
Oakley	*15–
Parsons	*39
Phillipsburg	*22–
Pittsburg	7+ and 14
Pratt	*32+
Randall	
Salina	18+, 34–, 44
Sedan	*28
Topeka	*11, 13+, 27, 43, 49
Wichita	3–, 10–, *15+, 24–, 33, *42

KENTUCKY

	Channel No.
Ashland	*25–, 61+
Beattyville	65
Bowling Green	13, 40+, *53–
Campbellsville	34
Covington	*54+
Danville	56
Elizabethtown	*23+
Hazard	*35+, 57–
Hopkinsville	51
Lexington	18+, 27–, 36, *46, 62
Louisville	3–, 11, *15, 21–, 32–, 41+, *68+

Madisonville	*19–, *35–
Morehead	*38+
Murray	*21+
Owensboro	31–, 48
Owenton	*52+
Paducah	6+, 29
Paintsville	69+
Pikeville	*22–, 51+
Somerset	16, *29+

LOUISIANA

	Channel No.
Alexandria	5, *25+, 31+, 41+
Baton Rouge	2, 9–, *27+, 33–
De Ridder	*23–
Houma	11
Lafayette	3+, 10, 15, *24
Lake Charles	7–, *18, 29–
Monroe	8+, *13, 14–, 39+
Morgan City	*14+
Natchitoches	*28–
New Iberia	36–
New Orleans	4+, 6, 8–, *12, 20–, 26, *32+, 38+
Shreveport	3–, 12, *24–, 33
Tallulah	*19

MAINE

	Channel No.
Augusta	*10–
Bangor	2–, 5+, 7–
Calais	*13–
Fort Kent	*46+
Fryeburg	*18+
Houlton	*25+
Kittery	*39
Lewiston	8–, 35–
Millinocket	*44–
Orono	*12–
Portland	6–, 13+, *26–, 51
Presque Island	8, *10+
Rumford	*43+

MARYLAND

	Channel No.
Annapolis	*22+
Baltimore	2+, 11–, 13+, 24+, 45, 54, *67–
Cumberland	52+, 65
Frederick	*62
Hagerstown	25–, *31
Oakland	*36+
Salisbury	16+, *28–, 47–
Waldorf	*58+

MASSACHUSETTS

	Channel No.
Boston	*2+, 4–, 5–, 7+, 25+, 38, *44+, 56, 68+
Greenfield	32+
Middleton	62
New Bedford	6+, 28–, *34
North Adams	19, *35
Norwell	46+
Pittsfield	51+
Springfield	22, 40, *57+
Vineyard Haven	58+
Worcester	14, [1]27, *48+, 66

MICHIGAN

	Channel No.
Alpena	*6, 11
Ann Arbor	31+, *58+
Bad Axe	*[1]15–
Battle Creek	41+
Bay City	5–, *19+, 61+
Cadillac	9, *27
Calumet	5–, *22–
Cheboygan	4+
Detroit	2+, 4, 7–, 20+, 50–, *56, 62
East Lansing	*23–, *69–
Escanaba	3+
Flint	12–, *28–, 66–
Grand Rapids	8+, 13+, 17, *35+
Iron Mountain	8–, *17+
Ironwood	*15–, 24+
Jackson	18+
Kalamazoo	3–, *52+, 64
Lansing	6–, 47, 53–
Manistee	*21
Manistique	*15+
Marquette	6–, *13, 19
Mount Clemens	38+
Mount Pleasant	*14
Muskegon	54+
Parma	10–
Petoskey	*23+
Port Huron	46+
Saginaw	25–, 49–
Sault Ste. Marie	8, 10+, *32–
Traverse City	7+, 29–
West Branch	*24

MINNESOTA

	Channel No.
Alexandria	7, *24
Appleton	*10–
Austin	6–, *15–
Bemidji	*9, 26+
Brainerd	*22
Crookston	*33
Duluth	3, *8, 10+, 21+, 27–
Ely	*17–
Fairmont	*16+
Hibbing	13–, *18–
International Falls	11, *35+
Mankato	12, *26–
Marshall	*30–
Minneapolis-St. Paul	*2–, 4, 5–, 9+, 11–, *17, 23+, 29+
Rochester	10, 47–
St. Cloud	19, *25–, 41
St. James	32+
Thief River Falls	10
Wadena	*20–
Walker	12–
Wilmar	*14–
Winona	*35+, 44–
Worthington	*20

MISSISSIPPI

	Channel No.
Biloxi	13+, *19+, 25–
Booneville	*12–
Bude	*17+
Clarksdale	*21

Cleveland	*31 –
Columbia	*45
Columbus	4 –, 27, *43
Greenville	15 –, 44
Greenwood	6 +, *23 +
Hattiesburg	22, *47
Houston	45 +
Jackson	3, 12 +, 16, *29 +, 40 +
Laurel	7, 18 +
Magee	34 +
Meridian	11 –, *14, 24 –, 30 –
Mississippi State	*2 +
Natchez	*42 +, 48
Oxford	*18
Senatobia	*34 –
Tupelo	9 –
Vicksburg	35 –
Yazoo City	*32 –

MISSOURI

	Channel No.
Birchtree	*20 –
Bowling Green	*35 +
Cape Girardeau	12, 23, *39 –
Carrollton	*18
Columbia	8 +, 17 – *23 +
Flat River	*22
Hannibal	7 –
Jefferson City	13, 25, *36 –
Joplin	12 +, 16, *22 –
Kansas City	4, 5 +, 9 +, *19 +, 41 –, 50 –, 62 +, *68 –
King City	*28 –
Kirksville	3 –
LaPlata	*21 +
Lowry City	*15 –
Poplar Bluff	15 +, *26 +
Rolla	*28
St. Joseph	2 –, 16 –, 22
St. Louis	2, 4 –, 5 –, *9, 11 –, 24 +, 30 +, *40 –, *46
Sedalia	6
Springfield	3 +, 10, *21 –, 27 –, 33–

MONTANA

	Channel No.
Anaconda	2 +
Billings	2, 8, *11, 14, 20 +
Bozeman	7 –, *9
Butte	*2 +, 4, 6 +, 18, 24
Cut Bank	*14 –
Dillon	*14 +
Glendive	5 +, 13 +, *16 –
Great Falls	3 +, 5 +, 16, 26, *32
Hardin	4 +
Havre	9 +, 11 +, *18 –
Helena	10 +, 12, *15 +
Joplin	35 –, 48, 54 –
Kalispell	9 –, *29 –
Lewistown	13
Miles City	3 –, *6, 10
Missoula	8 –, *11 –, 13 –, 17 –, 23 –
Sikeston	45
Wolf Point	*17 +

NEBRASKA

	Channel No.
Albion	8 +, *21 +
Alliance	*13 –
Bassett	*7 –
Beatrice	23 +
Falls City	*24
Grand Island	11 –, 17 –
Hastings	5 –, *29 +
Hayes Center	6
Hay Springs-Scottsbluff	4 +
Kearney	13
Lexington	*3 +
Lincoln	10 +, *12 –, 45, 51
McCook	8 –
Merriman	*12
Norfolk	*19 +
North Platte	235, 246, 278
Omaha	3, 6 +, 7, 15, *26, 42 +, *48 –
Orchard	16
Scottsbluff	10 –, 16
Superior	4 +

NEVADA

	Channel No.
Boulder City	5 +
Elko	10 –, *14 +
Ely	3 –, 6 +
Fallon	*25
Goldfield	2 –
Las Vegas	3, 8 –, *10 +, 13 –, 15 +, 21 +
McGill	*13
Pawnee City	*33 +
Reno	2, 4, *5, 8, 21 +, 27 +
Tonopah	9 –, *17 +
Winnemucca	7 +, *15 –
Yerington	*16 +

NEW HAMPSHIRE

	Channel No.
Berlin	*40 –
Concord	21 +
Durham	*11
Hanover	*15 +, 31
Keene	*52 +
Littleton	*49 +
Manchester	9 –, 50 –, 60 +
Portsmouth	[1]17 –

NEW JERSEY

	Channel No.
Asbury Park	[5]58
Atlantic City	*[1]18, *36, 53 +
Burlington	48 –
Camden	*23 +
Little Falls	*50 +
Newark	13 –, 68
New Brunswick	*[1]19 –, 47 +, *58
Peterson	41 –
Trenton	*52 –
Vineland	65 –
Wildwood	40

NEW MEXICO

	Channel No.
Alamogordo	*18 –
Albuquerque	4 +, *5 +, 7 +, 13 +, 14 –, 23 –, *32 +
Carlsbad	6 –, *15 +, 25 –
Clayton	*17

Clovis	12—
Deming	*16
Farmington	12+, *15+
Gallup	3, *8–, 10
Hobbs	29+
Las Cruces	*22–, 48+
Lovington	*19
Portales	*3+
Raton	*18–
Roswell	8, 10–, 21–, 27–, *33+
Santa Fe	2+, *9+, 11–, 19–
Silver City	6, *10+, *12
Socorro	*15–
Tucumcari	*15

New York

	Channel No
Albany-Schenectady	6, 10–, 13, *17+, 23–, *29+, 45
Amsterdam	*39+, 55
Binghamton	12–, 34, 40–, *46+
Buffalo	2, 4–, 7+, 17, *23, 29–, 49–
Carthage	7–
Corning	*30
Elmira	18+, 36–
Glens Falls	*58–
Ithaca	52, *65+
Jamestown	26+, *46
Kingston	63
Lake Placid	5, *34+
Levittown	*21–
Massena	*18
New York	2, 4, 5+, 7, 9+, 11+, *25, 31–
Oneonta	[1]15, *42
Patchogue	67
Plattsburg	*57
Poughkeepsie	54+
Riverhead	55+
Rochester	8, 10+, 13–, *21, 31+, *61+
Syracuse	3–, 5–, 9–, *24+, 43+, 62+
Utica	2–, 20+, 33, *59
Watertown	*16, 50+

North Carolina

	Channel No.
Andrews	*59
Asheville	13–, 21+, *33, 62+
Bryson City	*67—
Burlington	16
Canton	*27
Chapel Hill	*4+
Charlotte	3, 9+, 18, 36, *42+
Columbia	*2
Concord	*58
Durham	11+, 28+
Fayetteville	40+, 62
Forest City	66+
Franklin	*56+
Goldsboro	17–
Greensboro	2–, 48–, 61
Greenville	9–, 14, *25
Hickory	14–
High Point	8–, *32+, 67+
Jacksonville	*19
Kannapolis	64–
Laurel Hill	59+
Lexington	20
Linville	*17
Lumberton	*31
Morganton	23–
New Bern	12+
Raleigh	5, 22, *34–
Roanoke Rapids	*36–
Rockingham	*53
Rocky Mount	47+
Washington	7
Waynesville	59
Wilmington	3–, 6, 29+, *39–
Wilson	30–
Winston-Salem	12, *26+, 45

North Dakota

	Channel No.
Bismarck	*3, 5, 12–, 17–, 26+
Devils Lake	8+, *22+
Dickinson	2+, *9–, 7
Ellendale	*19–
Fargo	6, 11+, *13, 15–
Grand Forks	*2, 14+, 27+
Jamestown	7–, *23
Minot	*6+, 10–, 13–, 14–, 24
Pembina	12
Valley City	4–
Williston	*4, 8–, 11–, *15–

Ohio

	Channel No.
Akron	23+, *49+, 55–
Alliance	*45+
Ashtabula	[1]15
Athens	20*
Bowling Green	*27+
Cambridge	*44–
Canton	17–, 67
Chillicothe	53
Cincinnati	5–, 9, 12, 19+, *48–, 64–
Cleveland	3, 5+, 8, 19, *25+, 61
Columbus	4–, 6+, 10+, 28–, *34, *56–
Dayton	2, 7+, *16+, 22+, 45
Defiance	65+
Hillsboro	*24+
Lima	35–, 44+, *57+
Lorain	43
Mansfield	*47+, 68–
Newark	*31–, 51
Oxford	*14+
Portsmouth	30, *42–
Sandusky	52
Springfield	26+, *66
Steubenville	9+, *62+
Toledo	11–, 13, 24–, *30+, 36–
Youngstown	21–, 27, 33, *58
Zanesville	18–

Oklahoma

	Channel No.
Ada	10+, *22
Altus	*19–
Ardmore	12–, [1]*17, *28–
Bartlesville	17+
Cheyenne	12+
Elk City	8+, *15–
Enid	20–, *26+
Eufaula	*3
Guymon	*16
Hugo	42+, [1]*15+, *48+
Lawton	7+, *36–, 16–, 45
McAlester	*32–
Miami	*18–
Muskogee	19
Oklahoma City	4–, 5, 9–, *13, 14–, 25–,

	34–, 43+, 52
Shawnee	30
Tulsa	2+, 6+, 8–, *11–, 23, *35–, 41+, 47
Woodward	*17–

Oregon

	Channel No.
Astoria	*21
Bend	*3+, *15, 21+
Brookings	*14–
Burns	*18
Corvallis	*7–
Eugene	9+, 13, 16+, *28–
Grants Pass	*18+
Klamath Falls	2–, *22+
LaGrande	*13+, 16
Medford	5, *8+, 10+, 12+
North Bend	11, *17+
Portland	2, 6+, 8–, *10, 12, 24+, *30, 40–
Roseburg	4+
Salem	22, 32
The Dalles	*17–

Pennsylvania

	Channel No.
Allentown	*39, 69
Altoona	10–, 23–, 47, *57+
Bethlehem	60–
Clearfield	*3+
Erie	12, 24, 35+, *54+, 66+
Greensburg	40+
Harrisburg	21+, 27–, *33+
Hazleton	56
Johnstown	6, 8–, 19+, *28+
Lancaster	8+, 15+
Lebanon	59–
Philadelphia	3, 6–, 10, 17–, 29, *35–, 57
Pittsburgh	2–, 4+, 11, *13–, *16, 22, 53+
Reading	51
Scranton	16–, 22–, 38+, *44–, 64
State College	29+, *55+
Wilkes Barre	28
Williamsport	¹20–
York	43, 49+

Rhode Island

	Channel No.
Providence	10+, 12+, ¹16, *36, 64+

South Carolina

	Channel No.
Aiken	*44
Allendale	*14
Anderson	40
Beaufort	*16–
Charleston	2+, 4, 5+, *7–, 24
Columbia	10–, 19+, 25–, *35+, 57–
Conway	*23+
Florence	13+, 15–, 21, *33+
Georgetown	*41–
Greenville	4–, 16+, *29
Greenwood	*38, 48+
Myrtle Beach	43+
Rock Hill	30+, *55–
Spartanburg	7+, 49
Sumter	*27–

South Dakota

	Channel No.
Aberdeen	9–, *16–
Allen	22+
Brookings	*8
Eagle Butte	*13
Huron	12+
Lead	5–, 11+
Lowry	*11–, 56, 62+, 68–
Martin	*8–
Mitchell	5+
Pierre	4, *10+
Rapid City	3+, 7+, *9, 15–, 21–
Reliance	6–
Seneca	*2–
Sioux Falls	11, 13+, 17–, *23, 36+
Vermillion	*2+
Watertown	3–

Tennessee

	Channel No.
Athens	*24
Chattanooga	3+, 9, 12+, *45, 61–
Cookeville	*22, 28+
Crossville	20+, *55+
Fayetteville	*52–
Greeneville	39–
Jackson	7+, 16+, *32+
Johnson City	11–, *41
Kingsport	19
Knoxville	6, 8, 10+, *15–, 26–, 43+
Lexington	*11+
Memphis	3–, 5+, *10+, 13+, *14+, 24, 30
Murfreesboro	39+
Nashville	2–, 4+, 5, *8+, 17+, 30+, *42
Sneedville	*2+
Tullahoma	64+

Texas

	Channel No.
Abilene	9+, 15, *26+, 32+
Alpine	12–
Alvin	67
Amarillo	*2–, 4, 7, 10, 14+
Austin	7+, *18+, 24, 36, 42–
Bay City	*43+
Beaumont	6–, 12–, 21, *34–
Big Spring	4–, *14
Boquillas	8–
Brady	13
Brownsville	23
Bryan	3, *15–
Childress	*21
Corpus Christi	3–, 6, 10–, *16, 28–, 38+
Dallas	4+, 8, *13+, 27–, 33+, 39
Del Rio	10, *24+
Denton	*2
El Paso	4, 7, 9, *13, 14, 26+, *38–
Fort Stockton	5+
Fort Worth	5+, 11–, 21–, *31+
Galveston	*22, 48
Harlingen	4+, *44, 60
Houston	2–, *8, 11+, 13–, *14, 20, 26, 39–, and 61
Irving	49

Kennville	35+
Laredo	8, 13, 27−, *39
Longview	[1] 16+, 51−
Lubbock	*5−, 11, 13−, 28, 34−
Lufkin	9
McAllen	48
Marfa	3
Marshall	*22−, 35+
Midland	2+, 18
Monahans-Odessa	9−
Nacogdoches	19−, *32
Odessa	7−, 24−, 30, *36+
Paris-Hugo (Oklahoma)	42+
Port Arthur	4−
Presidio	7+
Richardson	23
Rio Grande City	40
Rosenberg	45
San Angelo	3−, 6, 8+, *21+
San Antonio	4, 5, *9−, 12+, *23−, 29+, 41+
Sherman	20−, *26−
Sonora	11+
Sweetwater	12
Temple	6+, 46−
Texarkana	6, 17−, *34
Tyler	7, 14+, *38
Victoria	19+, 25, 31
Waco	10+, 25+, *34+, 44−
Weslaco	5−
Wichita Falls	3+, 6−, 18−, *24

UTAH

	Channel No.
Cedar City	4, *16+
Logan	12−, *22
Moab	*14+
Monticello	*16−
Ogden	*9+, *18−, 24, 30
Price	3+, *15
Provo	*11+, 16
Richfield	8+, *19
Salt Lake City	2−, 4−, 5+, *7−, 13+, 14−, 20+, *26−
St. George	*18−
Vernal	6, *17+

VERMONT

	Channel No.
Burlington	3, 22+, *33−
Rutland	*28+
St. Johnsbury	*20−
Windsor	*41

VIRGINIA

	Channel No.
Blacksburg	*43
Bristol	5+, *28−
Bluefield	*63+
Charlottesville	29−, *41−, 64+
Courtland	*52
Danville	24−, 44+, *56
Farmville	*31−
Fairfax	*56−
Fredericksburg	*53, 69+
Front Royal	*42
Hampton	13−, *15
Harrisonburg	3−
Lynchburg	13, 21−, *54+
Manassas	66+
Marion	*52−
Norfolk-Portsmouth-Newport News	3+, 10+, 27, 33, 49−, *55+
Norton	*47−
Onancock	*25+
Petersburg	8
Richmond	6+, 12−, *23, 35+, *57−, 63
Roanoke	7−, 10, *15+, 27+, 38−
Staunton	*51−
Virginia Beach	43+
West Point	*46

WASHINGTON

	Channel No.
Anacortes	24
Bellingham	12+, *34, 64
Centralia	*15+
East Wenatchee	249A
Everett	16−
Kennewick	42+
Pasco	19−
Pullman	*10−
Richland	25, *31
Seattle	4, 5+, 7, *9, 22+, and *62
Spokane	2−, 4−, 6−, *7+, 22, 28−
Tacoma	11+, 13−, 20, *28, and *56
Vancouver	*14, 49
Walla Walla	14−
Wenatchee	*18+, 27
Yakima	23+, 29+, 35, *47

WEST VIRGINIA

	Channel No.
Beckley	4
Bluefield	6−, 40−
Charleston	8+, 11+, 23, 29, *49−
Clarksburg	12+, 46−
Fairmont	66−
Grandview	*9−
Huntington	3+, 13+, *33+
Keyser	*48+
Martinsburg	*44
Morgantown	*24−
Parkersburg	15−, 39+, *57
Weirton	*50+
Weston	5
Wheeling	7, [1] 14, *41
Williamson	*31+

WISCONSIN

	Channel No.
Appleton	32+
Bloomington	*49
Colfax	*28−
Eau Claire	13+, 18
Fond du Lac	34+
Green Bay	2+, 5+, 11+, 26+, *38
Highland	*51
Janesville	57+
Kenosha	55−
Kieler	*46+
LaCrosse	8+, 19+, 25, *31
Madison	3, 15, *21−, 27+, 47+
Manitowoc	16+
Milwaukee	4−, 6, *10+, 12, 18−, 24+, 30, *36
Oshkosh	22+
Park Falls	*36+
Racine	49+
Rhinelander	12+
Sheboygan	28
Superior	6+, 40

Suring	14−
Tomah	43
Wausau	7−, 9, *20+, 33−

WYOMING

	Channel No.
Casper	2+, *6+, 14−, 20−
Cheyenne	5+, *17, 27−, 33−
Lander	*4, 5
Laramie	*8+
Rawlins	11−
Riverton	10+
Rock Springs	13
Sheridan	7, 12+

U.S. TERRITORIES AND POSSESSIONS

	Channel No.
Guam: Agana	*4, 8, 10, *12
Puerto Rico:	
Aguadilla	*32, 44
Arecibo-Aguadilla	12+
Arecibo	54, 80
Bayamon	36
Caguas	11−, *58
Carolina	52
Cayey	76
Fajardo	13+, *40
Guayama	46
Humacoa	68
Mayaguez	3+, 5−, 16, 22
Ponce	7+, 9−, 14, 20, *26, 48
San Juan	2+, 4−, *6+, 18, 24, 30, *74
San Sebastian	38
Utuado	*70
Vega Baja	64
Yauco	42
Virgin Islands:	
Charlotte Amalie	10−, 17, *23, 43
Christiansted	8+, 15, *21, 27
Charlotte Amalie-Christiansted	*3, [6]*12

[1] Following the decision in Docket No. 18261, channels so indicated will not be available for television use until further action by the Commission.

[2] Operation on this channel is subject to the conditions, terms, and requirements set out in the Report and Order in Docket No. 19075, RM-1645, adopted January 5, 1972, released January 7, 1972, FCC 72-19.

[3] Channel 15 will not be available for television use until further action by the Commission.

[4] This channel is not available for use at Elgin unless and until it is determined by the Commission that it is not needed for use at Joliet, Ill.

[5] This channel is not available for use at Asbury Park unless and until it is determined by the Commission that it is not needed for educational use at New Brunswick, N.J.

[6] Stations using these assignments shall limit radiation toward stations on the same channel in Puerto Rico, to no more than the effective radiated power which would be radiated by an omnidirectional radio station using maximum permissible effective radiated power for antenna height above average terrain, at the minimum distances from such stations specified in § 73.610(b). The Commission shall consider the status of the negotiations with the appropriate British authorities concerning these assignments when the applications for construction permits come before the Commission.

(Sec. 5, 48 Stat. 1068; 47 U.S.C. 155)

EDITORIAL NOTE: § 73.606 was published in its entirety at 28 FR 13660, Dec. 14, 1963. FEDERAL REGISTER citations to subsequent amendments appear in the List of CFR Sections Affected in the Finding Aids section of this volume.

EFFECTIVE DATE NOTES: 1. AT 47 FR 34571, AUG. 10, 1982, THE ENTRIES FOR BEND AND SALEM, OREGON WERE ADDED TO § 73.606(B) EFFECTIVE OCT. 13, 1982.

2. At 47 FR 37901, Aug. 27, 1982, the entries for Mountain Home, Arkansas; Green Valley, Arizona; Carroll, Iowa; Magee, Mississippi; Scottsbluff, Nebraska were added to § 73.606(b) effective Oct. 19, 1982.

3. At 47 FR 38903, Sept. 3, 1982, the entires for Seattle and Tacoma, Washington were added to § 73.606(b) effective Oct. 26, 1982.

4. At 47 FR 43389, Oct. 1, 1982, the entry for Las Vegas, Nevada was added to § 73.606(b), effective Nov. 19, 1982.

Glossary

aspect ratio—In the NTSC system for America, the aspect ratio for any picture tube is four units wide versus three units high, and said to equal 4:3. In high definition TV, however, this ratio will probably become 5:3, and you'll see a broader expanse of picture in any particular scene.

attenuation—Signal or energy reduction usually produced by analog or digital information traveling through transmission lines, waveguides, or through a space medium. Deliberate attenuation is normally induced by specifically designed attenuators, with results expressed in decibels (dB).

audio carrier—*Always* 4.5 MHz *above* the video carrier, it extends to within 0.25 MHz of the assigned channel frequency. For channel 2, its active frequency is 59.75 MHz.

audio modulation—Formerly ± 25 kHz, has now been broadend to ± 50 kHz for multichannel stereo sound. It remains an FM carrier and must be detected accordingly.

bandwidth—The range of frequencies passing through some transmission medium from dc to gigahertz (GHz). Bandpass filters are often used to limit or protect such transmissions as signals progress through transmit or receiving equipment. In television, rf channels are 6 MHz wide, with baseband limited to 4.5 MHz.

baseband—As applied to television means pure audio and video, no rf modulation.

burst—Appearing on the back porch of the horizontal sync pulse, this 3.579545 MHz frequency is used to sync the receiver's regenerated subcarrier color oscillator.

cathode-ray tubes—Made of glass, with either dot or vertical stripe phosphors and excited through a grill shadow mask, CRTs are the display devices for most visual images now available. Flat panel devices already coming out of the laboratories, however, are expected to supplant CRTs in the not- too-distant future, removing the high-voltage problem and permitting more reasonable drive voltages.

chroma—Color produced by I and Q sidebands having bandwidths of 1.5 and 0.5 mHz, respectively.

color—Light information producing hue, saturation and brightness. Brightness is color intensity, hue becomes color difference, and saturation defines the degree of color purity. The human eye recognizes 30,000 different colors, all within a spectrum of 400 to 700 nanometers. Primary red, blue, and greens are seen at 680, 470, and 540 nanometers.

color modulation—Consists of 500 kHz of Q information and 1.5 MHz of I information, modulated on 90° out-of-phase suppressed subcarriers. Still AM, it is then "folded" into video at odd multiples of 1/2 the line scanning frequencies, but is decoded (detected) separately from luminance.

color purity—An area (or field) of pure color with no contaminations resulting from the absence of stray magnetic fields and the presence of adequate color beam landings on cathode-ray tube phosphors.

color subcarrier—The 3.579545 MHz color subcarrier oscillator always suppressed in transmission but reconstituted in the receiver to synchronously time color sync for correct color demodulation.

conductors—Often described as metal or wire that permits the easy flow of electrons such as copper, silver, aluminum, etc.

convergence—Red, blue, and green colors blended in every given area to produce white. Crosshatch, dothatch, and dots are used to insure correct color beam phosphor landings as adjusted by the various types of convergence devices and magnets as an integral part of the deflection system. Satisfactory convergence on today's CRTs should be within at least 95%. There continue to be both dynamic (operational control) and static (fixed) beam convergence systems in modern TV receivers, but with vastly more effective and simpler methods.

cross coupling—Similar to crosstalk, where energy is coupled from one circuit to another, producing undesirable interference.

cross modulation—Where interfering signals from one circuit or system modulate another.

crosstalk—Stray energy coupling between circuits producing obvious interference.

decibel—Listed as 1/10 of a Bel and used to define the ratio of two powers, voltages or currents in terms of gains or losses. It is 10 × the log of the power ratio and 20 × that of either voltage or current.

decoupling—Preventing the transfer of energy from one circuit or subsystem to another. Capacitors, for instance, are often used as rf decouplers.

DIAC—A two-electrode, three-layer bidirectional diode that will switch between *on* and *off* for either polarity of applied voltage.

digital television—At the moment, by use of A/D converters, analog signals following the video detector are digitized for both audio and video, then returned to baseband once again by D/A converters for sound and the cathode-ray tube. Eventually, with introduction of LCD or other flat panel displays, completely digitized receivers should appear and eventually become less expensive and more reliable than analog types.

diode—Simple two-terminal devices with a cathode and anode devoted principally to ac rectification in power supplies. In some cheap radios they are still used for nonlinear audio detectors and formerly in outmoded DTL, or diode-transistor logic circuits.

distortion—Where phase, amplitude, delay, and interference undesirably distort otherwise acceptable received or transmit signals throughout the audio, visible, and radio frequency spectrums.

electrons and holes—Current carriers in semiconductors that cross electrode barriers and result in active devices devoted usually to switching or amplification.

EMI—Usually encountered at rf, it is electromagnetic interference to any electrical signal in either transmit or receive modes. At rf it is usually called radio-frequency interference (RFI).

equalizing pulses—Permit sync circuits to recognize and lock on vertical sync. Their durations are 2.54 μsec each. Sequences are in sets of six before and after the six vertical pulses and each sequence occupies 3H, or a period of three horizontal lines.

FETs—Dating back to the 1920's, junction field-effect transistors are single-channel devices with source, gate, and drain electrodes. A reverse bias applied to the gate *depletes* electron flow.

field sync pulses—Have durations of 27.1 μsec for each of the six.

flat-square—Means cathode-ray tubes that are both full- square and have relatively flat screen surfaces also.

full-square—Refers to the new sharply-rectangular cathode-ray tubes in sizes 14, 20, 26, and 27 inches.

horizontal blanking interval—Nominally occupies 11 microseconds and contains both the horizontal blanking pulse and color burst.

horizontal sync—Appearing during each horizontal scan line and lasting for 4.76 microseconds. On its "back porch" you will find color burst information of 8 or 9 cycles of 2.24 microseconds duration.

insulators—Materials such as glass, wood, ceramics, etc. that actually block the flow of electrons.

interlaced scanning—The sequential display of odd and even fields to form a flicker-free, linear picture. Normally, the better the interlace the better the picture because there are few video dropouts, resulting in less information loss.

line scan—The forward scan time of 52.4 μsc required to trace one of the 525 horizontal scan lines from left to right on the cathode-ray tube.

luminance—Identified as brightness, is actually the sum of all color signal voltages and is the algebraic addition of reds, blues, and greens. Y (luminance) = 30% Red, 59% Green and 11% Blue.

MAC-B—Stands for mixed analog component signals, and places TV sound into the horizontal line-blanking period, then separates color and luminance for periods of 20 and 40 microseconds each during horizontal scan. In the process, luminance and chroma are compressed in transmission and expanded during reception, enlarging their bandwidths considerably. Transmitted as FM, this system could be used extensively in forthcoming satellite transmission for considerably better TV definition and resolution. Its present parameters are within the existing NTSC format.

MOSFETS—Are insulated gate FETs with two modes of operation: depletion and enhancement. In the latter instance, forward biasing opens the single channel causing current flow.

multichannel sound—Approved for transmission in early 1984 by the FCC, it is AM doublesideband for stereo L-R and operates on a 15,734 Hz pilot carrier, doubled to 31,468 Hz for stereo. Multichannel sound also contains higher-frequency carriers for SAP—second audio program, and professional channel(s).

multiplexing—Used in the sense of time-division multiplexing where samples of information are transmitted during separate time intervals. Will be used in the MAC system for companding luminance and chroma.

PAL (phase alternation line)—So called because chroma phase is alternated 180° from one line to the next to cancel differential phase errors. This is the German system of colorcasting and was developed by Walter Bruch in 1961.

picture frames—Two interlaced fields make a frame, and there are 30 frames/second in the U.S. NTSC television transmit and receive systems.

professional channel(s)—Are usually 6.5 × the pilot carrier, or may be interspersed between the stereo position and 102 kHz if there is no SAP conflict.

SAP—Second audio program transmitted in FM, is used for dual-language purposes or other audio at a bandwidth of only 10 kHz.

satellites—Limited here to geosynchronous space vehicles launched from earth and positioned at specific slots in an arc directly above the equator at 22.3 kilomiles above earth. Being synchronous with earth movements, they remain in fixed positions with only minor station-keeping thruster propulsion during their 8 to 10-year lifetimes.

SCR—A silicon-controlled rectifier is a four-layer PNPN semiconductor with cathode, anode, and gate and is actually a reverse-blocking thyristor.

SECAM (sequential with memory)—Invented by Henri de France in 1957, it transmits R-Y and B-Y in sequence, depending on frequency modulation for video and amplitude modulation for audio—just the reverse of our NTSC. It's the French method of broadcasting color.

semiconductors—Not a conductor nor an insulator, a semiconductor can be a two or multiterminal device used for amplification or switching, and usually called a diode, transistor, or thyristor, but all belonging to the same semiconductor family and usually derived from purified sand.

television—The transmission or reception of a televised image, accompanied usually by V/H sync, sound, and color (chroma), delivered on a common carrier called a channel which is 6 MHz wide.

thyrister semiconductors—Are similar to thyratron tubes that have two conducting states, *on* and *off*.

TRIAC—A forward-blocking, forward-conducting thyristor that operates on either voltage polarity applied to its main terminals, of which it has two.

tv channels—Total 12 VHF and 56 UHF, beginning with channel 2 and ending with channel 69. Former UHF channels 70 through 83 are now either used by repeaters or directly assigned to land mobile. Frequency ranges begin at 54 MHz and extend to 806 MHz.

vertical-field blanking interval—Occupies 16.667 milliseconds and occurs at the end of each 262.5 field scan lines. There are 21 usable lines in this interval devoted to 6 vertical sync pulses, 12 equalizing pulses, as well as additional horizontal pulse intervals.

video—Composite video contains color and luminance (brightness information) as well as horizontal and vertical sync pulses during horizontal and vertical blanking intervals. One line of video, for instance, amounts to approximately 52 microseconds in horizontal scan time, and 63.5 microseconds for a complete line, including color burst and horizontal sync.

video carrier—Always assigned 1.25 MHz above the lower edge of the assigned 6 MHz frequency band. At channel 2, for instance, with a bandwidth of 54-60 MHz, the video carrier rests at 55.25 MHz.

video modulation—In our American NTSC format, video modulation derives from amplitude modulation (AM), including both color and monochrome information within a bandwidth of 4.5 MHz.

Index

Edited by Roland S. Phelps